D0221405

The Facts On File

DICTIONARY
of
CHEMISTRY

The Facts On File DICTIONARY of CHEMISTRY

Fourth Edition

Edited by
John Daintith

The Facts On File Dictionary of Chemistry
Fourth Edition

Checkmark Books
An imprint of Facts On File, Inc.
132 West 31st Street
New York NY 10001

For Library of Congress Cataloging-in-Publication Data,
please contact Facts On File, Inc.
ISBN 0-8160-5650-1

Checkmark Books are available at special discounts when purchased in bulk quantities for businesses, associations, institutions, or sales promotions. Please call our Special Sales Department in New York at (212) 967-8800 or (800) 322-8755.

You can find Facts On File on the World Wide Web at
http://www.factsonfile.com

Compiled and typeset by Market House Books Ltd, Aylesbury, UK

Printed in the United States of America

MP PKG 10 9 8 7 6 5 4 3 2 1

This book is printed on acid-free paper.

PREFACE

This dictionary is one of a series designed for use in schools. It is intended for students of chemistry, but we hope that it will also be helpful to other science students and to anyone interested in science. Facts On File also publishes dictionaries in a variety of disciplines, including biology, physics, mathematics, forensic science, weather and climate, marine science, and space and astronomy.

The Facts On File Dictionary of Chemistry was first published in 1980 and the third edition was published in 1999. This fourth edition of the dictionary has been extensively revised and extended. The dictionary now contains over 3,000 headwords covering the terminology of modern chemistry. A totally new feature of this edition is the inclusion of over 1,700 pronunciations for terms that are not in everyday use. A number of appendixes have been included at the end of the book containing useful information, including a list of chemical elements and a periodic table. There is also a list of Web sites and a bibliography. A guide to using the dictionary has also been added to this latest version of the book.

We would like to thank all the people who have cooperated in producing this book. A list of contributors is given on the acknowledgments page. We are also grateful to the many people who have given additional help and advice.

ACKNOWLEDGMENTS

Contributors

Julia Brailsford B.Sc.
John Clark B.Sc.
John Connor B.Sc.
Derek Cooper Ph.D. F.R.I.C.
Rich Cutler B.Sc.
D.E. Edwards B.Sc. M.Sc.
Richard Hadfield B.Sc.
Valerie Illingworth B.Sc. M.Phil.
Alan Isaacs B.Sc. Ph.D.
Elizabeth Martin M.A.
P.R. Mercer B.A. T.C.
R.S. Smith B.Sc. C.Chem. M.R.I.C.
Derek Stefaniw B.Sc. Ph.D.
Elizabeth Tootil M.Sc.
J. Truman B.Sc.
David Eric Ward B.Sc. M.Sc. Ph.D.

Pronunciations

William Gould B.A.

Note

Unless otherwise stated, the melting and boiling points given in the dictionary are at standard pressure. Relative densities of liquids are at standard pressure with the liquid at 20°C relative to water at 4°C. Relative densities of gases are relative to air, both gases being at standard temperature and pressure.

The following abbreviations are used in the text:

p.n.	proton number (atomic number)
r.a.m.	relative atomic mass (atomic weight)

CONTENTS

GUIDE TO USING THE DICTIONARY

The main features of dictionary entries are as follows.

Headwords

The main term being defined is in bold type:

> **acid** A substance that gives rise to hydrogen ions when dissolved in water.

Plurals

Irregular plurals are given in brackets after the headword.

> **quantum** (*pl.* **quanta**) A definite amount of energy released or absorbed in a process.

Variants

Sometimes a word has a synonym or alternative spelling. This is placed in brackets after the headword, and is also in bold type:

> **promoter** (**activator**) A substance that improves the efficiency of a catalyst.

Here, 'activator' is another word for promoter. Generally, the entry for the synonym consists of a simple cross-reference:

> **activator** *See* promoter.

Abbreviations

Abbreviations for terms are treated in the same way as variants:

> **electron spin resonance** (ESR) A similar technique to nuclear magnetic resonance, but applied to unpaired electrons ...

The entry for the synonym consists of a simple cross-reference:

> **ESR** *See* electron spin resonance.

Formulas

Chemical formulas are also placed in brackets after the headword. These are in normal type:

> **potassium iodide** (KI) A white ionic solid....

Here, KI is the chemical formula for potassium iodide.

Multiple definitions
Some terms have two or more distinct senses. These are numbered in bold type

> **abundance** **1.** The relative amount of a given element among others; for example, the abundance of oxygen in the Earth's crust is approximately 50% by mass.
> **2.** The amount of a nuclide (stable or radioactive) relative to other nuclides of the same element in a given sample.

Cross-references
These are references within an entry to other entries that may give additional useful information. Cross-references are indicated in two ways. When the word appears in the definition, it is printed in small capitals:

> **boron nitride** (BN) A compound formed by heating BORON in nitrogen...

In this case the cross-reference is to the entry for `boron'.

Alternatively, a cross-reference may be indicated by 'See', 'See also', or 'Compare', usually at the end of an entry:

> **boron trifluoride** (BF_3) A colorless fuming gas made by... See boron trichloride.

Hidden entries
Sometimes it is convenient to define one term within the entry for another term:

> **charcoal** An amorphous form of carbon made by... *Activated charcoal* is charcoal heated to...

Here, 'activated charcoal' is a hidden entry under charcoal, and is indicated by italic type. The entry for 'activated charcoal' consists of a simple cross-reference:

> **activated charcoal** *See* charcoal.

Pronunciations
Where appropriate pronunciations are indicated immediately after the headword, enclosed in forward slashes:

> **deuteride** /dew-ter-ÿd/ A compound of deuterium...

Note that simple words in everyday language are not given pronunciations. Also headwords that are two-word phrases do not have pronunciations if the component words are pronounced elsewhere in the dictionary.

Pronunciation Key

A consonant is sometimes doubled to prevent accidental mispronunciation of a syllable resembling a familiar word; for example, /**ass**-id/ /acid/,rather than /**as**-id/ and /ul-tră-/**sonn**-iks// /ultrasonics/, rather than /ul-tră-**son**-iks/. An apostrophe is used: (a) between two consonants forming a syllable, as in /**den**-t'l/ /dental/,and (b) between two letters when the syllable might otherwise be mispronounced through resembling a familiar word, as in /**th'e**-ră-pee/ /therapy/ and /tal'k/ /talc/. The symbols used are:

/a/ *as in* back /bak/, active /**ak**-tiv/
/ă/ *as in* abduct /ăb-**dukt**/, gamma /**gam**-ă/
/ah/ *as in* palm /pahm/, father /**fah**-*th*er/,
/air/ *as in* care /kair/, aerospace /**air**-ŏ-spays/
/ar/ *as in* tar /tar/, starfish /**star**-fish/, heart /hart/
/aw/ *as in* jaw /jaw/, gall /gawl/, taut /tawt/
/ay/ *as in* mania /**may**-niă/ ,grey /gray/
/b/ *as in* bed /bed/
/ch/ *as in* chin /chin/
/d/ *as in* day /day/
/e/ *as in* red /red/
/ĕ/ *as in* bowel /**bow**-ĕl/
/ee/ *as in* see /see/, haem /heem/, caffeine //**ka**f-een/,/ baby /**bay**-bee/
/eer/ *as in* fear /feer/, serum /**seer**-ŭm/
/er/ *as in* dermal /**der**-măl/, labour /**lay**-ber/
/ew/ *as in* dew /dew/, nucleus /**new**-klee-ŭs/
/ewr/ *as in* epidural /ep-i-**dewr**-ăl/
/f/ *as in* fat /fat/, phobia /**foh**-biă/, rough /ruf/
/g/ *as in* gag /gag/
/h/ *as in* hip /hip/
/i/ *as in* fit /fit/, reduction /ri-**duk**-shăn/
/j/ *as in* jaw /jaw/, gene /jeen/, ridge /rij/
/k/ *as in* kidney /**ki**d-nee/, chlorine /**klor**-een/, crisis /**krÿ**-sis/
/ks/ *as in* toxic /**toks**-ik/
/kw/ *as in* quadrate /**kwod**-rayt/
/l/ *as in* liver /**liv**-er/, seal /seel/
/m/ *as in* milk /milk/
/n/ *as in* nit /nit/
/ng/ *as in* sing /sing/
/nk/ *as in* rank /rank/, bronchus /**bronk**-ŭs/
/o/ *as in* pot /pot/
/ô/ *as in* dog /dôg/
/o/ *as in* buttock /**but**-ŏk/
/oh/ *as in* home /hohm/, post /pohst/
/oi/ *as in* boil /boil/
/oo/ *as in* food /food/, croup /kroop/, fluke /flook/
/oor/ *as in* pruritus /proor-**ÿ**-tis/
/or/ *as in* organ /**or**-găn/, wart /wort/
/ow/ *as in* powder /**pow**-der/, pouch /powch/
/p/ *as in* pill /pil/
/r/ *as in* rib /rib/
/s/ *as in* skin /skin/, cell /sel/
/sh/ *as in* shock /shok/, action /**ak**-shŏn/
/t/ *as in* tone /tohn/
/th/ *as in* thin /thin/, stealth /stelth/
/*th*/ as in then /*th*en/, bathe /bay*th*/
/u/ *as in* pulp /pulp/, blood /blud/
/ŭ/ *as in* typhus /**tÿ**-fŭs/
/û/ *as in* pull /pûl/, hook /hûk/
/v/ *as in* vein /vayn/
/w/ *as in* wind /wind/
/y/ *as in* yeast /yeest/
/ÿ/ *as in* bite /bÿt/, high /hÿ/, hyperfine /**hÿ**-per-fÿn/
/yoo/ *as in* unit /**yoo**-nit/, formula /**form**-yoo-lă/
/yoor/ *as in* pure /pyoor/, ureter /yoor-**ee**-ter/
/ÿr/ *as in* fire /fÿr/
/z/ *as in* zinc /zink/, glucose /**gloo**-kohz/
/zh/ *as in* vision /**vizh**-ŏn/

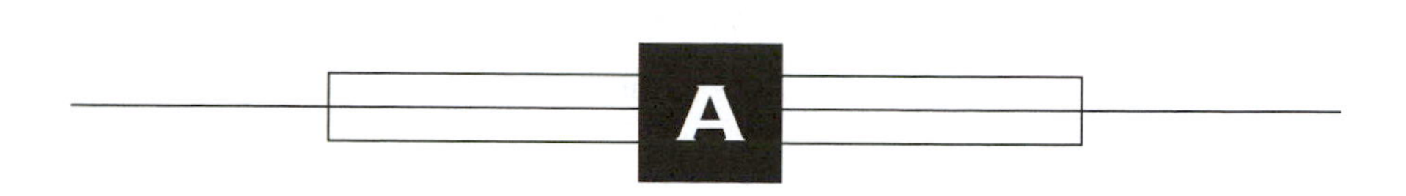

AAS *See* atomic absorption spectroscopy.

absolute alcohol Pure alcohol (ethanol).

absolute configuration A particular molecular configuration of a CHIRAL molecule, as denoted by comparison with a reference molecule or by some sequence rule. There are two systems for expressing absolute configuration in common use: the *D–L convention* and the *R–S convention*. *See* optical activity.

absolute temperature Symbol: T A temperature defined by the relationship:

$$T = \theta + 273.15$$

where θ is the Celsius temperature. The absolute scale of temperature was a fundamental scale based on Charles' law applied to an ideal gas:

$$V = V_0(1 + \alpha\theta)$$

where V is the volume at temperature θ, V_0 the volume at 0, and α the thermal expansivity of the gas. At low pressures (when real gases show ideal behavior) α has the value 1/273.15. Therefore, at $\theta = -273.15$ the volume of the gas theoretically becomes zero. In practice, of course, substances become solids at these temperatures. However, the extrapolation can be used for a scale of temperature on which –273.15°C corresponds to 0° (absolute zero). The scale is also known as the *ideal-gas scale*; on it temperature intervals were called *degrees absolute* (°A) or *degrees Kelvin* (°K), and were equal to the Celsius degree. It can be shown that the absolute temperature scale is identical to the thermodynamic temperature scale (on which the unit is the kelvin).

absolute zero The zero value of thermodynamic temperature; 0 kelvin or –273.15°C.

absorption A process in which a gas is taken up by a liquid or solid, or in which a liquid is taken up by a solid. In absorption, the substance absorbed goes into the bulk of the material. Solids that absorb gases or liquids often have a porous structure. The absorption of gases in solids is sometimes called *sorption*. *Compare* adsorption.

absorption indicator (**adsorption indicator**) An indicator used for titrations that involve a precipitation reaction. The method depends upon the fact that at the equivalence point there is a change in the nature of the ions absorbed by the precipitate particles. Fluorescein – a fluorescent compound – is commonly used. For example, in the titration of sodium chloride solution with added silver nitrate, silver chloride is precipitated. Sodium ions and chloride ions are absorbed in the precipitate. At the end point, silver ions and nitrate ions are in slight excess and silver ions are then absorbed. If fluorescein is present, negative fluorescein ions absorb in preference to nitrate ions, producing a pink complex.

absorption spectrum *See* spectrum.

abundance **1.** The relative amount of a given element among others; for example, the abundance of oxygen in the Earth's crust is approximately 50% by mass.
2. The amount of a nuclide (stable or radioactive) relative to other nuclides of the same element in a given sample. The *natural abundance* is the abundance of a nuclide as it occurs naturally. For instance, chlorine has two stable isotopes of masses

35 and 37. The abundance of ^{35}Cl is 75.5% and that of ^{37}Cl is 24.5%. For some elements the abundance of a particular nuclide depends on the source.

accelerator A catalyst added to increase the rate of cross-linking reactions in polymers.

acceptor /ak-**sep**-ter, -tor/ The atom or group to which a pair of electrons is donated in a coordinate bond. Pi-acceptors are compounds or groups that accept electrons into pi, p or d orbitals.

accumulator (**secondary cell; storage battery**) An electric cell or battery that can be charged by passing an electric current through it. The chemical reaction in the cell is reversible. When the cell begins to run down, current in the opposite direction will convert the reaction products back into their original forms. The most common example is the LEAD-ACID ACCUMULATOR, used in vehicle batteries.

acenaphthene /as-ĕ-**naf**-th'een, **-nap-**/ ($C_{12}H_{10}$) A colorless crystalline derivative of naphthalene, used in producing some dyes.

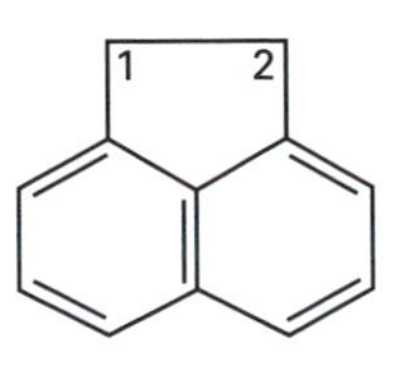

Acenaphthene

acetal /**ass**-ĕ-tal/ A type of organic compound formed by addition of an alcohol to an aldehyde. Addition of one alcohol molecule gives a *hemiacetal*. Further addition yields the full acetal. Similar reactions occur with ketones to produce hemiketals and ketals.

acetaldehyde /ass-ĕ-**tal**-dĕ-hÿd/ *See* ethanal.

acetamide /ass-ĕ-**tam**-ÿd, -id; ă-**set**-ĕ-mÿd, -mid/ *See* ethanamide.

acetate /**ass**-ĕ-tayt/ *See* ethanoate.

acetic acid /ă-**see**-tik, ă-**set**-ik/ *See* ethanoic acid.

acetone /**ass**-ĕ-tohn/ *See* propanone.

acetonitrile /ass-ĕ-toh-**nÿ**-trăl, ă-see-toh-, -tril, -trÿl/ *See* methyl cyanide.

acetophenone /ass-ĕ-toh-**fee**-nohn, ă-see-toh-/ *See* phenyl methyl ketone.

acetylation /ă-set'l-**ay**-shŏn/ *See* acylation.

acetyl chloride /**ass**-ĕ-t'l, ă-**see**-t'l/ *See* ethanoyl chloride.

acetylene /ă-**set**-ă-leen, -lin/ *See* ethyne.

acetyl group *See* ethanoyl group.

acetylide /ă-**set**-ă-lÿd/ *See* carbide.

acetylsalicylic acid /**ass**-ĕ-t'l-sal-ă-**sil**-ik, ă-**see**-t'l-/ *See* aspirin.

Acheson process /**ach**-ĕ-s'n/ *See* carbon.

achiral /ă-**kÿr**-ăl/ Describing a molecule that does not exhibit optical activity.

acid A substance that gives rise to hydrogen ions when dissolved in water. Strictly, these ions are hydrated, known as *hydroxonium* or *hydronium ions*, and are usually given the formula H_3O^+. An acid in solution will have a pH below 7. This definition does not take into account the competitive behavior of acids in solvents and it refers only to aqueous systems. The *Lowry–Brønsted theory* defines an acid as a substance that exhibits a tendency to release a proton, and a base as a substance that tends to accept a proton. Thus, when an acid releases a proton, the ion formed is the *conjugate base* of the acid. Strong acids (e.g. HNO_3) react completely with water to give H_3O^+, i.e. HNO_3 is stronger than H_3O^+ and the conjugate base NO_3^- is weak. Weak acids (e.g. CH_3COOH and C_6H_5COOH) are only partly dissociated because H_3O^+ is a stronger acid than the free acids and the ions CH_3COO^- and $C_6H_5COO^-$ are moderately strong bases. The Lowry–Brønsted theory is named for the English chemist Thomas Martin Lowry (1874–1936) and the Danish physical chemist Johannes Nicolaus Brønsted (1879–1947). *See also* Lewis acid.

acid anhydride A type of organic compound of general formula RCOOCOR′, where R and R′ are alkyl or aryl groups. They are prepared by reaction of an acyl halide with the sodium salt of a carboxylic acid, e.g.:

$$RCOCl + R'COO^-Na^+ \rightarrow RCOOCOR' + NaCl$$

Like the acyl halides, they are very reactive acylating agents. Hydrolysis is to carboxylic acids:

$$RCOOCOR' + H_2O \rightarrow RCOOH + R'COOH$$

See also acylation.

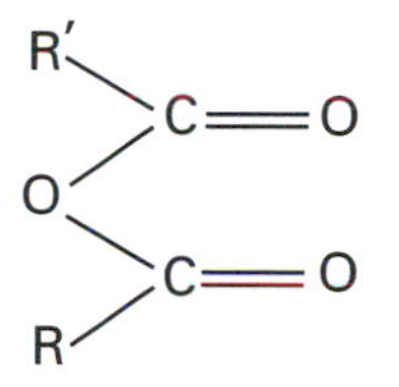

Acid anhydride structure

acid–base indicator An indicator that is either a weak base or a weak acid and whose dissociated and undissociated forms differ markedly in color. The color change must occur within a narrow pH range. Examples are methyl orange and phenolphthalein.

acid dyes The sodium salts of organic acids used in the dyeing of silks and wool. They are so called because they are applied from a bath acidified with dilute sulfuric or ethanoic acid.

acid halide *See* acyl halide.

acidic Having a tendency to release a proton or to accept an electron pair from a donor. In aqueous solutions the pH is a measure of the acidity, i.e. an acidic solution is one in which the concentration of H_3O^+ exceeds that in pure water at the same temperature; i.e. the pH is lower than 7. A pH of 7 is regarded as being neutral.

acidic hydrogen A hydrogen atom in a molecule that enters into a dissociation equilibrium when the molecule is dissolved in a solvent. For example, in ethanoic acid (CH_3COOH) the acidic hydrogen is the one on the carboxyl group, –COOH.

acidic oxide An oxide of a nonmetal that reacts with water to produce an acid or with a base to produce a salt and water. For example, sulfur(VI) oxide (sulfur trioxide) reacts with water to form sulfuric(VI) acid:

$$SO_3 + H_2O \rightarrow H_2SO_4$$

and with sodium hydroxide to produce sodium sulfate and water:

$$SO_3 + NaOH \rightarrow Na_2SO_4 + H_2O$$

See also amphoteric; basic oxide.

acidimetry /ass-ă-**dim**-ĕ-tree/ Volumetric analysis or acid-base titration in which a standard solution of an acid is added to the unknown (base) solution plus the indicator. *Alkalimetry* is the converse, i.e. the base is in the buret.

acidity constant *See* dissociation constant.

acid rain *See* pollution.

acid salt (**acidic salt**) A salt in which there is only partial replacement of the acidic hydrogen of an acid by metal or other cations. For polybasic acids the formulae are of the type $NaHSO_4$ (sodium hydrogensulfate) and $Na_3H(CO_3)_2.2H_2O$ (sodium sesquicarbonate). For monobasic acids such as HF the acid salts are of the form KHF_2 (potassium hydrogen fluoride). Although the latter were at one time formulated as a normal salt plus excess acid (i.e. KF.HF) it is preferable to treat these as hydrogen-bonded systems of the type K^+ $(F–H–F)^-$.

acid value A measure of the free acid present in fats, oils, resins, plasticizers, and solvents, defined as the number of milligrams of potassium hydroxide required to neutralize the free acids in one gram of the substance.

acridine /**ak**-ri-deen/ ($C_{12}H_9N$) A colorless crystalline heterocyclic compound with three fused rings. Derivatives of acridine are used as dyes and biological stains.

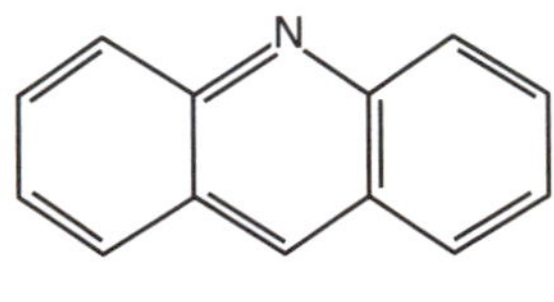

Acridine

Acrilan /**ak**-ră-lan/ (*Trademark*) A synthetic fiber that consists of a copolymer of 1-cyanoethene (acrylonitrile, vinyl cyanide) and ethenyl ethanoate (vinyl acetate).

acrolein /ă-**kroh**-lee-in/ *See* propenal.

acrylic acid /ă-**kril**-ik/ *See* propenoic acid.

acrylic resin A synthetic resin made by polymerizing an amide or ester derivative of 2-propenoic acid (acrylic acid). Examples of acrylic materials are Acrilan (from propenonitrile) and Plexiglas (polymethylmethacrylate). Acrylic resins are also used in paints.

acrylonitrile /ak-ră-loh-**nÿ**-trăl, -tril, -trÿl, ă-kril-oh-/ *See* propenonitrile.

actinic radiation /ak-**tin**-ik/ Radiation that can cause a chemical reaction; for example, ultraviolet radiation is actinic.

actinides /**ak**-tă-nÿdz/ *See* actinoids.

actinium /ak-**tin**-ee-ŭm/ A soft silvery-white radioactive metallic element that is the first member of the actinoid series. It occurs in minute quantities in uranium ores and the metal can be obtained by reducing the trifluoride with lithium. It can be produced by neutron bombardment of radium and is used as a source of alpha particles. The metal glows in the dark; it reacts with water to produce hydrogen.

Symbol: Ac; m.p. 1050±50°C; b.p. 3200±300°C; r.d. 10.06 (20°C); p.n. 89; most stable isotope ^{227}Ac (half-life 21.77 years).

actinoids /**ak**-tă-noidz/ (**actinides**) A group of 15 radioactive elements whose electronic configurations display filling of the 5f level. As with the lanthanoids, the first member, actinium, has no f electrons (Ac [Rn]$6d^17s^2$) but other members also show deviations from the smooth trend of f-electron filling expected from simple considerations, e.g. thorium Th [Rn]$6d^27s^2$, berkelium Bk [Rn]$5f^86d^17s^2$. The actinoids are all radioactive and their chemistry is often extremely difficult to study. In general, artificial methods using high-energy bombardment are used to generate them. *See also* transuranic elements.

activated charcoal *See* charcoal.

activated complex The partially bonded system of atoms in the transition state of a chemical reaction.

activation energy Symbol: E_a The minimum energy a particle, molecule, etc., must acquire before it can react; i.e. the energy required to *initiate* a reaction regardless of whether the reaction is exothermic or endothermic. Activation energy is often represented as an energy barrier that must be overcome if a reaction is to take place. *See* Arrhenius equation.

activator *See* promoter.

active mass *See* mass action; law of.

active site **1.** a site on the surface of a catalyst at which catalytic activity occurs or at which the catalyst is particularly effective.

2. The position on the molecule of an enzyme that binds to the substrate when the ENZYME acts as a catalyst.

activity 1. Symbol: *a* Certain thermodynamic properties of a solvated substance are dependent on its concentration (e.g. its tendency to react with other substances). Real substances show departures from ideal behavior and a corrective concentration term – the activity – has to be introduced into equations describing real solvated systems.
2. Symbol: *A* The average number of atoms disintegrating per unit time in a radioactive substance.

activity coefficient Symbol: *f* A measure of the degree of deviation from ideality of a solvated substance, defined as:

$$a = fc$$

where *a* is the activity and *c* the concentration. For an ideal solute $f = 1$; for real systems *f* can be less or greater than unity.

acyclic /ay-**sÿ**-klik, **-sik**-lik/ Describing a compound that is not cyclic (i.e. a compound that does not contain a ring in its molecules).

acyl anhydride /**ass**-ăl, **ay**-săl/ *See* acid anhydride.

acylating agent /**ass**-ă-layt-ing/ *See* acylation.

acylation /ass-ă-**lay**-shŏn/ A reaction that introduces the acyl group (RCO–). *Acylating agents* are acyl halides (R.CO.X) and acid anhydrides (R.CO.O.CO.R), which react with such nucleophiles as H_2O, ROH, NH_3, and RNH_2. In these compounds a hydrogen atom of a hydroxyl or amine group is replaced by the RCO– group. In *acetylation* the acetyl group (CH_3CO–) is used. In *benzoylation* the benzoyl group (C_6H_5CO–) is used. Acylation is used to prepare crystalline derivatives of organic compounds to identify them (e.g. by melting point) and to protect –OH groups in synthetic reactions.

acyl group The group of atoms RCO–.

acyl halide (**acid halide**) A type of organic compound of the general formula RCOX, where X is a halogen (acyl chloride, acyl bromide, etc.).

Acyl halides can be prepared by the reaction of carboxylic acid with a halogenating agent. Commonly, phosphorus halides are used (e.g. PCl_5) or a sulfur dihalide oxide (e.g. $SOCl_2$):

$$RCOOH + PCl_5 \rightarrow RCOCl + POCl_3 + HCl$$

$$RCOOH + SOCl_2 \rightarrow RCOCl + SO_2 + HCl$$

The acyl halides have irritating vapors and fume in moist air. They are very reactive to the hydrogen atom of compounds containing hydroxyl (–OH) or amine (–NH_2) groups. *See* acylation.

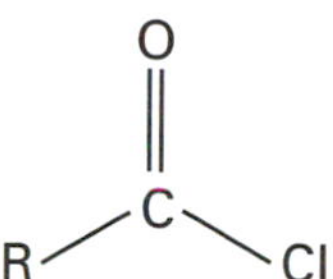

Acyl halide: an acyl chloride

addition polymerization *See* polymerization.

addition reaction A reaction in which additional atoms or groups of atoms are introduced into an unsaturated compound, such as an alkene or ketone. A simple example is the addition of bromine across the double bond in ethene:

$$H_2C{:}CH_2 + Br_2 \rightarrow BrH_2CCH_2Br$$

Addition reactions can be induced either by electrophiles or by nucleophiles. *See also* electrophilic addition; nucleophilic addition.

adduct /ă-**dukt**/ *See* coordinate bond.

adenine /**ad**-ĕ-neen, -nin, -nÿn/ A nitrogenous base found in DNA and RNA. It is also a constituent of certain coenzymes and when combined with the sugar ribose it forms the nucleoside adenosine found in AMP, ADP, and ATP. Adenine has a purine ring structure.

adenosine /ă-**den**-ŏ-seen, -sin, **ad**-'n-ŏ-seen/ (**adenine nucleoside**) A NUCLEOSIDE formed from adenine linked to D-ribose with a β-glycosidic bond. It is widely found in all types of cell, either as the free nucleoside or in combination in nucleic acids. Phosphate esters of adenosine, such as ATP, are important carriers of energy in biochemical reactions.

adenosine diphosphate /dÿ-**fos**-fayt/ *See* ADP.

adenosine monophosphate /mon-oh-**fos**-fayt/ *See* AMP.

adenosine triphosphate /trÿ-**fos**-fayt/ *See* ATP.

adiabatic change /ad-ee-ă-**bat**-ik/ A change during which no energy enters or leaves the system.

In an adiabatic expansion of a gas, mechanical work is done by the gas as its volume increases and the gas temperature falls. For an ideal gas undergoing a reversible adiabatic change it can be shown that

$$pV^{\gamma} = K_1$$
$$T^{\gamma}p^{1-\gamma} = K_2$$
$$\text{and } TV^{\gamma-1} = K_3$$

where K_1, K_2, and K_3 are constants and γ is the ratio of the principal specific heat capacities. *Compare* isothermal change.

adipic acid /ă-**dip**-ik/ *See* hexanedioic acid.

ADP (**adenosine diphosphate**) A NUCLEOTIDE consisting of adenine and ribose with two phosphate groups attached. *See also* ATP.

adsorbate /ad-**sor**-bayt, **-zor-**/ A substance that is adsorbed on a surface. *See* adsorption.

adsorbent /ad-**sor**-bĕnt, **-zor-**/ The substance on whose surface ADSORPTION takes place.

adsorption /ad-**sorp**-shŏn, **-zorp-**/ A process in which a layer of atoms or molecules of one substance forms on the surface of a solid or liquid. All solid surfaces take up layers of gas from the surrounding atmosphere. The adsorbed layer may be held by chemical bonds (*chemisorption*) or by weaker van der Waals forces (*physisorption*). *Compare* absorption.

adsorption indicator *See* absorption indicator.

aerobic /air-**oh**-bik/ Describing a biochemical process that takes place only in the presence of free oxygen. *Compare* anaerobic.

aerosol *See* sol.

AES *See* atomic emission spectroscopy.

affinity The extent to which one substance is attracted to or reacts with another.

afterdamp *See* firedamp.

agate /**ag**-it, -ayt/ A hard microcrystalline form of the mineral chalcedony (a variety of quartz). Typically it has greenish or brownish bands of coloration, and is used for making ornaments. *Moss agate* is not banded, but has mosslike patterns resulting from the presence of iron and manganese oxides. Agate is used in instrument bearings because of its resistance to wear.

agent orange A herbicide consisting of a mixture of two weedkillers (2,4-D and 2,4,5-T), which was formerly used in warfare to defoliate trees where an enemy may be hiding or to destroy an enemy's crops. It also contains traces of the highly toxic chemical dioxin, which may cause cancers and birth defects.

air The mixture of gases that surrounds the Earth. The composition of dry air, by volume, is:

nitrogen 78.08%
oxygen 20.95%
argon 0.93%

carbon dioxide 0.03%
neon 0.0018%
helium 0.0005%
krypton 0.0001%
xenon 0.00001%

Air also contains a variable amount of water vapor, as well as particulate matter (e.g. dust and pollen), and small amounts of other gases.

air gas *See* producer gas.

alabaster A mineral form of gypsum ($CaSO_4.2H_2O$).

alanine /**al**-ă-neen, -nÿn/ *See* amino acids.

albumen /al-**byoo**-mĕn/ The white of an egg, which consists mainly of albumin. *See* albumin.

albumin /al-**byoo**-min/ A soluble protein that occurs in many animal fluids, such as blood serum and egg white.

alchemy An ancient pseudoscience that was the precursor of chemistry, dating from early Christian times until the 17th century. It combined mysticism and experimental techniques. Many ancient alchemists searched for the *Philosopher's stone* – a substance that could transmute base metals into gold and produce the *elixir of life*, a universal remedy for all ills.

alcohol A type of organic compound of the general formula ROH, where R is a hydrocarbon group. Examples of simple alcohols are methanol (CH_3OH) and ethanol (C_2H_5OH).

Alcohols have the –OH group attached to a carbon atom that is not part of an aromatic ring: C_6H_5OH, in which the –OH group is attached to the ring, is thus a phenol. Phenylmethanol ($C_6H_5CH_2OH$) does have the characteristic properties of alcohols.

Alcohols can have more than one –OH group; those containing two, three or more such groups are described as *dihydric*, *trihydric*, and *polyhydric* respectively (as opposed to those containing one –OH group, which are *monohydric*). Examples are ethane-1,2-diol (ethylene glycol; ($HOCH_2$-CH_2OH) and propane-1,2,3-triol (glycerol; $HOCH_2CH(OH)CH_2OH$).

Alcohols are further classified according to the environment of the –C–OH grouping. If the carbon atom is attached to two hydrogen atoms, the compound is a *primary alcohol*. If the carbon atom is attached to one hydrogen atom and two other groups, it is a *secondary alcohol*. If the carbon atom is attached to three other groups, it is a *tertiary alcohol*. Alcohols can be prepared by:

1. Hydrolysis of HALOALKANES using aqueous potassium hydroxide:
$$RI + OH^- \rightarrow ROH + I^-$$
2. Reduction of aldehydes by nascent hydrogen (from sodium amalgam in water):
$$RCHO + 2[H] \rightarrow RCH_2OH$$

The main reactions are:

1. Oxidation by potassium dichromate(VI) in sulfuric acid. Primary alcohols give

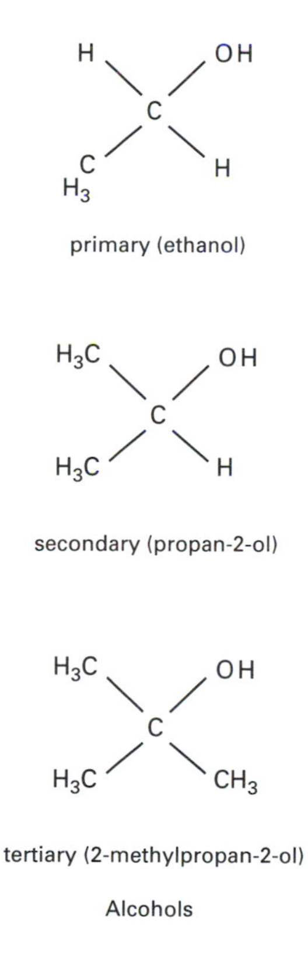

Alcohols

aldehydes, which are further oxidized to carboxylic acids:

$RCH_2OH \rightarrow RCHO \rightarrow RCOOH$

Secondary alcohols are oxidized to ketones.

2. Formation of esters with acids. The reaction, which is reversible, is catalyzed by H^+ ions:

   $ROH + R'COOH \rightleftharpoons R'COOR + H_2O$
3. Dehydration over hot pumice (400°C) to alkenes:

   $RCH_2CH_2OH - H_2O \rightarrow RCH{:}CH_2$
4. Reaction with sulfuric acid. Two types of reaction are possible. With excess acid at 160°C dehdyration occurs to give an alkene:

   $RCH_2CH_2OH + H_2SO_4 \rightarrow H_2O + RCH_2CH_2.HSO_4$

   $RCH_2CH_2.HSO_4 \rightarrow RCH{:}CH_2 + H_2SO_4$

   With excess alcohol at 140°C an ether is formed:

   $2ROH \rightarrow ROR + H_2O$

See also acylation.

aldehyde /**al**-dĕ-hȳd/ A type of organic compound with the general formula RCHO, where the –CHO group (the aldehyde group) consists of a carbonyl group attached to a hydrogen atom. Simple examples of aldehydes are methanal (formaldehyde, HCHO) and ethanal (acetaldehyde, CH_3CHO).

Aldehydes are formed by oxidizing a primary alcohol. In the laboratory potassium dichromate(VI) is used in sulfuric acid. They can be further oxidized to carboxylic acids. Reduction (using a catalyst or nascent hydrogen from sodium amalgam in water) produces the parent alcohol.

Aldehydes undergo a number of reactions:

1. They act as reducing agents, being oxidized to carboxylic acids in the process. These reactions are used as tests for aldehydes using such reagents as Fehling's solution and Tollen's reagent (silver-mirror test).
2. They form addition compounds with hydrogen cyanide to give 'cyanohydrins'. For example, propanal gives 2-hydroxybutanonitrile:

   $C_2H_5CHO + HCN \rightarrow C_2H_5CH(OH)CN$
3. They form addition compounds (*bisulfite addition compounds*) with the hydrogensulfate(IV) ion (hydrogensulfite; HSO_3^-):

   $RCHO + HSO_3^- \rightarrow RCH(OH)(HSO_3)$
4. They undergo CONDENSATION REACTIONS with such compounds as hydrazine, hydroxylamine, and their derivatives.
5. With alcohols they form hemiacetals and ACETALS.
6. They polymerize readily. Polymethanal or methanal trimer can be formed from methanal depending on the conditions. Ethanal gives ethanal trimer or ethanal tetramer.

See also Cannizzaro reaction; ketone.

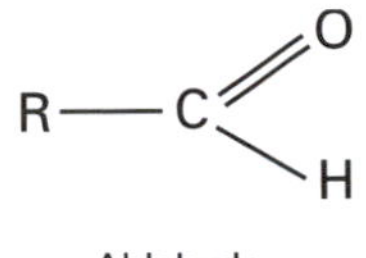

Aldehyde

aldohexose /al-doh-**heks**-ohs/ An aldose SUGAR with six carbon atoms.

aldol /**al**-dol, -dohl/ *See* aldol reaction.

aldol reaction A reaction in which two molecules of aldehyde combine to give an *aldol* – i.e. a compound containing both aldehyde and alcohol functional groups. The reaction is base-catalyzed; the reaction of ethanal refluxed with sodium hydroxide gives:

$2CH_3CHO \rightarrow CH_3CH(OH)CH_2CHO$

The mechanism is similar to that of the CLAISEN CONDENSATION: the first step is removal of a proton to give a carbanion, which subsequently attacks the carbon of the carbonyl group on the other molecule:

$CH_3CHO + OH^- \rightarrow {}^-CH_2CHO + H_2O$.

aldopentose /al-doh-**pen**-tohs/ An aldose SUGAR with five carbon atoms.

aldose /**al**-dohs/ A SUGAR containing an aldehyde (CHO) or potential aldehyde group.

alginic acid /al-**jin**-ik/ (**algin**; $(C_6H_8O_6)_n$) A yellow-white organic solid that is found in brown algae. It is a complex polysaccharide and produces, in even very dilute solutions, a viscous liquid. Alginic acid has various uses, especially in the food industry as a stabilizer and texture agent.

alicyclic compound /al-ă-**sÿ**-klik, -**sik**-lik/ An aliphatic cyclic compound, such as cyclohexane.

aliphatic compound /al-ă-**fat**-ik/ An organic compound with properties similar to those of the alkanes, alkenes, and alkynes and their derivatives. Most aliphatic compounds have an open chain structure but some, such as cyclohexane and sucrose, have rings. The term is used in distinction to aromatic compounds, which are similar to benzene. *Compare* aromatic compound.

alizarin /ă-**liz**-ă-rin/ (**1,2-dihydroxyanthraquinone**) An important orange-red organic compound used in the dyestuffs industry to produce red lakes. It occurs naturally in the root of the plant madder and may also be synthesized from anthraquinone.

alkali /**al**-kă-lÿ/ A water-soluble strong base. Strictly the term refers to the hydroxides of the alkali metals (group 1) only, but in common usage it refers to *any* soluble base. Thus borax solution may be described as mildly alkaline.

alkali metals (**group 1 elements**) A group of soft reactive metals, each representing the start of a new period in the periodic table and having an electronic configuration consisting of a rare-gas structure plus one outer electron. The alkali metals are lithium (Li), sodium (Na), potassium (K), rubidium (Rb), cesium (Cs), and francium (Fr). They formerly were classified in subgroup IA of the periodic table.

The elements all easily form positive ions M^+ and consequently are highly reactive (particularly with any substrate that is oxidizing). As the group is descended there is a gradual decrease in ionization potential and an increase in the size of the atoms; the group shows several smooth trends which follow from this. For example, lithium reacts in a fairly controlled way with water, sodium ignites, and potassium explodes. There is a general decrease in the following: melting points, heats of sublimation, lattice energy of salts, hydration energy of M^+, ease of decomposition of nitrates and carbonates, and heat of formation of the '-ide' compounds (fluoride, hydride, oxide, carbide, chloride).

Lithium has the smallest ion and therefore the highest charge/size ratio and is polarizing with a tendency towards covalent character in its bonding; the remaining elements form typical ionic compounds in which ionization, M^+X^-, is regarded as complete. The slightly anomalous position of lithium is illustrated by the similarity of its chemistry to that of magnesium. For example, lithium hydroxide is much less soluble than the hydroxides of the other group 1 elements; lithium perchlorate is soluble in several organic solvents. Because of the higher lattice energies associated with smaller ions lithium hydride and nitride are fairly stable compared to NaH, which decomposes at 345°C. Na_3N, K_3N etc., are not obtained pure and decompose below room temperature.

The oxides also display the trend in properties as lithium forms M_2O with only traces of M_2O_2, sodium forms M_2O_2 and at high temperatures and pressures MO_2, potassium, rubidium, and cesium form M_2O_2 if oxygen is restricted but MO_2 if burnt in air. Hydrolysis of the oxides or direct reaction of the metal with water leads to the formation of the hydroxide ion.

Salts of the bases MOH are known for all acids and these are generally white crystalline solids. The ions M^+ are hydrated in water and remain unchanged in most reactions of alkali metal salts.

Because of the ease of formation of the ions M^+ there are very few coordination compounds of the type ML_n^+ apart from solvated species of very low correlation times. The group 1 elements form a variety of organometallic compounds; the bonding in lithium alkyls and aryls is essentially covalent but the heavier elements form ionic compounds. Organo-alkali metal compounds – particularly the lithium compounds – are widely used in synthetic organic chemistry.

Francium is formed only by radioactive decay and in nuclear reactions; all the isotopes of francium have short half-lives, the longest of which is 21 minutes (francium-223). The few chemical studies which have been carried out indicate that it would have similar properties to those of the other alkali metals.

alkalimetry /al-kă-**lim**-ĕ-tree/ *See* acidimetry.

alkaline-earth metals (**group 2 elements**) A group of moderately reactive metals, harder and less volatile than the alkali metals. They were formerly classified in subgroup IIA of the periodic table. The term *alkaline earth* strictly refers to the oxides, but is often used loosely for the elements themselves. The electronic configurations are all those of a rare-gas structure with an additional two electrons in the outer s orbital. The elements are beryllium (Be), magnesium (Mg), calcium (Ca), strontium (Sr), barium (Ba), and radium (Ra). The group shows an increasing tendency to ionize to the divalent state M^{2+}. The first member, beryllium has a much higher ionization potential than the others and the smallest atomic radius. Thus it has a high charge/size ratio and consequently the bonding in beryllium compounds is largely covalent. The chemistry of the heavier members of the group is largely that of divalent ions.

The group displays a typical trend towards metallic character as the group is descended. For example, beryllium hydroxide is amphoteric; magnesium hydroxide is almost insoluble in water and is slightly basic; calcium hydroxide is sparingly soluble and distinctly basic; and strontium and barium hydroxides are increasingly soluble in water and strongly basic. The group also displays a smooth trend in the solubilities of the sulfates ($MgSO_4$ is soluble, $CaSO_4$ sparingly soluble, and $BaSO_4$ very insoluble). The trend to increasing metallic character is also shown by the increase in thermal stabilities of the carbonates and nitrates with increasing relative atomic mass.

The elements all burn in air (beryllium must be finely powdered) to give the oxide MO (covalent in the case of beryllium) and for barium the peroxide, BaO_2 in addition to BaO. The heavier oxides, CaO, SrO, and BaO, react with water to form hydroxides, $M(OH)_2$; magnesium oxide reacts only at high temperatures and beryllium oxide not at all. The metals Ca, Sr, and Ba all react readily with water to give the hydroxide:

$$M + 2H_2O \rightarrow M^{2+} + 2OH^- + H_2$$

In contrast, magnesium requires dilute acids in order to react (to the salt plus hydrogen), and beryllium is resistant to acid attack. A similar trend is seen in the direct reaction of hydrogen: under mild conditions calcium, strontium, and barium give ionic hydrides, high pressures are required to form magnesium hydride, and beryllium hydride can not be prepared by direct combination.

Because of its higher polarizing power, beryllium forms a range of complexes in which the beryllium atom should be treated as an electron acceptor (i.e. the vacant p orbitals are being used). Complexes such as etherates, acetylethanoates, and the tetrafluoride (BeF_4^{2-}) are formed, all of which are tetrahedral. In contrast Mg^{2+}, Ca^{2+}, Sr^{2+}, and Ba^{2+} have poor acceptor properties and form only weak complexes, even with donors such as ammonia or edta. Magnesium forms Grignard reagents (RMgX), which are important in organic synthesis, and related compounds $R_2Mg.MgX_2$ and R_2Mg are known. The few organic compounds of Ca, Sr, and Ba are ionic. All isotopes of radium are radioactive and radium was once widely used for radiotherapy. The half-life of ^{226}Ra (formed by decay of ^{238}U) is 1600 years.

alkaloid /**al**-kă-loid/ One of a group of natural organic compounds found in plants. They contain oxygen and nitrogen atoms; most are poisonous. However, they include a number of important drugs with characteristic physiological effects, e.g.

morphine, codeine, caffeine, cocaine, and nicotine.

alkane /**al**-kayn/ A type of hydrocarbon with general formula C_nH_{2n+2}. Alkanes are saturated compounds, containing no double or triple bonds. Methane (CH_4) and ethane (C_2H_6) are typical examples. The alkanes are fairly unreactive (their former name, the *paraffins*, means 'small affinity'). In ultraviolet radiation they react with chlorine to give a mixture of substitution products. There are a number of ways of preparing alkanes:

1. From a sodium salt of a carboxylic acid treated with soda lime:
   $RCOO^-Na^+ + NaOH \rightarrow RH + Na_2CO_3$
2. By reduction of a haloalkane with nascent hydrogen from the action of ethanol on a zinc–copper couple:
   $RX + 2[H] \rightarrow RH + HX$
3. By the Wurtz reaction – i.e. sodium in dry ether on a haloalkane:
   $2RX + 2Na \rightarrow 2NaX + RR$
4. By the Kolbé electrolytic method:
   $RCOO^- \rightarrow RR$
5. By refluxing a haloalkane with magnesium in dry ether to form a Grignard reagent:
   $RI + Mg \rightarrow RMgI$
   With acid this gives the alkane:
   $RMgI + H \rightarrow RH$

The main source of lower molecular weight alkanes is natural gas (for methane) and crude oil.

alkene /**al**-keen/ A type of hydrocarbon with the general formula C_nH_{2n}. The alkenes (formerly called *olefins*) are unsaturated compounds containing double carbon–carbon bonds. They can be obtained from crude oil by cracking alkanes. Important examples are ethene (C_2H_4) and propene (C_3H_6), both of which are used in plastics production and as starting materials for the manufacture of many other organic chemicals.

The methods of synthesizing alkenes are:

1. The elimination of HBr from a haloalkane using an alcoholic solution of potassium hydroxide:
   $RCH_2CH_2Br + KOH \rightarrow KBr + H_2O + RCH{:}CH_2$
2. The dehydration of an alcohol by passing the vapor over hot pumice (400°C):
   $RCH_2CH_2OH \rightarrow RCH{:}CH_2 + H_2O$

The reactions of alkenes include:

1. Hydrogenation using a catalyst (usually nickel at about 150°C):
   $RCH{:}CH_2 + H_2 \rightarrow RCH_2CH_3$
2. Addition reactions with halogen acids to give haloalkanes:
   $RCH{:}CH_2 + HX \rightarrow RCH_2CH_2X$
   The addition follows Markovnikoff's rule.
3. Addition reactions with halogens, e.g.
   $RCH{:}CH_2 + Br_2 \rightarrow RCHBrCH_2Br$
4. Hydration using concentrated sulfuric acid, followed by dilution and warming:
   $RCH{:}CH_2 + H_2O \rightarrow RCH(OH)CH_3$
5. Oxidation by cold potassium permanganate solutions to give diols:
   $RCH{:}CH_2 + H_2O + [O] \rightarrow RCH(OH)CH_2OH$
   Ethene can be oxidized in air using a silver catalyst to the cyclic compound epoxyethane (C_2H_4O).
6. Polymerization to polyethene (by the Ziegler or Phillips process).

See also oxo process; ozonolysis.

alkoxide /al-**koks**-ÿd/ An organic compound containing an ion of the type RO^-, where R is an alkyl group. Alkoxides are made by the reaction of metallic sodium on an alcohol. Sodium ethoxide ($C_2H_5O^-Na^+$) is a typical example.

alkoxyalkane /al-koks-ee-**al**-kayn/ (**Diethyl ether.**) *See* ether.

alkylbenzene /al-kăl-**ben**-zeen/ A type of organic hydrocarbon containing one or more alkyl groups substituted onto a benzene ring. Methylbenzene ($C_6H_5CH_3$) and 1,3-dimethylbenzene are simple examples. Alkylbenzenes can be made by a Friedel-Crafts reaction or by the Fittig reaction. Industrially, large quantities of methylbenzene are made by the hydroforming of crude oil.

Substitution of alkylbenzenes can occur at the benzene ring. The alkyl group directs the substituent into the 2- or 4-position.

Substitution of hydrogen atoms on the alkyl group can also occur.

alkyl group /**al**-kăl/ A group obtained by removing a hydrogen atom from an alkane or other aliphatic hydrocarbon.

alkyl halide *See* haloalkane.

alkyl sulfide A thioether with the general formula RSR′, where R and R′ are alkyl groups.

alkyne /**al**-kÿn/ A type of hydrocarbon with the general formula C_nH_{2n-2}. The alkynes are unsaturated compounds containing triple carbon–carbon bonds. The simplest member of the series is ethyne (C_2H_2), which can be prepared by the action of water on calcium dicarbide.

$$CaC_2 + 2H_2O \rightarrow Ca(OH)_2 + C_2H_2$$

The alkynes were formerly called the *acetylenes*.

In general, alkynes can be made by the cracking of alkanes or by the action of a hot alcoholic solution of potassium hydroxide on a dibromoalkane, for example:

$$BrCH_2CH_2Br + KOH \rightarrow KBr + CH_2{:}CHBr + H_2O$$

$$CH_2{:}CHBr + KOH \rightarrow CHCH + KBr + H_2O$$

The main reactions of the alkynes are:

1. Hydrogenation with a catalyst (usually nickel at about 150°C):
$$C_2H_2 + H_2 \rightarrow C_2H_4$$
$$C_2H_4 + H_2 \rightarrow C_2H_6$$
2. Addition reactions with halogen acids:
$$C_2H_2 + HI \rightarrow H_2C{:}CHI$$
$$H_2C{:}CHI + HI \rightarrow CH_3CHI_2$$
3. Addition of halogens; for example, with bromine in tetrachloromethane:
$$C_2H_2 + Br_2 \rightarrow BrHC{:}CHBr$$
$$BrHC{:}CHBr + Br_2 \rightarrow Br_2HCCHBr_2$$
4. With dilute sulfuric acid at 60–80°C and mercury(II) catalyst, ethyne forms ethanal (acetaldehyde):
$$C_2H_2 + H_2O \rightarrow H_2C{:}C(OH)H$$
This enol form converts to the aldehyde:
$$CH_3COH$$
5. Ethyne polymerizes if passed through a hot tube to produce some benzene:
$$3C_2H_2 \rightarrow C_6H_6$$
6. Ethyne forms unstable dicarbides (acetylides) with ammoniacal solutions of copper(I) and silver(I) chlorides.

allotropy /ă-**lot**-rŏ-pee/ The ability of certain elements to exist in more than one physical form. Carbon, sulfur, and phosphorus are the most common examples. Allotropy is more common in groups 14, 15, and 16 of the periodic table than in other groups. *See also* enantiotropy; monotropy.

alloy A mixture of two or more metals (e.g. bronze or brass) or a metal with small amounts of non-metals (e.g. steel). Alloys may be completely homogeneous mixtures or may contain small particles of one phase in the other phase.

allyl group /**al**-ăl/ *See* propenyl group.

Alnico /**al**-mă-koh/ (*Trademark*) Any of a group of very hard brittle alloys used to make powerful permanent magnets. They contain nickel, aluminum, cobalt, and copper in various proportions. Iron, titanium, and niobium can also be present. They have a high remanence and coercive force.

alpha particle A He^{2+} ion emitted with high kinetic energy by a radioactive substance. Alpha particles are used to cause nuclear disintegration reactions.

alternating copolymer *See* polymerization.

alum A type of double salt. Alums are double sulfates obtained by crystallizing mixtures in the correct proportions. They have the general formula:

$$M_2SO_4.M'_2(SO_4)_3.24H_2O$$

Where M is a univalent metal or ion, and M′ is a trivalent metal. Thus, aluminum potassium sulfate (called *potash alum*, or simply *alum*) is

$$K_2SO_4.Al_2(SO_4)_3.24H_2O$$

Aluminum ammonium sulfate (called *ammonium alum*) is

$$(NH_4)_2SO_4.Al_2(SO_4)_3.24H_2O$$

The name 'alum' originally came from the presence of Al^{3+} as the trivalent ion, but is also applied to other salts containing trivalent ions, thus, Chromium(III) potassium sulfate (*chrome alum*) is

$$K_2SO_4.Cr_2(SO_4)_3.24H_2O$$

alumina *See* aluminum oxide.

aluminate *See* aluminum hydroxide.

aluminosilicate /ă-loo-mă-noh-**sil**-ă-kayt/ *See* silicates.

aluminum A soft moderately reactive metal; the second element in group 3 of the periodic table. It was formerly classified in subgroup IIIA. Aluminum has the electronic structure of neon plus three additional outer electrons. There are numerous minerals of aluminum; it is the most common metallic element in the Earth's crust (8.1% by weight) and the third in order of abundance. Commercially important minerals are bauxite (hydrated Al_2O_3), corundum (anhydrous Al_2O_3), cryolite (Na_3AlF_6), and clays and mica (aluminosilicates).

The metal is produced on a massive scale by the Hall–Heroult method in which alumina, a non-electrolyte, is dissolved in molten cryolite and electrolyzed. The bauxite contains iron, which would contaminate the product, so the bauxite is dissolved in hot alkali, the iron oxide is removed by filtration, and the pure alumina then precipitated by acidification. Molten aluminum is tapped off from the base of the cell and oxygen evolved at the anode. The aluminum atom is much bigger than boron (the first member of group 3) and its ionization potential is not particularly high. Consequently aluminum forms positive ions Al^{3+}. However, it also has non-metallic chemical properties. Thus, it is amphoteric and also has a number of covalently bonded compounds.

Unlike boron, aluminum does not form a vast range of hydrides – AlH_3 and Al_2H_6 may exist at low pressures, and the only stable hydride, $(AlH_3)_n$, must be prepared by reduction of aluminum trichloride. The ion AlH_4^- is widely used in the form of $LiAlH_4$ as a vigorous reducing agent.

The reaction of aluminum metal with oxygen is very exothermic but at ordinary temperatures an impervious film of the oxide protects the bulk metal from further attack. This oxide film also protects aluminum from oxidizing acids. There is only one oxide, Al_2O_3 (alumina), but a variety of polymorphs and hydrates are known. It is relatively inert and has a high melting point, and for this reason is widely used as a furnace lining and for general refractory brick. Aluminum metal will react with alkalis releasing hydrogen and producing initially $Al(OH)_3$ then $Al(OH)_4^-$.

Aluminum reacts readily with the halogens; in the case of chlorine thin sheets will burst into flame. The fluoride has a high melting point (1290°C) and is ionic. The other halides are dimers in the vapor phase (two halogen bridges). Aluminum also forms a sulfide (Al_2S_3), nitride (AlN), and carbide (Al_4C), the latter two at extremely high temperatures.

Because of aluminum's ability to expand its coordination number and tendency towards covalence it forms a variety of complexes such as AlF_6^{2-} and $AlCl_4^-$. A number of very reactive aluminum alkyls are also known, some of which are important as polymerization catalysts.

Symbol: Al; m.p. 660.37°C; b.p. 2470°C; r.d. 2.698 (20°C); p.n. 13; r.a.m. 26.981539.

aluminum acetate *See* aluminum ethanoate.

aluminum bromide ($AlBr_3$) A white solid soluble in water and many organic solvents.

aluminum chloride ($AlCl_3$) A white covalent solid that fumes in moist air:

$$AlCl_3 + 3H_2O \rightarrow Al(OH)_3 + 3HCl$$

It is prepared by heating aluminum in dry chlorine or dry hydrogen chloride. Vapor-density measurements show that its structure is a dimer; it consists of Al_2Cl_6 molecules in the vapor. The $AlCl_3$ structure would be electron-deficient. Aluminum chloride is used in Friedel–Crafts reactions in organic preparations.

aluminum ethanoate (**aluminum acetate**; $Al(OOCCH_3)_3$) A white solid soluble in water. It is usually obtained as the dibasic salt, *basic aluminum ethanoate*, $Al(OH)(CH_3COO)_2$. It is prepared by dissolving aluminum hydroxide in ethanoic acid and is used extensively as a mordant in dyeing and as a size for paper and cardboard products. The solution is hydrolyzed and contains various complex aluminum-hydroxyl species and colloidal aluminum hydroxide.

aluminum fluoride (AlF_3) A white crystalline solid that is slightly soluble in water but insoluble in most organic solvents. Its primary use is as an additive to the cryolite (Na_3AlF_6) electrolyte in the production of aluminum.

aluminum hydroxide ($Al(OH)_3$) A white powder prepared as a colorless gelatinous precipitate by adding ammonia solution or a small amount of sodium hydroxide solution to a solution of an aluminum salt. It is an amphoteric hydroxide and is used as a foaming agent in fire extinguishers and as a mordant in dyeing.

Its amphoteric nature causes it to dissolve in excess sodium hydroxide solution to form the *aluminate* ion (systematic name tetrahydroxoaluminate(III)):

$$Al(OH)_3 + OH^- \rightarrow Al(OH)_4^- + H_2O$$

When precipitating from solution, aluminum hydroxide readily absorbs colored matter from dyes to form lakes.

aluminum nitrate ($Al(NO_3)_3.9H_2O$) A hydrated white crystalline solid prepared by dissolving freshly prepared aluminum hydroxide in nitric acid. It cannot be prepared by the action of dilute nitric acid on aluminum since the metal is rendered passive by a thin surface layer of oxide.

aluminum oxide (**alumina**; Al_2O_3) A white powder that is almost insoluble in water. Because of its amphoteric nature it will react with both acids and alkalis. Aluminum oxide occurs naturally as bauxite, corundum, and white sapphire; it is manufactured by heating aluminum hydroxide. It is used in the extraction by electrolysis of aluminum, as an abrasive (corundum), in furnace linings (because of its refractory properties), and as a catalyst (e.g. in the dehydration of alcohols).

aluminum potassium sulfate (**potash alum**; $Al_2(SO_4)_3.K_2SO_4.24H_2O$) A white solid, soluble in water but insoluble in alcohol, prepared by mixing equimolecular quantities of solutions of ammonium and aluminum sulfate followed by crystallization. It is used as a mordant for dyes, as a waterproofing agent, and as a tanning additive.

aluminum sulfate ($Al_2(SO_4)_3.18H_2O$) A white crystalline solid. It is used as a size for paper, a precipitating agent in sewage treatment, a foaming agent in fire control, and as a fireproofing agent. Its solutions are acidic by hydrolysis, containing such species as $Al(H_2O)_5(OH)^{2+}$.

aluminum trimethyl /trÿ-**meth**-ăl/ *See* trimethylaluminum.

amalgam /ă-**mal**-găm/ An alloy of mercury with one or more other metals. Amalgams may be liquid or solid. An amalgam of sodium (Na/Hg) with water is used as a source of nascent hydrogen.

amatol /**am**-ă-tol, -tohl/ A high explosive that consists of a mixture of ammonium nitrate and TNT (trinitrotoluene).

ambidentate *See* isomerism.

americium /am-ĕ-**rish**-ee-ŭm/ A highly toxic radioactive silvery element of the actinoid series of metals. A transuranic element, it is not found naturally on Earth but is synthesized from plutonium. The element can be obtained by reducing the trifluoride with barium metal. It reacts with oxygen, steam, and acids. ^{241}Am has been used in gamma-ray radiography.

Symbol: Am; m.p. 1172°C; b.p. 2607°C; r.d. 13.67 (20°C); p.n. 95; most stable isotope ^{243}Am (half-life 7.37×10^3 years).

amethyst A purple form of the mineral

quartz (silicon(IV) oxide, SiO_2) used as a semiprecious gemstone. The color comes from impurities such as oxides of iron.

amide /**am**-ÿd, -id/ **1.** A type of organic compound of general formulae $RCONH_2$ (primary), $(RCO)_2NH$ (secondary), and $(RCO)_3N$ (tertiary). Amides are white, crystalline solids and are basic in nature, some being soluble in water. Amides can be formed by reaction of ammonia with acid chlorides or anhydrides:

$$(RCO)_2O + 2NH_3 \rightarrow RCONH_2 + RCOO^-NH_4^+$$

Reactions of amides include:

1. Reaction with hot acids to give carboxylic acids:
   $$RCONH_2 + HCl + H_2O \rightarrow RCOOH + NH_4Cl$$ 2.
2. Reaction with nitrous acid to give carboxylic acids and nitrogen:
   $$RCONH_2 + HNO_2 \rightarrow RCOOH + N_2 + H_2O$$ 3.
3. Dehydration by phosphorus(V) oxide to give a nitrile:
   $$RCONH_2 - H_2O \rightarrow RCN$$

See also Hofmann degradation.

2. An inorganic salt containing the NH_2^- ion. They are formed by the reaction of ammonia with certain metals (such as sodium and potassium). *See* sodamide.

amination /am-ă-**nay**-shŏn/ The introduction of an amino group ($-NH_2$) into an organic compound. An example is the conversion of an aldehyde or ketone into an amide by reaction with hydrogen and ammonia in the presence of a catalyst:

$$RCHO + NH_3 + N_2 \rightarrow RCH_2NH_2 + H_2O$$

amine /ă-**meen, am**-in/ A compound containing a nitrogen atom bound to hydrogen atoms or hydrocarbon groups. They have the general formula R_3N, where R can be hydrogen or an alkyl or aryl group. Amines can be prepared by reduction of amides or nitro compounds.

An amine is classified according to the number of organic groups bonded to the nitrogen atom: one, primary; two, secondary; three, tertiary. Since amines are basic in nature they can form the quater-

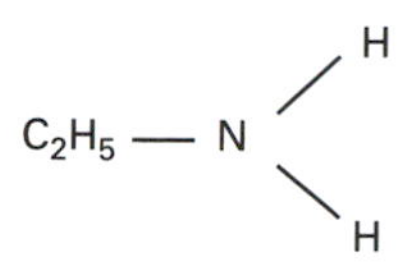

primary (ethylamine)

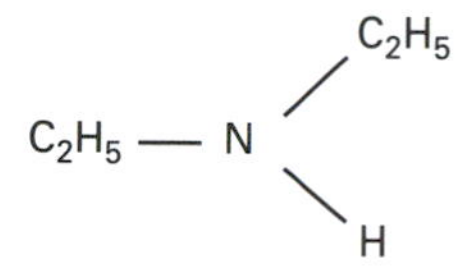

secondary (diethylamine)

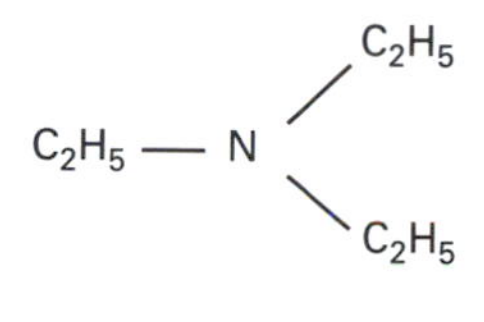

tertiary (triethylamine)

Amines

nium ion, R_3NH^+. All three classes, plus a quaternium salt, can be produced by the Hofmann reaction (which occurs in a sealed vessel at 100°C):

$$RX + NH_3 \rightarrow RNH_3^+ X^-$$
$$RNH_3^+ X^- + NH_3 \rightleftharpoons RNH_2 + NH_4X$$
$$RNH_2 + RX \rightarrow R_2NH_2^+ X^-$$
$$R_2NH_2^+ X^- + NH_3 \rightleftharpoons R_2NH + NH_4X$$
$$R_2NH + RX \rightarrow R_3NH^+ X^-$$
$$R_3NH^+ X^- + NH_3 \rightleftharpoons R_3N + NH_4X$$
$$R_3N + RX \rightarrow R_4N^+X^-$$

Reactions of amines include:

1. Reaction with acids to form salts:
   $$R_3N + HX \rightarrow R_3NH^+X^-$$
2. Reaction with acid chlorides to give *N*-substituted acid amides (primary and secondary amines only):
   $$RNH_2 + R'COCl \rightarrow R'CONHR + HX$$

amine salt A salt similar to an ammonium salt, but with organic groups attached to the nitrogen atom. For example, triethylamine ($(C_2H_5)_3N$) will react with hydrogen chloride to give triethylammonium chloride:

$$(C_2H_5)_3N + HCl \rightarrow (C_2H_3)_3NH^+ Cl^-$$

Sometimes amine salts are named using the suffix '-ium'. For instance, aniline ($C_6H_5NH_2$) forms anilinium chloride $C_6H_5NH_3^+ Cl^-$. Often insoluble alkaloids are used in medicine in the form of their amine salt (sometimes referred to as the 'hydrochloride').

It is also possible for amine salts of this type to have four groups on the nitrogen atom. For example, with chloroethane, tetraethylammonium chloride can be formed:

$$(C_2H_5)_3N + C_2H_5Cl \rightarrow (C_2H_5)_4N^+Cl^-$$

amino acids /ă-**mee**-noh, **am**-ă-/ Derivatives of carboxylic acids in which a hydrogen atom in an aliphatic acid has been replaced by an amino group. Thus, from ethanoic acid, the amino acid 2-aminoethanoic acid (glycine) is formed. All are white, crystalline, soluble in water (but not in alcohol), and with the sole exception of the simplest member, all are optically active.

In the body the various proteins are assembled from the necessary amino acids and it is important therefore that all the amino acids should be present in sufficient quantities. In humans, twelve of the twenty amino acids can be synthesized by the body itself. Since these are not required in the diet they are known as *nonessential amino acids*. The remaining eight cannot be synthesized by the body and have to be supplied in the diet. They are known as *essential* amino acids.

The amino acids that occur in proteins all have the $-NH_2$ group and the $-COOH$ group attached to the same carbon atom. They are thus *alpha amino acids*, the carbon atom being the alpha carbon. They have complex formulae and are usually referred to by their common names, rather than systematic names:

alanine $CH_3CH(NH_2)COOH$
arginine $NH_2C(NH)NH(CH_2)_3CH(NH_2)$-$COOH$
asparagine $NH_2COCH_2CH(NH_2)COOH$
aspartic acid $COOHCH_2CH(NH_2)COOH$
cysteine $SHCH_2CH(NH_2)COOH$
cystine $[HOOCCH(NH_2)CH_2S]_2$
glutamic acid $COOH(CH_2)_2CH(NH_2)$-$COOH$
glutamine $NH_2CH(CH_2)_2(CONH_2)$-$COOH$
glycine $CH_2(NH_2)COOH$
histidine $C_3H_3N_2CH_2CH(NH_2)COOH$
isoleucine $(CH_3)CH_2CH(CH_3)CH(NH_2)$-$COOH$
leucine $(CH_3)_2CHCH_2CH(NH_2)COOH$
lysine $NH_2(CH_2)_4CH(NH_2)COOH$
methionine $CH_3S(CH_2)_2CH(NH_2)COOH$
phenylalanine $C_6H_5CH_2CH(NH_2)COOH$
proline $NH(CH_2)_3CHCOOH$
serine $CH_2OHCH(NH_2)COOH$
threonine $CH_3CHOHCH(NH_2)$-$COOH$
tryptophan $C_6H_4NHC_2HCH_2CH(NH_2)$-$COOH$
tyrosine $C_6H_4OHCH_2CH(NH_2)COOH$
valine $(CH_3)_2CHCH(NH_2)COOH$

Note that proline is in fact a cyclic *imino acid*, with the nitrogen atom bonded to the alpha carbon.

See also optical activity.

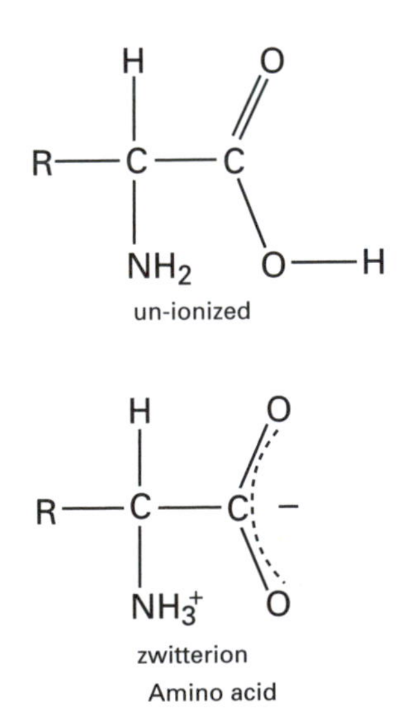

Amino acid

aminobenzene /ă-mee-noh-**ben**-zeen, am-ă-/ *See* aniline.

aminoethane /a-mee-noh-**eth**-ayn, am-ă-/ *See* ethylamine.

amino group The group $-NH_2$.

aminotoluine /ă-mee-noh-**tol**-yoo-een, am-ă-/ *See* toluidine.

ammine /am-een, ă-**meen/** A complex in which ammonia molecules are coordinated to a metal ion; e.g. $[Cu(NH_3)_4]^{2+}$.

ammonia /ă-**moh**-nee-ă/ (NH_3) A colorless gas with a characteristic pungent odor. On cooling and compression it forms a colorless liquid, which becomes a white solid on further cooling. Ammonia is very soluble in water (a saturated solution at 0°C contains 36.9% of ammonia): the aqueous solution is alkaline and contains a proportion of free ammonia. Ammonia is also soluble in ethanol. It occurs naturally to a small extent in the atmosphere, and is usually produced in the laboratory by heating an ammonium salt with a strong alkali. Ammonia is synthesized industrially from hydrogen and atmospheric nitrogen by the Haber process.

The compound does not burn readily in air but ignites, giving a yellowish-brown flame, in oxygen. It will react with atmospheric oxygen in the presence of platinum or a heavy metal catalyst – a reaction used as the basis of the commercial manufacture of nitric acid, which involves the oxidation of ammonia to nitrogen monoxide and then to nitrogen dioxide. Ammonia coordinates readily to form ammines and reacts with sodium or potassium to form inorganic amides and with acids to form ammonium salts; for example, it reacts with hydrogen chloride to form ammonium chloride:

$$NH_3(g) + HCl(g) \rightarrow NH_4Cl(g)$$

Ammonia is also used commercially in the manufacture of fertilizers, mainly ammonium nitrate, urea, and ammonium sulfate. It is used to a smaller extent in the refrigeration industry. Liquid ammonia is an excellent solvent for certain substances, which ionize in the solutions to give ionic reactions similar to those occurring in aqueous solutions. Ammonia is marketed as the liquid, compressed in cylinders ('anhydrous ammonia'), or as aqueous solutions of various strengths. *See also* ammonium hydroxide.

ammoniacal /am-ŏ-**nÿ**-ă-kăl/ Describing a solution in aqueous ammonia.

ammonia-soda process *See* Solvay process.

ammonium alum *See* alum.

ammonium carbonate (**sal volatile**; $(NH_4)_2CO_3$) A white solid that crystallizes as plates or prisms. It is very soluble in water and readily decomposes on heating to ammonia, carbon dioxide, and water. The white solid sold commercially as ammonium carbonate is actually a double salt of both ammonium hydrogencarbonate (NH_4HCO_3) and ammonium aminomethanoate ($NH_2CO_2NH_4$). This salt is manufactured from ammonium chloride and calcium carbonate. It decomposes on exposure to air into ammonium hydrogencarbonate and ammonia, and it reacts with ammonia to give the true ammonium carbonate. Commercial ammonium carbonate is used in baking powders, smelling salts, and in the dyeing and wool-scouring industries.

ammonium chloride (**sal ammoniac**; NH_4Cl) A white crystalline solid with a characteristic saline taste. It is very soluble in water (37 g per 100 g of water at 20°C). Ammonium chloride can be manufactured by the action of ammonia on hydrochloric acid. It sublimes on heating because of the equilibrium:

$$NH_4Cl(s) \rightleftharpoons NH_3(g) + HCl(g)$$

Ammonium chloride is used in galvanizing, as a flux for soldering, in dyeing and calico printing, and in the manufacture of Leclanché and 'dry' cells.

ammonium hydroxide (**ammonia solution**; NH_4OH) An alkali that is formed when ammonia dissolves in water. It probably contains hydrated ammonia molecules as well as some NH_4^+ and OH^- ions. A saturated aqueous solution of ammonia has a relative density of 0.88 g cm^{-3}, and is known as 880 ammonia. Ammonia solution is a useful reagent and cleansing agent.

ammonium ion The ion NH_4^+, formed by coordination of NH_3 to H^+. *See also* quaternary ammonium compound.

ammonium nitrate (NH_4NO_3) A colorless crystalline solid that is very soluble in water (871 g per 100 g of water at 100°C). It is usually manufactured by the action of ammonia on nitric acid. It is used in the manufacture of explosives and, because of its high nitrogen content, as a fertilizer.

ammonium phosphate (**triammonium phosphate (V)**; $(NH_4)_3PO_4$) A colorless crystalline salt made from ammonia and phosphoric(V) acid, used as a fertilizer to add both nitrogen and phosphorus to the soil.

ammonium sulfate ($(NH_4)_2SO_4$) A colorless crystalline solid that is soluble in water. When heated carefully it gives ammonium hydrogensulfate, which on stronger heating yields nitrogen, ammonia, sulfur(IV) oxide (sulfur dioxide), and water. Ammonium sulfate is manufactured by the action of ammonia on sulfuric acid. It is the most important ammonium salt because of its widespread use as a fertilizer. Its only drawback as a fertilizer is that it tends to leave an acidic residue in the soil.

amorphous /ă-**mor**-fŭs/ Describing a solid substance that has no 'long-range' regular arrangement of atoms; i.e. is not crystalline. Amorphous materials can consist of minute particles that possess order over a very short distance. Glasses are also amorphous; the atoms in the solid have a random arrangement. X-ray analysis has shown that many substances that were once described as amorphous are composed of very small crystals. For example, charcoal, coke, and soot (all forms of carbon) are made up of small graphite-like crystals.

amount of substance Symbol: *n* A measure of the number of entities present in a substance. *See* mole.

AMP (**adenosine monophosphate**) A NUCLEOTIDE consisting of adenine, ribose, and phosphate. *See* ATP.

ampere /**am**-pair/ Symbol: A The SI base unit of electric current, defined as the constant current that, maintained in two straight parallel infinite conductors of negligible circular cross section placed one meter apart in vacuum, would produce a force between the conductors of 2×10^{-7} newton per meter. The unit is named for the French physicist and mathematician André Marie Ampère (1775–1836).

amphiprotic /am-fă-**proh**-tik/ *See* solvent.

ampholyte ion /**am**-fŏ-lÿt/ *See* zwitterion.

amphoteric /am-fŏ-**te**-rik/ A material that can display both acidic and basic properties. The term is most commonly applied to the oxides and hydroxides of metals that can form both cations and complex anions. For example, zinc oxide dissolves in acids to form zinc salts and also dissolves in alkalis to form zincates, $[Zn(OH)_4]^{2-}$.

amu /ay-em-**yoo**/ *See* atomic mass unit.

amyl group /**am**-ăl/ *See* pentyl group.

amyl nitrite ($C_5H_{11}ONO$) A pale brown volatile liquid organic compound; a nitrous acid ester of 3-methylbutanol (isoamyl alcohol). It is used in medicine as an inhalant to dilate the blood vessels (and thereby prevent pain) in patients with angina pectoris.

amylopectin /am-ă-lo-**pek**-tin/ The water-insoluble fraction of STARCH.

amylose /**am**-ă-lohs/ A polymer of GLUCOSE, a polysaccharide sugar that is found in starch.

anabolism /ă-**nab**-ŏ-liz-ăm/ All the metabolic reactions that synthesize complex molecules from more simple molecules. *See also* metabolism.

anaerobic /an-air-**oh**-bik/ Describing a biochemical process that takes place in the absence of free oxygen. *Compare* aerobic.

analysis The process of determining the constituents or components of a sample. There are two broad major classes of analysis, *qualitative analysis* – essentially answering the question 'what is it?' – and *quantitative analysis* – answering the question 'how much of such and such a component is present?' There is a vast number of analytical methods which can be applied, depending on the nature of the sample and the purpose of the analysis. These include gravimetric, volumetric, and systematic qualitative analysis (classical wet methods); and instrumental methods, such as chromatographic, spectroscopic, nuclear, fluorescence, and polarographic techniques.

Andrews' experiment An investigation (1861) into the relationship between pressure and volume for a mass of carbon dioxide at constant temperature. The resulting isothermals showed clearly the existence of a critical point and led to greater understanding of the liquefaction of gases. The experiment is named for the Irish physical chemist Thomas Andrews (1813–1885).

ångstrom /**ang**-strŏm/ Symbol: Å A unit of length defined as 10^{-10} meter. The ångstrom is sometimes still used for expressing wavelengths of light or ultraviolet radiation or for the sizes of molecules, although the nanometer is preferred. The unit is named for the Swedish physicist Anders Jonas Ångstrom (1814–74).

anhydride /an-**hÿ**-drÿd/ A compound formed by removing water from an acid or, less commonly, a base. Many non-metal oxides are anhydrides of acids: for example CO_2 is the anhydride of H_2CO_3 and SO_3 is the anhydride of H_2SO_4. Organic anhydrides are formed by removing H_2O from two carboxylic-acid groups, giving compounds with the functional group –CO.O.CO–. These form a class of organic compounds called ACID ANHYDRIDES (or *acyl anhydrides*).

anhydrite /an-**hÿ**-drÿt/ *See* calcium sulfate.

anhydrous /an-**hÿ**-drŭs/ Describing a substance that lacks moisture, or a salt with no water of crystallization. For example, on strong heating, blue crystals of copper(II) sulfate pentahydrate, $CuSO_4.5H_2O$, form white anhydrous copper(II) sulfate, $CuSO_4$.

aniline /**an**-ă-lin, -lÿn/ (**aminobenzene; phenylamine**; $C_6H_5NH_2$) A colorless oily substance made by reducing nitrobenzene ($C_6H_5NO_2$). Aniline is used for making

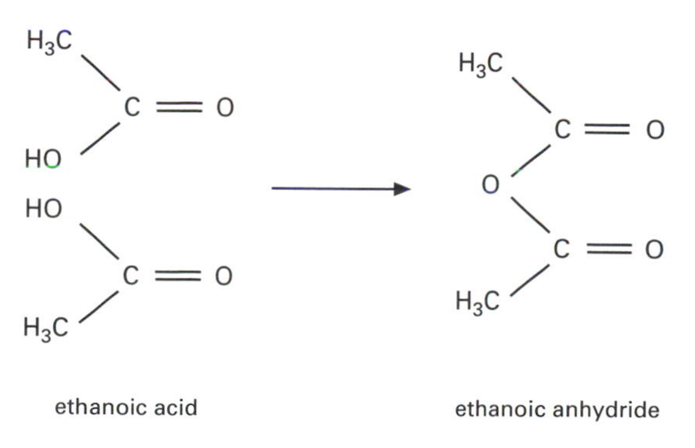

Anhydride

dyes, pharmaceuticals, and other organic compounds.

anion /**an**-ÿ-ŏn, -on/ A negatively charged ion, formed by addition of electrons to atoms or molecules. In electrolysis anions are attracted to the positive electrode (the anode). *Compare* cation.

anionic detergent /an-ÿ-**on**-ik/ *See* detergent.

anionic resin An ION-EXCHANGE material that can exchange anions, such as Cl^- and OH^-, for anions in the surrounding medium. Such resins are used for a wide range of analytical and purification purposes.

They are often produced by addition of a quaternary ammonium group ($-N(CH_3)_3^+$) or a phenolic group ($-OH^-$) to a stable polyphenylethene resin. A typical exchange reaction is:

$$\text{resin}-N(CH_3)_3^+Cl^- + KOH \rightleftharpoons \text{resin}-N(CH_3)_3^+OH^- + KCl$$

Anionic resins can be used to separate mixtures of halide ions. Such mixtures can be attached to the resin and recovered separately by elution.

anisotropic /an-ÿ-sŏ-**trop**-ik/ A term descriptive of certain substances which have one or more physical properties that differ according to direction. Most crystals are anisotropic.

annealing /ă-**neel**-ing/ A type of heat treatment applied to metals to change their physical properties. The metal is heated to, and held at, an appropriate temperature before being cooled at a suitable rate to produce the desired grain structure. Annealing is most commonly used to remove the stresses that have arisen during rolling, to increase the softness of the metal, and to make it easier to machine. Objects made of glass can also be annealed to remove strains.

annulene /**an**-yŭ-leen/ A ring compound containing alternating double and single C–C bonds. The compound C_8H_8, having an 8-membered ring of carbon atoms, is the first annulene larger than benzene. It is not an AROMATIC COMPOUND because it is not planar and does not obey the Hückel rule. C_8H_8 is called *cyclo-octatetrane*. Higher annulenes are designated by the number of carbon atoms in the ring. [10]-annulene obeys the Hückel rule but is not aromatic because it is not planar as a result of interactions of the hydrogen atoms inside the ring. There is some evidence that [18]-annulene, which is a stable red solid, has aromatic properties.

anode /**an**-ohd/ In electrolysis, the electrode that is at a positive potential with respect to the cathode. In any electrical system, such as a discharge tube or electronic device, the anode is the terminal at which electrons flow out of the system.

anode sludge *See* electrolytic refining.

anodizing /**an**-ŏ-dÿz-ing/ An industrial process for protecting aluminum with an oxide layer formed in an electrolytic cell containing an oxidizing acid (e.g. sulfuric(VI) acid). The layer of Al_2O_3 is porous and can be colored with certain dyes.

anomer /**an**-ŏ-mer/ Either of two isomeric forms of a cyclic form of a sugar that differ in the disposition of the –OH group on the carbon next to the O atom of the ring (the *anomeric carbon*). Anomers are diastereoisomers. They are designated α– or β– according to whether the –OH is above or below the ring respectively.

anthracene /**an**-thră-seen/ ($C_{14}H_{10}$) A white crystalline solid used extensively in the manufacture of dyes. Anthracene is found in the heavy- and green-oil fractions of crude oil and is obtained by fractional crystallization. Its structure is benzene-like, having three six-membered rings fused to-

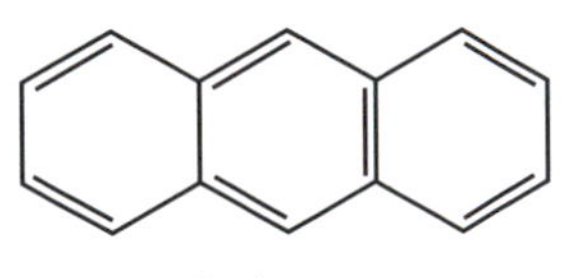

Anthracene

gether. The reactions are characteristic of AROMATIC COMPOUNDS.

anthracite /**an**-thră-sÿt/ The highest grade of coal, with a carbon content of between 92% and 98%. It burns with a hot blue flame, gives off little smoke and leaves hardly any ash.

anthraquinone /an-thră-kwi-**nohn**, an-thră-**kwin**-ohn/ ($C_6H_4(CO)_2C_6H_4$) A colorless crystalline quinone used in producing dyestuffs such as alizarin.

antibonding orbital *See* orbital.

anti-isomer /an-tee-**ÿ**-sŏ-mer/ *See* isomerism.

antiknock agent /an-tee-**nok**/ A substance added to gasoline to inhibit preignition or 'knocking'. A common example is lead tetraethyl.

antimonic /an-tă-**mon**-ik/ Designating an antimony(IV) compound.

antimonous /**an**-tă-moh-nŭs/ Designating an antimony(III) compound.

antimony /**an**-tă-moh-nee/ A metalloid element existing in three allotropic forms; the most stable is a brittle silvery metal. Antimony belongs to group 15 (formerly VB) of the periodic table. It is found in many minerals, principally stibnite (Sb_2S_3). It is used in alloys – small amounts of antimony can harden other metals. It is also used in semiconductor devices.

Symbol: Sb; m.p. 630.74°C; b.p. 1635°C; r.d. 6.691; p.n. 51; r.a.m. 112.74.

antimony(III) chloride (**antimony trichloride**; $SbCl_3$) A white deliquescent solid, formerly known as *butter of antimony*. It is prepared by direct combination of antimony and chlorine. It is readily hydrolyzed by cold water to form a white precipitate of *antimony(III) chloride oxide* (antimonyl chloride, SbOCl):

$$SbCl_3 + H_2O = SbOCl + 2HCl$$

antimony(III) chloride oxide *See* antimony(III) chloride.

antimonyl chloride /**an**-tă-mŏ-nil, an-**tim**-ŏ-nil/ *See* antimony(III) chloride.

antimony(III) oxide (**antimony trioxide**; Sb_2O_3) A white insoluble solid. It is an amphoteric oxide with a strong tendency to act as a base. It can be prepared by direct oxidation by air, oxygen, or steam and is formed when antimony(III) chloride is hydrolyzed by excess boiling water.

antimony(V) oxide (**antimony pentoxide**; Sb_2O_5) A yellow solid. It is usually formed by the action of concentrated nitric acid on antimony or by the hydrolysis of antimony(V) chloride. Although an acidic oxide, it is only slightly soluble in water.

antimony pentoxide /pen-**toks**-ÿd/ *See* antimony(V) oxide.

antimony trichloride /trÿ-**klor**-ÿd, -**kloh**-rÿd/ *See* antimony(III) chloride.

antimony trioxide /trÿ-**oks**-ÿd/ *See* antimony(III) oxide.

antioxidant /an-tee-**oks**-ă-dănt/ A substance that inhibits oxidation. Antioxidants are added to such products as foods, paints, plastics, and rubber to delay their oxidation by atmospheric oxygen. Some work by forming chelates with metal ions, thus neutralizing the catalytic effect of the ions in the oxidation process. Other types remove intermediate oxygen free radicals. Naturally occurring antioxidants can limit tissue or cell damage in the body. These include vitamin E and β-carotene.

antiparallel spins /an-tee-**pa**-ră-lel/ Spins of two neighboring particles in which the magnetic moments associated with the spin are aligned in opposite directions.

apatite /**ap**-ă-tÿt/ A naturally occurring phosphate of calcium, $CaF_2.Ca_3(PO_4)_3$.

aprotic /ă-**prot**-ik, -**proh**-tik/ *See* solvent.

aqua fortis /**a**-kwă **for**-tis/ An old name for nitric acid, HNO_3.

aqua regia /**ree**-jee-ă/ A mixture of concentrated nitric acid and three to four parts of hydrochloric acid. It dissolves all metals including gold, hence the name. The mixture contains chlorine and NOCl (nitrosyl chloride).

aqueous /**ay**-kwee-ŭs, **ak**-wee-/ Describing a solution in water.

aragonite /ă-**rag**-ŏ-nÿt, **a**-ră-gŏ-nÿt/ An anhydrous mineral form of calcium carbonate, $CaCO_3$, which occurs associated with limestone and in some metamorphic rocks. It is also the main ingredient of pearls. It is not as stable as calcite, into which it may change over time.

arene /ă-**reen**/ An organic compound containing a benzene ring; i.e. an aromatic hydrocarbon or derivative.

argentic oxide /ar-**jen**-tik/ *See* silver(II) oxide.

argentous oxide /ar-**jen**-tŭs/ *See* silver(I) oxide.

arginine /**ar**-jă-nÿn/ *See* amino acids.

argon /**ar**-gon/ An inert colorless odorless monatomic element of the rare-gas group. It forms 0.93% by volume of air. Argon is used to provide an inert atmosphere in electric and fluorescent lights, in welding, and in extracting titanium and silicon. The element forms no known compounds.

Symbol: Ar; m.p. −189.37°C; b.p. −185.86°C; d. 1.784 kg m^{-3} (0°C); p.n. 18; r.a.m. 39.95.

aromatic compound An organic compound containing benzene rings in its structure. Aromatic compounds, such as benzene, have a planar ring of atoms linked by alternate single and double bonds. The characteristic of aromatic compounds is that their chemical properties are not those expected for an unsaturated compound; they tend to undergo nucleophilic substitution of hydrogen (or other groups) on the ring, and addition reactions only occur under special circumstances.

The explanation of this behavior is that the electrons in the double bonds are delocalized over the ring, so that the six bonds are actually all identical and intermediate between single bonds and double bonds. The pi electrons are thus spread in a molecular orbital above and below the ring. The evidence for this delocalization in benzene is that: The bond lengths between carbon atoms in benzene are all equal and intermediate between single and double bond lengths. Also, if two hydrogen atoms attached to adjacent carbon atoms are substituted by other groups, the compound has only one structure. If the bonds were different two isomers would exist. Benzene has a stabilization energy of 150 kJ mol^{-1} over the Kekulé structure.

The delocalization of the electrons in the pi orbitals of benzene accounts for the properties of benzene and its derivatives, which differ from the properties of alkenes and other aliphatic compounds. The phenomenon is called *aromaticity*. A definition of aromaticity is that it occurs in compounds that obey the *Hückel rule*: i.e. that there should be a planar ring with a total of $(4n + 2)$ pi electrons (where n is any integer). Using this rule as a criterion certain non-benzene rings show aromaticity. Such compounds are called *nonbenzenoid aromatics*. Other compounds that have a ring of atoms with alternate double and single bonds, but do not obey the rule (e.g. cyclooctotetraene, which has a non-planar ring of alternating double and single bonds) are called *pseudoaromatics*. The rule is named for the German chemist Erich Armand Arthur Joseph Hückel (1896–1980).

Compare aliphatic compound. *See also* resonance.

aromaticity /ă-roh-mă-**tis**-ă-tee/ *See* aromatic compound.

Arrhenius equation /ah-**ren**-ee-ŭs/ An equation relating the rate constant of a chemical reaction and the temperature at which the reaction is taking place:

$$k = A\exp(-E_a/RT)$$

where A is a constant, k the rate constant, T the thermodynamic temperature in kelvins, R the gas constant, and E_a the activation energy of the reaction.

Reactions proceed at different rates at different temperatures, i.e. the magnitude of the rate constant is temperature dependent. The Arrhenius equation is often written in a logarithmic form, i.e.

$$\log_e k = \log_e A - E/2.3RT$$

This equation enables the activation energy for a reaction to be determined. The equation is named for the Swedish physical chemist Svante August Arrhenius (1859–1927).

arsenate(III) /**ar**-sĕ-nayt/ (**arsenite**) A salt of the hypothetical arsenic(III) acid, formed by reacting arsenic(III) oxide with alkalis. Arsenate(III) salts contain the ion AsO_3^{3-}. Copper arsenate(III) is used as an insecticide.

arsenate(V) A salt of arsenic(V) acid, made by reacting arsenic(III) oxide, As_2O_3, with nitric acid. Arsenate(V) salts contain the ion AsO_4^{3-}. Disodiumhydrogenarsenate(V) is used in printing calico.

arsenic /**ar**-sĕ-nik, **ars**-nik; *adj.* ars-**sen**-ik/ A toxic metalloid element existing in several allotropic forms; the most stable is a brittle gray metal. It belongs to group 15 (formerly VA) of the periodic table. Arsenic is found native and in several ores including mispickel (FeSAs), realgar (As_4S_4), and orpiment (As_2S_3). The element reacts with hot acids and molten sodium hydroxide but is unaffected by water and acids and alkalis at normal temperatures. It is used in semiconductor devices, alloys, and gun shot. Various compounds are used in medicines and agricultural insecticides and poisons.

Symbol: As; m.p. 817°C (gray) at 3 MPa pressure; sublimes at 616°C (gray); r.d. 5.78 (gray at 20°C); p.n. 33; r.a.m. 74.92159.

arsenic(III) chloride (**arsenious chloride**; $AsCl_3$) A poisonous oily liquid. It fumes in moist air due to hydrolysis with water vapor:

$$AsCl_3 + 3H_2O = As_2O_3 + 6HCl$$

Arsenic(III) chloride is covalent and exhibits nonmetallic properties.

arsenic hydride *See* arsine.

arsenic(III) oxide (**white arsenic**; **arsenious oxide**; As_2O_3) A colorless crystalline solid that is very poisonous (0.1 g would be a lethal dose). Analysis of the solid and vapor states suggests a dimerized structure of As_4O_6. An amphoteric oxide, arsenic(III) oxide is sparingly soluble in water, producing an acidic solution. It is formed when arsenic is burned in air or oxygen.

arsenic(V) oxide (**arsenic oxide**; As_2O_5) A white amorphous deliquescent solid. It is an acidic oxide prepared by dissolving arsenic(III) oxide in hot concentrated nitric acid, followed by crystallization then heating to 210°C.

arsenide /**ar**-sĕ-nÿd/ A compound of arsenic and another metal. For example, with iron arsenic forms iron(III) arsenide, $FeAs_2$, and gallium arsenide, GaAs, is an important semiconductor.

arsenious chloride /ar-**sen**-ee-ŭs/ *See* arsenic(III) chloride.

arsenious oxide *See* arsenic(III) oxide.

arsenite /**ar**-sĕ-nÿt/ *See* arsenate(III).

arsine /ar-**seen**, **ar**-seen, -sin/ (**arsenic hydride**; AsH_3) A poisonous colorless gas with an unpleasant smell. It decomposes to arsenic and hydrogen at 230°C. It is produced in the analysis for arsenic (*Marsh's test*).

artificial radioactivity Radioactivity induced by bombarding stable nuclei with high-energy particles. For example:

$$^{27}_{13}Al + ^{1}_{0}n \rightarrow ^{24}_{11}Na + ^{4}_{2}He$$

represents the bombardment of aluminum with neutrons to produce an isotope of sodium. All the transuranic elements (atomic numbers 93 and above) are artificially radioactive since they do not occur in nature.

aryl group /**a**-răl/ An organic group derived by removing a hydrogen atom from an aromatic hydrocarbon or derivative.

asbestos /ass-**best**-ŏs/ A fibrous variety of various rock-forming silicate minerals, such as the amphiboles and chrysotile. It has many uses that employ its properties of heat-resistance and chemical inertness. Prolonged exposure to asbestos dust may cause asbestosis – a form of lung cancer.

asparagine /ă-**spa**-ră-jeen, -jin/ *See* amino acids.

aspartic acid /ă-**spar**-tik/ *See* amino acids.

aspirator An apparatus for sucking a gas or liquid from a vessel or body cavity.

aspirin (**acetylsalicylic acid**; $C_9H_8O_4$) A colorless crystalline compound made by treating salicylic acid with ethanoyl hydride. It is used as an analgesic and antipyretic drug, and small doses are prescribed for patients at risk of heart attack or stroke. It should not be given to young children.

association The combination of molecules of a substance with those of another to form more complex species. An example is a mixture of water and ethanol (which are termed *associated liquids*), the molecules of which combine via hydrogen bonding.

astatine /**ass**-tă-teen, -tin/ A radioactive element belonging to the halogen group. It occurs in minute quantities in uranium ores. Many short-lived radioisotopes are known, all alpha-particle emitters.

Symbol: At; m.p. 302°C (est.); b.p. 337°C (est.); p.n. 85; most stable isotope ^{210}At (half-life 8.1 hours).

asymmetric atom *See* chirality; isomerism; optical activity.

atactic polymer *See* polymerization.

atmolysis /at-**mol**-ă-sis/ The separation of gases by using their different rates of diffusion.

atmosphere A unit of pressure defined as 101 325 pascals (atmospheric pressure). The atmosphere is used in chemistry only for rough values of pressure; in particular, for stating the pressures of high-pressure industrial processes.

atom The smallest part of an element that can exist as a stable entity. Atoms consist of a small dense positively charged nucleus, made up of neutrons and protons, with electrons in a cloud around this nucleus. The chemical reactions of an element are determined by the number of electrons (which is equal to the number of protons in the nucleus). All atoms of a given element have the same number of protons (the proton number). A given element may have two or more isotopes, which differ in the number of neutrons in the nucleus.

The electrons surrounding the nucleus are grouped into *shells* – i.e. main orbits around the nucleus. Within these main orbits there may be sub-shells. These correspond to atomic orbitals. An electron in an atom is specified by four quantum numbers:

1. The *principal quantum number* (n), which specifies the main energy levels. n can have values 1, 2, etc. The corresponding shells are denoted by letters K, L, M, etc., the K shell (n = 1) being the nearest to the nucleus. The maximum number of electrons in a given shell is $2n^2$.
2. The *orbital quantum number* (l), which specifies the angular momentum. For a given value of n, *1* can have possible values of n–1, n–2, ... 2, 1, 0. For instance, the M shell (n = 3) has three sub-shells with different values of l (0, 1, and 2). Sub-shells with angular momentum 0, 1, 2, and 3 are designated by letters s, p, d, and f.

3. The *magnetic quantum number* (m). This can have values $-l$, $-(l - 1)$ ... 0 ... + $(l + 1)$, + l. It determines the orientation of the electron orbital in a magnetic field.
4. The *spin quantum number* (m_s), which specifies the intrinsic angular momentum of the electron. It can have values +½ and –½.

Each electron in the atom has four quantum numbers and, according to the Pauli exclusion principle, no two electrons can have the same set of quantum numbers. This explains the electronic structure of atoms. *See also* Bohr theory.

atomic absorption spectroscopy (**AAS**) A technique in chemical analysis in which a sample is vaporized and an absorption spectrum is taken of the vapor. The elements present are identified by their characteristic absorption lines.

atomic emission spectroscopy (**AES**) A technique in chemical analysis that involves vaporizing a sample of material at high temperature. Atoms in excited states decay to the ground state, emitting electromagnetic radiation at particular frequencies characteristic of that type of atom.

atomic force microscope (**AFM**) An instrument used to investigate surfaces. A small probe consisting of a very small chip of diamond is held just above a surface of a sample by a spring-loaded cantilever. As the probe is slowly moved over the surface the force between the surface and the tip is measured and the probe is automatically raised and lowered to keep this force constant. Scanning the surface in this way enables a contour map of the surface to be generated with the help of a computer. An atomic force microscope closely resembles a SCANNING TUNNELLING MICROSCOPE (STM) in some ways, although it uses forces rather than electrical signals to investigate the surface. Like a STM, it can resolve individual molecules. Unlike a STM, it can be used to investigate nonconducting materials, a feature that is useful in investigating biological samples.

atomic heat *See* Dulong and Petit's law.

atomicity /at-ŏ-**mis**-ă-tee/ The number of atoms per molecule of an element. Helium, for example, has an atomicity of one, nitrogen two, and ozone three.

atomic mass unit (**amu; dalton**) Symbol: u A unit of mass used for atoms and molecules, equal to 1/12 of the mass of an atom of carbon-12. It is equal to $1.660\ 33 \times 10^{-27}$ kg.

atomic number *See* proton number.

atomic orbital *See* orbital.

atomic weight *See* relative atomic mass (r.a.m.).

ATP (**adenosine triphosphate**) The universal energy carrier of living cells. Energy from respiration or, in photosynthesis, from sunlight is used to make ATP from ADP. It is then reconverted to ADP in various parts of the cell by enzymes known as *ATPases*, the energy released being used to drive three main cellular processes: mechanical work (muscle contraction and cellular movement); the active transport of molecules and ions; and the biosynthesis of biomolecules. It can also be converted to light, electricity, and heat.

ATP is a NUCLEOTIDE consisting of adenine and ribose with three phosphate groups attached. Hydrolysis of the terminal phosphate bond releases energy (30.6 kJ mol^{-1}) and is coupled to an energy-requiring process. Further hydrolysis of ADP to AMP sometimes occurs, releasing more energy.

atto- Symbol: a A prefix denoting 10^{-18}. For example, 1 attometer (am) = 10^{-18} meter (m).

Aufbau principle /**owf**-bow/ A principle that governs the order in which the atomic orbitals are filled in elements of successive proton number; i.e. a statement of the order of increasing energy. The order is as follows:

$1s^2$, $2s^2$, $2p^6$, $3s^2$, $3p^6$, $4s^2$, $3d^{10}$, $4p^6$, $5s^2$, $4d^{10}$, $5p^6$, $6s^2$, $4f^{14}$, $5d^{10}$, $6p^6$, $7s^2$, $5f^{14}$, $6d^{10}$
(the superscript indicates the maximum number of electrons for each level).
Note

1. Hund's rule of maximum multiplicity applies; i.e. degenerate orbitals are occupied singly before spin pairing occurs.
2. the unexpected position of the d-levels, which give rise to the first, second, and third transition series.
3. the unusual position of the f-levels, giving rise to the lanthanoids and actinoids.

'Aufbau' is a German word meaning 'building up'.

Auger effect /oh-**zhay**/ An effect in which an excited ion decays by emission of an electron (rather than a photon). For example, if a substance is bombarded by high-energy electrons or gamma rays, an electron from an inner shell may be ejected. The result is a positive ion in an excited state. This ion will decay to its ground state by a transition of an outer electron to an inner shell. The energy released in the transition may result in the emission of a photon in the x-ray region of the electromagnetic spectrum (this is x-ray fluorescence). Alternatively, the energy may be released in the form of a second electron ejected from the atom to give a doubly charged ion. The emitted electron (known as an *Auger electron*) has a characteristic energy corresponding to the difference in energy levels in the ion. The Auger effect is a form of autoionization. The effect is named for the French physicist Pierre Auger (1899–93).

auric /**ô**-rik/ Designating a compound of gold(III).

auric chloride /**or**-ik/ *See* gold(III) chloride.

aurous /**ôr**-ŭs/ Designating a compound of gold(I).

autocatalysis /aw-toh-kă-**tal**-ă-sis/ *See* catalyst.

autoclave /**aw**-tŏ-klayv/ An apparatus consisting of an airtight container whose contents are heated by high-pressure steam; the contents may also be agitated. Autoclaves are used for reactions between gases under pressure in industrial processing and for sterilizing objects.

autoionization /aw-toh-ÿ-ŏ-ni-**zay**-shŏn/ The spontaneous ionization of excited atoms, ions or molecules as in the Auger effect

Avogadro constant /ah-vŏ-**gah**-droh/ (**Avogrado's number**) Symbol: N_A The number of particles in one mole of a substance. Its value is 6.022 52 × 10^{23} mol^{-1}. The constant is named for the Italian physicist and chemist Lorenzo Romano Amedeo Carlo Avogadro, Count of Quaregna and Cerreto (1776–1856).

Avogadro's law The principle that equal volumes of all gases at the same temperature and pressure contain equal numbers of molecules. It is often called *Avogadro's hypothesis.* It is strictly true only for ideal gases.

Avogadro's number *See* Avogadro constant.

azeotrope /ă-zee-ŏ-**trop**-ik/ (**azeotropic mixture**) A mixture of liquids for which the vapor phase has the same composition as the liquid phase. It therefore boils without change in composition and consequently without progressive change in boiling point.

The composition and boiling points of azeotropes vary with pressure, indicating that they are not chemical compounds. Azeotropes may be broken by distillation in the presence of a third liquid, by chemical reactions, adsorption, or fractional crystallization. *See* constant-boiling mixture.

azeotropic distillation A method used to separate mixtures of liquids that cannot be separated by simple distillation. Such a mixture is called an *azeotrope.* A solvent is added to form a new azeotrope with one of

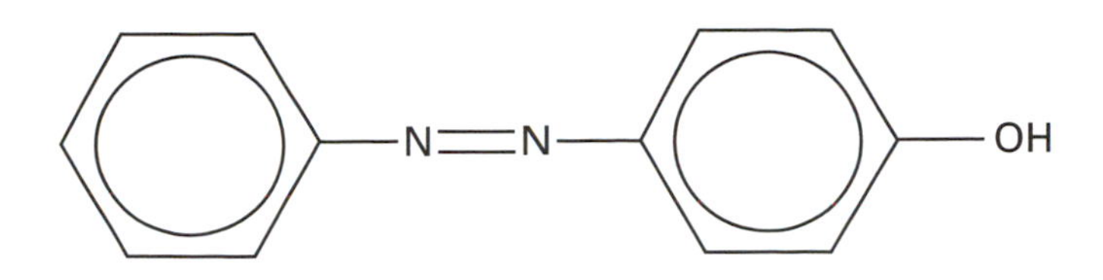

Azo compound

the components, and this is then removed and subsequently separated in a second column. An example of the use of azeotropic distillation is the dehydration of 96% ethanol to absolute ethanol. Azeotropic distillation is not widely used because of the difficulty of finding inexpensive non-toxic non-corrosive solvents that can easily be removed from the new azeotrope.

azeotropic mixture *See* azeotrope.

azide /**az**-ÿd, -id, **ay**-zÿd/ **1.** An inorganic compound containing the ion N_3^-.
2. An organic compound of general formula RN_3.

azine /**az**-een, -in/ An organic heterocyclic compound that has a hexagonal ring containing carbon and nitrogen atoms. Pyridine (C_5H_5N) is the simplest example.

azo compound /**az**-oh, **ay**-zoh/ A type of organic compound of the general formula RN:NR′, where R and R′ are aromatic groups. Azo compounds can be formed by coupling a diazonium compound with an aromatic phenol or amine. Most are colored because of the presence of the *azo group* –N:N–.

azo dye An important type of dye used in acid dyes for wool and cotton. The dyes are azo compounds; usually sodium salts of sulfonic acids.

azo group *See* azo compound.

azoimide /az-oh-**im**-ÿd, -id, ay-zoh-/ *See* hydrazoic acid.

azulene /**az**-yŭ-leen/ ($C_{10}H_8$) A blue crystalline compound having a seven-membered ring fused to a five-membered ring. It converts to naphthalene on heating.

azurite /**azh**-ŭ-rÿt/ *See* copper(II) carbonate.

Babo's law /**bah**-bohz/ The principle that if a substance is dissolved in a liquid (solvent) the vapor pressure of the liquid is reduced; the amount of lowering is proportional to the amount of solute dissolved. *See also* Raoult's law. The law is named for the German chemist Lambert Heinrich Clemens von Babo (1818–99).

back e.m.f. An e.m.f. that opposes the normal flow of electric charge in a circuit or circuit element. In some electrolytic cells a back e.m.f. is caused by the layer of hydrogen bubbles that builds up on the cathode as hydrogen ions pick up electrons and form gas molecules (i.e. as a result of polarization of the electrode).

Bakelite /**bay**-kĕ-lÿt/ (*Trademark*) A common thermosetting synthetic polymer formed by the condensation of phenol and methanal.

baking powder A mixture of sodium hydrogencarbonate (sodium bicarbonate, baking soda) and a weak acid such as tartaric acid. The addition of moisture or heating causes a reaction that produces bubbles of carbon dioxide gas, which make dough or cake mixture rise.

baking soda *See* sodium hydrogencarbonate.

ball mill A device commonly used in the chemical industry for reducing the size of solid material. Ball mills usually have slowly rotating steel-lined drums, which contain steel balls. The material is crushed by the tumbling action of the contents of the drum. *Compare* hammer mill.

Balmer series /**bahl**-mer/ A series of lines in the spectrum of radiation emitted by excited hydrogen atoms. The lines correspond to the atomic electrons falling into the second lowest energy level, emitting energy as radiation. The wavelengths (λ) of the radiation in the Balmer series are given by:

$$1/\lambda = R(1/2^2 - 1/n^2)$$

where n is an integer and R is the Rydberg constant. The series is named for the Swiss mathematician Johann Jakob Balmer (1825–98). *See* Bohr theory. *See also* spectral series.

banana bond (**bent bond**) **1.** In strained-ring compounds the angles assumed on the basis of hybridization are often not equal to the angles obtained by joining the atomic centers. In these cases it is sometimes assumed that the bonding orbital is bent or banana-like in shape. Cyclopropane is an example, in which geometric considerations imply a bond angle of 60° while sp^3 hybridization implies an inter-orbital angle of around 100°, giving a banana bond.
2. A multicentre bond of the type present in such compounds as diborane (B_2H_6).

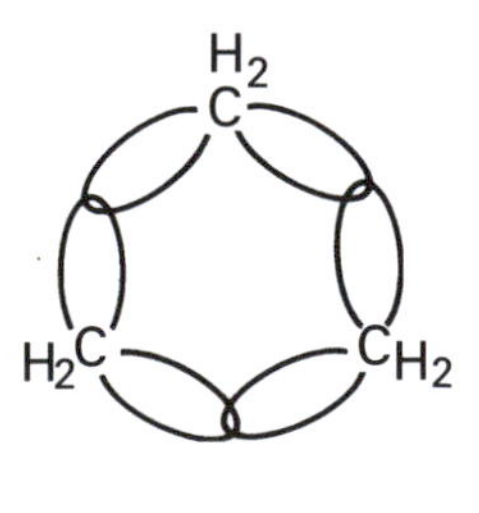

Banana bond

band spectrum A spectrum that appears as a number of bands of emitted or absorbed radiation. Band spectra are characteristic of molecules. Often each band can be resolved into a number of closely spaced lines. The bands correspond to changes of electron orbit in the molecules. The close lines seen under higher resolution are the result of different vibrational states of the molecule. *See also* spectrum.

bar A unit of pressure defined as 10^5 pascals. The *millibar* (mb) is more common; it is used for measuring atmospheric pressure in meteorology.

Barft process A process formerly used for protecting iron from corrosion by heating it in steam, to form a layer of tri-iron tetroxide (Fe_3O_4).

barites /bă-**rȳ**-teez/ *See* barium sulfate.

barium /**bair**-ee-ŭm/ A dense, low-melting reactive metal; the fifth member of group 2 (formerly IIA) of the periodic table and a typical alkaline-earth element. The electronic configuration is that of xenon with two additional outer 6s electrons. Barium is of low abundance; it is found as witherite ($BaCO_3$) and barytes ($BaSO_4$). The metal is obtained by the electrolysis of the fused chloride using a cooled cathode which is slowly withdrawn from the melt. Because of its low melting point barium is readily purified by vacuum distillation. Barium metal is used as a 'getter', i.e., a compound added to a system to seek out the last traces of oxygen; and as an alloy constituent for certain bearing metals.

Barium has a low ionization potential and a large radius. It is therefore strongly electropositive and its properties, and those of its compounds, are very similar to those of the other alkaline-earth elements calcium and strontium. Notable differences in the chemistry of barium from the rest of the group are:

1. The much higher stability of the carbonate.
2. The formation of the peroxide below 800°C. Barium peroxide decomposes on strong heating to give oxygen and barium oxide:

$$BaO_2 \rightleftharpoons BaO + O$$

This equilibrium was the basis of the totally obsolete *Brin process* for the commercial production of oxygen.

Barium is also notable for the very low solubility of the sulfate, which permits its application to gravimetric analysis for either barium or sulfate. Barium compounds give a characteristic green color to flames which is used in qualitative analysis. Barium salts are all highly toxic with the exception of the most insoluble materials. Metallic barium has the body-centered cubic structure.

Symbol: Ba; m.p. 729°C; b.p. 1640°C; r.d. 3.594 (20°C); p.n. 56; r.a.m. 137.327.

barium bicarbonate *See* barium hydrogencarbonate.

barium carbonate ($BaCO_3$) A white insoluble salt that occurs naturally as the mineral witherite. Barium carbonate can be readily precipitated by adding an alkali carbonate to a barium salt solution. On heating it decomposes reversibly with the formation of the oxide and carbon dioxide:

$$BaCO_3 \rightleftharpoons BaO + CO_2$$

It is used as a rat poison.

barium chloride ($BaCl_2$) A white solid that can be prepared by dissolving barium carbonate in hydrochloric acid and crystallizing out the dihydrate ($BaCl_2.2H_2O$). Barium chloride is used as the electrolyte in the extraction of barium, as a rat poison, and in the leather industry.

barium hydrogencarbonate (**barium bicarbonate**; $Ba(HCO_3)_2$) A compound that occurs only in aqueous solution. It is formed by the action of cold water containing carbon dioxide on barium carbonate, to which it reverts on heating:

$$BaCO_3 + CO_2 + H_2O = Ba(HCO_3)_2$$

barium hydroxide (**baryta**; $Ba(OH)_2$) A white solid usually obtained as the octahydrate, $Ba(OH)_2.8H_2O$. Barium hydroxide is the most soluble of the group 2 hydroxides and can be used in volumetric analysis for the estimation of weak acids using phenolphthalein as an indicator.

barium oxide (BaO) A white powder prepared by heating barium in oxygen or by thermal decomposition of barium carbonate. It has been used in the manufacture of lubricating-oil additives.

barium peroxide (BaO_2) A dense off-white powder that can be prepared by carefully heating barium oxide in oxygen. Barium peroxide is used for bleaching straw and silk and in the laboratory preparation of hydrogen peroxide.

barium sulfate ($BaSO_4$) A white solid that occurs naturally as the mineral barytes. Barium sulfate is very insoluble in water and can be prepared easily as a precipitate by adding sulfuric acid to barium chloride. Barium sulfate is an important industrial chemical. Under the name of 'blanc fixe' it is used as a pigment extender in surface coating compositions. It is also used in the glass and rubber industries and medically (taken orally) to allow radiographs to be taken of the digestive system.

barn Symbol: b A unit of area defined as 10^{-28} square meter. The barn is sometimes used to express the effective cross-sections of atoms or nuclei in the scattering or absorption of particles.

barrel A measurement of volume often used in the chemical industry. It is equal to 159 liters (about 29 US gallons).

baryta /bă-**rȳ**-tă/ *See* barium hydroxide.

barytes /bă-**rȳ**-teez/ (**heavy spar**) A mineral form of barium sulfate ($BaSO_4$).

basalt /bă-**sawlt**, **bass**-awlt/ A dark-colored basic igneous rock derived from solidified volcanic lava. It consists mainly of fine crystals of pyroxine and plagioclase feldspar.

base A compound that releases hydroxyl ions, OH^-, in aqueous solution. Basic solutions have a pH greater than 7. In the Lowry-Brønsted treatment a base is a substance that tends to accept a proton. Thus OH^- is basic as it accepts H^+ to form water, but H_2O is also a base (although somewhat weaker) because it can accept a further proton to form H_3O^+. In this treatment the ions of classical mineral acids such as SO_4^{2-} and NO_3^- are weak *conjugate bases* of their respective acids. Electron-pair donors, such as trimethylamine and pyridine, are examples of organic bases.

base-catalyzed reaction A reaction catalyzed by bases. Typical base-catalyzed reactions are the Claisen condensation and the aldol reaction, in which the first step is abstraction of a proton to give a carbanion.

base metal A metal such as iron, lead, or copper, distinguished from a noble metal such as gold or silver.

base unit A unit that is defined in terms of reproducible physical quantities. *See also* SI units.

basic Having a tendency to release OH^- ions. Thus any solution in which the concentration of OH^- ions is greater than that in pure water at the same temperature is described as basic; i.e. the pH is greater than 7.

basic aluminum ethanoate *See* aluminum ethanoate.

basic oxide An oxide of a metal that reacts with water to form a base, or with an acid to form a salt and water. For example, calcium oxide reacts with water to form calcium hydroxide:

$$CaO + H_2O \rightarrow Ca(OH)_2$$

and with hydrochloric acid to produce calcium chloride and water:

$$CaO + 2HCl \rightarrow CaCl_2 + H_2O$$

See also acidic oxide.

basic oxygen process *See* Bessemer process.

basic salt A compound intermediate between a normal salt and a hydroxide or oxide. The term is often restricted to hydroxyhalides (such as $Pb(OH)Cl$, $Mg_2(OH)_3Cl_2.4H_2O$, and $Zn(OH)F$) and hydroxy-oxy salts (for example, $2PbCO_3$.

$Pb(OH)_2$, $Cu(OH)_2.CuCl_2$, $Cu(OH)_2.CuCo_3$). The hydroxyhalides are essentially closely packed assemblies of OH^- with metal or halide ions in the octahedral holes. The hydroxy-oxy salts have more complex structures than those implied by the formulae.

basic slag *See* slag.

batch process A manufacturing process in which the reactants are fed into the process in fixed quantities (batches), rather than in a continuous flow. At any particular instant all the material, from its preparation to the final product, has reached a definite stage in the process. Baking a cake is an example of a batch process. Such processes present problems of automation and instrumentation and tend to be wasteful of energy. For this reason, batch processing is only economically operated on an industrial scale when small quantities of valuable or strategic materials are required, e.g. drugs, titanium metal. *Compare* continuous process.

battery A number of similar units, such as electric cells, working together. Many dry 'batteries' used in radios, flashlights, etc., are in fact single cells. If a number of identical cells are connected in series, the total e.m.f. of the battery is the sum of the e.m.f.s of the individual cells. If the cells are in parallel, the e.m.f. of the battery is the same as that of one cell, but the current drawn from each is less (the total current is split among the cells).

bauxite // A mineral hydrated form of aluminum hydroxide; the principal ore of aluminum.

b.c.c. *See* body-centered cubic crystal.

Beckmann rearrangement /**bek**-măn/ A type of reaction in which the OXIME of a ketone is converted into an amide using a sulfuric acid catalyst. First discovered by the German chemist Ernst Otto Beckmann (1853–1923), it is used in the manufacture of polyamides.

Beckmann thermometer /**bek**-măn/ A type of thermometer designed to measure differences in temperature rather than scale degrees. Beckmann thermometers have a larger bulb than common thermometers and a stem with a small internal diameter, so that a range of 5°C covers about 30 centimeters in the stem. The mercury bulb is connected to the stem in such a way that the bulk of the mercury can be separated from the stem once a particular 5° range has been attained. The thermometer can thus be set for any particular range of working temperature. It is named for the German chemist Ernst Otto Beckmann (1853–1923).

becquerel /bek-ĕ-**rel**/ Symbol: Bq The SI unit of activity equal to the activity of a radioactive substance that has one spontaneous nuclear change per second; 1 Bq = 1 s^{-1}. The unit is named for the French physicist Antoine Henri Becquerel (1852–1908).

beneficiation /ben-ĕ-fish-ee-**ay**-shŏn/ The process of separating an ore into a useful component and waste material (known as *gangue*). The process is sometimes described as *ore dressing*.

bent bond *See* banana bond.

bentonite /**ben**-tŏ-nÿt/ A type of clay that is used as an adsorbent in making paper. The gelatinous suspension it forms with water is used to bind together the sand for making iron castings. Chemically bentonite is an aluminosilicate of variable composition.

bent sandwich compound *See* sandwich compound.

benzaldehyde /ben-**zal**-dĕ-hÿd/ *See* benzenecarbaldehyde.

benzene /**ben**-zeen, ben-**zeen**/ (C_6H_6) A colorless liquid hydrocarbon with a characteristic odor. Benzene is a highly toxic compound and continued inhalation of the vapor is harmful. It was originally isolated from coal tar and for many years this was the principal source of the compound.

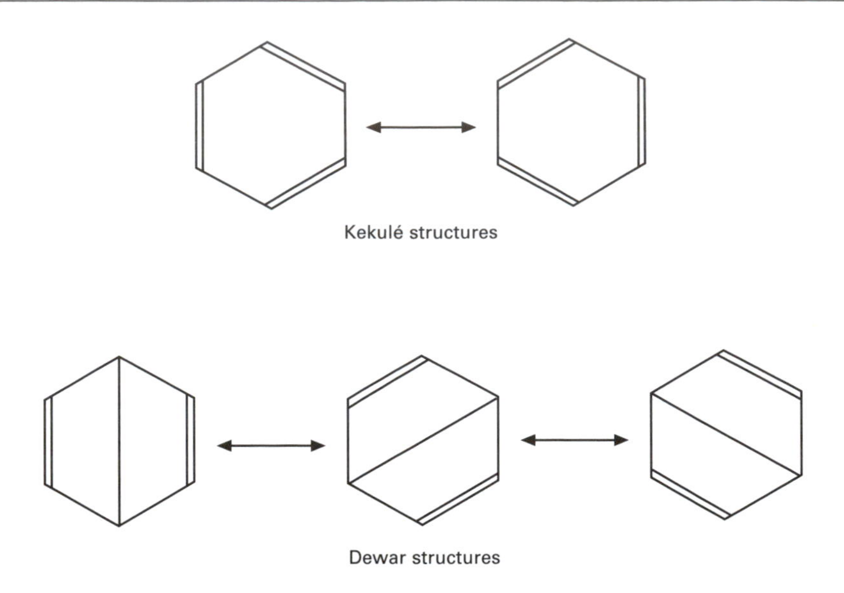
Kekulé structures

Dewar structures

Benzene structures

Contemporary manufacture is from hexane; petroleum vapor is passed over platinum at 500°C and at 10 atmospheres pressure:

$$C_6H_{14} \rightarrow C_6H_6 + 4H_2$$

Benzene is the simplest aromatic hydrocarbon. It shows characteristic electrophilic substitution reactions, which are difficult to explain assuming a simple unsaturated structure (such as Kekulé's (1865) or Dewar's (1867) formulae). This anomolous behavior can now be explained by assuming that the six pi electrons are delocalized and that benzene is, therefore, a resonance hybrid. It consists of two Kekulé structures and three Dewar structures, the former contributing 80% and the latter 20% to the hybrid. Benzene is usually represented by either a Kekulé structure or a hexagon containing a circle (which denotes the delocalized electrons).

benzenecarbaldehyde /ben-zeen-kar-**bal**-dĕ-hÿd/ (**benzaldehyde**; C_6H_5CHO) A yellow organic oil with a distinct almond-like odor. Benzenecarbaldehyde undergoes the reactions characteristic of aldehydes and may be synthesized in the laboratory by the usual methods of aldehyde synthesis. It is used as a food flavoring and in the manufacture of dyes and antibiotics, and can be readily manufactured by the chlorination of methylbenzene and the subsequent hydrolysis of (dichloromethyl) benzene:

$$C_6H_5CH_3 + Cl_2 \rightarrow C_6H_5CHCl_2$$

$$C_6H_5CHCl_2 + 2H_2O \rightarrow C_6H_5CH(OH)_2 + 2HCl$$

$$C_6H_5CH(OH)_2 \rightarrow C_6H_5CHO + H_2O$$

benzenecarbonyl chloride /ben-zeen-**kar**-bŏ-nil/ (**benzoyl chloride**; C_6H_5COCl) A liquid acyl chloride used as a benzoylating agent.

benzenecarbonyl group (**benzoyl group**) The group $C_6H_5.CO–$.

benzenecarboxylic acid /ben-zeen-kar-boks-**il**-ik/ (**benzoic acid**; C_6H_5COOH) A white crystalline carboxylic acid. It is used as a food preservative. The carboxyl group (–COOH) directs further substitution onto the benzene ring in the 3 position.

benzene-1,2-dicarboxylic acid (**phthalic acid**; $C_6H_4(COOH)_2$) A white crystalline aromatic acid. On heating it loses water to form *phthalic anhydride*, which is used to make dyestuffs and polymers for plasticizers and plastics.

benzene-1,4-dicarboxylic acid (**terephthalic acid**; $C_6H_4(COOH)_2$) A colorless crystalline organic acid used to produce Dacron and other polyesters.

benzene-1,3-diol (**resorcinol**; $C_6H_4(OH)_2$) A white crystalline phenol used in the manufacture of dyestuffs and celluloid.

benzene-1,4-diol (**hydroquinone, quinol**; $C_6H_4(OH)_2$) A white crystalline phenol used in making dyestuffs. *See also* quinone.

benzene ring The cyclic hexagonal arrangement of six carbon atoms that are characteristic of AROMATIC COMPOUNDS, i.e. of BENZENE and its derivatives.

benzenesulfonic acid /ben-zeen-sul-**fon**-ik/ ($C_6H_5SO_2OH$) A white crystalline acid made by sulfonation of benzene. Any further substitution onto the benzene ring is directed into the 3 position.

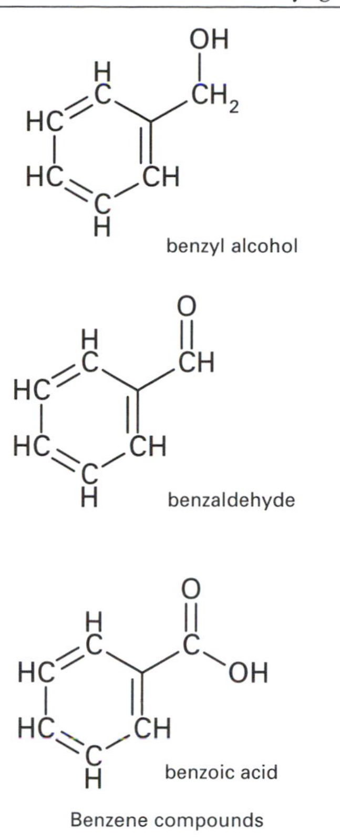

Benzene compounds

benzfuran /benz-**fyoo**-ran, -fyoo-**ran**/ (**coumarone**; C_8H_6O) A crystalline compound having a benzene ring fused to a furan ring.

benzil *See* Benzilic acid rearrangement.

benzilic acid rearrangement /ben-**zil**-ik/ A reaction in which *benzil* (1,2-diphenylethan-1,2-dione) is treated with hydroxide and then with acid to give *benzilic acid* (2-hydroxy-2,2-diphenylethanoic acid):

$$C_6H_5.CO.CO.C_6H_5 \rightarrow (C_6H_5)_2C(OH).COOH$$

The reaction, which involves migration of a phenyl group (C_6H_5–) from one carbon atom to another, was the first rearrangement reaction to be described (by Justus von Liebig in 1828).

benzoic acid /ben-**zoh**-ik/ *See* benzenecarboxylic acid.

benzole /**ben**-zohl/ A mixture of mainly aromatic hydrocarbons obtained from coal. *See also* benzene.

benzopyrene /ben-zoh-**pȳ**-reen/ *See* benzpyrene.

benzoquinone /ben-zoh-kwi-**nohn**/ *See* quinone.

benzoylation /ben-zoh-ă-**lay**-shŏn/ *See* acylation.

benzoyl chloride /**ben**-zoh-ăl/ *See* benzenecarbonyl chloride.

benzoyl group *See* benzenecarbonyl group.

benzpyrene /benz-**pȳ**-reen/ (**benzopyrene**; $C_{20}H_{12}$) A cyclic aromatic hydrocarbon with a structure consisting of five fused benzene rings. It occurs in coal tar and tobacco smoke and has strong carcinogenic properties.

benzpyrrole /benz-**pi**-rohl/ *See* indole.

benzyl alcohol /**ben**-zăl/ *See* phenylmethanol.

benzyl group The group $C_6H_5CH_2$–.

benzyne /**ben**-zÿn/ (C_6H_4) A short-lived intermediate present in some reactions. The ring of six carbon atoms contains two double bonds and one triple bond (the systematic name is *1,2-didehydrobenzene*).

Bergius process /**ber**-gee-ŭs/ A process formerly used for making hydrocarbon fuels from coal. A powdered mixture of coal, heavy oil, and a catalyst was heated with hydrogen at high pressure. The process is named for the German industrial chemist Friedrich Karl Rudolph Bergius (1884–1949).

berkelium /**ber**-klee-ŭm, ber-**kee**-lee-ŭm/ A silvery radioactive transuranic element of the actinoid series of metals, not found naturally on Earth. Several radioisotopes have been synthesized. The metal reacts with oxygen, steam, and acids.

Symbol: Bk; m.p. 1050°C; p.n. 97; r.d. 14.79 (20°C); most stable isotope ^{247}Bk (half-life 1400 years).

beryl A mineral, $3BeO.Al_2O_3.6SiO_2$, used as a source of beryllium. There are several varieties including the gemstones emerald (colored green by traces of chromium oxide) and aquamarine (a blue-green color).

beryllate /bĕ-**ril**-ayt/ *See* beryllium; beryllium hydroxide.

beryllium /bĕ-**ril**-ee-ŭm/ A light metallic element, similar to aluminum but somewhat harder; the first element in group 2 (formerly IIA) of the periodic table. It has the electronic configuration of helium with two additional outer 2s electrons.

Beryllium occurs in a number of minerals such as beryllonite ($NaBePO_4$), chrysoberyl ($Be(AlO_2)_2$), bertrandite ($4BeO.2SiO_2$), and beryl ($3BeO.Al_2O_3.6SiO_2$). The element accounts for only 0.0006% by mass of the Earth's crust. The metal is obtained by conversion of the ore to the sulfate at high temperature and pressure with concentrated sulfuric acid, then to the chloride, followed by electrolysis of the fused chloride. Alternatively, extraction by hydrogen fluoride followed by electrolysis of the fused fluoride may be employed. The metal has a much lower general reactivity than lithium or other elements in group 2. It is used as an antioxidant and hardener in some alloys, such as copper and phosphor bronzes.

Beryllium has the highest ionization potential of group 2 and the smallest size. Consequently it is less electropositive and more polarizing than other members of the group. Thus, Be^{2+} ions do not exist as such in either solids or solutions, and even with the most electronegative elements there is partial covalent character in the bonds. The metal reacts directly with oxygen, nitrogen, sulfur, and the halogens at various elevated temperatures, to form the oxide BeO, nitride Be_3N_2, sulfide BeS, and halides BeX_2, all of which are covalent. Beryllium does not react directly with hydrogen but a polymeric hydride $(BeH_2)_n$ can be prepared by reduction of $(CH_3)_2Be$ using lithium tetrahydridoaluminate.

Beryllium is amphoteric forming *beryllate* species, such as $[Be(OH)_4]^{2-}$ and $[Be(OH)]_3^{3+}$. The hydroxide is only weakly basic. The element does not form a true carbonate; the basic beryllium carbonate, $BeCO_3.Be(OH)_2$ is formed when sodium carbonate is added to solutions of beryllium compounds.

Beryllium hydride, chloride, and dimethylberyllium form polymeric bridged species but, whereas the bridging in the chloride is via an electron pair on chlorine atoms and can be regarded as an electron-pair donor bond, the bonding in the hydride and in the methyl compound involves two-electron three-centre bonds. Coordination compounds are quite common with beryllium; some examples include $[BeCl_4]^{2-}$, $(R_2O)_2BeCl_2$, and $[Be(NH_3)_4]Cl_2$. Beryllium also forms a number of alkyl compounds, some of which can be stabilized by coordination. Beryllium is extremely toxic.

Symbol: Be; m.p. 1278±5°C; b.p. 2970°C (under pressure); r.d. 1.85 (20°C); p.n. 4; r.a.m. 9.012182.

beryllium bicarbonate *See* beryllium hydrogencarbonate.

beryllium bronze *See* bronze.

beryllium carbonate ($BeCO_3$) An unstable solid prepared by prolonged treatment of a suspension of beryllium hydroxide with carbon dioxide. The resulting solution is evaporated and filtered in an atmosphere of carbon dioxide.

beryllium chloride ($BeCl_2$) A white solid obtained by passing chlorine over a heated mixture of beryllium oxide and carbon. It is a poor conductor in the fused state and in the solid form consists of a covalent polymeric structure. It is readily soluble in organic solvents but is hydrolyzed in the presence of water to give the hydroxide. The anhydrous salt is used as a catalyst.

beryllium hydrogencarbonate (**beryllium bicarbonate**; $Be(HCO_3)_2$) A compound formed in solution by the action of carbon dioxide on a suspension of the carbonate, to which it reverts on heating:

$$BeCO_3 + CO_2 + H_2O = Be(HCO_3)_2$$

beryllium hydroxide ($Be(OH)_2$) A solid that can be precipitated from beryllium-containing solutions by an alkali. In an excess of alkali, beryllium hydroxide dissolves to give a beryllate, $Be(OH)_4^{2-}$. In this way, beryllium and aluminum hydroxides are similar and the reaction is an indication that beryllium hydroxide is amphoteric.

beryllium oxide (BeO) A white solid formed by heating beryllium in oxygen or by the decomposition of beryllium hydroxide or carbonate. Beryllium oxide is insoluble in water but it shows basic properties by dissolving in acids to form beryllium salts:

$$BeO + 2H^+ \rightarrow Be^{2+} + H_2O$$

However, beryllium oxide also resembles acidic oxides by reacting with alkalis to form beryllates:

$$BeO + 2OH^- + H_2O \rightarrow Be(OH)_4^{2-}$$

It is thus an amphoteric oxide. Beryllium oxide is used for the production of beryllium and beryllium-copper refractories, for high-output transistors, and for printed circuits. The chemical properties of beryllium oxide are similar to those of aluminum oxide.

beryllium sulfate ($BeSO_4$) An insoluble salt prepared by the reaction of beryllium oxide with concentrated sulfuric acid. On heating, it breaks down to give the oxide. When heated in the presence of water, it readily dissolves to give an acidic solution.

Bessemer process /**bess**-ĕ-mer/ A process for making steel from pig iron. A vertical cylindrical steel vessel (*converter*) is used, lined with a refractory material. Air is blown through the molten iron and carbon is oxidized and thus removed from the iron. Other impurities – silicon, sulfur, and phosphorus – are also oxidized. Irons containing large amounts of phosphorus are treated in a converter lined with a basic material, so that a phosphate slag is formed. The required amount of carbon is then added to the iron to produce the desired type of steel. The process is named for the British inventor and engineer Sir Henry Bessemer (1813–98).

beta particle An electron emitted by a radioactive substance as it decays.

bi- Prefix used formerly in naming acid salts. The prefix indicates the presence of hydrogen; for instance, sodium bisulfate ($NaHSO_4$) is sodium hydrogensulfate, etc.

bicarbonate /bÿ-**kar**-bŏ-nayt,-nit/ *See* hydrogencarbonate.

bimolecular /bÿ-mŏ-**lek**-yŭ-ler/ Describing a reaction or step that involves two molecules, ions, etc. A common example of this type of reaction is the decomposition of hydrogen iodide,

$$2HI \rightarrow H_2 + I_2$$

takes place between two molecules and is therefore a bimolecular reaction. Other common examples are:

$$H_2O_2 + I_2 \rightarrow OH^- + HIO$$

$$OH^- + H^+ \rightarrow H_2O$$

All bimolecular reactions are second order, but some second-order reactions are not bimolecular.

binary compound A chemical compound formed from two elements; e.g. Fe_2O_3 or NaCl.

biochemical oxygen demand (BOD) The amount of oxygen taken from natural water by microorganisms that decompose organic waste matter in the water. It is therefore a measure of the quantity of organic pollutants present. It is found by measuring the amount of oxygen in a sample of water, keeping the sample and then making the measurement again five days later.

biochemistry The study of chemical compounds and reactions occurring in living organisms.

biodegradable /bÿ-oh-di-**gray**-dă-băl/ *See* pollution.

bioinorganic chemistry /bÿ-oh-in-or-**gan**-ik/ The study of biological molecules that contain metal atoms or ions. Many enzymes have active metal atoms and bioinorganic compounds are important in other roles such as protein folding, oxygen transport, and electron transfer. Two important examples of bioinorganic compounds are hemoglobin (containing iron) and chlorophyll (containing magnesium).

biosynthesis /bÿ-oh-**sin**-th'ĕ-sis/ The reactions by which organisms obtain the various compounds needed for life.

biphenyl /bÿ-**fen**-ăl, -**fee**-năl/ ($C_6H_5C_6H_5$) An organic compound having a structure in which two phenyl groups are joined by a C–C bond. *See also* polychlorinated biphenyl.

bipyridyl /bÿ-**pÿ**-ră-dăl/ *See* dipyridyl.

Birkeland–Eyde process /**berk**-lănd ÿ-dĕ/ An industrial process for fixing nitrogen (as nitrogen monoxide) by passing air through an electric arc:

$$N_2 + O_2 \rightarrow 2NO$$

The process is named for the Norwegian physicist and chemist Kristian Olaf Bernhard Birkeland (1867–1917) and the Norwegian engineer and industrialist Samuel Eyde (1866–1940). *See also* nitrogen fixation.

bismuth /**biz**-mŭth/ A brittle pinkish metallic element belonging to group 15 (formerly VA) of the periodic table. It occurs native and in the ores Bi_2S_3 and Bi_2O_3. The element does not react with oxygen or water under normal temperatures. It can be dissolved by concentrated nitric acid. Bismuth is widely used in alloys, especially low-melting alloys. The element has the property of expanding when it solidifies. Compounds of bismuth are used in cosmetics and medicines.

Symbol: Bi; m.p. 271.35°C; b.p. 1560±5°C; r.d. 9.747 (20°C); p.n. 83; r.a.m. 208.98037.

bismuth(III) carbonate dioxide /dÿ-**oks**-ÿd, -id/ (**bismuthyl carbonate**; $Bi_2O_2CO_3$) A white solid prepared by mixing solutions of bismuth nitrate and ammonium carbonate. It contains the $(BiO)^+$ ion.

bismuth(III) chloride (**bismuth trichloride**; $BiCl_3$) A white deliquescent solid. It can be prepared by direct combination of bismuth and chlorine. Bismuth(III) chloride dissolves in excess dilute hydrochloric acid to form a clear liquid, but if diluted it produces a white precipitate of *bismuth(III) chloride oxide* (*bismuthyl chloride*, BiOCl):

$$BiCl_3 + H_2O = BiOCl + 2HCl$$

This reaction is often used as an example of a reversible reaction and as a confirmatory test for bismuth during quantitative analysis. Bismuth(V) chloride does not form.

bismuth(III) chloride oxide *See* bismuth(III) chloride.

bismuth(III) nitrate oxide (**bismuthyl nitrate**; $BiONO_3$) A white insoluble solid, often referred to as *bismuth subnitrate*. It is precipitated when bismuth(III) nitrate is diluted and contains the $(BiO)^+$ ion. Bismuth(III) nitrate oxide is used in pharmaceutical preparations.

bismuth trichloride /trÿ-**klor**-ÿd, -**kloh**-rÿd/ *See* bismuth(III) chloride.

bismuthyl carbonate /**biz**-mŭ-thăl/ *See* bismuth(III) carbonate dioxide.

bismuthyl chloride *See* bismuth(III) chloride.

bismuthyl compound A compound containing the $(BiO)^+$ ion or BiO grouping.

bismuthyl ion *See* bismuth(III) carbonate dioxide.

bismuthyl nitrate *See* bismuth(III) nitrate oxide.

bisulfate /bÿ-**sul**-fayt/ *See* hydrogensulfate.

bisulfite /bÿ-**sul**-fÿt/ *See* hydrogensulfite.

bisulfite addition compound *See* aldehyde.

bittern /**bit**-ern/ The liquid that is left after sodium chloride has been crystallized from sea water.

bitumen /bă-**tew**-mĕn, **bich**-û-mĕn/ A mixture of solid or semisolid hydrocarbons obtained from coal, oil, etc. *See* tar.

bituminous coal /bă-**tew**-mĕ-nŭs/ A type of second-grade coal, containing more than 65% carbon but also quantities of gas, coal tar, and water. It is the common type of coal used for domestic and industrial purposes. *See also* anthracite.

biuret /bÿ-yû-**ret**, bÿ-**yoo**-ret/ ($H_2NCONHCONH_2$) A colorless crystalline organic compound made by heating UREA (carbamide). It is used in a chemical test for PROTEINS.

bivalent /bÿ-**vay**-lĕnt, **biv**-ă-/ (**divalent**) Having a valence of two.

blackdamp *See* firedamp.

blanc fixe /blahng-**feeks**/ *See* barium sulfate.

blast furnace A furnace for producing iron from iron(III) oxide. A mixture of the ore with coke and a flux is heated by preheated air blown into the bottom of the furnace. The flux is often calcium oxide (from limestone). Molten pig iron is run off from the bottom of the furnace.

bleach Any substance used to remove color from materials such as cloth and paper. All bleaches are oxidizing agents, and include chlorine, sodium chlorate(I) solution (NaClO, sodium hypochlorite), hydrogen peroxide, and sulfur(IV) oxide. Sunlight also has a bleaching effect.

bleaching powder A white solid that can be regarded as a mixture of calcium chlorate(I) (calcium hypochlorite), calcium chloride, and calcium hydroxide. It is prepared on a large scale by passing a current of chlorine through a tilted cylinder down which is passed calcium hydroxide. Bleaching powder has been used for bleaching paper pulps and fabrics and for sterilizing water. Its bleaching power arises from the formation, in the presence of air containing carbon dioxide, of the oxidizing agent chloric(I) acid (hypochlorous acid, HClO):

$$Ca(ClO)_2.Ca(OH)_2.CaCl_2 + 2CO_2 \rightarrow 2CaCO_3 + CaCl_2 + 2HClO$$

blende /blend/ A naturally occurring sulfide ore.

block copolymer *See* polymerization.

blue vitriol /**vit**-ree-ŏl/ Copper(II) sulfate pentahydrate ($CuSO_4.5H_2O$).

boat conformation *See* conformation.

BOD *See* biochemical oxygen demand.

body-centered cubic crystal (b.c.c.) A crystal structure in which the unit cell has an atom, ion, or molecule at each corner of a cube and at the center of the cube. In this type of structure the coordination number is 8. It is less close-packed than the face-centered cubic structure. The alkali metals form crystals with body-centered cubic structures.

bohrium /**bor**-ee-ŭm, **boh**-ree-/ A synthetic radioactive element first detected by bombarding a bismuth target with chromium nuclei. Only a small number of atoms have ever been produced.

Symbol: Bh; p.n. 107; most stable isotope ^{262}Bh (half life 0.1s).

Bohr magneton /bor, bohr/ *See* magneton.

Bohr theory A theory introduced in 1911 to explain the spectrum of atomic hydrogen. The model he used was that of a nucleus with charge +e, orbited by an electron with charge –e, moving in a circle of radius r. If v is the velocity of the electron, the centripetal force, mv^2/r is equal to the force of electrostatic attraction, $e^2/4\pi\varepsilon_0 r^2$. Using this, it can be shown that the total energy of the electron (kinetic and potential) is $-e^2/8\pi\varepsilon_0 r$.

If the electron is considered to have wave properties, then there must be a whole number of wavelengths around the orbit, otherwise the wave would be a progressive wave. For this to occur

$$n\lambda = 2\pi r$$

where n is an integer, 1, 2, 3, 4, ... The wavelength, λ, is h/mv, where h is the Planck constant and mv the momentum. Thus for a given orbit:

$$nh/2\pi = mvr$$

This means that orbits are possible only when the angular momentum (mvr) is an integral number of units of $h/2\pi$. Angular momentum is thus quantized. In fact, Bohr in his theory did not use the wave behavior of the electron to derive this relationship. He assumed from the beginning that angular momentum was quantized in this way. Using the above expressions it can be shown that the electron energy is given by

$$E = -me^4/8\varepsilon_0^2 n^2 h^2$$

Different values of n (1, 2, 3, etc.) correspond to different orbits with different energies; n is the principal quantum number. In making a transition from an orbit n_1 to another orbit n_2 the energy difference ΔW is given by:

$$\Delta W = W_1 - W_2 = me^4(1/n_2^2 - 1/n_1^2)/8\varepsilon_0^2 h^2$$

This is equal to hv where v is the frequency of radiation emitted or absorbed. Since $v\lambda = c$, then

$$1/\lambda = me^4(1/n_1^2 - 1/n_2^2)/8\varepsilon_0^2 ch^3$$

The theory is in good agreement with experiment in predicting the wavelengths of lines in the hydrogen spectrum, although it is less successful for larger ATOMS. Different values of n_1 and n_2 correspond to different spectral series, with lines given by the expression:

$$1/\lambda = R(1/n_1^2 - 1/n_2^2)$$

R is the *Rydberg constant.* Its experimental value is $1.096\ 78 \times 10^7$ m^{-1}. The value from Bohr theory ($me^4/8\pi\varepsilon_1^2 ch^2$) is $1.097\ 00 \times 10^7$ m^{-1}.

The theory is named for the Danish physicist Niels Bohr (1885–1962).

boiling The process by which a liquid is converted into a gas or vapor by heating at its boiling point. At this temperature the vapor pressure of the liquid is equal to the external pressure, and bubbles of vapor can form within the liquid.

boiling point The temperature at which the vapor pressure of a liquid is equal to atmospheric pressure. This temperature is always the same for a particular liquid at a given pressure (for reference purposes usually taken as standard pressure). *See also* elevation of boiling point.

boiling-point-composition diagram A diagram for a two-component liquid system representing both the variation of the boiling point and the composition of the vapor phase as the liquid-phase composition is varied. A mixture of A and B at composition L_1 would have a boiling point T_1 and a vapor composition V_1, which when condensed would give a liquid at L_2. L_2 would boil at T_2 to give vapor V_2 equivalent to liquid of composition L_3, and so on. Thus the whole process of either distillation or fractionation of this system will lead to progressive enrichment in component A.

For perfect solutions obeying Raoult's law the curves would coincide but in real cases there will be sufficient intermolecular attraction to cause deviation from this. The separation of the curves as well as the difference in boiling points determines the performance of fractionation columns. *See also* constant-boiling mixture.

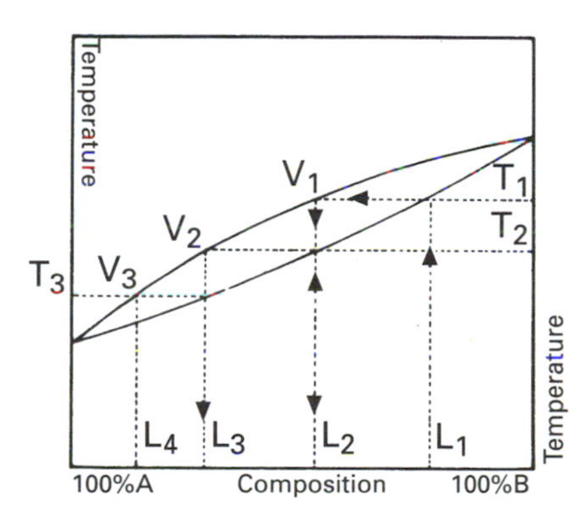

Boiling point–composition diagram

Boltzmann constant /bohlts-măn, -mahn/ Symbol: k The constant 1.380 54 J K^{-1}, equal to the gas constant (R) divided by the Avogadro constant (N_A). The constant is named for the Austrian theoretical physicist Ludwig Edward Boltzmann (1844–1906). *See also* degrees of freedom.

Boltzmann formula A fundamental result in statistical mechanics stating that the entropy S of a system is related to the number W of distinguishable ways in which the system can exist by the equation: $S = k \ln W$, where k is the Boltzmann constant. This formula is a quantitative expression of the idea that the entropy of a system is a measure of its disorder. It was discovered by Ludwig Boltzmann in the late 19th century in the course of his investigations into the foundations of statistical mechanics.

bomb calorimeter A sealed insulated container, used for measuring energy released during combustion of substances (e.g. foods and fuels). A known amount of the substance is ignited inside the calorimeter in an atmosphere of pure oxygen, and undergoes complete combustion at constant volume. The resultant rise in temperature is related to the energy released by the reaction. Such energy values (*calorific values*) are often quoted in joules per kilogram (J kg^{-1}).

bond The interaction between atoms, molecules, or ions holding these entities together. The forces giving rise to bonding may vary from extremely weak intermolecular forces, about 0.1 kJ mol^{-1}, to very strong chemical bonding, about 10^3 kJ mol^{-1}. *See* bond energy; bond length; coordinate bond; covalent bond; electrovalent bond; hydrogen bond; metallic bond.

bond dissociation energy *See* bond energy.

bond energy The energy involved in forming a bond. For ammonia, for instance, the energy of the N–H bond is one third of the energy involved for the process

$$NH_3 \rightarrow N + 3H$$

It is thus one third of the heat of atomization. Strictly speaking *bond enthalpy* should be used.

The *bond dissociation energy* is a different quantity to the bond energy. It is the energy required to break a particular bond in a compound, e.g.:

$$NH_3 \rightarrow NH_2 + H$$

bond enthalpy *See* bond energy.

bonding orbital *See* orbital.

bond length The length of a chemical BOND, the distance between the centers of the nuclei of two atoms joined by a chemical bond. Bond lengths may be measured by electron or x-ray diffraction.

boracic acid /bŏ-**rass**-ik/ *See* boric acid.

boranes /**bor**-aynz, **boh**-raynz/ *See* boron hydrides.

borate /**bor**-ayt, **boh**-rayt/ *See* boron.

borax /**bor**-aks, **boh**-raks/ *See* disodium tetraborate decahydrate.

borax-bead test A preliminary test in qualitative inorganic analysis that can be a guide to the presence of certain metals. A borax bead is formed by heating a little borax on a loop in a platinum wire. A minute sample is introduced into the bead and the color observed in both the oxidizing and reducing areas of a Bunsen-burner flame. The color is also noted when the bead is cold.

boric acid /**bor**-ik, **boh**-rik/ (**boracic acid;** H_3BO_3) A white crystalline solid soluble in water; in solution it is a very weak acid. Boric acid is used as a mild antiseptic eye lotion and was formerly used as a food preservative. It is used in glazes for enameled objects and is a constituent of Pyrex glass.

Trioxoboric(III) acid is the full systematic name for the solid acid and it exists in this form in its dilute solutions. However, in more concentrated solutions polymerization occurs to give *polydioxoboric(III) acid.*

boride /**bor**-ÿd, **boh**-rÿd/ A compound of boron, especially one with a more electropositive element.

Born–Haber cycle /**born hay**-ber/ A cycle used in calculating the lattice energies of solids. The steps involved are:

Atomization of sodium:

$$Na(s) \rightarrow Na(g)\ \Delta H_1$$

Ionization of sodium:

$$Na(g) \rightarrow Na^+(g)\ \Delta H_2$$

Atomization of chlorine:

$$Cl_2(g) \rightarrow 2Cl(g)\ \Delta H_3$$

Ionization of chlorine:

$$Cl(g) + e^- \rightarrow Cl^-(g)\ \Delta H_4$$

Formation of solid:

$$Na^+(g) + Cl^-(g) \rightarrow NaCl(s)\ \Delta H_5$$

This last step involves the lattice energy ΔH_5. The sum of all these enthalpies is equal to the heat of the reaction:

$$Na(s) + \tfrac{1}{2}Cl_2(g) \rightarrow NaCl(s)$$

The cycle is named for the German physicist Max Born (1882–1970) and the German physical chemist Fritz Haber (1868–1934). *See also* Hess's law.

borohydride ions /bor-oh-**hÿ**-drÿd, boh-roh-/ *See* boron hydride.

boron /**bor**-on, **boh**-ron/ A hard rather brittle metalloid element of group 3 (formerly IIIA) of the periodic table. It has the electronic structure $1s^22s^22p^1$. Boron is of low abundance (0.0003%) but the natural minerals occur in very concentrated forms as either borax ($Na_2B_4O_7.10H_2O$) or colemanite ($Ca_2B_6O_{11}$). The element is obtained by conversion to boric acid followed

BORAX-BEAD COLORS (H = hot; C = cold)

Metal	*Oxidizing flame*	*Reducing flame*
chromium	green H+C	green H+C
cobalt	blue H+C	blue H+C
copper	green H, blue C	often opaque
iron	brown-red H, yellow C	green H+C
manganese	violet H+C	colourless H+C
nickel	red–brown C	gray–black C

by dehydration to B_2O_3 then reduction with magnesium. High-purity boron for semiconductor applications is obtained by conversion to boron trichloride, which can be purified by distillation, then reduction using hydrogen. Only small quantities of elemental boron are needed commercially; the vast majority of boron supplied by the industry is in the form of borax or boric acid.

As boron has a small atom and has a relatively high ionization potential its compounds are predominantly covalent; the ion B^{3+} does not exist. Boron does not react directly with hydrogen to form boron hydrides (boranes) but the hydrolysis of magnesium boride does produce a range of boranes such as B_4H_{10}, B_5H_9, and B_6H_{10}. Thermal decomposition of these higher boranes produces, among other things, the simplest borane, B_2H_6 (diborane). The species BH_3 is only a short-lived reaction intermediate.

Finely divided boron burns in oxygen above 600°C to give the oxide, B_2O_3, an acidic oxide which will dissolve slowly in water to give boric acid ($B(OH)_3$) and rapidly in alkalis to give borates such as $Na_2B_4O_7$. A number of polymeric species with B–B and B–O links are known, e.g. a lower oxide $(BO)_x$, and a polymeric acid $(HBO_2)_n$. Although the parent acid is weak, many salts containing *borate* anions are known but their stoichiometry gives little indication of their structure, many of which are cyclic or linear polymers. These contain both BO_3 planar groups and BO_4 tetrahedra. Boric acid and the borates give a range of glassy substances on heating; these contain cross-linked B–O–B chains and nets. In the molten state these materials react with metal ions to form borates, which on cooling give characteristic colors to the glass. *See* borax-bead test.

Boron reacts with nitrogen on strong heating (1000°C) to give boron nitride, a slippery white solid with a layer structure similar to that of graphite, i.e., hexagonal rings of alternating B and N atoms. The material has an extremely high melting point and is thermally very stable but there is sufficient bond polarity in the B–N links to permit slow hydrolysis by water to give ammonia. There is also a 'diamond-like' form of B–N which is claimed to be even harder than diamond.

Elemental boron reacts directly with fluorine and chlorine but for practical purposes the halides are obtained via the BF_3 route from boric acid:

$$B_2O_3 + 3CaF_2 + 3H_2SO_4 \rightarrow 3CaSO_4 + 3H_2O + 2BF_3$$

$$BF_3 + AlCl_3 \rightarrow AlF_3 + BCl_3$$

These halides are all covalent molecules, which are all planar and trigonal in shape. Boron halides are industrially important as catalysts or promoters in a variety of organic reactions including polymerization and Friedel–Crafts type alkylations. The decomposition of boron halides in atmospheres of hydrogen at elevated temperatures is also used to deposit traces of pure boron in semiconductor devices.

Boron forms a range of compounds with elements that are less electronegative than itself, called *borides*. Borides such as ZrB_2 and TiB_2 are hard refractory substances, which are chemically inert and have remarkably high electrical conductivities. Borides have a wide range of stoichiometries, from M_4B through to MB_6, and can exist in close-packed arrays, chains, and two-dimensional nets. Natural boron consists of two isotopes, ^{10}B (18.83%) and ^{11}B (81.17%). These percentages are sufficiently high for their detection by splitting of infrared absorption or by n.m.r. spectroscopy.

Both borax and boric acid are used as mild antiseptics and are not regarded as toxic; boron hydrides are however highly toxic.

Symbol: B; m.p. 2300°C; b.p. 2658°C; r.d. 2.34 (20°C); p.n. 5; r.a.m. 10.811.

boron hydrides (**boranes**) Hydrides of boron ranging from B_2H_6 to $B_{20}H_{16}$, which have complex molecular structures. These compounds are notable for containing B–H–B bonds. Investigations have shown that the boron hydrides are electron deficient and thus the bonding can only be rationalized in terms of MULTICENTER BONDS. The compound B_2H_6 can be prepared by reacting boron trichloride with hydrogen.

A complex ion, tetrahydridoborate(III) (borohydride, BH_4^-), can be formed; it has very strong reducing properties. Other, more complex, *borohydride ions* exist.

boron nitride (BN) A compound formed by heating BORON in nitrogen (1000°C). It has two crystalline forms.

boron oxide (B_2O_3) A glassy hygroscopic solid that eventually forms boric acid. It forms various salts but also exhibits some amphoteric properties.

boron tribromide (BBr_3) A colorless liquid. *See* boron trichloride.

boron trichloride /trÿ-**klor**-ÿd, -**kloh**-rÿd/ (BCl_3) A fuming liquid made by passing dry chlorine over heated boron. It is rapidly hydrolysed by water:

$$BCl_3 + 3H_2O \rightarrow 3HCl + H_3BO_3$$

As there are only three pairs of shared electrons in the outer shell of the boron atom, boron halides form very stable addition compounds with ammonia by the acceptance of a lone electron pair in a coordinate bond to complete a shared octet.

boron trifluoride (BF_3) A colorless fuming gas made by heating a mixture of boron oxide, calcium fluoride, and concentrated sulfuric acid. *See* boron trichloride.

borosilicates /bor-ŏ-**sil**-ă-kayts/ Complex compounds similar to silicates, but containing BO_3 and BO_4 units in addition to the SiO_4 units. Certain crystalline borosilicate minerals are known. In addition, borosilicate glasses can be made by using boron oxide in addition to silicon(IV) oxide. These tend to be tougher and more heat resistant than 'normal' silicate glass.

Bosch process /bosh/ The reaction

$$CO + H_2O \rightarrow CO_2 + H_2$$

using water gas over a hot catalyst. It has been used to make hydrogen for the Haber process. The process is named for the German industrial chemist Carl Bosch (1874–1940).

bowl classifier A device that separates solid particles in a mixture of solids and liquid into fractions according to particle size. Feed enters the center of a shallow bowl, which contains revolving blades. The coarse solids collect on the bottom, fine solids at the periphery.

Boyle's law At a constant temperature, the pressure of a fixed mass of a gas is inversely proportional to its volume: i.e.

$$pV = K$$

where K is a constant. The value of K depends on the temperature and on the nature of the gas. The law holds strictly only for ideal gases. Real gases follow Boyle's law at low pressures and high temperatures. The law is named for the British chemist and physicist Robert Boyle (1627–91). *See* gas laws.

Brackett series /**brak**-it/ *See* hydrogen atom spectrum.

Brady's reagent /**bray**-dee/ *See* 2,4-dinitrophenylhydrazine.

Bragg equation /brag/ An equation used to deduce the crystal structure of a material using data obtained from x-rays directed at its surface. The conditions under which a crystal will reflect a beam of x-rays with maximum intensity is:

$$n\lambda = 2d\sin\theta$$

where θ is the angle of incidence and reflection (*Bragg angle*) that the x-rays make with the crystal planes, n is a small integer, λ is the wavelength of the x-rays, and d is the distance between the crystal planes. The equation is named for the British physicist Sir William Lawrence Bragg (1890–1971).

branched chain *See* chain.

brass Any of a group of copper–zinc alloys containing up to 50% of zinc. The color of brass changes from red-gold to golden to silvery-white with increasing zinc content. Brasses are easy to work and resist corrosion well. Brasses with up to 35% zinc can be worked cold and are specially suited for rolling into sheets, drawing into wire, and making into tubes. Brasses with 35–46% zinc are harder and stronger but less ductile; they require hot working (e.g. forging). The properties of brass can be improved by the addition of other elements; lead improves its ability to be machined, while aluminum and tin increase its corrosion resistance. *See also* bronze; nickel–silver.

bremsstrahlung /**brem**-shtrah-lûng/ *See* x-radiation.

brine A concentrated solution of sodium chloride.

Brin process *See* barium.

bromic(I) acid /**broh**-mik/ (**hypobromous acid**; HBrO) A pale yellow liquid made by reacting mercury(II) oxide with bromine water. It has strong bleaching powers. Bromic(I) acid will donate protons to only a small extent and hence is a weak acid in aqueous solution. It is a strong oxidizing agent. *See also* bromine water.

bromic(V) acid ($HBrO_3$) A colorless liquid made by the addition of dilute sulfuric acid to barium bromate. It dissociates extensively in aqueous solution and is a strong acid.

bromide /**broh**-mÿd/ *See* halide.

bromination /broh-mă-**nay**-shŏn/ *See* halogenation.

bromine /**broh**-meen, -min/ A deep red, moderately reactive element belonging to the halogens; i.e. group 17 (formerly VIIA) of the periodic table. Bromine is a liquid at room temperature (mercury is the only other element with this property). It occurs in small amounts in seawater, salt lakes, and salt deposits but is much less abundant than chlorine. Bromine reacts with most metals but generally with less vigor than chlorine. It has less oxidizing power than chlorine and consequently can be released from solutions of bromides by reaction with chlorine gas. The laboratory method is the more convenient oxidation by manganese dioxide. Industrial methods of production utilize oxidation by chlorine or electrolysis with removal from the solution by purging with air. Bromine and its compounds are used in pharmaceuticals, photography, chemical synthesis, fumigants, and in significantly large quantities as 1,2-dibromoethane (which is added to gasoline to combine with lead produced from the decomposition of the antiknock agent lead tetraethyl).

The electropositive elements form ionic bromides and the non-metals form fully covalent bromides. Like chlorine, bromine forms oxides, Br_2O and BrO_2, both of which are unstable. The related oxo-acid anions hypobromite (BrO^-) and bromate (BrO_3^-) are formed by the reaction of bromine with cold aqueous alkali and hot aqueous alkali respectively, but the bromine analogs of chlorite and perchlorate are not known.

Bromine and the interhalogens are highly toxic. Liquid bromine and bromine solutions are also very corrosive and goggles and gloves should always be worn when handling such compounds.

Symbol: Br; m.p. –7.25°C; b.p. 58.78°C; r.d. 3.12 (20°C); p.n. 35; r.a.m. 79.904.

bromine trifluoride (BrF_3) A colorless fuming liquid made by direct combination of fluorine and bromine. It is a very reactive compound, its reactions being similar to those of its component halides.

bromine water A yellow solution of bromine in water, which contains bromic(I) acid (hypobromous acid, HBrO). It is a weak acid and a strong oxidizing agent, and decomposes to give a mixture of bromide (Br^-) and bromate(V) (BrO_3^-) ions.

bromobutyl /broh-moh-**byoo**-t'l/ *See* butyl rubber.

bromoethane /broh-moh-**eth**-ayn/ (**ethyl bromide**; C_2H_5Br) A colorless volatile compound, used as a refrigerant. It can be made from ethene and hydrogen bromide.

bromoform /**broh**-mŏ-form/ *See* tribromomethane.

bromomethane /broh-moh-**meth**-ayn/ (**methyl bromide**; CH_3Br) A colorless volatile compound used as a solvent. It can be made from methane and bromine.

Brønsted–Lowry theory /**bron**-sted **low**-ree/ *See* acid.

bronze Any of a group of copper-tin alloys usually containing 0.5–10% of tin. They are generally harder, stronger in compression, and more corrosion resistant than brass. Zinc is often added, as in *gunmetal* (2–4% zinc), to increase strength and corrosion-resistance; bronze coins often contain more zinc (2.5%) than tin (0.5%). The presence of lead improves its machining qualities.

Some copper-rich alloys containing no tin are also called bronzes. *Aluminum bronzes*, for example, with up to 10% aluminum, are strong, resistant to corrosion and wear, and can be worked cold or hot; *silicon bronzes*, with 1–5% silicon, have high corrosion-resistance; *beryllium bronzes*, with about 2% beryllium, are very hard and strong.

brown coal *See* lignite.

Brownian movement (**Brownian motion**) The random motion of small particles in a fluid – for example, smoke particles in air. The particles, which may be large enough to be visible with a microscope, move because they are continuously bombarded by the molecules of the fluid. Brownian movement is named for the Scottish botanist Robert Brown (1773–1858).

brown-ring test A qualitative test used for the detection of nitrate. A freshly prepared solution of iron(II) sulfate is mixed with the sample and concentrated sulfuric acid is introduced slowly to the bottom of the tube using a dropping pipette so that two layers are formed. A brown ring formed where the liquids meet indicates the presence of nitrate. The brown color is $[Fe(NO)]SO_4$, which breaks down on shaking.

buckminsterfullerene /buk-min-ster-**fûl**-ĕ-reen/ An allotrope of carbon containing clusters of 60 carbon atoms bound in a highly symmetric polyhedral structure. The C_{60} polyhedron has a combination of pentagonal and hexagonal faces similar to the panels on a soccer ball. The molecule was named for the American architect Richard Buckminster Fuller (1895–1983) because its structure resembles a geodesic dome (invented by Fuller). The C_{60} polyhedra are informally called *bucky balls*. The original method of making the allotrope was to fire a high-power laser at a graphite target. This also produces less stable carbon clusters, such as C_{70}. It can be produced more conveniently using an electric arc between graphite electrodes in an inert gas. The allotrope is soluble in benzene, from which it can be crystallized to give yellow crystals. This form of carbon is also known as *fullerite*.

The discovery of buckminsterfullerene led to a considerable amount of research into its properties and compounds. Particular interest has been shown in trapping metal ions inside the carbon cage to form enclosure compounds. Buckminsterfullerene itself is often simply called *fullerene*. The term also applies to derivatives of buckminsterfullerene and to simi-

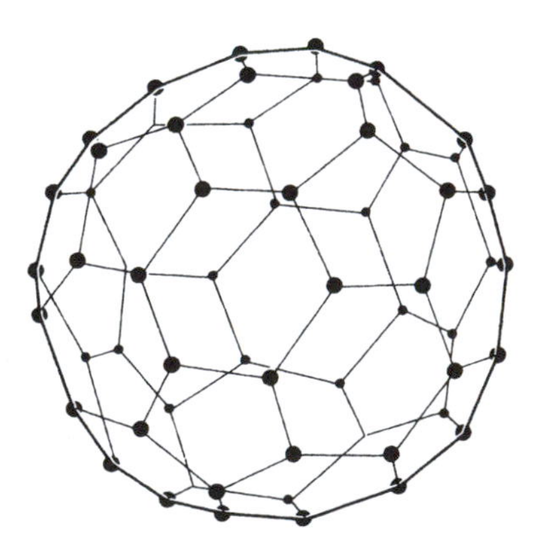

Buckminsterfullerene

lar cluster (e.g. C_{70}). Carbon structures similar to that in C_{60} can also form small tubes, known as *bucky tubes*.

bucky ball /**buk**-ee/ *See* buckminsterfullerene.

bucky tube *See* buckminsterfullerene.

buffer A solution in which the pH remains reasonably constant when acids or alkalis are added to it; i.e. it acts as a buffer against (small) changes in pH. Buffer solutions generally contain a weak acid and one of its salts derived from a strong base; e.g. a solution of ethanoic acid and sodium ethanoate. If an acid is added, the H^+ reacts with the ethanoate ion (from dissociated sodium ethanoate) to form undissociated ethanoic acid; if a base is added the OH^- reacts with the ethanoic acid to form water and the ethanoate ion. The effectiveness of the buffering action is determined by the concentrations of the acid–anion pair:

$$K = [H^+][CH_3COO^-]/[CH_3COOH]$$

where K is the dissociation constant.

Phosphate, oxalate, tartrate, borate, and carbonate systems can also be used for buffer solutions.

bumping Violent boiling of a liquid caused when bubbles form at a pressure above atmospheric pressure.

Bunsen burner /**bun**-sĕn/ A gas burner consisting of a vertical metal tube with an adjustable air-inlet hole at the bottom. Gas is allowed into the bottom of the tube and the gas–air mixture is burnt at the top. With too little air the flame is yellow and sooty. Correctly adjusted, the burner gives a flame with a pale blue inner cone of incompletely burnt gas, and an almost invisible outer flame where the gas is fully oxidized and reaches a temperature of about 1500°C. The inner region is the reducing part of the flame and the outer region the oxidizing part (*see* borax-bead test). The Bunsen burner is named for the German chemist Wilhelm Bunsen (1811–99). He did not invent the Bunsen burner, but used it to great effect in pioneering work on spectroscopy.

Bunsen cell A type of primary cell in which the positive electrode is formed by carbon plates in a nitric acid solution and the negative electrode consists of zinc plates in sulfuric acid solution.

buret /byû-**ret**/ A piece of apparatus used for the addition of variable volumes of liquid in a controlled and measurable way. The buret is a long cylindrical graduated tube of uniform bore fitted with a stopcock and a small-bore exit jet, enabling a drop of liquid at a time to be added to a reaction vessel. Burets are widely used for titrations in volumetric analysis. Standard burets permit volume measurement to 0.005 cm^3 and have a total capacity of 50 cm^3; a variety of smaller *microburettes* is available. Similar devices are used to introduce measured volumes of gas at regulated pressure in the investigation of gas reactions.

buta-1,3-diene /**byoo**-tă **dÿ**-een/ (**butadiene**; $CH_2{:}CHCH{:}CH_2$) A colorless gas made by catalytic dehydrogenation of butane. It is used in the manufacture of synthetic rubber. Buta-1,3-diene is conjugated and to some extent the pi electrons are delocalized over the whole of the molecule.

butanal /**byoo**-tă-nal/ (**butyraldehyde**; C_3H_7CHO) A colorless liquid aldehyde.

butane /**byoo**-tayn, byoo-**tayn**/ (C_4H_{10}) A gaseous alkane obtained either from the gaseous fraction of crude oil or by the 'cracking' of heavier fractions. Butane is easily liquefied and its main use is as a portable supply of fuel (bottle gas).

Butane is the fourth member of the homologous series of alkanes.

butanedioic acid /byoo-tayn-dÿ-**oh**-ik/ (**succinic acid**) A crystalline carboxylic acid, $HOOC(CH_2)_2COOH$, that occurs in amber and certain plants. It forms during the fermentation of sugar (sucrose).

butanoic acid /byoo-tă-**noh**-ik/ (**butyric acid**; C_3H_7COOH) A colorless liquid carboxylic acid. Esters of butanoic acid are present in butter.

butanols /**byoo**-tă-nolz, -nohlz/ (**butyl alcohols**; C_4H_9OH) Two alcohols that are derived from butane: the primary alcohol butan-1-ol ($CH_3(CH_2)_2CH_2OH$) and the secondary alcohol butan-2-ol ($CH_3CH(OH)CH_2CH_3$). Both are colorless volatile liquids used as solvents.

butanone /**byoo**-tă-nohn/ (**methyl ethyl ketone**; $CH_3COC_2H_5$) A colorless volatile liquid ketone. It is manufactured by the oxidation of butane and used as a solvent.

butenedioic acid /byoo-teen-dȳ-**oh**-ik/ Either of two isomers. *Transbutenedioic acid* (fumaric acid) is a crystalline compound found in certain plants. *Cis-butenedioic acid* (maleic acid) is used in the manufacture of synthetic resins. It can be converted into the trans isomer by heating at 120°C.

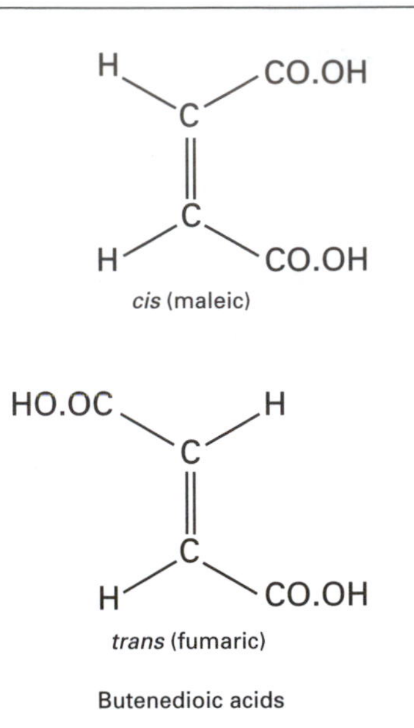

Butenedioic acids

butyl alcohol /**byoo**-t'l/ *See* butanols.

butyl group The straight chain alkyl group $CH_3(CH_2)_2CH_2$–.

butyl rubber A type of synthetic rubber made by copolymerizing isobutylene (2-methylpropene, $CH_3{:}C(CH_3)_2$) with small amounts of isoprene (methylbuta-1,3-diene, $CH_3{:}C(CH_3)CH{:}CH_2$).

Before the introduction of tubeless tires butyl rubber was used for inner tubes because it is impervious to air. Subsequently halogenated butyl rubbers were developed (*halobutyls*), which could be cured at higher temperature and vulcanized with other rubbers. Both *chlorobutyls* and *bromobutyls* are manufactured. These types of rubber are used in tubeless tires bonded to the inner surface of the tire. Other uses are in sealants, hoses, and pond liners.

butyraldehyde /byoo-t'l-**al**-dĕ-hȳd/ *See* butanal.

butyric acid /byoo-**ti**-rik/ *See* butanoic acid.

by-product A substance obtained during the manufacture of a main chemical product. For example, propanone is not now manufactured from propan-1-ol, but is obtained as a by-product in the manufacture of phenol by the cumene process. Calcium chloride is a by-product of the Solvay process for making sodium carbonate. Some metals are obtained as by-products in processes to extract other metals. Cadmium, for instance, is a by-product of the extraction of zinc.

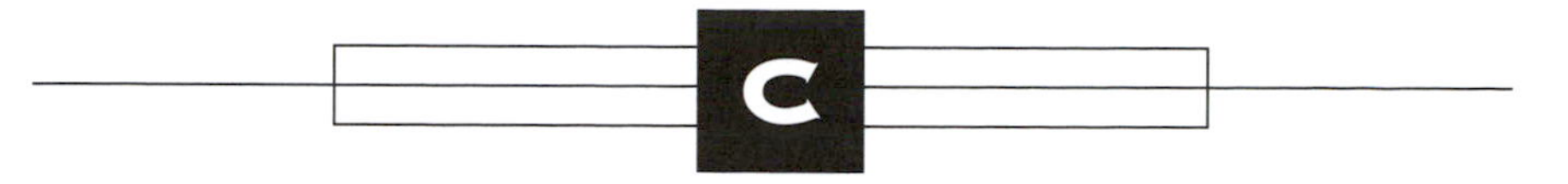

cadmium /**kad**-mee-ŭm/ A transition metal obtained as a by-product during the extraction of zinc. It is used to protect other metals from corrosion, as a neutron absorber in nuclear reactors, in alkali batteries, and in certain pigments. It is highly toxic.

Symbol: Cd; m.p. 320.95°C; b.p. 765°C; r.d. 8.65 (20°C); p.n. 48; r.a.m. 112.411.

cadmium cell *See* Weston cadmium cell.

caesium /**see**-zee-ŭm, **see**-see-/ *See* cesium.

cage compounds *See* clathrate.

calamine /**kal**-ă-mÿn, -min/ **1.** A mineral consisting of hydrated zinc silicate ($2ZnO.SiO_2.H_2O$). It is also known as *hemimorphite*.
2. Zinc carbonate ($ZnCO_3$), which occurs naturally as a mineral also known as *smithsonite*. The carbonate is used medicinally in suspension as a soothing lotion for sunburn and skin complaints. Formerly, the natural mineral was used for this but now medicinal calamine is made by precipitating a basic zinc carbonate from a solution of a zinc salt. It is usually colored pink by the addition of small quantities of iron(III) oxide.

calcination /kal-să-**nay**-shŏn/ The formation of a calcium carbonate deposit from hard water.

calcite /**kal**-sÿt/ A mineral form of calcium carbonate occurring in limestone, chalk, and marble.

calcium /**kal**-see-ŭm/ A moderately soft, low-melting reactive metal; the third element in group 2 (formerly IIA) of the periodic table. The electronic configuration is that of argon with an additional pair of 4s electrons.

Calcium is widely distributed in the Earth's crust and is the third most abundant element. Large deposits occur as chalk or marble, $CaCO_3$; gypsum, $CaSO_4.2H_2O$; anhydrite, $CaSO_4$; fluorspar, CaF; and apatite, $CaF_2.Ca_3(PO_4)_3$. However sufficiently large quantities of calcium chloride are available as waste from the Solvay process to satisfy industrial requirements for the metal, which is produced by electrolysis of the fused salt. Large quantities of lime, $Ca(OH)_2$, and quicklime, CaO, are produced by decomposition of the carbonate for use in both building and agriculture. Several calcium minerals are mined as a source of other substances. Thus, limestone is a cheap source of carbon dioxide, gypsum and anhydrite are used in the manufacture of sulfuric acid, phosphate rock for phosphoric acid, and fluorspar for a range of fluorochemicals.

Calcium has a low ionization potential and a relatively large atomic radius. It is therefore a very electropositive element. The metal is very reactive and the compounds contain the divalent ion Ca^{2+}. Calcium forms the oxide (CaO), a white ionic solid, on burning in air, but for practical purposes the oxide is best prepared by heating the carbonate, which decomposes at about 800°C. Both the oxide and the metal itself react with water to give the basic hydroxide ($Ca(OH)_2$). On heating with nitrogen, sulfur, or the halogens, calcium reacts to form the nitride (Ca_3N_2), sulfide (CaS), or the halides (CaX_2). Calcium also reacts directly with hydrogen to give the hydride CaH_2 and borides, arsenides, carbides, and silicides can be prepared in a similar way. Both the carbonate and sulfate are insoluble. Calcium salts impart a characteristic brick-red color to flames which is an aid to qualitative analysis. At ordinary temperatures calcium has the face-centered cubic structure with a transition at 450°C to the close-packed hexagonal structure.

Symbol: Ca; m.p. 839°C; b.p. 1484°C; r.d. 1.55 (20°C); p.n. 20; r.a.m. 40.0878.

calcium acetylide *See* calcium dicarbide.

calcium bicarbonate *See* calcium hydrogencarbonate.

calcium carbide *See* calcium dicarbide.

calcium carbonate ($CaCO_3$) A white solid that occurs naturally in two crystalline forms: calcite and aragonite. These minerals make up the bulk of such rocks as marble, limestone, and chalk. Calcium carbonate also occurs in the mineral dolomite ($CaCO_3.MgCO_3$). It is sparingly soluble in water but dissolves in rainwater containing carbon dioxide to form calcium hydrogencarbonate, which causes temporary hardness of water. Calcium carbonate is a basic raw material in the Solvay process and is used for making glass, mortar, and cement.

calcium chloride ($CaCl_2$) A white solid that occurs in a number of hydrated forms and is readily available as a by-product of the Solvay process. It is readily soluble in water and the solution, known as *brine*, is used in refrigerating plants. Other applications that depend on its water-absorbing property and the low freezing point of the aqueous solution include the suppression of dust on roads and in mines and the melting of snow. Calcium chloride is used as the electrolyte in the production of calcium.

calcium cyanamide ($CaCN_2$) A solid prepared by heating calcium dicarbide to a temperature in excess of 800°C in an atmosphere of nitrogen. It is used as a fertilizer because ammonia and calcium carbonate are slowly formed when water is added:

$$CaCN_2 + 3H_2O \rightarrow CaCO_3 + 2NH_3$$

Other uses include the defoliation of cotton plants and the production of melamine.

calcium dicarbide (**calcium acetylide; calcium carbide**; CaC_2) A colorless solid when pure. In countries where electricity is cheap, calcium dicarbide is produced on a large scale by heating calcium oxide with coke at a temperature in excess of 2000°C in an electric-arc furnace. Water is then added to give ETHYNE (C_2H_2), an important industrial organic chemical:

$$CaC_2 + 2H_2O \rightarrow C_2H_2 + Ca(OH)_2$$

The structure of calcium dicarbide is interesting because the carbon is present as carbide ions (C_2^{2-}).

calcium fluoride (CaF_2) A white crystalline compound found naturally as *fluorite* (*fluorspar*).

calcium-fluoride structure (**fluorite structure**) A form of crystal structure in which each calcium ion is surrounded by eight fluorine ions arranged at the corners of a cube and each fluorine ion is surrounded tetrahedrally by four calcium ions.

calcium hydrogencarbonate (calcium bicarbonate, $Ca(HCO_3)_2$) A solid formed when water containing carbon dioxide dissolves calcium carbonate:

$$CaCO_3 + H_2O + CO_2 \rightarrow Ca(HCO_3)_2$$

Calcium hydrogencarbonate is a cause of temporary hard water. The solid is unknown at room temperature. *See* hardness.

calcium hydroxide (**slaked lime**; $Ca(OH)_2$) A white solid that dissolves sparingly in water to give the alkali known as *limewater*. Calcium hydroxide is manufactured by adding water to the oxide, a process known as *slaking*, which evolves much heat. If just sufficient water is added so that the oxide turns to a fine powder, the product is *slaked lime*. If more water is added, a thick suspension called *milk of lime* is formed. Calcium hydroxide has many uses. As a base, it is used to neutralize acid soil and in industrial processes such as the Solvay process. It is also used in the manufacture of mortar, whitewash, and bleaching powder and for the softening of temporary hard water.

calcium nitrate ($Ca(NO_3)_2$) A deliquescent salt that is very soluble in water. It is usually crystallized as the tetrahydrate $Ca(NO_3)_2.4H_2O$. When the hydrate is heated, the anhydrous salt is first produced and this subsequently decomposes to give calcium oxide, nitrogen dioxide, and oxygen. Calcium nitrate is used as a nitrogenous fertilizer.

calcium octadecanoate /ok-tă-dek-ă-**noh**-ayt/ (**calcium stearate**; $Ca(CH_3(CH_2)_{16}COO)_2$) An insoluble salt of octadecanoic acid. It is formed as 'scum' when SOAP, containing the soluble salt sodium octadecanoate, is mixed with hard water containing calcium ions.

calcium oxide (**quicklime**; CaO) A white solid formed by heating calcium in oxygen or, more widely, by the thermal decomposition of calcium carbonate. On a large scale, limestone (calcium carbonate) is heated in a tall tower called a lime kiln to a temperature of 550°C. The reversible reaction:

$$CaCO_3 \rightleftharpoons CaO + CO_2$$

proceeds in a forward direction as the carbon dioxide is carried away by the upward current through the kiln. Calcium oxide is used in extractive metallurgy to produce a slag with the impurities in metal ores; it is also used as a drying agent and it is an intermediate for the production of calcium hydroxide.

calcium phosphate ($Ca_3(PO_4)_2$) A solid that occurs naturally in the mineral apatite ($CaF_2.Ca_3(PO_4)_3$) and in rock phosphate. It is the chief constituent of animal bones and is used extensively in fertilizers.

calcium silicate (Ca_2SiO_4) A white insoluble crystalline compound found in various cements and minerals. It is also a component of the slag produced in smelting iron and other metals in a blast furnace.

calcium stearate *See* calcium octadecanoate.

calcium sulfate (**anhydrite**; $CaSO_4$) A white solid that occurs abundantly as the mineral anhydrite and as the dihydrate ($CaSO_4.2H_2O$), known as *gypsum* or *alabaster*. When heated, gypsum loses water to form the hemihydrate ($2CaSO_4.H_2O$), which is *plaster of Paris*. If the water is replaced, gypsum reforms and sets as a solid. Plaster of Paris is therefore used for taking casts and for setting broken limbs. Calcium sulfate is sparingly soluble in water and is a cause of permanent hardness in water. It is used in ceramics, paint, and in paper making.

Calgon /**kal**-gon/ (*Trademark*) A substance often added to detergents to remove unwanted chemicals that have dissolved in water and would otherwise react with soap to form a scum. Calgon consists of complicated polyphosphate molecules, which absorb dissolved calcium and magnesium ions. The metal ions become trapped within the Calgon molecules.

caliche /kah-**lee**-chay/ An impure commercial form of sodium nitrate.

californium /kal-ă-**for**-nee-ŭm/ A silvery radioactive transuranic element of the actinoid series of metals, not found naturally on Earth. Several radioisotopes have been synthesized, including californium-252, which is used as an intense source of neutrons in certain types of portable detector and in the treatment of cancer.

Symbol: Cf; m.p. 900°C; p.n. 98; most stable isotope ^{251}Cf (half-life 900 years).

calixarene /kă-**liks**-ă-reen/ *See* host–guest chemistry.

calomel /**kal**-ŏ-mel/ *See* mercury(I) chloride.

calomel electrode A half cell having a mercury electrode coated with mercury(I) chloride (called *calomel*), in an electrolyte consisting of potassium chloride and (saturated) mercury(I) chloride solution. Its standard electrode potential against the hydrogen electrode is accurately known (–0.2415 V at 25°C) and it is a convenient secondary standard.

calorie /**kal**-ŏ-ree/ Symbol: cal A unit of energy approximately equal to 4.2 joules. It was formerly defined as the energy needed to raise the temperature of one gram of water by one degree Celsius. Because the specific thermal capacity of water changes with temperature, this definition is not precise. The mean or thermochemical calorie (cal_{TH}) is defined as 4.184 joules. The international table calorie (cal_{IT}) is defined as 4.1868 joules. Formerly the mean calorie was defined as one hundredth of the heat needed to raise one gram of water from 0°C to 100°C, and the 15°C calorie as the heat needed to raise it from 14.5°C to 15.5°C.

calorific value /kal-ŏ-**rif**-ik/ The energy content of a fuel, defined as the energy released in burning unit mass of the fuel.

calorimeter /kal-ŏ-**rim**-ĕ-ter/ A device or apparatus for measuring thermal properties such as specific heat capacity, calorific value, etc. *See* bomb calorimeter.

calx /kal'ks/ A metal oxide obtained by heating an ore to high temperatures in air.

camphor /**kam**-fer/ ($C_{10}H_{16}O$) A naturally-occurring white organic compound with a characteristic penetrating odor. It is a cyclic compound and a ketone, formerly obtained from the wood of the camphor tree but now made synthetically. Camphor is used as a platicizer for celluloid and as an insecticide against clothes moths.

candela Symbol: cd The SI base unit of luminous intensity, defined as the intensity (in the perpendicular direction) of the black-body radiation from a surface of 1/600 000 square meter at the temperature of freezing platinum and at a pressure of 101 325 pascals.

cane sugar *See* sucrose.

Cannizzaro reaction /kahn-need-**dzah**-roh/ The reaction of aldehydes to give alcohols and acid anions in the presence of strong bases. The aldehydes taking part in the Cannizzaro reaction lack hydrogen atoms on the carbon attached to the aldehyde group. For instance, in the presence of hot aqueous sodium hydroxide:

$$NaOH + 2C_6H_5CHO \rightarrow C_6H_5CH_2OH + C_6H_5COO^-Na^+$$

This reaction is a disproportionation, involving both oxidation (to acid) and reduction (to alcohol). It can also occur with methanal to give methanol and methanoate ions.

$$NaOH + 2HCHO \rightarrow CH_3OH + HCOO^-Na^+$$

The reaction is named for the Italian chemist Stanislao Cannizzaro (1826–1910) who first described it in 1853.

canonical form *See* resonance.

caproic acid /kă-**proh**-ik/ *See* hexanoic acid.

caprolactam /kap-roh-**lak**-tăm/ ($C_6H_{11}NO$) A white crystalline substance used in the manufacture of NYLON.

caprylic acid /kă-**pril**-ik/ *See* octanoic acid.

carbamide /**kar**-bă-mÿd, -mid, kar-**bam**-ÿd, -id/ *See* urea.

carbanion /kar-**ban**-ÿ-ŏn, -on/ An intermediate in an organic reaction in which one carbon atom is electron-rich and carries a negative charge. Carbanions are usually formed by abstracting a hydrogen ion from a C–H bond using a base, e.g. from ethanal to form $^-CH_2CHO$, or from organometallic compounds in which the carbon atom is bonded to an electropositive metal.

carbazole /**kar**-bă-zohl/ ($C_{12}H_9N$) A white crystalline compound used in the manufacture of dyestuffs.

carbene /**kar**-been/ A transient species of the form RR′C:, with two valence electrons that do not form bonds. The simplest example is *methylene*, H_2C:. Carbenes are short-lived intermediates in some organic reactions. They attack double bonds to form three-membered rings (i.e. cyclopropane derivatives). They also attack single bonds in *insertion reactions*, for example:

$$R–O–H + R_2C: \rightarrow R–O–C(R_2)H$$
$$R–H + R_2C: \rightarrow R–C(R_2)–H$$

Carbenes are important 'reagents' in organic synthesis and various methods exist for generating them in the reaction medium.

carbenium ion /kar-**bee**-nee-ŭm/ *See* carbocation.

carbide /**kar**-bÿd/ A compound of carbon with a more electropositive element. The carbides of the elements are classified into:

1. *Ionic carbides*, which contain the carbide ion C^{4-}. An example is aluminum carbide, Al_4C_3. Compounds of this type react with water to give methane (they were formerly also called *methanides*). The *dicarbides* are ionic carbon compounds that contain the dicarbide ion $^-C:C^-$. The best-known example is calcium dicarbide, CaC_2, also known as calcium carbide, or simply *carbide*. Compounds of this type give ethyne with water. They were formerly called *acetylides* or *ethynides*. Ionic carbides are formed with very electropositive metals. They are crystalline.
2. *Covalent carbides*, which have giant-molecular structures, as in silicon carbide (SiC) and boron carbide (B_4C_3). These are hard high-melting solids. Other covalent compounds of carbon (CO_2, CS_2, CH_4, etc.) have covalent molecules.
3. *Interstitial carbides*, which are interstitial compounds of carbon with transition metals. Titanium carbide (TiC) is an example. These compounds are all hard high-melting solids, with metallic properties. Some carbides (e.g. nickel carbide Ni_3C) have properties intermediate between those of interstitial and ionic carbides.

carbocation An ion with a positive charge in which the charge is mostly localized on a carbon atom. There are two types. *Carbonium ions* have five bonds to the carbon atom and a complete outer shell of 8 electrons. The simplest example would be the carbonium ion CH_5^+, which could be regarded as formed by adding H^+ to methane, CH_4, in the same way that the ammonium ion, NH_4^+, is formed from ammonia, NH_3. There is, however, a difference between ammonia and methane in that ammonia has a lone pair of electrons, which it can donate in forming the NH_4^+ ion. The carbonium ion CH_5^+ (and similar ions) is a transient species, produced in the gas phase by electron bombardment of organic compounds and detected in a mass spectrum. Its shape is that of a carbon atom with three hydrogens in a plane and one hydrogen above and one below (a trigonal bipyramid).

Carbenium ions have three bonds to the central carbon and are planar, with the bonds directed toward the corners of a triangle (sp^2 hybridization). They have six electrons in outer shell of carbon and a vacant p orbital. Carbenium ions are important intermediates in a number of organic reactions, notably the S_N1 mechanism of NUCLEOPHILIC SUBSTITUTION. It is possible to produce stable carbenium ions in salts of the type $(C_6H_5)_3C^+Cl^-$, which are orange-red solids. In these the triphenylmethyl cation is stabilized by delocalization over the three phenyl groups. It is also possible to produce carbenium ions using SUPERACIDS.

carbocyclic compound /kar-boh-**sÿ**-klik, -**sik**-lik/ A compound, such as benzene or cyclohexane, that contains a ring of carbon atoms in its structure.

carbohydrates /kar-boh-**hÿ**-dayts/ A class of compounds occurring widely in nature and having the general formula type $C_x(H_2O)_y$. (Note that although the name suggests a hydrate of carbon these com-

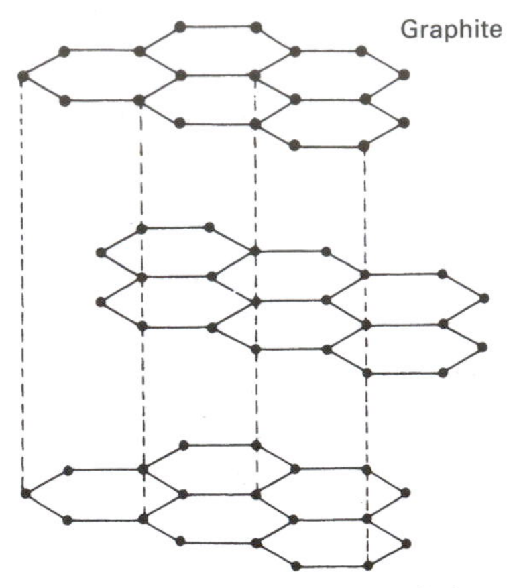

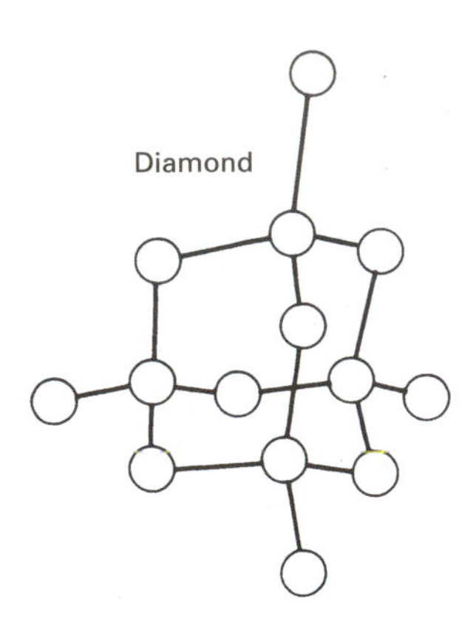

Carbon structures

pounds are in no way hydrates and have no similarities to classes of hydrates.) Carbohydrates are generally divided into two main classes: SUGARS and POLYSACCHARIDES.

Carbohydrates are both stores of energy and structural elements in living systems; plants having typically 15% carbohydrate and animals about 1% carbohydrate. The body is able to build up polysaccharides from simple units (anabolism) or break the larger units down to more simple units for releasing energy (catabolism).

carbolic acid /kar-**bol**-ik/ A former name for phenol (hydroxybenzene, C_6H_5OH).

carbon /**kar**-bŏn/ The first element of group 14 (formerly IVA) of the periodic table. Carbon is a universal constituent of living matter and the principal deposits of carbon compounds are derived from living sources; i.e., carbonates (chalk and limestone) and fossil fuels (coal, oil, and gas). It also occurs in the mineral dolomite. The element forms only 0.032% by mass of the Earth's crust. Minute quantities of elemental carbon also occur as the allotropes graphite and diamond. A third allotrope, buckminsterfullerene (C_{60}), also exists.

The industrial demand for graphite is such that it is manufactured in large quantities using the Acheson process in which coke and small amounts of asphalt or clay are raised to high temperatures. Large quantities of impure carbon are also consumed in the reductive extraction of metals. Apart from the demand for diamond as a gemstone there is a large industrial demand for low-grade small diamonds for drilling and grinding machinery. High-temperature studies show that graphite and diamond can be inter-converted at 3000°C and extremely high pressures but the commercial exploitation would not be viable.

Carbon burns in oxygen to form carbon dioxide and carbon monoxide, CO. Carbon dioxide is soluble in water forming the weakly acidic carbonic acid, the parent acid of the metal carbonates. In contrast CO is barely soluble in water but it will react with alkali to give the methanoate (formate) ion:

$$CO + OH^- \rightarrow HCO_2^-$$

Carbon will react readily with sulfur at red heat to form CS_2 but it does not react directly with nitrogen. Cyanogen, $(CN)_2$, must be prepared by heating covalent metal cyanides such as CuCN. Carbon will also react directly with many metals at elevated temperatures to give carbides. Carbides can also be obtained by heating the metal oxide with carbon or heating the metal with a hydrocarbon. There is a bewilderingly wide range of metal carbides, both salt-like with electropositive elements (CaC_2) and covalent with the metalloids (SiC), and there are also many interstitial carbides formed with metals such as Cr, Mn, Fe, Co, and Ni.

The compounds with C–N bonds form a significant branch of inorganic chemistry of carbon, these are hydrogen cyanide (HCN) and the cyanides, cyanic acid (HNCO) and the cyanates, and thiocyanic acid (HNCS) and the thiocyanates.

Naturally occurring carbon has the isotopic composition ^{12}C (98.89%), ^{13}C (1.11%) and ^{14}C (minute traces in the upper atmosphere produced by slow neutron capture by ^{14}N atoms). ^{14}C is used for radiocarbon dating because of its long half-life of 5730 years.

Symbol: C; m.p. 3550°C; b.p. 4830°C (sublimes); C_{60} sublimes at 530°C; r.d. 3.51 (diamond), 2.26 (graphite), 1.65 (C_{60}) (all at 20°C); p.n. 6; r.a.m. 12.011.

See also buckminsterfullerene.

carbonate /**kar**-bŏ-nayt, -nit/ A salt of carbonic acid (containing the ion CO_3^{2-}).

carbonation /kar-bŏ-**nay**-shŏn/ **1.** The solution of carbon dioxide in a liquid under pressure, as in carbonated soft drinks.
2. The addition of carbon dioxide to compounds, e.g. the insertion of carbon dioxide into Grignard reagents.

carbon black A finely divided form of carbon produced by the incomplete combustion of such hydrocarbon fuels as natural gas or petroleum oil. It is used as a black pigment in inks and as a filler for rubber in tire manufacture.

carbon cycle The circulation of carbon compounds in the environment, one of the major natural cycles of an element. Carbon dioxide in the air is used by green plants in photosynthesis (in which it is combined with water to form sugars and starches). Plants are eaten by animals which exhale carbon dioxide, or when plants and animals die their remains decompose with the production of carbon dioxide. Some plants are burned or converted to fossil fuels which are burned, again with a formation of carbon dioxide.

carbon dating (**radiocarbon dating**) A method of dating – measuring the age of (usually archaeological) materials that contain matter of living origin. It is based on the fact that ^{14}C, a beta emitter of half-life approximately 5730 years, is being formed continuously in the atmosphere as a result of cosmic-ray action.

The ^{14}C becomes incorporated into living organisms. After death of the organism the amount of radioactive carbon decreases exponentially by radioactive decay. The ratio of ^{12}C to ^{14}C is thus a measure of the time elapsed since the death of the organic material.

The method is most valuable for specimens of up to 20 000 years old, though it has been modified to measure ages up to 70 000 years. For ages of up to about 8000 years the carbon time scale has been calibrated by dendrochronology; i.e. by measuring the ^{12}C:^{14}C ratio in tree rings of known antiquity.

carbon dioxide /dȳ-**oks**-ȳd, -id/ (CO_2) A colorless odorless nonflammable gas formed when carbon burns in excess oxygen. It is also produced by respiration. Carbon dioxide is present in the atmosphere (0.03% by volume) and is converted in plants to carbohydrates by photosynthesis. In the laboratory it is made by the action of dilute acid on metal carbonates. Industrially, it is obtained as a by-product in certain processes, such as fermentation or the manufacture of lime. The main uses are as a refrigerant (solid carbon dioxide, called *dry ice*) and in fire extinguishers and carbonated drinks. Increased levels of carbon dioxide in the atmosphere from the combustion of fossil fuels are thought to contribute to the greenhouse effect.

Carbon dioxide is the anhydride of the weak acid carbonic acid, which is formed in water:

$$CO_2 + H_2O \rightarrow H_2CO_3$$

carbon disulfide (CS_2) A colorless poisonous flammable liquid made from methane (natural gas) and sulfur. It is a solvent. The pure compound is virtually odorless, but CS_2 usually has a revolting smell because of the presence of other sulfur compounds.

carbon fibers Fibers of graphite, which are used, for instance, to strengthen polymers. They are made by heating stretched textile fibers and have an orientated crystal structure.

carbonic acid /kar-**bon**-ik/ (H_2CO_3) A dibasic acid formed in small amounts in solution when carbon dioxide dissolves in water:

$$CO_2 + H_2O \rightleftharpoons H_2CO_2$$

It forms two series of salts: hydrogencarbonates (HCO_3^-) and carbonates (CO_3^{2-}). The pure acid cannot be isolated.

carbonium ion /kar-**boh**-nee-ŭm/ *See* carbocation.

carbonize /**kar**-bŏ-nÿz/ (**carburize**) To convert an organic compound into carbon by incomplete oxidation at high temperature.

carbon monoxide /mon-**oks**-ÿd, -id/ (CO) A colorless flammable toxic gas formed by the incomplete combustion of carbon. In the laboratory it can be made by dehydrating methanoic acid with concentrated sulfuric acid:

$$HCOOH - H_2O \rightarrow CO$$

Industrially, it is produced by the oxidation of carbon or of natural gas, or by the water-gas reaction. It is a powerful reducing agent and is used in metallurgy.

Carbon monoxide is neutral and only sparingly soluble in water. It is not the anhydride of methanoic acid, although under extreme conditions it can react with sodium hydroxide to form sodium methanoate. It forms metal carbonyls with transition metals, and its toxicity is due to its ability to form a complex with hemoglobin.

carbon tetrachloride /tet-ră-**klor**-ÿd, -id, -**kloh**-rÿd, -rid/ *See* tetrachloromethane.

carbonyl /**kar**-bŏ-nil/ A complex in which carbon monoxide ligands are coordinated to a metal atom. A common example is tetracarbonyl nickel(0), $Ni(CO)_4$.

carbonyl chloride (**phosgene**; $COCl_2$) A colorless toxic gas with a choking smell. It is used as a chlorinating agent and to make polyurethane plastics and insecticides; it was formerly employed as a war gas.

carbonyl group The group –C=O. It occurs in aldehydes (RCO.H), ketones (RR′CO), carboxylic acids (RCO.OH), and in carbonyl complexes of transition metals.

carborundum /kar-bŏ-**run**-dŭm/ *See* silicon carbide.

carboxylate ion /kar-**boks**-ă-layt/ The ion $-COO^-$, produced by ionization of a carboxyl group. In a carboxylate ion the negative charge is generally delocalized over the O–C–O grouping and the two C–O bonds have the same length, intermediate between that of a double C=O and a single C–O.

carboxyl group /kar-**boks**-ăl/ The organic group –CO.OH, present in carboxylic acids.

carboxylic acid /kar-boks-**il**-ik/ A type of organic compound with the general formula RCOOH. Many carboxylic acids occur naturally in plants and (in the form of esters) in fats and oils, hence the alternative name *fatty acids*. Carboxylic acids with one COOH group are called *monobasic*, those with two, *dibasic*, and those with

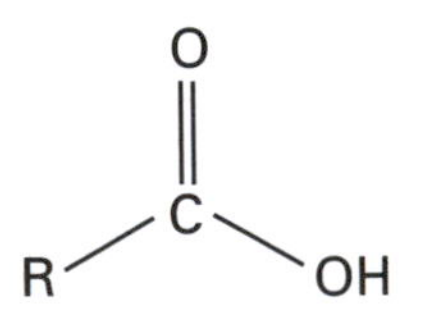

Carboxylic acid

three, *tribasic*. The methods of preparation are:

1. Oxidation of a primary alcohol or an aldehyde:
   $RCH_2OH + 2[O] \rightarrow RCOOH + H_2O$
2. Hydrolysis of a nitrile using dilute hydrochloric acid:
   $RCN + HCl + 2H_2O \rightarrow RCOOH + NH_4Cl$

The acidic properties of carboxylic acids are due to the carbonyl group, which attracts electrons from the C–O and O–H bonds. The carboxylate ion formed, $R\text{–}COO^-$, is also stabilized by delocalization of electrons over the O–C–O grouping.

Other reactions of carboxylic acids include the formation of esters and the reaction with phosphorus(V) chloride to form acyl halides.

carburizing /**kar**-bŭ-rÿz, -byŭ-/ *See* case hardening.

carbylamine reaction /kar-băl-ă-**meen**, -**am**-in/ *See* isocyanide test.

carcinogen /kar-**sin**-ŏ-jĕn/ An agent that causes cancer in animals, such as tobacco smoke or ionizing radiation.

Carius method /**kah**-ree-ŭs/ A method in quantitative analysis for identifying halogens, phosphorus, and sulfur in organic compounds. The compound is heated with concentrated nitric acid, which oxidizes the organic matter and leaves the element in a form in which it can be detected by normal qualitative analysis. The method is named for the German chemist Georg Ludwig Carius (1829–75).

carnallite /**kar**-nă-lÿt/ A mineral chloride of potassium and magnesium, $KCl.MgCl_2.6H_2O$.

Carnot cycle /kar-**noh**/ The idealized reversible cycle of four operations occurring in a perfect heat engine. These are the successive adiabatic compression, isothermal expansion, adiabatic expansion, and isothermal compression of the working substance. The cycle returns to its initial pressure, volume, and temperature, and transfers energy to or from mechanical work. The efficiency of the Carnot cycle is the maximum attainable in a heat engine. The cycle is named for the French physicist Nicolas Leonard Sadi Carnot (1796–1832). *See* Carnot's principle.

Carnot's principle (**Carnot theorem**) The efficiency of any heat engine cannot be greater than that of a reversible heat engine operating over the same temperature range. It follows directly from the second law of thermodynamics, and means that all reversible heat engines have the same efficiency, independent of the working substance. If heat is absorbed at temperature T_1 and given out at T_2, then the Carnot efficiency is $(T_1 - T_2)/T_1$.

Caro's acid /**kah**-roh/ *See* peroxosulfuric(VI) acid.

carrier gas The gas used to carry the sample in gas chromatography.

case hardening Processes for increasing the hardness of the surface or 'case' of the steel used to make such components as gears and crankshafts. The oldest method is *carburizing*, in which the carbon content of the surface layer is increased by heating in a carbon-rich environment. *Nitriding* involves the diffusion of nitrogen into the surface layer of steel forming intensely hard nitride particles in the structure. A combination of both carburizing and nitriding is sometimes employed.

casein /**kay**-seen, -see-in/ A phosphorus-containing protein that occurs in milk and cheese. It is easily digested by young mammals and is their major source of protein and phosphorus. It has been used in the manufacture of paper and buttons.

cast iron Alloys of iron and carbon made by remelting the crude iron produced in a blast furnace. The carbon content is usually between 2.4 and 4.0% and may be present as iron carbide (white cast iron) or as graphite (gray cast iron). Metals can be added to improve the properties of the alloy. Additional elements such as phosphorus, sulfur, and manganese are also present as impurities. Cast iron is a cheap metal with an extensive range of possible properties and it has been used on a very large scale.

catabolism /kă-**tab**-ŏ-liz-ăm/ All the metabolic reactions that break down complex molecules to simpler compounds. The function of catabolic reactions is to provide energy. *See also* metabolism.

catalyst /**kat**-ă-list/ A substance that alters the rate of a chemical reaction without itself being changed chemically in the reaction. The catalyst can, however, undergo physical change; for example, large lumps of catalyst can, without loss in mass, be converted into a powder. Small amounts of catalyst are often sufficient to increase the rate of reaction considerably. A *positive catalyst* increases the rate of a reaction and a *negative catalyst* reduces it. *Homogeneous catalysts* are those that act in the same phase as the reactants (i.e. in gaseous and liquid systems). For example, nitrogen(II) oxide gas will catalyze the reaction between sulfur(IV) oxide and oxygen in the gaseous phase. *Heterogeneous catalysts* act in a different phase from the reactants. For example, finely divided nickel (a solid) will catalyze the hydrogenation of oil (liquid).

The function of a catalyst is to provide a new pathway for which the rate-determining step has a lower activation energy than in the uncatalyzed reaction. A catalyst does not change the products in an equilibrium reaction and their concentration is identical to that in the uncatalyzed reaction; i.e. the position of the equilibrium remains unchanged. The catalyst simply increases the rate at which equilibrium is attained.

In *autocatalysis*, one of the products of the reaction itself acts as a catalyst. In this type of reaction the reaction rate increases with time to a maximum and finally slows down. For example, in the hydrolysis of ethyl ethanoate, the ethanoic acid produced catalyzes the reaction.

catalytic converter /kat-ă-**lit**-ik/ A device fitted to the exhaust system of gasoline-fuelled vehicles to remove pollutant gases from the exhaust. It consists of a honeycomb structure (to provide maximum area) coated with platinum, palladium, and rhodium catalysts. Such devices can convert carbon monoxide to carbon dioxide, oxides of nitrogen to nitrogen, and unburned fuel to carbon dioxide and water.

catalytic cracking The conversion, using a catalyst, of long-chain hydrocarbons from the refining of petroleum into more useful shorter-chain compounds such as those occurring in kerosene and gasoline.

catechol /**kat**-ĕ-chohl, -kohl/ (**1,2-dihydroxybenzene**) A colourless crystalline PHENOL containing two hydroxyl groups. It is used in photographic developing.

catecholamine /kat-ĕ-**chohl**-ă-meen, **-kohl-**/ Any of a group of important amines that contain a catechol ring in their molecules. They are neurotransmitters and hormones. Examples are epinephrine and norepinephrine.

catenation /kat-ĕ-**nay**-shŏn/ The formation of chains of atoms in molecules.

cathode /**kath**-ohd/ In electrolysis, the electrode that is at a negative potential with respect to the anode. In any electrical system, such as a discharge tube or electronic device, the cathode is the terminal at which electrons enter the system.

cation /**kat**-ÿ-ŏn, -on/ A positively charged ion, formed by removal of electrons from atoms or molecules. In electrolysis, cations are attracted to the negatively charged electrode (the cathode). *Compare* anion.

cationic detergent /kat-ÿ-**on**-ik/ *See* detergent.

cationic resin An ION-EXCHANGE material that can exchange cations, such as H^+ and Na^+, for ions in the surrounding medium. Such resins are used for a wide range of purification and analytical purposes.

They are often produced by adding a sulfonic acid group ($-SO_3^-H^+$) or a carboxylate group ($-COO^-H^+$) to a stable polyphenylethene resin. A typical exchange reaction is:

$$resin{-}SO_3^-H^+ + NaCl = resin{-}SO_3^-Na^+ + HCl$$

They have been used to great effect to separate mixtures of cations of similar size having the same charge. Such mixtures can be attached to cationic resins and progressive elution will recover them in order of decreasing ionic radius. Promethium was first isolated using this technique.

caustic lime *See* calcium hydroxide.

caustic potash *See* potassium hydroxide.

caustic soda *See* sodium hydroxide.

celestine /**sel**-ĕ-stÿn, -stin/ A mineral sulfate of strontium, $SrSO_4$.

cell A system having two plates (electrodes) in a conducting liquid (electrolyte). An *electrolytic cell* is used for producing a chemical reaction by passing a current through the electrolyte (i.e. by ELECTROLYSIS). A *voltaic* (or *galvanic*) *cell* produces an e.m.f. by chemical reactions at each electrode. Electrons are transferred to or from the electrodes, giving each a net charge.

There is a convention for writing cell reactions in voltaic cells. The DANIELL CELL, for instance, consists of a zinc electrode in a solution of Zn^{2+} ions connected (through a porous pot) to a solution of Cu^{2+} ions in which is placed a copper electrode. The reactions at the electrodes are

$$Zn \rightarrow Zn^{2+} + 2e$$

i.e. oxidation of the zinc to zinc(II), and

$$Cu^{2+} + 2e \rightarrow Cu$$

i.e. reduction of copper(II) to copper. A cell reaction of this type is written:

$$Zn|Zn^{2+}(aq)|Cu^{2+}(aq)|Cu$$

The e.m.f. is the potential of a lead on the right minus the potential of a lead on the left. Copper is positive in this case and the e.m.f. of the cell is stated as +1.10 volts. *See also* accumulator; Leclanché cell.

cellulose /**sel**-yŭ-lohs/ A polysaccharide $(C_6H_{10}O_5)_n$ of glucose, which is the main constituent of the cell walls of plants. It is obtained from wood pulp.

cellulose acetate (**cellulose ethanoate**) A polymeric substance made by acetylating cellulose. It is used in plastics, in acetate film, and in acetate rayon.

cellulose trinitrate /trÿ-**nÿ**-trayt/ (**gun cotton; nitrocellulose**) A highly flammable substance made by treating cellulose with a nitric–sulfuric acid mixture. Cellulose trinitrate is used in explosives and in lacquers. It is an ester of nitric acid (i.e. not a true nitro compound).

Celsius scale A TEMPERATURE SCALE in which the temperature of melting pure ice is taken as 0° and the temperature of boiling water 100° (both at standard pressure). The *degree Celsius* (°C) is equal to the kelvin. This was known as the *centigrade scale* until 1948, when the present name became official. Celsius' original scale was inverted (i.e. had 0° as the steam temperature and 100° as the ice temperature). The scale is named for the Swedish astronomer Anders Celsius (1701–44).

cement A powdered mixture of calcium silicates and aluminates, which is made by heating limestone ($CaCO_3$) with clay, and grinding the result. When mixed with water, reactions occur with the water (hence the name *hydraulic cement*) and a hard solid aluminosilicate is formed.

cementite /sĕ-**men**-tÿt/ (Fe_3C) A constituent of certain cast irons and steels. The presence of cementite increases the hardness of the metal.

centi- Symbol: c A prefix denoting 10^{-2}. For example, 1 centimeter (cm) = 10^{-2} meter (m).

centigrade scale *See* Celsius scale.

centrifugal pump A device commonly used for transporting fluids around a chemical plant. Centrifugal pumps usually have 6–12 blades rotating inside a fixed circular casing. As the blades rotate, the fluid is impelled out of the pump along a pipe. Centrifugal pumps do not produce high pressures but they have the advantage of being relatively cheap because they are simple in design, have no valves, and work at high speeds. In addition they are not damaged if a blockage develops. *Compare* displacement pump.

centrifuge /**sen**-tră-fyooj/ An apparatus for rotating a container at high speeds, used to increase the rate of sedimentation of suspensions or the separation of two immiscible liquids. *See also* ultracentrifuge.

ceramics /sĕ-**ram**-iks/ Useful high-melting inorganic materials. Ceramics include silicates and aluminosilicates, refractory metal oxides, and metal nitrides, borides, etc. Pottery and porcelain are examples of ceramics.

cerium /**seer**-ee-ŭm/ A ductile malleable gray element of the lanthanoid series of metals. It occurs in association with other lanthanoids in many minerals, including monazite and bastnasite. The metal is reactive; it reacts with water and tarnishes in air. It is used in several alloys (especially for lighter flints), as a catalyst, and in compound form in carbon-arc searchlights, etc., and in the glass industry.

Symbol: Ce; m.p. 799°C; b.p. 3426°C; r.d. 6.7 (hexagonal structure, 25°C); p.n. 58; r.a.m. 140.15.

cermet /**ser**-met/ A synthetic composite material made by combining a ceramic and a sintered metal. Cermets have better temperature and corrosion resistance than straight ceramics. For example, a chromium–alumina cermet is used to make blades for gas-turbine engines.

cerussite /**seer**-ŭ-sÿt/ A naturally occurring form of lead(II) carbonate that is an important lead ore. It forms orthorhombic crystals and is often found together with galena (PbS).

cesium /**see**-zee-ŭm, **see**-see-/ (**caesium**) A soft golden highly reactive low-melting element of the alkali-metal group. It is found in several silicate minerals, including pollucite ($CsAlSi_2O_6$). The metal oxidizes in air and reacts violently with water. Cesium is used in photocells, as a catalyst, and in the cesium atomic clock. The radioactive isotopes ^{134}Cs (half life 2.065 years) and ^{137}Cs (half life 30.3 years) are produced in nuclear reactors and are potentially dangerous atmospheric pollutants.

Symbol: Cs; m.p. 28.4°C; b.p. 678.4°C; r.d. 1.873 (20°C); p.n. 55; r.a.m. 132.91.

cesium-chloride structure A form of crystal structure that consists of alternate layers of cesium ions and chloride ions with the center of the lattice occupied by a cesium ion in contact with eight chloride ions (i.e. four chloride ions in the plane above and four in the plane below).

CFC Chlorofluorocarbon. *See* halocarbon.

c.g.s. system A system of units that uses the centimeter, the gram, and the second as the base mechanical units. Much early scientific work used this system, but it has now almost been abandoned.

chain When two or more atoms form bonds with each other in a molecule, a chain of atoms results. This chain may be a *straight chain*, in which each atom is added to the end of the chain, or it may be a *branched chain*, in which the main chain of atoms has one or more smaller SIDE CHAINS branching off it.

chain reaction A self-sustaining chemical reaction consisting of a series of steps, each of which is initiated by the one before it. An example is the reaction between hydrogen and chlorine:

$$Cl_2 \rightarrow 2Cl\bullet$$
$$H_2 + Cl\bullet \rightarrow HCl + H\bullet$$
$$H\bullet + Cl_2 \rightarrow HCl + Cl\bullet$$

$$2H\bullet \rightarrow H_2$$
$$2Cl\bullet \rightarrow Cl_2$$

The first stage, chain initiation, is the dissociation of chlorine molecules into atoms; this is followed by two chain propagation reactions. Two molecules of hydrogen chloride are produced and the ejected chlorine atom is ready to react with more hydrogen. The final steps, chain termination, stop the reaction.

Induced nuclear fission reactions depend on chain reactions; the fission reaction is maintained by the two or three neutrons set free in each fission.

chair conformation *See* conformation.

chalcogens /**chal**-kŏ-jĕnz/ *See* group 16 elements.

chalk A natural form of calcium carbonate ($CaCO_3$) formed originally by marine organisms. (Blackboard chalk is calcium sulfate, $CaSO_4$.)

chamber process *See* lead-chamber process.

chaotic reaction A chemical reaction in which the concentrations of the reactants show chaotic behavior, i.e. the evolution of the reaction may become unpredictable. Reactions of this type ususally involve a large number of complex interlinked steps.

charcoal An amorphous form of carbon made by heating wood or other organic material in the absence of air. *Activated charcoal* is charcoal that has been heated to drive off absorbed gas. It is used for absorbing gases and for removing impurities from liquids.

Charles' law /sharlz, **charl**-ziz/ For a given mass of gas at constant pressure, the volume increases by a constant fraction of the volume at 0°C for each Celsius degree rise in temperature. The constant fraction (α) has almost the same value for all gases – about 1/273 – and Charles' law can be written in the form

$$V = V_0(1 + \alpha_v\theta)$$

where V is the volume at temperature θ°C and V_0 the volume at 0°C. The constant α_v is the thermal expansivity of the gas. For an ideal gas its value is 1/273.15.

A similar relationship exists for the pressure of a gas heated at constant volume:

$$p = p_0(1 + \alpha_p\theta)$$

Here, α_p is the pressure coefficient. For an ideal gas

$$\alpha_p = \alpha_v$$

although they differ slightly for real gases. It follows from Charles' law that for a gas heated at constant pressure,

$$V/T = K$$

where T is the thermodynamic temperature and K is a constant. Similarly, at constant volume, p/T is a constant.

Charles' volume law is sometimes called *Gay-Lussac's law* after its independent discoverer. The law is named for both the French physicist and physical chemist Jacques Alexandre César Charles (1746–1823) and for the French physicist Joseph-Louis Gay-Lussac (1778–1850). *See also* absolute temperature; gas laws.

chelate /**kee**-layt/ A metal coordination complex in which one ligand coordinates at two or more points to the same metal ion. The resulting complex contains rings of atoms that include the metal atom. An example of a *chelating agent* is 1,2-diaminoethane ($H_2NCH_2CH_2NH_2$), which can coordinate both its amine groups to the same atom. It is an example of a *bidentate ligand* (having two 'teeth'). Edta, which can form up to six bonds, is another example of a chelating agent. The word chelate comes from the Greek word meaning 'claw'. *See also* sequestration.

chemical bond A link between atoms that leads to an aggregate of sufficient stability to be regarded as an independent molecular species. Chemical bonds include covalent bonds, electrovalent (ionic) bonds, coordinate bonds, and metallic bonds. Hydrogen bonds and van der Waals forces are not usually regarded as true chemical bonds.

chemical combination, laws of A group of chemical laws developed during the late 18th and early 19th centuries, which arose from the recognition of the importance of quantitative (as opposed to qualitative) study of chemical reactions. The laws are:

1. the law of conservation of mass (matter);
2. the law of constant (definite) proportions;
3. the law of multiple proportions;
4. the law of equivalent (or reciprocal) proportions,

These laws played a significant part in Dalton's development of his atomic theory (1808). *See* conservation of mass, law of; constant proportions, law of; equivalent proportions, law of; multiple proportions, law of.

chemical dating A method of using chemical analysis to find the age of an archaeological specimen in which compositional changes have taken place over time. For example, the determination of the amount of fluorine in bone that has been buried gives an indication of its age because phosphate in the bone has gradually been replaced by fluoride ions from groundwater. Another dating technique depends on the fact that, in living organisms, amino acids are optically active. After death a slow racemization reaction occurs and a mixture of L- and D-isomers forms. The age of bones can be accurately determined by measuring the relative amounts of L- and D-amino acids present.

chemical engineering The branch of engineering concerned with the design and maintenance of a chemical plant and its ability to withstand extremes of temperature and pressure, corrosion, and wear. It enables laboratory processes producing grams of material to be converted into a large-scale plant producing tonnes of material. Chemical engineers plan large-scale chemical processes by linking together the appropriate unit processes and by studying such parameters as heat and mass transfer, separations, and distillations.

chemical equation A method of representing a chemical reaction using chemical formulas (see formula). The formulas of the reactants are given on the left-hand side of the equation, with the formulas of the products on the right. The two halves are separated by a directional arrow or arrows (on an equals sign). A number preceding a formula (called a *stoichiometric coefficient*) indicates the number of molecules of that substance involved. The equation must balance – that is, the number of atoms of any one element must be the same on both sides of the equation. A simple example is the equation for the reaction between hydrogen and oxygen to form water:

$$2H_2 + O_2 \rightarrow 2H_2O$$

A more complex equation represents the reaction between disodium tetraborate (borax) and aqueous hydrochloric acid to give boric acid and sodium chloride:

$$Na_2B_4O_7 + 2HCl + 5H_2O \rightarrow 4H_3BO_3 + 2NaCl$$

chemical equilibrium *See* equilibrium.

chemical formula *See* formula.

chemical potential Symbol: μ. For the *i*th component of a mixture the chemical potential μ_i is defined by the partial derivative of the Gibbs free energy G of the system with respect to the amount n_i of the component, when the temperature, pressure, and amounts of other components are constant, i.e. $\mu i = \partial G/\partial n_i$. If the chemical potentials of components are equal then the components are in equilibrium. Also, in a one-component system with two phases it is necessary for the chemical potentials to be equal in the two phases for there to be equilibrium.

chemical reaction A process in which one or more elements or chemical compounds (the reactants) react to produce a different substance or substances (the products).

chemical shift An effect of chemical structure on the position of a spectral line. *See* nuclear magnetic resonance.

chemical symbol A letter or pair of letters that stand for a chemical element, as used in chemical formulas and equations. *See* chemical equation; formula.

chemiluminescence /kem-ă-loo-mă-**nes**-ĕns/ The emission of light during a chemical reaction.

chemisorption /kem-ă-**sorp**-shŏn, -**zorp**-/ *See* adsorption.

Chile saltpeter /**chil**-ee/ *See* sodium nitrate.

china clay (**kaolin**) A white powder obtained from the natural decomposition of granites. It is used as a filler in paper-making, in the pottery industries, and in pharmaceuticals. *See* kaolinite.

chiral /**kÿr**-ăl/ Having the property of chirality. For example, lactic acid is a chiral compound because it has two possible structures that cannot be superposed. *See* optical activity.

chirality /kÿr-**al**-ă-tee/ The property of existing in left- and right-handed forms; i.e. forms that are not superposable in three-dimensional space. In chemistry the term is applied to the existence of optical isomers. *See* optical activity.

chirality element A part of a molecule that causes it to display chirality. The most common type of element is a *chirality center*, which is an atom attached to four different atoms or groups. This is also referred to as an *asymmetric atom*. Less commonly a molecule may have a *chirality axis*, as in the case of certain substituted allenes of the type $R_1R_2C=C=CR_3R_4$. In this form of compound the R_1 and R_2 groups do not lie in the same plane as the R_3 and R_4 groups because of the nature of the double bonds. The chirality axis lies along the C=C=C chain. It is also possible to have molecules that contain a *chirality plane*. *See* optical activity.

chloral /**klor**-ăl, **kloh**-răl/ *See* trichloroethanal.

chloral hydrate *See* trichloroethanal.

chloramine /**klor**-ă-meen, **kloh**-ră-/ (NH_2Cl) A colorless liquid made by reacting ammonia with sodium chlorate(I) (NaOCl). It is formed as an intermediate in the production of hydrazine. Chloramine is unstable and changes explosively into ammonium chloride and nitrogen trichloride.

chlorate /**klor**-ayt, **kloh**-rayt/ A salt of chloric(V) acid.

chloric(I) acid (**hypochlorous acid**; HClO) A colorless liquid produced when chlorine is dissolved in water. It is a bleach and gives chlorine water its disinfectant properties. To increase the yield of acid, the chlorine water can be shaken with a small amount of mercury(II) chloride. The Cl–O bond is broken more easily than the O–H bond in aqueous solution; the acid is consequently a poor proton donor and hence a weak acid.

chloric(III) acid (**chlorous acid**; $HClO_2$) A pale yellow liquid produced by reacting chlorine dioxide with water. It is a weak acid and oxidizing agent.

chloric(V) acid (**chloric acid**; $HClO_3$) A colorless liquid with a pungent odor, formed by the action of dilute sulfuric acid on barium chlorate. It is a strong acid and has bleaching properties. Chloric(V) acid is a strong oxidizing agent and in concentrated solution it will ignite organic substances, such as paper and sugar.

chloric(VII) acid (**perchloric acid**; $HClO_4$) A colorless liquid that fumes strongly in moist air. It is made by vacuum distillation of a mixture of potassium perchlorate and concentrated sulfuric acid. In contact with organic material it is dangerously explosive.

The hydrate ($HClO_4.H_2O$) of chloric(VII) acid is a white crystalline solid at room temperature and has an ionic lattice structure of the form $(H_3O)^+(ClO_4)^-$.

chloric acid /**klor**-ik, **kloh**-rik/ *See* chloric(V) acid.

chloride /**klor**-ÿd, **kloh**-rÿd/ *See* halide; chlorine.

chlorination /klor-ă-**nay**-shŏn, kloh-ră-/ 1. Treatment with chlorine; for instance, the use of chlorine to disinfect water.
2. *See* halogenation.

chlorine /**klor**-een, -in, **kloh**-reen, -rin/ A green reactive gaseous element belonging to the halogens; i.e. group 17 (formerly VIIA) of the periodic table. It occurs in seawater, salt lakes, and underground deposits of halite, NaCl. It accounts for about 0.055% of the Earth's crust. Chlorine is strongly oxidizing and can be liberated from its salts only by strong oxidizing agents, such as manganese(IV) oxide, potassium permanganate(VII), or potassium dichromate; note that sulfuric acid is not sufficiently oxidizing to release chlorine from chlorides. Industrially, chlorine is prepared by the electrolysis of brine and in some processes chlorine is recovered by the high-temperature oxidation of waste hydrochloric acid. Chlorine is used in large quantities, both as the element, to produce chlorinated organic solvents, and for the production of polyvinyl chloride (PVC), the major thermoplastic in use today, and in the form of hypochlorites for bleaching.

Chlorine reacts directly and often vigorously with many elements; it reacts explosively with hydrogen in sunlight to form hydrogen chloride, HCl, and combines with the electropositive elements to form metal chlorides. The metals of main groups 1 and 2 form ionic chlorides but an increase in the metal charge/size ratio leads to the chlorides becoming increasingly covalent. For example, CsCl is totally ionic, $AlCl_3$ has a layer lattice, and $TiCl_4$ is essentially covalent. The electronegative elements form volatile molecular chlorides characterized by the single covalent bond to chlorine. With the exception of Pb^{2+}, Ag^+, and Hg_2^{2+}, the ionic chlorides are soluble in water, dissolving to give the hydrated metal ion and the chloride ion Cl^-. Chlorides of metals other than the most electropositive are hydrolyzed if aqueous solutions are evaporated; for example,

$$ZnCl_2 + H_2O \rightarrow Zn(OH)Cl + HCl$$
$$FeCl_3 + 3H_2O \rightarrow Fe(OH)_3 + 3HCl$$

Chlorine forms four oxides, chlorine monoxide, Cl_2O; chlorine dioxide, ClO_2; chlorine hexoxide, Cl_2O_6; and chlorine heptoxide, Cl_2O_7; all of which are highly reactive and explosive. Chlorine dioxide finds commercial application as an active oxidizing agent but because of its explosive nature it is usually diluted by air or other gases. The chloride ion is able to function as a ligand with a large variety of metal ions forming such species as $[FeCl_4]^-$, $[CuCl_4]^{3-}$, and $[Co(NH_3)_4Cl_2]^+$. The formation of anionic chloro-complexes is applied to the separation of metals by anion-exchange methods.

Because of the hydrolysis of many metal chlorides when solutions are evaporated, special techniques must be used to prepare anhydrous chlorides. These are:

1. reaction of dry chlorine with the hot metal;
2. reaction of dry hydrogen chloride with the hot metal (lower valences);
3. reaction of dry hydrogen chloride on the hydrated chloride.

The solubility of inorganic metal chlorides is such that they are not an environmental problem unless the metal ion itself is toxic but many organochlorine compounds are sufficiently stable for the accumulated residues of chlorine-containing pesticides to present a severe problem in some areas. This arises because they can accumulate in food chains and concentrate in the tissues of higher animals (*see* DDT). Chlorine and hydrogen chloride are both highly toxic. Thus chlorides that hydrolyze to release HCl should also be regarded as toxic. As organochlorine compounds are frequently even more toxic, they should not be handled without gloves and precautions should be taken against inhalation.

Symbol: Cl; m.p. −100.38°C; b.p. −33.97°C; d. 3.214 kg m^{-3} (0°C); p.n. 17; r.a.m. 35.4527.

chlorine dioxide /dÿ-**oks**-ÿd, -id/ (ClO_2) An orange gas formed by the action of concentrated sulfuric acid on potassium chlorate. It is a powerful oxidizing agent and its explosive properties in the presence of a reducing agent were used to make one of the first matches. It is widely used in the purification of water and as a bleach in the flour and wood-pulp industry. On an industrial scale an aqueous solution of chlorine dioxide is made by passing nitrogen dioxide up a tower packed with a fused mixture of aluminum oxide and clay, down which a solution of sodium chlorate flows.

chlorine monoxide /mon-**oks**-ÿd, -id/ *See* dichlorine oxide.

chlorine(I) oxide *See* dichlorine oxide.

chlorite /**klor**-ÿt, **kloh**-rÿt/ A chlorate(III) salt; i.e. a salt of chloric(III) acid (chlorous acid).

chloroacetic acid /klor-oh-ă-**see**-tik, -**set**-ik, kloh-roh-/ *See* chloroethanoic acid.

chlorobenzene /klor-oh-**ben**-zeen, klor-oh-ben-**zeen**, kloh-roh-/ (**monochlorobenzene**; C_6H_5Cl) A colorless liquid made by the catalytic reaction of chlorine with benzene. It can be converted to phenol by reaction with sodium hydroxide under extreme conditions (300°C and 200 atmospheres pressure). It is also used in the manufacture of other organic compounds.

2-chlorobuta-1,3-diene /klor-ŏ-byoo-tă-**dÿ**-een, kloh-roh- / (**chloroprene**; $H_2C{:}CH.CCl{:}CH_2$) A colorless liquid derivative of butadiene used in the manufacture of neoprene rubber.

chlorobutyl /klor-oh-**byoo**-t'l, kloh-roh-/ *See* butyl rubber.

chloroethane /klor-oh-**eth**-ayn, kloh-roh-/ (**ethyl chloride**; C_2H_5Cl) A gaseous compound made by the addition of hydrogen chloride to ethene. It is used as a refrigerant and a local anesthetic.

chloroethanoic acid /klor-oh-eth-ă-**noh**-ik, kloh-roh-/ (**chloroacetic acid**; $CH_2ClCOOH$) A colorless crystalline solid made by substituting one of the hydrogen atoms of the methyl group of ethanoic acid with chlorine, using red phosphorus. It is a stronger acid than ethanoic acid because of the electron-withdrawing effect of the chlorine atom. *Dichloroethanoic acid* (dichloroacetic acid, $CHCl_2COOH$) and *trichloroethanoic acid* (trichloroacetic acid, CCl_3COOH) are made in the same way. The acid strength increases with the number of chlorine atoms present.

chloroethene /klor-oh-**eth**-een, kloh-roh-/ (**vinyl chloride**; $H_2C{:}CHCl$) A gaseous organic compound used in the manufacture of PVC (polyvinyl chloride). Chloroethene is manufactured by the reaction between ethyne and hydrogen chloride using a mercury(II) chloride catalyst:

$$C_2H_2 + HCl \rightarrow H_2C{:}CHCl$$

An alternative source, making use of the ready supply of ethene, is via dichloroethane:

$$H_2C{:}CH_2 + Cl_2 \rightarrow CH_2Cl.CH_2Cl \rightarrow H_2C{:}CHCl$$

chlorofluorocarbon /klor-ŏ-floo-ŏ-rŏ-**kar**-bŏn, kloh-roh-/ *See* halocarbon.

chloroform /**klor**-ŏ-form, kloh-rŏ-/ *See* trichloromethane.

chloromethane /klor-ŏ-**meth**-ayn, kloh-roh-/ (**methyl chloride**; CH_3Cl)

chlorophylls /**klor**-ŏ-filz, **kloh**-rŏ-/ The pigments present in plants that act as catalysts in the photosynthesis of carbohydrates from carbon dioxide and water. There are four types, known as chlorophylls *a*, *b*, *c*, and *d*.

chloroplatinic acid /klor-ŏ-plă-**tin**-ik, kloh-rŏ-/ (**platinic chloride**; H_2PtCl_6) A reddish compound prepared by dissolving platinum in aqua regia. When crystallized from the resulting solution, crystals of the hexahydrate ($H_2PtCl_6.6H_2O$) are obtained. The crystals are needle-shaped and deliquesce on exposure to moist air. Chloroplatinic acid is a relatively strong acid, giving rise to the family of chloroplatinates.

chloroprene /**klor**-ŏ-preen, **kloh**-rŏ-/ *See* 2-chlorobuta-1,3-diene.

chlorous acid /**klor**-ŭs, **kloh**-rŭs/ *See* chloric(III) acid.

cholesteric crystal /kŏ-**less**-tĕ-rin/ *See* liquid crystal.

chromate /**kroh**-mayt/ Any oxygen-containing derivative of chromium. Usually the term is reserved for the chromate(VI) species.

chromatography /kroh-mă-**tog**-ră-fee/ A technique used to separate or analyze complex mixtures. A number of related techniques exist; all depend on two phases: a *mobile phase*, which may be a liquid or a gas, and a *stationary phase*, which is either a solid or a liquid held by a solid. The sample to be separated or analyzed is carried by the mobile phase through the stationary phase. Different components of the mixture are absorbed or dissolved to different extents by the stationary phase, and consequently move along at different rates. In this way the components are separated. There are many different forms of chromatography depending on the phases used and the nature of the partition process between mobile and stationary phases. The main classification is into *column chromatography* and *planar chromatography*.

A simple example of column chromatography is in the separation of liquid mixtures. A vertical column is packed with an absorbent material, such as alumina (aluminum oxide) or silica gel. The sample is introduced into the top of the column and washed down it using a solvent. This process is known as *elution*; the solvent used is the *eluent* and the sample being separated is the *eluate*. If the components are colored, visible bands appear down the column as the sample separates out. The components are separated as they emerge from the bottom of the column. In this particular example of chromatography the partition process is adsorption on the particles of alumina or silica gel. Column chromatography can also be applied to mixtures of gases. *See* gas chromatography. In the other main type of chromatography, planar chromatography, the stationary phase is a flat sheet of absorbent material. *See* paper chromatography; thin-layer chromatography.

Components of the mixture are held back by the stationary phase either by adsorption (e.g. on the surface of alumina) or because they dissolve in it (e.g. in the moisture within chromatography paper).

chrome alum /krohm/ *See* alum.

chromic anhydride /**kroh**-mik/ *See* chromium(VI) oxide.

chromic oxide *See* chromium(III) oxide.

chromite /**kroh**-mÿt/ (**chrome iron ore**; $FeCr_2O_4$) A mineral that consists of mixed oxides of chromium and iron, the principal ore of chromium. It occurs as black masses with a metallic luster.

chromium /**kroh**-mee-ŭm/ A transition metal that occurs naturally as chromite ($FeO.Cr_2O_3$), large deposits of which are found in Zimbabwe. The ore is converted into sodium dichromate(VI) and then reduced with carbon to chromium(III) oxide and finally to metallic chromium with aluminum. Chromium is used in strong alloy steels and stainless steel and for plating articles. It is a hard silvery metal that resists corrosion at normal temperatures. It reacts slowly with dilute hydrochloric and sulfuric acids to give hydrogen and blue chromium(II) compounds, which quickly oxidize in air to green chromium(III) ions. The oxidation states are +6 in chromates (CrO_4^{2-}) and dichromates ($Cr_2O_7^{2-}$), +3 (the most stable), and +2. In acidic solutions the yellow chromate(VI) ion changes to the orange dichromate(VI) ion. Dichromates are used as oxidizing agents in the laboratory; for example as a test for sulfur(IV) oxide (sulfur dioxide) and to oxidize alcohols.

Symbol: Cr; m.p. 1860±20°C; b.p. 2672°C; r.d. 7.19 (20°C); p.n. 24; r.a.m. 51.9961.

chromium(II) oxide (**chromous oxide**; CrO) A black powder prepared by the oxidation of chromium amalgam with dilute nitric acid. At high temperatures (around 1000°C) chromium(II) oxide is reduced by hydrogen.

chromium(III) oxide (**chromic oxide**; **chromium sesquioxide**; Cr_2O_3) A green powder that is almost insoluble in water. It is isomorphous with iron(III) oxide and aluminum(III) oxide. Chromium(III) oxide is prepared by gently heating chromium(III) hydroxide or by heating ammonium dichromate. Alternative preparations include the heating of a mixture of ammonium chloride and potassium dichromate or the decomposition of chromyl chloride by passing it through a red-hot tube. Chromium(III) oxide is used as a pigment in the paint and glass industries.

chromium(IV) oxide (**chromium dioxide**; CrO_2) A black solid prepared by heating chromium(III) hydroxide in oxygen at a temperature of 300–350°C. Chromium(IV) oxide is very unstable.

chromium(VI) oxide (**chromium trioxide**; **chromic anhydride**; CrO_3) A red crystalline solid formed when concentrated sulfuric acid is added to a cold saturated solution of potassium dichromate. The long prismatic needle-shaped crystals that are produced are extremely deliquescent. Chromium(VI) oxide is readily soluble in water, forming a solution that contains several of the polychromic acids. On heating it decomposes to give chromium(III) oxide. Chromium(VI) oxide is used as an oxidizing reagent.

chromophore /**kroh**-mŏ-for, -fohr/ A group of atoms in a molecule that is responsible for the color of the compound.

chromous oxide /**kroh**-mŭs/ *See* chromium(II) oxide.

chromyl chloride /**kroh**-măl/ (CrO_2Cl_2) A dark red covalent liquid prepared either by distilling a dry mixture of potassium dichromate and sodium chloride with concentrated sulfuric acid or by the action of concentrated sulfuric acid on chromium(VI) oxide dissolved in concentrated hydrochloric acid. Chromyl chloride is hydrolyzed by water, and with solutions of alkalis it undergoes immediate hydrolysis to produce chromate ions. It is used as an oxidizing agent in organic chemistry. Chromyl chloride oxidizes methyl groups at the ends of aromatic side chains to aldehyde groupings (Étard's reaction).

cinnabar /**sin**-ă-bar/ *See* mercury(II) sulfide.

cinnamic acid /să-**nam**-ik, **sin**-ă-mik/ *See* 3-phenylpropenoic acid.

CIP system (**Cahn–Ingold–Prelog system**) A method of producing a sequence rule used in the absolute description of stereoisomers in the *R–S* convention (*see* optical activity) or the E-Z CONVENTION. The rule is to consider the atoms that are bound directly to a chiral center (or to a double bond). The group in which this atom has the highest proton number has the highest priority. So, for example, in $HCClBr(NH_2)$, the order of priority is $Br > Cl > NH_2 > H$. If two atoms are the same, the substituents are considered, with the substituents of highest proton number taking precedence. So in

$$C(NH_2)(NO_2)(CH_3)(C_2H_6)$$

the order is $NO_2 > NH_2 > C_2H_6 > CH_3$. The system is named after the British chemists Robert Cahn (1899–1981) and Sir Christopher Ingold (1893–1970) and the Bosnian–Swiss chemist Vladimir Prelog (1906–).

cis- Designating an isomer with groups that are adjacent. *See* isomerism.

***cis-trans* isomerism** /sis-**tranz**/ *See* isomerism.

citric acid /**sit**-rik/ A white crystalline carboxylic acid important in plant and animal cells. It is present in many fruits. The systematic name is 2-hydroxypropane-1,2,3-tricarboxylic acid. The formula is:

$HOOCCH_2C(OH)(COOH)CH_2COOH$

Claisen condensation /**klÿ**-zěn/ A reaction in which two molecules of ester combine to give a keto-ester – a compound containing a ketone group and an ester group. The reaction is base-catalyzed by sodium ethoxide; the reaction of ethyl ethanoate refluxed with sodium ethoxide gives:

$$2CH_3.CO.OC_2H_5 \rightarrow CH_3.CO.CH_2.CO.OC_2H_5 + C_2H_5OH$$

The mechanism is similar to that of the aldol reaction, the first step being formation of a carbanion from the ester:

$$CH_3COC_2H_5 + {}^{-}OC_2H_5 \rightarrow {}^{-}CH_2COC_2H_5 + C_2H_5OH$$

This attacks the carbon atom of the carbonyl group on the other ester molecule, forming an intermediate anion that decomposes to the keto-ester and the ethanoate ion. The reaction is named for the German organic chemist Ludwig Claisen (1851–1930).

Clark cell A type of cell formerly used as a standard source of e.m.f. It consists of a mercury cathode coated with mercury sulfate, and a zinc anode. The electrolyte is zinc sulfate solution. The e.m.f. produced is 1.4345 volts at 15°C. The Clark cell has been superseded as a standard by the Weston (*Trademark*) cadmium cell. The cell is named for the English engineer Josiah Latimer Clark (1822–98).

clathrate /**klath**-rayt/ (**enclosure compound**) A substance in which small (guest) molecules are trapped within the lattice of a crystalline (host) compound. Clathrates are formed when suitable host compounds are crystallized in the presence of molecules of the appropriate size. Although the term 'clathrate compound' is often used, they are not true compounds; no chemical bonds are formed, and the guest molecules interact by weak van der Waals forces. The clathrate is maintained by the cagelike lattice of the host. The host lattice must be broken down, for example, by heating or dissolution in order to release the guest. This should be compared with zeolites, in which the holes in the host lattice are large enough to permit entrance or emergence of the guest without breaking bonds in the host lattice. Quinol forms many clathrates, e.g. with SO_2; water (ice) forms a clathrate with xenon.

clays Naturally occurring aluminosilicates which form pastes and gels with water.

cleavage The splitting of a crystal along planes of atoms, to form smooth surfaces.

close packing The arrangement of particles (usually atoms) in crystalline solids in which each particle has 12 nearest neighbors: six in the same layer (or plane) as itself and three each in the layer above and below. This arrangement provides the most economical use of space. The two principal types are cubic close packing and hexagonal close packing.

cluster compound A type of compound in which a cluster of metal atoms are joined by metal–metal bonds. Cluster compounds are formed by certain transition elements, such as molybdenum and tungsten.

coagulation /koh-ag-yŭ-**lay**-shŏn/ The association of particles (e.g. in colloids) into clusters. *See* flocculation.

coal A black mineral that consists mainly of carbon, used as a fuel and as a source of organic chemicals. It is the fossilized remains of plants that grew in the Carboniferous and Permian periods and were buried and subjected to high pressures underground. There are various types of coal, classified according to their increasing carbon content. *See* anthracite; bituminous coal; lignite.

coal gas A fuel gas made by heating coal in a limited supply of air. It consists mainly of hydrogen and methane, with some carbon monoxide (which makes the gas highly poisonous). COAL TAR and coke are formed as by-products.

coal tar Tar produced by heating coal in the absence of oxygen. It is a mixture of many organic compounds (e.g. benzene, toluene, and naphthalene) and also contains free carbon.

cobalt /**koh**-bawlt/ A lustrous silvery-blue hard ferromagnetic transition metal occurring in association with nickel. It is used in alloys for magnets, cutting tools, and electrical heating elements and in catalysts and some paints.

Symbol: Co; m.p. 1495°C; b.p. 2870°C; r.d. 8.9 (20°C); p.n. 27; r.a.m. 58.93320.

cobaltic oxide /koh-**bawl**-tik/ *See* cobalt(III) oxide.

cobaltous oxide /koh-**bawl**-tŭs/ *See* cobalt(II) oxide.

cobalt(II) oxide (**cobaltous oxide**; CoO) A green powder prepared by the action of heat on cobalt(II) hydroxide in the absence of air. Alternatively it may be prepared by the thermal decomposition of cobalt(II) sulfate, nitrate, or carbonate, also in the absence of air. Cobalt(II) oxide is a basic oxide, reacting with acids to give solutions of cobalt(II) salts. It is stable in air up to temperatures of around 600°C, after which it absorbs oxygen to form tricobalt tetroxide. Cobalt(II) oxide can be reduced to cobalt by heating in a stream of carbon monoxide or hydrogen. It is used in the pottery industry and in the production of vitreous enamels.

cobalt(III) oxide (**cobaltic oxide**; Co_2O_3) A dark gray powder formed by the thermal decomposition of either cobalt(II) nitrate or carbonate in air. If heated in air, cobalt(III) oxide undergoes further oxidation to give tricobalt tetroxide, Co_3O_4.

coenzymes Small organic molecules (compared to the size of an enzyme) that enable enzymes to carry out their catalytic activity. Examples include nicotinamide adenine dinucleotide (NAD) and ubiquinone (coenzyme Q). Some coenzymes are capable of catalyzing reactions in the absence of an enzyme but the rate of reaction is never as high as when a catalyst is present. A coenzyme is not a true catalyst because it undergoes chemical change during the reaction.

coherent units A system or sub-set of units (e.g. SI units) in which the derived units are obtained by multiplying or dividing together base units, with no numerical factor involved.

coinage metals A group of malleable metals forming group 11 of the periodic table. (They were formerly subgroup IB.) The metals all have an outer s^1 electronic configuration but they differ from the alkali metals (also outer s^1) in having inner d-electrons. They are copper (Cu), silver (Ag), and gold (Au). The coinage metals all have much higher ionization potentials than the alkali metals and high positive standard electrode potentials and are therefore much more difficult to oxidize. A further significant difference from the alkali metals is the variety of oxidation states observed for the coinage metals. Thus copper in aqueous chemistry is the familiar (hydrated) blue divalent Cu^{2+} ion but colorless copper(I) compounds can be prepared with groups that stabilize low valences such as CN^-, e.g. CuCN. A few compounds of copper(III) have also been prepared. The common form of silver is Ag(I), e.g. $AgNO_3$, with a few Ag(II) compounds stable in the solid only. The most common oxidation state of gold is Au(III), although cyanide ion again stabilizes Au(I) compounds, e.g., $[Au(CN)_4]^{3-}$. The metals all form a large number of coordination compounds (unlike group 1) and are generally treated with the other elements of the appropriate transition series.

coke A dense substance obtained from the carbonization of coal. It is used as a fuel and as a reducing agent.

colligative properties /kŏ-**lig**-ă-tiv/ A group of properties of solutions that depends on the number of particles present, rather than the nature of the particles. Colligative properties include:

1. The lowering of vapor pressure.

2. The elevation of boiling point.
3. The lowering of freezing point.
4. Osmotic pressure.

Colligative properties are all based upon empirical observation. The explanation of these closely related phenomena depends on intermolecular forces and the kinetic behavior of the particles, which is qualitatively similar to those used in deriving the kinetic theory of gases.

collimator /**kol**-ă-may-ter/ An arrangement for producing a parallel beam of radiation for use in a spectrometer or other instrument. A system of lenses and slits is utilized.

colloid /**kol**-oid/ A heterogeneous system in which the interfaces between phases, though not visibly apparent, are important factors in determining the system properties. The three important attributes of colloids are:

1. They contain particles, commonly made up of large numbers of molecules, forming the distinctive unit or *disperse phase*.
2. The particles are distributed in a continuous medium (the *continuous phase*).
3. There is a stabilizing agent, which has an affinity for both the particle and the medium; in many cases the stabilizer is a polar group.

Particles in the disperse phase typically have diameters in the range 10^{-6}–10^{-4} mm. Milk, rubber, and emulsion paints are typical examples of colloids. *See also* sol.

colorimetric analysis /kul-ŏ-ră-**met**-rik/ Quantitative analysis in which the concentration of a colored solute is measured by the intensity of the color. The test solution can be compared against standard solutions.

columbium /kŏ-**lum**-bee-ŭm/ The former name (in the USA) for niobium.

column chromatography *See* chromatography, gas chromatography.

combustion A reaction with oxygen with the production of heat and light. The combustion of solids and liquids occurs when they release flammable vapor, which reacts with oxygen in the gas phase. Combustion reactions usually involve a complex sequence of free-radical chain reactions. The light is produced by excited atoms, molecules, or ions. In highly luminous flames it comes from small incandescent particles of carbon.

Sometimes the term is also applied to slow reactions with oxygen, and also to reactions with other gases (for example, certain metals 'burn' in chlorine). *See also* spontaneous combustion.

common salt *See* sodium chloride.

complex (**coordination compound**) A type of compound in which molecules or ions form coordinate bonds with a metal atom or ion. The coordinating species (called *ligands*) have lone pairs of electrons, which they can donate to the metal atom or ion. They are molecules such as ammonia or water, or negative ions such as Cl^- or CN^-. The resulting complex may be neutral or it may be a *complex ion*. For example:

$$Cu^{2+} + 4NH_3 \rightarrow [Cu(NH_3)_4]^{2+}$$
$$Fe^{3+} + 6CN^- \rightarrow [Fe(CN)_6]^{3-}$$
$$Fe^{2+} + 6CN^- \rightarrow [Fe(CN)_6]^{4-}$$

The formation of such coordination complexes is typical of transition metals. Often the complexes contain unpaired electrons and are paramagnetic and colored. *See also* chelate. *See illustration overleaf.*

component /kŏm-**poh**-nĕnt/ One of the separate chemical substances in a mixture in which no chemical reactions are taking place. For example, a mixture of ice and water has one component; a mixture of nitrogen and oxygen has two components. When chemical reactions occur between the substances in a mixture, the number of components is defined as the number of chemical substances present minus the number of equilibrium reactions taking place. Thus, the system: $N_2 + 3H_2 \rightleftharpoons 2NH_3$ is a two-component system. *See also* phase rule.

compound A chemical combination of atoms of different elements to form a substance in which the ratio of combining atoms remains fixed and is specific to that substance. The constituent atoms cannot be separated by physical means; a chemical reaction is required for the compounds to be formed or to be changed. The existence of a compound does not necessarily imply that it is stable. Many compounds have lifetimes of less than a second. *See also* mixture.

concentrated Denoting a solution in which the amount of solute in the solvent is relatively high. The term is always relative; for example, whereas concentrated sulfuric acid may contain 96% H_2SO_4, concentrated potassium chlorate may contain as little as 10% $KClO_3$.
Compare dilute.

concentration The amount of substance per unit volume or mass in a solution. *Molar concentration* is amount of substance (in moles) per cubic decimeter (liter). *Mass concentration* is mass of solute per unit volume. *Molal concentration* is amount of substance (in moles) per kilogram of solute.

concerted reaction A reaction that takes place in a single stage rather than as a series of simple steps. In a concerted reaction there is a transition state in which bonds are forming and breaking at the same time. An example is the S_N2 mechanism in NUCLEOPHILIC SUBSTITUTION. *See also* pericyclic reaction.

condensation /kon-den-**say**-shŏn/ The conversion of a gas or vapor into a liquid or solid by cooling.

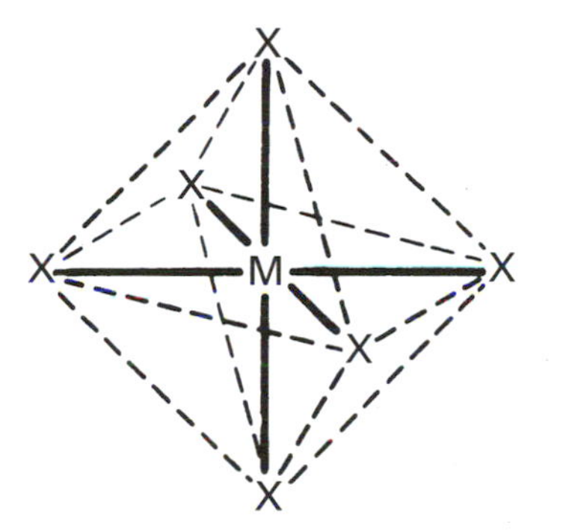

octahedral

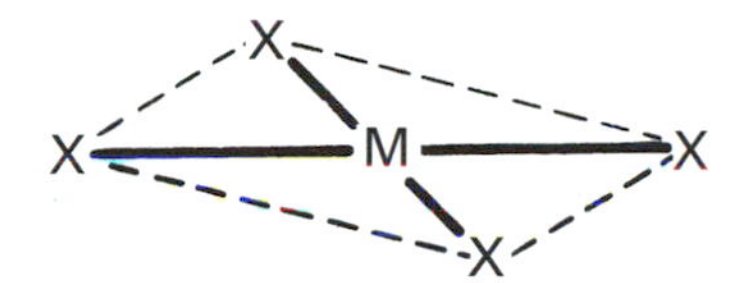

square-planar

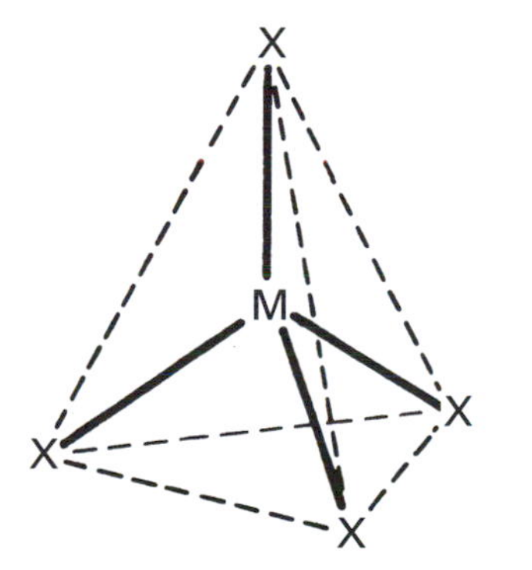

tetrahedral

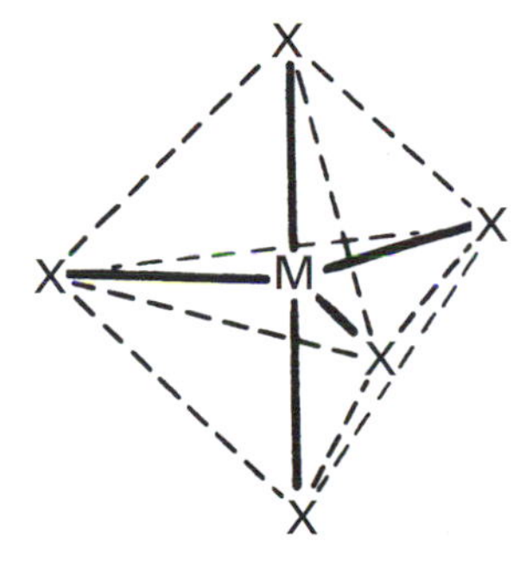

trigonal-bipyramid

Shapes of complexes

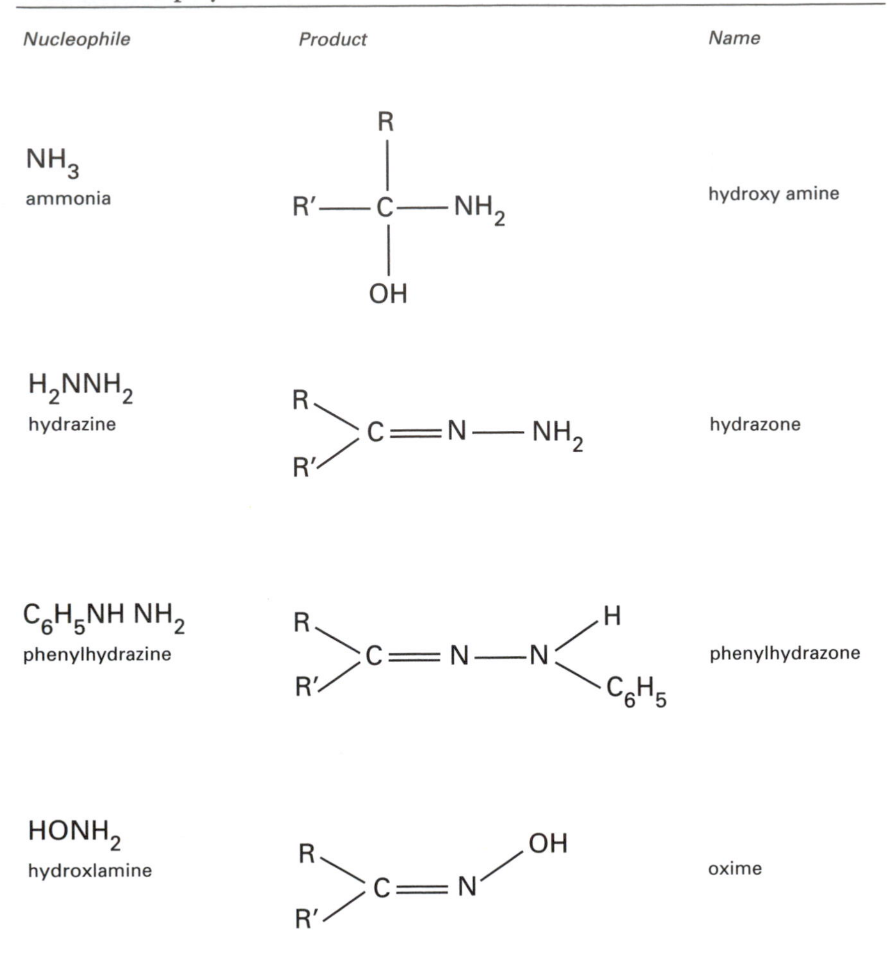

Condensation reactions with aldehydes and ketones

condensation polymerization *See* polymerization.

condensation reaction A reaction in which addition of two molecules occurs followed by elimination of a smaller molecule, usually water. Condensation reactions (addition–elimination reactions) are characteristic of aldehydes and ketones reacting with a range of nucleophiles. There is typically nucleophilic addition at the C atom of the carbonyl group followed by elimination of water.

conducting polymer A type of organic polymer that conducts electricity like a metal. Conducting polymers are crystalline substances containing conjugated unsaturated carbon–carbon bonds. In principle, they provide lighter and cheaper alternatives to metallic conductors.

conductiometric titration /kŏn-duk-tee-oh-**met**-rik/ A titration in which measurement of the electrical conductance is made continuously throughout the addition of the titrant and well beyond the equivalence

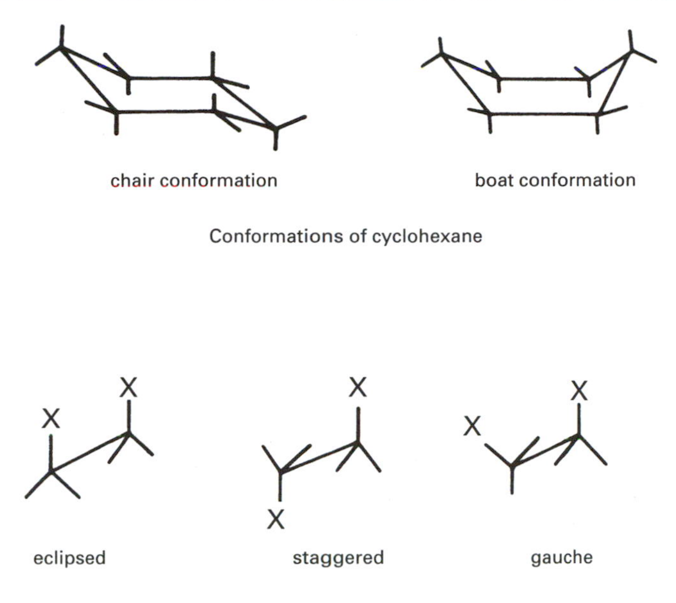

chair conformation

boat conformation

Conformations of cyclohexane

eclipsed

staggered

gauche

Conformations for rotation about a single bond

Conformations

point. This is in place of traditional end-point determination by indicators. The operation is carried out in a conductance cell, which is part of a resistance bridge circuit. The method depends on the fact that ions have different ionic mobilities, H^+ and OH^- having particularly high values. The method is especially useful for weak acid–strong base and strong acid–weak base titrations for which color-change titrations are unreliable.

configuration /kŏn-fig-yŭ-**ray**-shŏn/ **1.** The arrangement of electrons about the nucleus of an atom. Configurations are represented by symbols, which contain:

1. An integer, which is the value of the principal quantum number (shell number).
2. A lower-case letter representing the value of the azimuthal quantum number (l), i.e.

   s means $l = 0$, p means $l = 1$,
   d means $l = 2$, f means $l = 3$.
3. A numerical superscript giving the number of electrons in that particular set; for example, $1s^2$, $2p^3$, $3d^5$.

The ground state electronic configuration (i.e. the most stable or lowest energy state) may then be represented as follows, for example, He, $1s^2$; N, $1s^22s^22p^5$. These are identical to the 'electron box' models, which are often met. However, elements are commonly abbreviated by using an inert gas to represent the 'core', e.g. Zr has the configuration $[Kr]4d^25s^2$.

2. The arrangement of atoms or groups in a molecule.

conformation A particular shape of molecule that arises through the normal rotation of its atoms or groups about single bonds. Any of the possible conformations that may be produced is called a *conformer*, and there will be an infinite number of these possibilities, differing in the angle between certain atoms or groups on adjacent carbon atoms. Sometimes, the term 'conformer' is applied more strictly to possible conformations that have minimum energies – as in the case of the boat and chair conformations of CYCLOHEXANE.

conjugate acid /**kon**-jŭ-git/ *See* acid.

conjugate base *See* base.

conjugated /**kon**-jŭ-gay-tid/ Describing compounds that have alternating double and single bonds in their structure. For example, but-1,3-ene (H_2C:CHCH:CH_2) is a typical conjugated compound. In such compounds there is delocalization of the electrons in the double bonds.

conservation of energy, law of Enunciated by Helmholtz in 1847, this law states that in all processes occurring in an isolated system the energy of the system remains constant. The law does of course permit energy to be converted from one form to another (including mass, since energy and mass are equivalent).

conservation of mass, law of Formulated by Lavoisier in 1774, this law states that matter cannot be created or destroyed. Thus in a chemical reaction the total mass of the products equals the total mass of the reactants (the term 'mass' must include any solids, liquids, and gases – including air – that participate).

Constantan /**kon**-stăn-tan/ (*Trademark*) A copper-nickel (cupronickel) alloy containing 45% nickel. It has a high electrical resistivity and very low temperature coefficient of resistance and is therefore used in thermocouples and resistors.

constant-boiling mixture A general observation for most liquids is that the vapor phase above a liquid is richer in the more volatile component (a deviation from Raoult's law). Consequently most liquid mixtures show a regular increase in the boiling point as the liquid is progressively distilled. The boiling point-composition curve shows that distillation of the liquid of composition L_1 gives a vapor richer in A and represented by composition V_1. Further distillation leads to the liquid composition moving towards B. For certain mixtures in which there are strong intermolecular attractions the boiling point-composition curves show minima or maxima. Fractional distillation of the former leads to initial changes in the vapor until a distillate of composition L_2 is reached at which point a constant boiling mixture or AZEOTROPE distills over. Further attempts to fractionate the distillate do not lead to a change in composition. An example of an azeotropic mixture of minimum boiling point is water (b.p. 100°C) and ethanol (b.p. 78.3°C), the azeotrope being 4.4% water and boiling at 78.1°C.

Mixtures that display a maximum in the boiling point-composition curve can lead to initial separation of pure A on fractionation but as the composition of the liquid moves towards B and reaches the maximum, a constant boiling mixture L_3 is reached that will distill over unchanged. An example of an azeotropic mixture of maximum boiling point is water (b.p. 100°C) and hydrogen chloride (b.p. –80°C), the azeotrope being 80% water and boiling at 108.6°C.

constant composition, law of *See* constant proportions; law of.

constant proportions, law of Formulated by Proust in 1779 after the analysis of a large number of compounds, the principle that the proportion of each element in

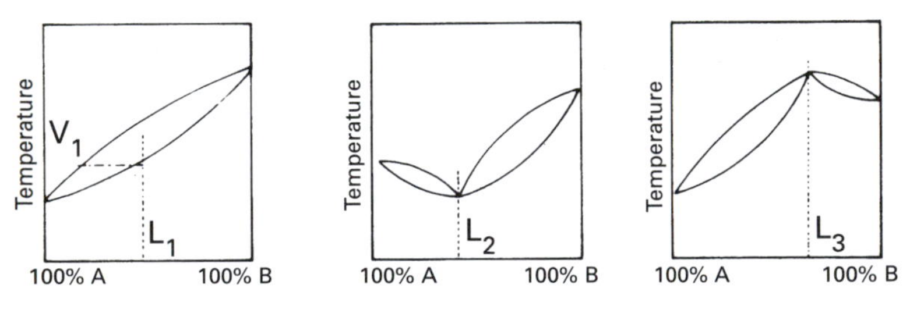

Constant-boiling mixtures

a compound is fixed or constant. It follows that the composition of a pure chemical compound is independent of the method of preparation. It is also called the *law of definite proportions* and the *law of constant composition.*

contact process An industrial process for the manufacture of sulfuric acid. Sulfur(IV) oxide and air are passed over a heated catalyst (vanadium(V) oxide or platinum) and sulfur(VI) oxide is produced:

$$2SO_2 + O_2 \rightarrow 2SO_3$$

The sulfur(VI) oxide is dissolved in sulfuric acid

$$SO_3 + H_2SO_4 \rightarrow H_2S_2O_7$$

The resulting oleum is diluted to give sulfuric acid.

continuous phase *See* colloid.

continuous process A manufacturing process in which the raw materials are constantly fed into the plant. These react as they flow through the equipment to give a continuing flow of product. At any point, only a small amount of material is at a particular stage in the process but material at all stages of the reaction is present. The fractional distillation of crude oil is an example of a continuous process. Such processes are relatively easy to automate and can therefore manufacture a product cheaply. The disadvantages of continuous processing are that it usually caters for a large demand and the plant is expensive to install and cannot normally be used to make other things. *Compare* batch process.

continuous spectrum A SPECTRUM composed of a continuous range of emitted or absorbed radiation. Continuous spectra are produced in the infrared and visible regions by hot solids.

converter *See* Bessemer process.

coordinate bond /koh-**or**-dă-nit/ (**dative bond**) A covalent bond in which the bonding pair is visualized as arising from the donation of a lone pair from one species to another species, which behaves as an electron acceptor. The definition includes such examples as the 'donation' of the lone pair of the ammonia molecule to H^+ (an acceptor) to form NH_4^+ or to Cu^{2+} to form $[Cu(NH_3)_4]^{2+}$.

The donor groups are known as Lewis bases and the acceptors are either hydrogen ions or Lewis acids. Simple combinations, such as $H_3N \rightarrow BF_3$, are known as *adducts*. *See also* complex.

coordination compound /koh-or-dă-**nay**-shŏn/ *See* complex.

coordination number The number of coordinate bonds formed to a metal atom or ion in a complex.

copolymerization /koh-pol-ă-mer-i-**zay**-shŏn/ *See* polymerization.

copper /**kopp**-er/ A transition metal occurring in nature principally as the sulfide. It is extracted by roasting the ore in a controlled air supply and purified by electrolysis of copper(II) sulfate solution using impure copper as the anode and pure copper as the cathode. Copper is used in electrical wires and in such alloys as brass and bronze.

The metal itself is golden-red in color; copper(I) compounds are white (except the oxide, which is red), and copper(II) compounds are blue in solution. Copper is unreactive to dilute acids (except nitric acid). Copper(I) compounds are unstable in solution and decompose to copper and copper(II) ions. Both copper(I) and copper(II) ions form complexes, copper(II) ions being identified by the dark blue complex $[Cu(NH_3)_4]^{2+}$ formed with excess ammonia solution.

Symbol: Cu; m.p. 1083.5°C; b.p. 2567°C; r.d. 8.96 (20°C); p.n. 29; r.a.m. 63.546.

copper(II) carbonate ($CuCO_3$) A green crystalline compound that occurs in mineral form as the basic salt in *azurite* and *malachite*. *See also* verdigris.

copper(I) chloride (**cuprous chloride**;

CuCl) A white solid, insoluble in water, prepared by heating copper(II) chloride in concentrated hydrochloric acid with excess copper turnings. When the solution is colorless, it is poured into air-free water (or water containing sulfur(IV) oxide) and a white precipitate of copper(I) chloride is obtained. On exposure to air this precipitate turns green due to the formation of basic copper(II) chloride. Copper(I) chloride absorbs carbon monoxide gas. It is used as a catalyst in the rubber industry. The chloride is essentially covalent in structure. In the vapor phase both dimeric and trimeric forms exist. It is used in organic chemistry in the Sandmeyer reactions.

copper(II) chloride (**cupric chloride**; $CuCl_2$) A compound prepared by dissolving excess copper(II) oxide or copper(II) carbonate in dilute hydrochloric acid. On crystallization, emerald green crystals of the dihydrate ($CuCl_2.2H_2O$) are obtained. The anhydrous chloride may be prepared as a brown solid by burning copper in excess chlorine. Alternatively the dihydrate may be dehydrated using concentrated sulfuric acid. Dilute solutions of copper(II) chloride are blue, concentrated solutions are green, and solutions in the presence of excess hydrochloric acid are yellow.

copper(II) nitrate (**cupric nitrate**; $Cu(NO_3)_2$) A compound prepared by dissolving either excess copper(II) oxide or copper(II) carbonate in dilute nitric acid. On crystallization, deep blue crystals of the trihydrate ($Cu(NO_3)_2.3H_2O$) are obtained. The crystals are prismatic in shape and are extremely deliquescent. On heating, copper(II) nitrate will decompose to give copper(II) oxide, nitrogen(IV) oxide, and oxygen. The white anhydrous salt is prepared by adding a solution of dinitrogen pentoxide in nitric acid to crystals of the trihydrate.

copper(I) oxide (**cuprous oxide**; CuO) An insoluble red powder prepared by the heating of copper with copper(II) oxide or the reduction of an alkaline solution of copper(II) sulfate. Copper(I) oxide is easily reduced by hydrogen when heated; it is oxidized to copper(II) oxide when heated in air. It is used in the glass industry. Copper(I) oxide undergoes disproportionation in acid solutions, producing copper(II) ions and copper. The oxide will dissolve in concentrated hydrochloric acid due to the formation of the complex ion $[CuCl_2]^-$. Copper(I) oxide is a covalent solid.

copper(II) oxide (**cupric oxide**; CuO) A black solid prepared by the action of heat on copper(II) nitrate, hydroxide, or carbonate. It is a basic oxide and reacts with dilute acids to form solutions of copper(II) salts. Copper(II) oxide can be reduced to copper by heating in a stream of hydrogen or carbon monoxide. It can also be reduced by mixing with carbon and heating the mixture. Copper(II) oxide is stable up to its melting point, after which it decomposes to give oxygen, copper(I) oxide, and eventually copper.

copper(II) sulfate (**cupric sulfate**; $CuSO_4$) A compound prepared as the hydrate by the action of dilute sulfuric acid on copper(II) oxide or copper(II) carbonate. On crystallization, blue triclinic crystals of the pentahydrate (blue vitriol, $CuSO_4.5H_2O$) are formed. Industrially copper(II) sulfate is prepared by passing air through a hot mixture of dilute sulfuric acid and scrap copper. The solution formed is recycled until the concentration of the copper(II) sulfate is sufficient. Copper(II) sulfate is readily soluble in water. The monohydrate ($CuSO_4.H_2O$) is formed at 100°C and the anhydrous salt at 250°C. Anhydrous copper(II) sulfate is white; it is extremely hygroscopic and turns blue on absorption of water. It decomposes on heating to give copper(II) oxide and sulfur(VI) oxide.

Copper(II) sulfate is used as a wood preservative, a fungicide (in Bordeaux mixture), and in the dyeing and electroplating industries.

coral A form of calcium carbonate that is secreted by various invertebrate marine animals (such as certain members of the class Anthozoa) for support and living space.

corn rule *See* optical activity.

corrosion Reaction of a metal with an acid, oxygen, or other compound with destruction of the surface of the metal. Rusting is a common form of corrosion.

corundum /kŏ-**run**-dŭm/ (**emery**; Al_2O_3) A naturally occurring form of aluminum oxide that sometimes contains small amounts of iron and silicon(IV) oxide. Ruby and sapphire are impure crystalline forms. It is used in various polishes, abrasives, and grinding wheels.

coulomb /koo-**lom**/ Symbol: C The SI unit of electric charge, equal to the charge transported by an electric current of one ampere flowing for one second. 1 C = 1 A s. The unit is named for the French physicist Charles Augustin de Coulomb (1736–1806).

coulombmeter /koo-**lom**-ĕ-ter/ (**coulometer; voltameter**) A device for determining electric charge or electric current using electrolysis. The mass m of material released in time t is measured and this can be used to calculate the charge (Q) and the current (I) from the electrochemical equivalent (z) of the element, using $Q = m/z$ or $I = m/zt$.

coulometer /koo-**lom**-ĕ-ter/ *See* coulombmeter.

coumarin /**koo**-mă-rin/ (**1,2-benzopyrone**; $C_9H_6O_2$) A colorless crystalline compound with a pleasant odor, used in making perfumes. On hydrolysis with sodium hydroxide it forms *coumarinic acid.*

coumarone /**koo**-mă-rohn/ *See* benzfuran.

coupling A chemical reaction in which two groups or molecules join together. An example is the formation of azo compounds.

covalent bond /koh-**vay**-lĕnt/ A bond formed by the sharing of an electron pair between two atoms. The covalent bond is conventionally represented as a line, thus H–Cl indicates that between the hydrogen atom and the chlorine atom there is an electron pair formed by electrons of opposite spin implying that the binding forces are strongly localized between the two atoms. Molecules are combinations of atoms bound together by covalent bonds; covalent bonding energies are of the order 10^3 kJ mol^{-1}.

Modern bonding theory treats the electron pairing in terms of the interaction of electron (atomic) orbitals and describes the covalent bond in terms of both 'bonding' and 'anti-bonding' molecular orbitals.

covalent crystal A crystal in which the atoms present are covalently bonded. They are sometimes referred to as giant lattices or macromolecules. The best known completely covalent crystal is diamond.

covalent radius The radius an atom is assumed to have when involved in a covalent bond. For homonuclear diatomic molecules (e.g. Cl_2) this is simply half the measured internuclear distance. For heteroatomic molecules substitutional methods are used. For example, the internuclear distance of bromine fluoride (BrF) is about 180 pm, therefore using 71 pm for the covalent radius of fluorine (from F_2) we get 109 pm for bromine. The accepted value is 114 pm.

cracking A process whereby petroleum distillates are thermally decomposed to lighter fractions. The process can be purely thermal, although *catalytic cracking* is by far the most widely used method.

cream of tartar *See* potassium hydrogentartrate.

creosote /**kree**-ŏ-soht/ A colorless oily liquid containing phenols and distilled from wood tar, used as a disinfectant. The name is also given to *creosote oil*, a dark brown liquid distilled from coal tar and used for preserving timber. It also consists of phenols, mixed with some methylphenols.

cresols /**kree**-solz, -sohlz/ *See* methylphenols.

critical point The conditions of temperature and pressure under which a liquid being heated in a closed vessel becomes indistinguishable from the gas or vapor phase. At temperatures below the critical temperature (T_c) the substance can be liquefied by applying pressure; at temperatures above T_c this is not possible. For each substance there is one critical point; for example, for carbon dioxide it is at 31.1°C and 73.0 atmospheres.

critical pressure The lowest pressure needed to bring about liquefaction of a gas at its critical temperature.

critical temperature The temperature below which a gas can be liquefied by applying pressure and above which no amount of pressure is sufficient to bring about liquefaction. Some gases have critical temperatures above room temperature (e.g. carbon dioxide 31.1°C and chlorine 144°C) and have been known in the liquid state for many years. Liquefaction proved much more difficult for those gases (e.g. oxygen –118°C and nitrogen –146°C) that have very low critical temperatures.

critical volume The volume of one mole of a substance at its critical point.

crossed-beam reaction A chemical reaction performed with two molecular beams intersecting at an angle. Crossed-beam reactions usually involve very simple reactions at low pressures. The products are detected by mass spectroscopy and fundamental information can be obtained about reaction mechanisms.

cross linkage An atom or short chain joining two longer chains in a polymer.

crown ether A compound that has a large ring composed of $–CH_2–CH_2–O–$ units. For example, 18-crown-6 has the formula $C_{12}H_{24}O_6$ (six CH_2CH_2O units). The rings of these compounds are not planar – the name comes from the shape of the molecule. The oxygen atoms of these cyclic ethers can coordinate to central metal ions or to other positive ions (e.g. NH_4^+). The crown ethers have a number of uses in analysis, separation of mixtures, and as catalysts. *Cryptands* are similar compounds in which the ether chains are linked by nitrogen atoms to give a three-dimensional cage structure. They are similar in action to crown ethers but generally form more strongly bound complexes. *See also* host–guest chemistry.

crude oil *See* petroleum.

cryohydrate /krȳ-oh-**hȳ**-drayt/ A EUTECTIC mixture of ice and certain salts.

cryolite /**krȳ**-ŏ-lȳt/ *See* sodium hexafluoroaluminate.

cryoscopic constant /krȳ-ŏ-**skop**-ik/ *See* depression of freezing point.

cryptands /**krip**-tandz/ *See* crown ethers.

crystal A solid substance that has a definite geometric shape. A crystal has fixed angles between its faces, which have distinct edges. The crystal will sparkle if the faces are able to reflect light. The constant angles are caused by the regular arrangements of particles (atoms, ions, or molecules) in the crystal. If broken, a large crystal will form smaller crystals.

In crystals, the atoms, ions, or molecules of the substance form a distinct regular array in the solid state. The faces and their angles bear a definite relationship to the arrangement of these particles.

crystal-field theory A theory of the properties of metal COMPLEXES. Originally it was introduced by Hans Bethe in 1929 to account for the properties of transition elements in ionic crystals. In the theory, the ligands surrounding the metal atom or ion are thought of as negative charges, and the effect of these charges on the energies of

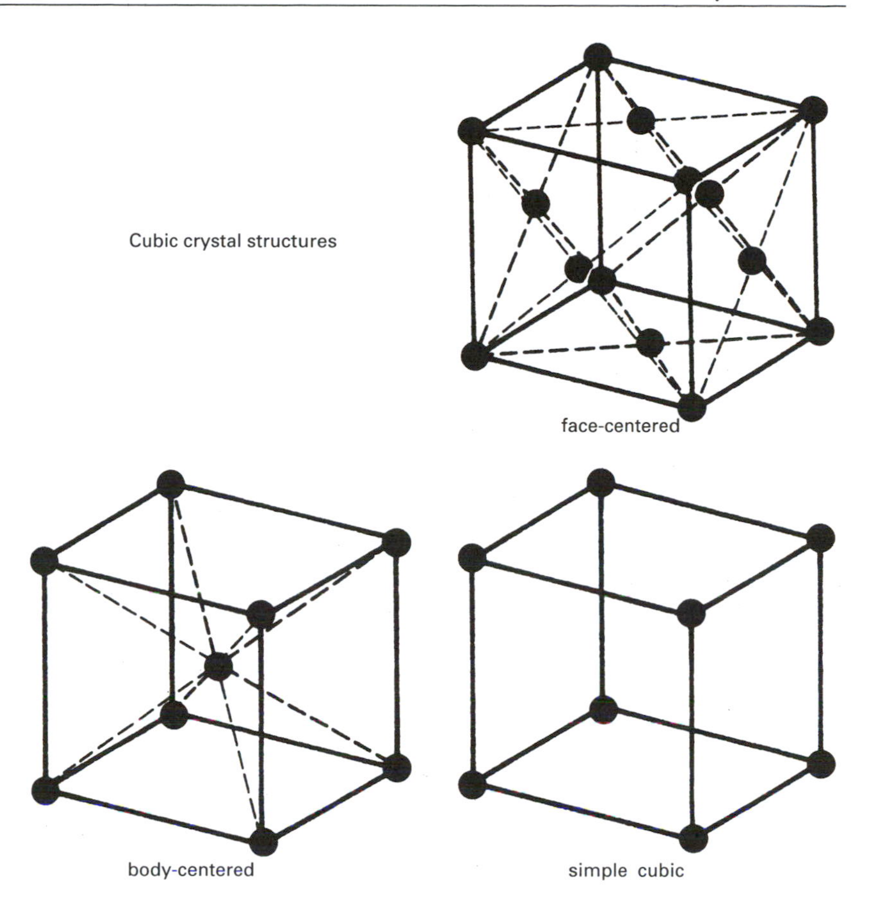

Cubic crystal structures

the d orbitals of the central metal ion is considered. Crystal-field theory was successful in describing the spectroscopic, magnetic, and other properties of complexes. It has been superseded by *ligand-field theory*, in which the overlap of the d orbitals on the metal ion with orbitals on the ligands is taken into account.

crystal habit The shape of a crystal. The habit depends on the way in which the crystal has grown; i.e. the relative rates of development of different faces.

crystalline /**kris**-tă-lin, -lÿn/ Denoting a substance that forms crystals. Crystalline substances have a regular internal arrangement of atoms, even though they may not exist as geometrically regular crystals. For instance, lead (and other metals) are crystalline. Such substances are composed of accumulations of tiny crystals. *Compare* amorphous.

crystallite /**kris**-tă-lÿt/ A small crystal that has the potential to grow larger. It is often used in mineralogy to describe specimens that contain accumulations of many minute crystals of unknown chemical composition and crystal structure.

crystallization /kris-tă-li-**zay**-shŏn/ The process of forming crystals. When a substance cools from the gaseous or liquid state to the solid state, crystallization occurs. Crystals will also form from a solution saturated with a solute.

crystallography /kris-tă-**log**-ră-fee/ The study of the formation, structure, and properties of crystals. *See also* x-ray crystallography.

crystalloid /**kris**-tă-loid/ A substance that is not a colloid and which will therefore not pass through a semipermeable membrane. *See* colloid; semipermeable membrane.

crystal structure The particular repeating arrangement of atoms (molecules or ions) in a crystal. 'Structure' refers to the internal arrangement of particles, not the external appearance.

crystal system A classification of crystals based on the shapes of their unit cell. If the unit cell is a parallelopiped with lengths *a*, *b*, and *c* and the angles between these edges are α (between *b* and *c*), β (between *a* and *c*), and γ (between *a* and *b*), then the classification is: cubic: $a = b = c$; $\alpha = \beta = \gamma = 90°$ tetragonal: $a = b \neq c$; $\alpha = \beta = \gamma = 90°$ orthorhombic: $a \neq b \neq c$; $\alpha = \beta = \gamma = 90°$ hexagonal: $a = b \neq c$; $\alpha = \beta = 90°$; $\gamma = 120°$ trigonal: $a = b \neq c$; $\alpha = \beta = \gamma \neq 90°$ monoclinic: $a \neq b \neq c$; $\alpha = \gamma = 90° \neq \beta$ triclinic: $a \neq b \neq c$; $\alpha \neq \beta \neq \gamma$
The orthorhombic system is also called the *rhombic* system.

CS gas (**(2-chlorobenzylidine-)-malanonitrile**; $C_6H_4ClCH{:}C(CN)_2$) A white organic compound that is a nasal irritant used in powder form as a tear gas for riot control.

cubic Denoting a crystal in which the unit cell is a cube. In a *simple cubic* crystal the particles are arranged at the corners of a cube. *See also* body-centered cubic crystal; face-centered cubic crystal; crystal system.

cubic close packed *See* face-centered cubic crystal.

cumene process /**kyoo**-meen/ An industrial process for the manufacture of phenol from isopropylbenzene (cumene), which is itself made by passing benzene vapor and propene over a phosphoric acid catalyst (250°C and 30 atmospheres):

$$C_6H_6 + CH_2{:}CH(CH_3) \rightarrow C_6H_5CH(CH_3)_2$$

The isopropylbenzene is oxidized by air to a 'hydroperoxide':

$$C_6H_5C(CH_3)_2\text{–O–O–H}$$

This is hydrolyzed by dilute acid to phenol (C_6H_5OH) and propanone (CH_3COCH_3), which is a valuable by-product.

cupellation /kyoo-pe-**lay**-shŏn/ A technique used for the separation of silver and/or gold from impurities by the use of an easily oxidizable metal, such as lead. The impure metal is heated in a furnace and a blast of hot air is directed upon it. The lead or other metal forms an oxide, leaving the silver or gold behind.

cupric chloride /**kyoo**-prik/ *See* copper(II) chloride.

cupric nitrate *See* copper(II) nitrate.

cupric oxide *See* copper(II) oxide.

cupric sulfate *See* copper(II) sulfate.

cuprite /**kyoo**-prÿt/ A red mineral form of copper(I) oxide (Cu_2O), a principal ore of copper.

cupronickel /kyoo-prŏ-**nik**-ăl/ A series of strong copper-nickel alloys containing up to 45% nickel. Those containing 20% and 30% nickel are very malleable, can be worked cold or hot, and are very corrosion-resistant. They are used, for example, in condenser tubes in power stations. Cupronickel with 25% nickel is widely used in coinage. *See also* Constantan.

cuprous chloride /**kyoo**-prŭs/ *See* copper(I) chloride.

cuprous oxide *See* copper(I) oxide.

curie /**kyoo**-ree, kyoo-**ree**/ Symbol: Ci A unit of radioactivity, equivalent to the amount of a given radioactive substance that produces 3.7×10^{10} disintegrations per second, the number of disintegrations produced by one gram of radium.

curium /**kyoo**-ree-ŭm/ A highly toxic radioactive silvery element of the actinoid series of metals. A transuranic element, it is not found naturally on Earth but is synthesized from plutonium. Curium-244 and curium-242 have been used in thermoelectric power generators.

Symbol: Cm; m.p. 1340±40°C; b.p. ≅ 3550°C; r.d. 13.3 (20°C); p.n. 96; most stable isotope ^{247}Cm (half-life 1.56×10^7 years). The unit is named for the French physicist Pierre Curie (1859–1906).

cyanamide process /sȳ-**an**-ă-mȳd, -mid, sȳ-ă-**nam**-ȳd, -id/ An industrial process for fixing nitrogen by heating calcium dicarbide in air.

$$CaC_2 + N_2 \rightarrow CaCN_2$$

The product, calcium cyanamide, hydrolyzes to ammonia and can be used as a fertilizer. The process is expensive due to the high cost of calcium dicarbide. *See also* nitrogen fixation.

cyanide /**sȳ**-ă-nȳd, -nid/ **1.** A salt of hydrogen cyanide, containing the cyanide ion (CN^-). *See* hydrocyanic acid.
2. *See* nitrile.

cyanide process A technique used for the extraction of gold from its ores. After crushing the gold ore to a fine powder it is agitated with a very dilute solution of potassium cyanide. The gold is dissolved by the cyanide to form potassium dicyanoaurate(I) (potassium aurocyanide, $KAu(CN)_2$). This complex is reduced with zinc, filtered, and melted to obtain the pure gold metal.

cyanoferrate /sȳ-an-oh-**fe**-rayt/ A compound containing the ion $[Fe(CN)_6]^{4-}$ (the hexacyanoferrate(II) ion) or the ion $[Fe(CN)_6]^{3-}$ (the hexacyanoferrate(III) ion). These ions are usually encountered as their potassium salts. Potassium hexacyanoferrate(II) ($K_4Fe(CN)_6$, potassium ferrocyanide) is a yellow crystalline compound. Potassium hexacyanoferrate(III) ($K_3Fe(CN)_6$, potassium ferricyanide) is an orange crystalline compound. Its solution gives a deep blue precipitate with iron(II) ions, and is used as a test for iron(II) (ferrous) compounds. *Prassian blue* is a blue pigment containing hexacyanoferrate ions.

cyanogen /sȳ-**an**-ŏ-jen/ (C_2N_2) A toxic flammable gas prepared by heating mercury cyanide. *See also* pseudohalogens.

cyanohydrin /sȳ-an-oh-**hȳ**-drin/ An addition compound formed between an aldehyde or ketone and hydrogen cyanide. The general formula is RCH(OH)(CN) (from an aldehyde) or RR′C(OH)(CN) (from a ketone). Cyanohydrins are easily hydrolyzed to hydroxycarboxylic acids. For instance, the compound 2-hydroxypropanonitrile ($CH_3CH(OH)(CN)$) is hydrolyzed to 2-hydroxypropanoic acid ($CH_3CH(OH)(COOH)$).

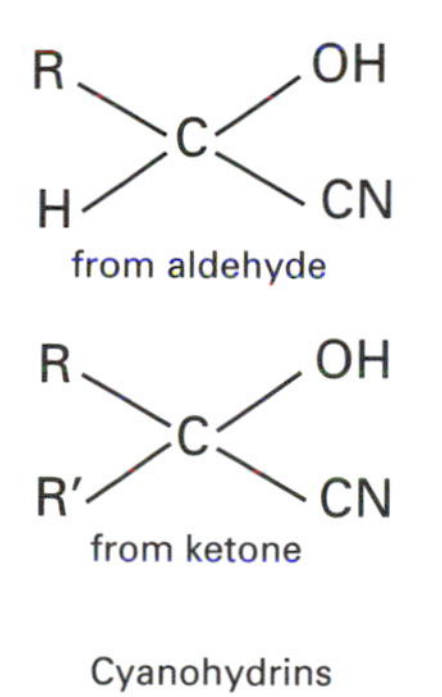

Cyanohydrins

cyclic compound A compound containing a ring of atoms. If the atoms forming the ring are all the same the compound is *homocyclic*; if different atoms are involved it is *heterocyclic*.

cyclization /sȳ-kli-**zay**-shŏn/ A reaction in which a straight-chain compound is converted into a cyclic compound.

cycloaddition /sȳ-kloh-ă-**dish**-ŏn/ *See* pericyclic reaction.

cycloalkane /sȳ-kloh-**al**-kayn/ A saturated cyclic hydrocarbon comprising a ring of carbon atoms, each carrying two hydrogen atoms, general formula C_nH_{2n}. Cyclopropane (C_3H_6), and cyclobutane (C_4H_8) both have strained rings and are highly reactive. Other cycloalkanes have similar properties to the alkanes, although they are generally less reactive.

cyclohexadiene-1,4-dione /sÿ-kloh-heks-ă-**dÿ**-een, -**dÿ**-ohn/ *See* quinone.

cyclohexane /sÿ-kloh-**heks**-ayn/ (C_6H_{12}) A colorless liquid alkane that is commonly used as a solvent and in the production of hexanedioic acid (adipic acid) for the manufacture of nylon. Cyclohexane, itself, is manufactured by the reformation of longer chain hydrocarbons present in crude-oil fractions. It is also interesting from a structural point of view, existing as a 'puckered' six-membered ring, having all bonds between carbon atoms at 109.9° (the tetrahedral angle). The molecule undergoes rapid interconversion between two 'chair-like' CONFORMATIONS, which are energetically equivalent, passing through a 'boat-like' structure of higher energy. It is commonly represented by a hexagon.

cyclonite /**sÿ**-klŏ-nÿt, **sik**-lŏ/ A high explosive made from hexamine.

cyclo-octatetraene /sÿ-kloh-ok-tă-**tet**-ră-een/ *See* annulene.

cyclopentadiene /sÿ-kloh-pen-tă-**dÿ**-een/ A cyclic hydrocarbon made by cracking petroleum. The molecules have a five-membered ring containing two carbon-carbon double bonds and one CH_2 group. It forms the negative *cyclopentadienyl ion* $C_5H_5^-$, present in sandwich compounds, such as ferrocene.

cyclopentadienyl ion /sÿ-kloh-pen-tă-dÿ-**en**-ăl/ *See* cyclopentadiene.

cysteine /**sis**-tee-een/ *See* amino acids.

cytidine /**sÿ**-tă-din, -deen, -**sit**-ă-/ (**cytosine nucleoside**) A NUCLEOSIDE formed when cytosine is linked to D-ribose via a β-glycosidic bond.

cytosine /**sÿ**-tŏ-sin, -seen/ A nitrogenous base found in DNA and RNA. Cytosine has the pyrimidine ring structure.

D

dalton /**dawl**-tŏn/ *See* atomic mass unit.

Dalton's atomic theory /**dawl**-tŏnz/ A theory explaining the formation of compounds by elements first published by John Dalton in 1803. It was the first modern attempt to describe chemical behavior in terms of atoms. The theory was based on certain postulates:

1. All elements are composed of small particles, which Dalton called atoms.
2. All atoms of the same element are identical.
3. Atoms can neither be created nor destroyed.
4. Atoms combine to form 'compound atoms' (i.e. molecules) in simple ratios.

Dalton also suggested symbols for the elements. The theory was used to explain the law of conservation of mass and the laws of chemical combination. The theory is named for the British chemist and physicist John Dalton (1766–1844).

Dalton's law (of partial pressures) The pressure of a mixture of gases is the sum of the partial pressures of each individual constituent. (The *partial pressure* of a gas in a mixture is the pressure that it would exert if it alone were present in the container.) Dalton's law is strictly true only for ideal gases.

Daniell cell /**dan**-yĕl/ A type of primary cell consisting of two electrodes in different electrolytes separated by a porous partition. The positive eletrode is copper immersed in copper(II) sulfate solution. The negative electrode is zinc–mercury amalgam in either dilute sulfuric acid or zinc sulfate solution. The porous pot prevents mixing of the electrolytes, but allows ions to pass. With sulfuric acid the e.m.f. is about 1.08 volts; with zinc sulfate it is about 1.10 volts.

At the copper electrode copper ions in solution gain electrons from the metal and are deposited as copper atoms:

$$Cu^{2+} + 2e^{-} \rightarrow Cu$$

The copper electrode thus gains a positive charge. At the zinc electrode, zinc atoms from the electrode lose electrons and dissolve into solution as zinc ions, leaving a net negative charge on the electrode

$$Zn \rightarrow 2e^{-} + Zn^{2+}$$

The cell is named for the British chemist and meteorologist John Frederic Daniell (1790–1845).

darmstadtium /darm-**stat**-ee-ŭm. -**shtat**-/ A radioactive metallic element that does not occur naturally on the Earth. It is made either by bombarding a lead target with nickel nuclei or by bombarding a plutonium target with sulfur nuclei. There are several isotopes; the most stable is ^{281}Ds, with a half-life of about 1.6 minutes. The chemical properties of darmstadtium should be similar to those of platinum. The element is named for Darmstadt, the place in Germany where it was discovered.

Symbol Ds; p.n. 110.

dative bond /**day**-tiv/ *See* coordinate bond.

Davy lamp /**day**-vee/ *See* safety lamp.

d-block elements The TRANSITION ELEMENTS of the first, second, and third long periods of the periodic table, i.e. Sc to Zn, Y to Cd, and La to Hg. They are so called because in general they have inner d-levels with configurations of the type $(n-1)d^{x}ns^{2}$ where $x = 1–10$.

DDT /dee-dee-**tee**/ (**dichlorodiphenyltrichloroethane**; $(ClC_6H_4)_2CH(CCl_3)$) A colorless crystalline organic compound, once widely used as an insecticide. Its use is now restricted or banned in many countries because it is stable and remains unchanged in the soil, and passes up the food chain to accumulate in the fatty tissues of carnivorous animals.

deactivation A process in which the reactivity of a substance is lessened, or even totally removed . Usually this deactivation is unwanted, as in the poisoning of catalysts.

de Broglie wave /dĕ-**broh**-lyee/ A wave associated with a particle, such as an electron or proton. In 1924, Louis de Broglie suggested that, since electromagnetic waves can be described as particles (photons), particles of matter could also have wave properties. The wavelength (λ) has the same relationship to momentum (p) as in electromagnetic radiation:

$$\lambda = h/p$$

where h is the Planck constant. The wave is named for the French physicist Prince Louis Victor Pierre Raymond de Broglie (1882–1987). *See also* quantum theory.

debye /dĕ-**bay**-ĕ/ Symbol: D A unit of electric dipole moment equal to 3.335 64 × 10^{-30} coulomb meter. It is used in expressing the dipole moments of molecules. the unit is named for the Dutch–American physicist Peter Joseph William Debye (1884–1966).

Debye–Hückel theory /dĕ-**bay**-ĕ **hoo**-kĕl/ A theory to predict the conductivity of ions in dilute solutions of strong ELECTROLYTES. It assumes that electrolytes in dilute solution are completely dissociated into ions but takes into account interionic attraction and repulsion. Agreement between the theory and experiment occurs only with very dilute solutions (less than 10^{-3}M). *See* dissociation.

deca- Symbol: da A prefix denoting 10. For example, 1 decameter (dam) = 10 meters (m).

decahydrate /dek-ă-**hy̆**-drayt/ A crystalline solid containing ten molecules of water of crystallization per molecule of compound.

decant /di-**kant**/ To pour off the clear liquid above a sediment.

decay **1.** The spontaneous breakdown of a radioactive isotope. *See* half-life; radioactivity.
2. The transition of excited atoms, ions, molecules, etc., to the ground state.

deci- Symbol: d A prefix denoting 10^{-1}. For example, 1 decimeter (dm) = 10^{-1} meter (m).

decomposition A chemical reaction in which a compound is broken down into compounds with simpler molecules.

decrepitation /di-krep-ă-**tay**-shŏn/ Heating a crystalline solid so that it emits a crackling noise, usually as a result of loss of water of crystallization.

defect An irregularity in the ordered arrangement of particles in a crystal lattice. There are two main types of defect in crystals. *Point defects* occur at single lattice points and there are two types. A *vacancy* is a missing atom; i.e. a vacant lattice point. Vacancies are sometimes called *Schottky defects* (named for the Swiss–German physicist Walter Schottky (1886–1976). An *interstitial* is an atom that is in a position that is not a normal lattice point. If an atom moves off its lattice point to an interstitial position the result (vacancy plus interstitial) is a *Frenkel defect* (named for the Russian mathematical physicist Jacov Il'ich Frenkel (1894–1952). All solids above absolute zero have a number of point defects, the concentration of defects depending on the temperature. Point defects can also be produced by strain or by irradiation.

Dislocations (or *line defects*) are also produced by strain in solids. They are irregularities extending over a number of lattice points along a line of atoms.

definite proportions, law of *See* constant proportions; law of.

degassing Removal of dissolved or absorbed gases from liquids or solids.

degenerate Describing different quantum states that have the same energy. For instance, the five d orbitals in a transition-metal atom all have the same energy but different values of the magnetic quantum number *m*. Differences in energy occur if a magnetic field is applied or if the arrangement of ligands around the atom is not symmetrical. The degeneracy is then said to be 'lifted'.

degradation A chemical reaction involving the decomposition of a molecule into simpler molecules, usually in stages. The Hofmann degradation of amides is an example.

degrees of freedom **1.** The independent ways in which particles can take up energy. In a monatomic gas, such as helium or argon, the atoms have three translational degrees of freedom (corresponding to motion in three mutually perpendicular directions). The mean energy per atom for each degree of freedom is $kT/2$, where k is the Boltzmann constant and T the thermodynamic temperature; the mean energy per atom is thus $3kT/2$.

A diatomic gas also has two rotational degrees of freedom (about two axes perpendicular to the bond) and one vibrational degree (along the bond). The rotations also each contribute $kT/2$ to the average energy. The vibration contributes kT ($kT/2$ for kinetic energy and $kT/2$ for potential energy). Thus, the average energy per molecule for a diatomic molecule is $3kT/2$ (translation) + kT (rotation) + kT (vibration) = $7kT/2$.

Linear triatomic molecules also have two significant rotational degrees of freedom; non-linear molecules have three. For non-linear polyatomic molecules, the number of vibrational degrees of freedom is $3N - 6$, where N is the number of atoms in the molecule.

The molar energy of a gas is the average energy per molecule multiplied by the Avogadro constant. For a monatomic gas it is $3RT/2$, etc.

2. The independent physical quantities (e.g. pressure, temperature, etc.) that define the state of a given system. *See* phase rule.

dehydration /dee-hÿ-**dray**-shŏn/ **1.** Removal of water from a substance.

2. Removal of the elements of water (i.e. hydrogen and oxygen in a 2:1 ratio) to form a new compound. An example is the dehydration of propanol to propene over hot pumice:

$$C_3H_7OH \rightarrow CH_3CH{:}CH_2 + H_2O$$

deionization /dee-ÿ-ŏ-ni-**zay**-shŏn/ A method of removing ions from a solution using ION EXCHANGE. The term is commonly applied to the purification of tap water; deionized water has superseded distilled water in many chemical applications.

deliquescent /del-ă-**kwess**-ĕnt/ Describing a solid compound that absorbs water from the atmosphere, eventually forming a solution. *See also* hygroscopic.

delocalization /dee-loh-kă-li-**zay**-shŏn/ A spreading out of bonding electrons in a molecule over the molecule.

delocalized bond /dee-**loh**-kă-lÿzd/ (**non-localized bond**) A type of bonding in molecules that occurs in addition to sigma bonding. The electrons forming the delocalized bond are no longer regarded as remaining between two atoms; i.e. the electron density of the delocalized electrons is spread over several atoms and may spread over the whole molecule.

The electron density of the delocalized bond is spread by means of a delocalized molecular orbital and may be regarded as a series of pi bonds extending over several atoms, for example the pi bonds in butadiene and the C–O pi bonds in the carbonate ion.

delta metal (**delta brass**) A strong alloy of copper and zinc, containing also a little iron. Its main use is for making cartridge cases.

denaturation /dee-nay-chŭ-**ray**-shŏn/ The changes in structure that occur when a protein is heated. These changes are irreversible and affect the properties of the protein.

denaturing /dee-**nay**-chŭ-ring/ The process of adding small amounts of contaminants to ethanol to make it unfit for drinking. Denatured alcohol (such as methylated spirits) is not subject to the excise duty on alcoholic drinks.

dendritic growth /den-**drit**-ik/ Growth of crystals in a branching ('tree-like') habit.

dendritic polymer *See* supramolecular polymer.

denitrification /dee-nÿ-tră-fă-**kay**-shŏn/ *See* nitrogen cycle.

density Symbol: ρ The mass per unit volume of a given substance. The units are g dm^{-3}, etc. *See also* relative density.

density functional theory A method of calculating the electronic structure of molecules using the electron density distribution in atoms. The theory was developed by Walter Kohn (1923–) and his colleagues in the mid-1960s and has been used extensively in chemistry and solid state physics.

deoxyribonucleic acid /dee-oks-ă-rÿ-boh-new-**klee**-ik/ *See* DNA.

deoxyribose /dee-oks-ă-**rÿ**-bohs/ *See* ribose.

depolarizer A substance used in a voltaic cell to prevent polarization. Hydrogen bubbles forming on the electrode can be removed by an oxidizing agent, such as manganese(IV) oxide.

depression of freezing point A colligative property of solutions in which the freezing point of a given solvent is lowered by the presence of a solute. The amount of the reduction is proportional to the molal concentration of the solute. The depression depends only on the concentration and is independent of solute composition. The proportionality constant, K_f, is called the *freezing point constant* or sometimes the *cryoscopic constant*. $\Delta t = K_f C_M$, where Δt is the lowering of the temperature and C_M is the molal concentration; the unit of K_f is kelvin kilogram mole^{-1} (K kg mol^{-1}). Although closely related to the property of boiling-point elevation, the cryogenic method can be applied to measurement of relative molecular mass with considerable precision. A known weight of pure solvent is slowly frozen, with stirring, in a suitable cold bath and the freezing temperature measured using a Beckmann thermometer. A known weight of solute of known molecular mass is introduced, the solvent thawed out, and the cooling process and measurement repeated. The addition is repeated several times and an average value of K_f for the solvent obtained by plotting Δt against C_M. The whole process is then repeated using the unknown solute and its relative molecular mass determined using the value of K_f previously obtained.

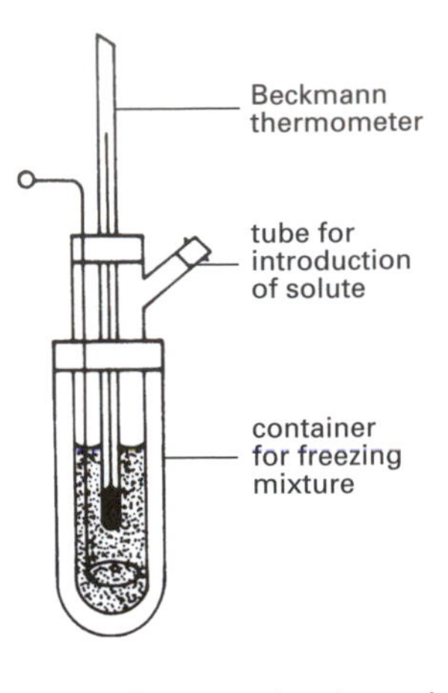

Apparatus for measuring depression of freezing point

The effect is applied to more precise measurement of relative molecular mass by using a pair of Dewar flasks (pure solvent and solution) and measuring Δt by means of thermocouples. The theoretical explanation is similar to that for LOWERING OF VAPOR PRESSURE. The freezing point of the solvent is that point at which the curve representing the vapor pressure above the liquid phase intersects the curve representing the vapor pressure above the frozen solvent. The addition of solute depresses the former curve but as the solid phase that separates is always pure solvent (above the eutectic point), there is no attendant depression of the latter curve. Consequently the point of intersection is depressed, resulting in a lowering of the freezing point.

derivative A compound obtained by reaction from another compound. The term is most often used in organic chemistry of compounds that have the same general structure as the parent compound.

derived unit A unit defined in terms of base units, and not directly from a standard value of the quantity it measures. For example, the newton is a unit of force defined as a kilogram meter second^{-2} (kg m s^{-2}). *See also* SI units.

desalination /dee-sal-ă-**nay**-shŏn/ Any of various techniques for removing the salts (mainly sodium chloride) from sea water to make it fit for drinking, irrigation, use in water-cooled engines, and for making steam for steam turbines. There are various methods. Those based on distillation depend on a cheap source of heat energy, of which solar energy is the most promising, especially in hot climates. Flash evaporation (evaporation under reduced pressure) and freezing to make pure ice are other methods, as are electrodialysis, ion exchange, reverse osmosis, and the use of molecular sieves.

desiccation /dess-ă-**kay**-shŏn/ Removal of moisture from a substance.

desiccator /**dess**-ă-kay-ter/ A laboratory apparatus for drying solids or for keeping solids free of moisture. It is a container in which is kept a hygroscopic material (e.g. calcium chloride or silica gel).

destructive distillation The process of heating an organic substance such that it wholly or partially decomposes in the absence of air to produce volatile products, which are subsequently condensed. The destructive distillation of coal was the process for manufacturing coal gas and coal tar. Formerly, methanol was made by the destructive distillation of wood.

detergents A group of substances that improve the cleansing action of solvents, particularly water. The majority of detergents, including soap, have the same basic structure. Their molecules have a hydrocarbon chain (tail) that does not attract water molecules. The tail is said to be hydrophobic (water hating). Attached to this tail is a small group (head) that readily ionizes and attracts water molecules. It is said to be hydrophilic (water loving). Detergents reduce the surface tension of water and thus improve its wetting power. Because the detergent ions have their hydrophilic heads anchored in the water and their hydrophobic tails protruding above it, the water surface is broken up, enabling the water to spread over the material to be cleaned and penetrate between the material and the dirt. With the assistance of agitation, the dirt can be floated off. The hydrophobic tails of the detergent molecules 'dissolve' in grease and oils. The protruding hydrophilic heads repel each other causing the oil to roll up and form a drop, which floats off into the water as an emulsion. More recently synthetic detergents, often derived from petrochemicals, have been developed. Unlike SOAPS these detergents do not form insoluble scums with hard water.

Synthetic detergents are of three types. *Anionic detergents* form ions consisting of a hydrocarbon chain to which is attached either a sulfonate group, $-SO_2-O^-$, or a sulfate group, $-O-SO_2-O^-$. The corresponding metal salts are soluble in water. *Cationic detergents* have organic positive ions of the type RNH_3^+, in which R has a long hydrocarbon chain. Non-ionic detergents are complex chemical compounds called ethoxylates. They owe their detergent properties to the presence of a number of oxygen atoms in one part of the molecule, which are capable of forming hydrogen bonds with the surface water molecules, thus reducing the surface tension of the water.

deuterated compound A compound in which one or more ^1H atoms have been replaced by deuterium (^2H) atoms.

deuteride /**dew**-ter-ÿd/ A compound of deuterium with other elements; i.e. a hydride in which the deuterium isotope is present rather than ^1H. *See* hydride.

deuterium /dew-**teer**-ee-ŭm/ Symbol: D, ^2H A naturally occurring stable isotope of hydrogen in which the nucleus contains one proton and one neutron. The atomic mass is thus approximately twice that of ^1H; deuterium is known as 'heavy hydrogen'. Chemically it behaves almost identically to hydrogen, forming analogous compounds, although reactions of deuterium compounds are often slower than those of the corresponding ^1H compounds. This is made use of in kinetic studies where the rate of a reaction may depend on transfer of a hydrogen atom (i.e. a kinetic isotope effect).

deuterium oxide (D_2O) The chemical name of heavy water. *See* deuterium.

deuteron /**dew**-ter-on/ The nucleus of the deuterium atom, $^2H^+$.

Dewar flask /**dew**-er/ (**vacuum flask**) A double-walled container of thin glass with the space between the walls evacuated and sealed to stop conduction and convection of energy through it. The glass is often silvered to reduce radiation.

Dewar structure A representation of the structure of BENZENE in which there is a single bond between two opposite corners of the hexagonal ring and two double bonds at the sides of the ring. The Dewar structures contribute to the resonance hybrid of benzene.

dextrin /**deks**-trin/ A polysaccharide sugar produced by the action of amylase enzymes on or the chemical hydrolysis of starch. Dextrins are used as adhesives.

dextro-form /**deks**-troh-form/ *See* optical activity.

dextronic acid /deks-**tron**-ik/ *See* gluconic acid.

dextrorotatory /deks-troh-**roh**-tă-tor-ee. -toh-ree/ *See* optical activity.

dextrose /**deks**-trohs/ (**grape-sugar**) Naturally occurring GLUCOSE belongs to the stereochemical series D and is dextrorotatory, indicated by the symbol (+). Thus the term *dextrose* is used to indicate D-(+)-glucose. As other stereochemical forms of glucose have no significance in biological systems the term 'glucose' is often used interchangeably with dextrose in biology.

***d*-form** *See* optical activity.

***D*-form** *See* optical activity.

diagonal relationship There is a general trend in the periodic table for electronegativity to increase from left to right along a period and decrease down a group. Thus a move of 'one across and one down', i.e. a diagonal move in the table, gives rise to effects that tend to cancel each other. There is a similar combined effect for size, which decreases along a period but increases down a group. This diagonal relationship gives rise to similarities in chemical properties, which are particularly noticeable for the following pairs: Li–Mg; Be–Al; B–Si.

Li–Mg:

1. both have carbonates that give CO_2 on heating;
2. both burn in air to give the normal oxide only;
3. both form a nitride;
4. both form hydrated chlorides that hydrolyze slowly.

Be–Al:

1. both give hydrogen with alkalis;
2. both give water-insoluble hydroxides that dissolve in alkali;

3. both form complex ions of the type MCl_3^-;
4. both have covalently bridged chlorides.

B–Si:
1. both form acidic oxides of a giant covalent-molecule type with high melting points;
2. both form low-stability hydrides that ignite in air;
3. both form readily hydrolyzable chlorides that fume in air;
4. both have amorphous and crystalline forms and both form glasses with basic oxides, such as Na_2O.

diamagnetism /dȳ-ă-**mag**-nĕ-tiz-ăm/ *See* magnetism.

1,6-diaminohexane /dȳ-am-ă-noh-**heks**-ayn/ (**hexamethylene diamine;** $H_2N(CH_2)_6NH_2$) An organic compound used as a starting material in the production of NYLON. It is manufactured from cyclohexane.

diamond An allotrope of CARBON. It is the hardest naturally occurring substance and is used for jewelry and, industrially, for cutting and drilling equipment. Each carbon atom is surrounded by four equally spaced carbon atoms arranged tetrahedrally. The carbon atoms form a three-dimensional network with each carbon–carbon bond equal to 0.154 nm and at an angle of 109.5° with its neighbors. In diamonds millions of atoms are covalently bonded to form a giant molecular structure, the great strength of which results from the strong covalent bonds. Diamonds can be formed synthetically from graphite in the presence of a catalyst and under extreme temperature and pressure; although small, such diamonds are of adequate size for many industrial uses.

diastereoisomer *See* isomerism.

diatomaceous earth /dȳ-ă-tŏ-**may**-shŭs/ *See* diatomite.

diatomic /dȳ-ă-**tom**-ik/ Describing a molecule that consists of two atoms. Hydrogen (H_2), oxygen (O_2), nitrogen (N_2), and the halogens are examples of diatomic elements.

diatomite /dȳ-**at**-ŏ-mȳt/ (**diatomaceous earth; kieselguhr**) A whitish powdered mineral consisting mainly of silica (silicon(IV) oxide) derived from the shells of diatoms. It is used to make fireproof cements and as an absorbent in the manufacture of DYNAMITE.

diazine /**dȳ**-ă-zeen, -zin/ *See* pyrazine.

diazole /**dȳ**-ă-zohl, dȳ-**ay**-/ *See* pyrazole.

diazonium salt /dȳ-ă-**zoh**-nee-ŭm/ A compound of the type $RN_2^+X^-$, where R is an aromatic group and X^- a negative ion. Diazonium salts are made by diazotization. They can be isolated but are very unstable, and are usually prepared in solution. The $-N_2^+$ group renders the benzene ring susceptible to nucleophilic substitution (rather than electrophilic substitution). Typical reactions are:

1. Reaction with water on warming the solution:

$$RN_2^+ + H_2O \rightarrow ROH + N_2 + H^+$$

2. Reaction with halogen ions (CuCl catalyst for chloride ions):

$$RN_2^+ + I^- \rightarrow RI + N_2$$

Diazonium ions can also act as electrophiles and substitute other benzene rings (*diazo coupling*). *See* azo compound.

diazotization /dȳ-a-zŏ-ti-**zay**-shŏn/ The reaction of an aromatic amine (e.g. aniline) with nitrous acid at low temperatures (below 5°C).

$$C_6H_5NH_2 + HNO_2 \rightarrow C_6H_5N^+N + OH^- + H_2O$$

The acid is prepared *in situ* by reaction between nitric acid and sodium nitrite. The resulting diazonium ion is susceptible to attack by nucleophiles and provides a method of nucleophilic substitution onto the benzene ring.

dibasic acid /dȳ-**bay**-sik/ An acid which has two active protons, such as sulfuric acid. Dibasic acids can give rise to two se-

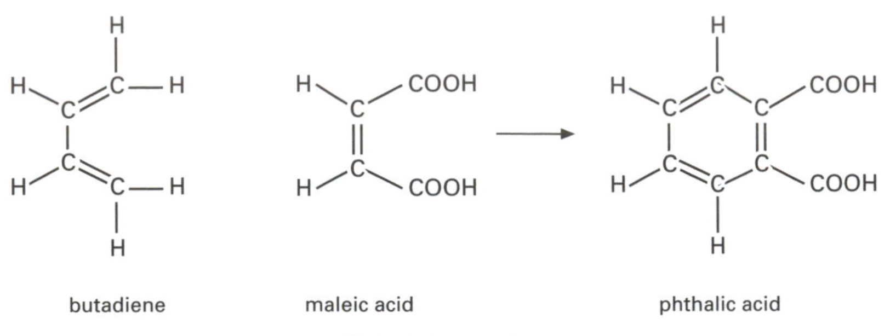

Diels–Alder reaction

ries of salts. For example, sulfuric acid (H_2SO_4) forms sulfates (SO_4^{2-}) and hydrogensulfates (HSO^-_4).

1,2-dibromoethane /dÿ-broh-moh-**eth**-ayn/ (**ethylene dibromide**; $BrCH_2CH_2Br$) A colorless volatile organic liquid, made by reacting bromine with ethene. It is used as a fuel additive to remove lead (as lead bromide, which is also volatile). *See also* lead tetraethyl.

dicarbide /dÿ-**kar**-bÿd/ *See* carbide.

dicarboxylic acid /dÿ-kar-boks-**il**-ik/ An organic acid that has two carboxyl groups (–COOH). An example is hexanedioic acid, $HOOC(CH_2)_4COOH$ (adipic acid).

dichlorine oxide /dÿ-**klor**-een, -in, **-kloh**-reen, -rin/ (**chlorine monoxide**; Cl_2O) An orange gas made by passing chlorine over mercury(II) oxide. It is a strong oxidizing agent and dissolves in water to give chloric(I) acid.

dichloroacetic acid /dÿ-klor-oh-ă-**see**-tik, **-set**-ik, -kloh-roh-/ *See* chloroethanoic acid.

dichloroethanoic acid /dÿ-kor-oh-eth-ă-**noh**-ik, -kloh-roh-/ *See* chloroethanoic acid.

dichromate(VI) /dÿ-**kroh**-mayt/ A salt containing the ion $Cr_2O_7^-$. Dichromates are strong oxidizing agents. *See* potassium dichromate.

dielectric constant /dÿ-i-**lek**-trik/ (**relative permittivity**) Symbol: ε_r A quantity that characterizes how a medium reduces the electric field strength associated with a distribution of electric charges. If two point charges Q_1 and Q_2 are a distance d apart in a medium with a *permittivity* ε this means that the force F between the charges is given by $F = (1/4\pi\varepsilon)(Q_1Q_2/d^2)$.
The dielectric constant of the medium is given by $\varepsilon_r = \varepsilon/\varepsilon_0$, where ε_0 is the permittivity of free space. The dielectric constant of air is very slightly greater than 1, while that of water is about 80. The value of the dielectric constant of a medium has important physical and chemical consequences, particularly for ions in solutions.

Diels–Alder reaction /**deelz awl**-der/ An organic addition reaction used to make six-membered ring compounds. A CONJUGATED compound (a compound with two double bonds separated by a single bond, such as a diene) adds to a compound with a single double bond to form the six-membered ring. The reaction is named for the German organic chemists Otto Paul Herman Diels (1876–1954) and Kurt Alder (1902–58).

diene /**dÿ**-een/ An organic compound containing two carbon–carbon double bonds.

diesel fuel /**dee**-zĕl/ A petroleum fraction consisting of various alkanes in the boiling range 200–350°C, used as a fuel for diesel (compression-ignition) engines.

diethyl ether /dÿ-**eth**-ăl/ *See* ethoxyethane.

diffusion /di-**fyoo**-zhŏn/ Movement of a

gas, liquid, or solid as a result of the random thermal motion of its particles (atoms or molecules). A drop of ink in water, for example, will slowly spread throughout the liquid. Diffusion in solids occurs very slowly at normal temperatures. *See also* Graham's law.

dihydrate /dÿ-**hÿ**-drayt/ A crystalline compound with two molecules of water of crystallization per molecule of compound.

dihydric alcohol /dÿ-**hÿ**-drik/ *See* diol.

2,3-dihydroxybutanedioic acid /dÿ-hÿ-**droks**-ăl-byoo-tayn-dÿ-**oh**-ik/ *See* tartaric acid.

dihydroxypurine /dÿ-hÿ-droks-ă-**poor**-een, -in/ *See* xanthine.

diiodine hexachloride /dÿ-**ÿ**-ŏ-deen heks-ă-**klor**-ÿd, -id, -kloh-rÿd, -kloh-rid/ (**iodine trichloride**; I_2Cl_6) A yellow crystalline solid made by reacting excess chlorine with iodine. It is a strong oxidizing agent and at 70°C it dissociates into iodine monochloride and chlorine.

dilead(II) lead(IV) oxide /**dÿ**-led/ (**red lead**; Pb_3O_4) A powder made by heating lead(II) oxide or lead(II) carbonate hydroxide at 400°C. It is black when hot and red or orange when cold. On strong heating it decomposes to lead(II) oxide and oxygen. Dilead(II) lead (IV) oxide is used as a pigment and in glass making. In practice it tends to have less oxygen than denoted in the formula.

diluent /**dil**-yoo-ĕnt/ A solvent that is added to reduce the strength of a solution.

dilute Denoting a solution in which the amount of solute is low relative to that of the solvent. The term is always relative and includes dilution at trace level as well as the common term 'bench dilute acid', which usually means a 2M solution. *Compare* concentrated.

dimensionless units The radian and steradian in SI units. *See* SI units.

dimer /**dÿ**-mer/ A compound (or molecule) formed by combination or association of two molecules of a monomer. For instance, aluminum chloride ($AlCl_3$) is a dimer (Al_2Cl_6) in the vapor.

dimesoiodic (VII) acid /dÿ-mess-oh-ÿ-**od**-ik/ *See* iodic (VII) acid.

dimethylbenzene /dÿ-meth-ăl-**ben**-zeen, ben-**zeen**/ (**xylene**; $C_6H_4(CH_3)_2$) An organic hydrocarbon present in the light-oil fraction of crude oil. It is used extensively as a solvent. There are three isomeric compounds with this name and formula, distinguished as 1,2-, 1,3-, and 1,4-dimethylbenzene according to the positions of the methyl groups on the benzene ring.

dimorphism /dÿ-**mor**-fiz-ăm/ *See* polymorphism.

dinitrogen oxide /dÿ-**nÿ**-trŏ-jĕn/ (**nitrous oxide**; N_2O) A colorless gas with a faintly sweet odor and taste. It is appreciably soluble in water (1.3 volumes in 1 volume of water at 0°C) but more soluble in ethanol. It is prepared commercially by the *careful* heating of ammonium nitrate:

$$NH_4NO_3(s) = N_2O(g) + 2H_2O(g)$$

Dinitrogen oxide is fairly easily decomposed on heating to temperatures above 520°C, giving nitrogen and oxygen. The gas is used as a mild anesthetic in medicine and dentistry, being marketed in small steel cylinders. It is sometimes called *laughing gas* because it induces a feeling of elation when inhaled.

dinitrogen tetroxide /te-**troks**-ÿd/ (N_2O_4) A colorless gas that becomes a pale yellow liquid below 21°C and solidifies below –11°C. On heating, the gas dissociates to nitrogen dioxide molecules:

$$N_2O_4(g) = 2NO_2(g)$$

This dissociation is complete at 140°C. Liquid dinitrogen tetroxide has good solvent properties and is used as a nitrating agent.

2,4-dinitrophenylhydrazine /dÿ-nÿ-troh-fen-ăl-**hÿ**-dră-zeen/ (**Brady's reagent**;

$C_6H_6N_4O_4$) An orange solid commonly used in solution with methanol and sulfuric acid to produce crystalline derivatives by condensation with aldehydes and ketones. The derivatives, known as 2,4-dinitrophenylhydrazones, can easily be purified by recrystallization and have characteristic melting points, used to identify the original aldehyde or ketone.

diol /**dȳ**-ôl/ (**dihydric alcohol; glycol**) An alcohol that has two hydroxyl groups (–OH) per molecule of compound.

1,4-dioxan /dȳ-**oks**-ăn/ ($(CH_2)_2O_2$) A colorless liquid cyclic ether. It is an inert compound miscible with water used as a solvent.

dioxins /dȳ-**oks**-inz/ A related group of highly toxic chlorinated compounds. Particularly important is the compound 2,3,7,8-tetrachlorodibenzo-*p*-dioxin (TCDD), which is produced as a by-product in the manufacture of 2,4,5-T, and may consequently occur as an impurity in certain types of weedkiller. The defoliant known as agent orange used in Vietnam contained significant amounts of TCDD. Dioxins cause a skin disease (chloracne) and birth defects. Dioxins have been released into the atmosphere as a result of explosions at herbicide manufacturing plants, most notably at Seveso, Italy, in 1976.

diphosphane /dȳ-**fos**-fayn/ (**diphosphine**; P_2H_4) A yellow liquid that can be condensed out from phosphine in a freezing mixture. It ignites spontaneously in air.

diphosphine /dȳ-**fos**-feen, -fin/ *See* diphosphane.

dipolar bond /dȳ-**pohl**-er/ *See* coordinate bond.

dipole /**dȳ**-pohl/ A system in which two equal and opposite electric charges are separated by a finite distance. Polar molecules have permanent dipoles. Induced dipoles can also occur. *See also* dipole moment.

dipole moment Symbol: μ A quantitative measure of polarity in either a bond (bond moment) or a molecule as a whole (molecular dipole moment). The unit is the debye (equivalent to 3.34×10^{-30} coulomb meter). Molecules such as HF, H_2O, NH_3, and $C_6H_5NH_2$ possess dipole moments; CCl_4, N_2, C_6H_6, and PF_5 do not.

The molecular dipole moment can be estimated by vector addition of individual bond moments if the bond angles are known. The possession of a dipole moment permits direct interaction with electric fields or interaction with the electric component of radiation.

dipyridyl /dȳ-**pȳ**-ră-dăl/ (**bipyridyl**) A compound formed by linking two pyridine rings. There are various isomers, some of which are used in herbicides.

direct dyes A group of dyes that are mostly azo-compounds derived from benzidene or benzidene derivatives. They are used to dye cotton, viscose rayon, and other cellulose fibers directly, using a neutral bath containing sodium chloride or sodium sulfate as a mordant.

disaccharide /dȳ-**sak**-ă-rȳd, -rid/ A SUGAR with molecules composed of two monosaccharide units. Sucrose and maltose are examples. These are linked by a –O– linkage (*glycosidic link*).

disconnection *See* retrosynthetic analysis.

dislocation *See* defect.

disodium hydrogen phosphate(V) /dȳ-**soh**-dee-ŭm/ (**disodium hydrogen orthophosphate**; Na_2HPO_4) A white solid prepared by titrating phosphoric acid with sodium hydroxide solution using phenolphthalein as the indicator. On evaporation the solution yields efflorescent monoclinic crystals of the dodecahydrate, $Na_2HPO_4.12H_2O$. The effloresced salt contains $7H_2O$. Disodium hydrogen phosphate is used in the textile industry.

disodium oxide (**sodium monoxide;**

Na_2O) A highly reactive whitish deliquescent solid that combines violently with water to form sodium hydroxide.

disodium tetraborate decahydrate (**borax**; $Na_2B_4O_7.10H_2O$) A white crystalline solid, sparingly soluble in cold water but readily soluble in hot water. It occurs naturally as salt deposits in dry lake beds, especially in California. It is an important industrial material, being used in the manufacture of enamels and heat-resistant glass, as a paper glaze, etc.

Its full systematic name is sodium(I) heptaoxotetraborate(III)-10-water. In solution hydrolysis occurs:

$$B_4O_7^{2-} + 7H_2O = 2OH^- + 4H_3BO_3$$

The borax-bead test gives characteristic colors with certain cations.

disperse dyes Water-insoluble dyes, which, when held in fine suspension, can be applied to acetate rayon fabrics. The dye, together with a dispersing agent, is warmed to a temperature of 45–50°C and the fabric added. By modifying the method of application it is possible to dye polyacrylic and polyester fibers. The yellow/orange shades are nitroarylamine derivatives and the green to bluish shades are derivatives of 1-amino anthraquinone. Certain azo compounds are disperse dyes and these give a range of colors.

disperse phase *See* colloid.

dispersing agent A compound used to assist emulsification or dispersion.

dispersion force A weak type of intermolecular force. *See* van der Waals force.

displacement pump A commonly used device for transporting liquids and gases around chemical plants. It works on the principle of the bicycle pump: a piston raises the pressure of the fluid and, when it is high enough, a valve opens and the fluid is discharged through an outlet pipe. As the piston moves back the pressure falls and the cycle continues. Displacement pumps can be used to generate very high pressures (e.g. in the synthesis of ammonia) but because of the system of valves, they are more expensive than other types of pump. *Compare* centrifugal pump.

displacement reaction A chemical reaction in which an atom or group displaces another atom or group from a molecule. A common example is the displacement of hydrogen from acids by metals:

$$Zn + 2HCl \rightarrow ZnCl_2 + H_2$$

disproportionation /dis-prŏ-por-shŏ-**nay**-shŏn/ A chemical reaction in which there is simultaneous oxidation and reduction of the same compound. An example is the reaction of copper(I) chloride to copper and copper(II) chloride:

$$2CuCl \rightarrow Cu + CuCl_2$$

in which there is simultaneous oxidation (to Cu(II)) and reduction (to Cu(0)).

Another example is the reaction of chlorine with water:

$$Cl_2 + H_2O \rightarrow 2H^+ + Cl^- + ClO^-$$

in which there is reduction to Cl^- and oxidation to ClO^-.

dissociation Breakdown of a molecule into two molecules, atoms, radicals, or ions. Often the reaction is reversible, as in the ionic dissociation of weak acids in water:

$$CH_3COOH + H_2O \rightleftharpoons CH_3COO^- + H_3O^+$$

dissociation constant The equilibrium constant of a dissociation reaction. For example, the dissociation constant of a reaction:

$$AB \rightleftharpoons A + B$$

is given by:

$$K = [A][B]/[AB]$$

where the brackets denote concentration (activity).

Often the degree of dissociation is used – the fraction (α) of the original compound that has dissociated at equilibrium. For an original amount of AB of n moles in a volume V, the dissociation constant is given by:

$$K = \alpha^2 n/(1 - \alpha)V$$

Note that this expression is for dissociation into two molecules.

Acid dissociation constants (or *acidity constants*, symbol: K_a) are dissociation constants for the dissociation into ions in solution:

$$HA + H_2O \rightleftharpoons H_3O^+ + A^-$$

The concentration of water can be taken as unity, and the acidity constant is given by:

$$K_a = [H_3O^+][A^-]/[HA]$$

The acidity constant is a measure of the strength of the acid. Base dissociation constants (K_b) are similarly defined. The expression:

$$K = \alpha^2 n/(1 - \alpha)V$$

applied to an acid is known as *Ostwald's dilution law* (named for the German chemist Friedrich Wilhelm Ostwald (1853–1932).. In particular if α is small (a weak acid) then $K = \alpha^2 n/V$, or $\alpha = C\sqrt{V}$, where C is a constant. The degree of dissociation is then proportional to the square root of the dilution.

distillation The process of boiling a liquid and condensing the vapor. Distillation is used to purify liquids or to separate components of a liquid mixture. *See also* destructive distillation; fractional distillation; steam distillation; vacuum distillation.

distilled water Water that has been purified by distillation, perhaps several times.

disulfur dichloride /dÿ-**sul**-fer/ (**sulfur monochloride**; S_2Cl_2) A red fuming liquid with a strong smell. It is prepared by passing chlorine over molten sulfur and is used to harden rubber.

diterpene *See* terpene.

dithionate /dÿ-**th'ÿ**-ŏ-nayt/ A salt of dithionic acid.

dithionic acid /dÿ-th'ÿ-**on**-ik/ (**sulfinic acid; hyposulfuric acid**; $H_2S_2O_6$) A strong sulfur oxyacid that decomposes slowly on standing or heating.

dithionite /dÿ-**th'ÿ**-ŏ-nÿt/ (**sulfinate**) A salt of dithionous acid. Dithionites are powerful reducing agents.

dithionous acid /dÿ-**th'ÿ**-ŏ-nŭs/ (**sulfinic acid; hyposulfurous acid**; $H_2S_2O_4$) An unstable sulfur oxyacid that is known only in solution. Its salts (*dithionites* or *sulfinates*) are powerful reducing agents.

divalent /dÿ-**vay**-lĕnt/ (**bivalent**) Having a valence of two.

***dl*-form** *See* optical activity.

***D-L* convention** *See* optical activity.

DNA (**deoxyribonucleic acid**) A nucleic acid, mainly found in the chromosomes, that contains the hereditary information of organisms. The molecule is made up of two antiparallel helical polynucleotide chains coiled around each other to give a *double helix*. It is also known as the *Watson–Crick model* after James Watson and Francis Crick who first proposed this model in 1953. Phosphate molecules alternate with deoxyribose sugar molecules along both chains and each sugar molecule is also joined to one of four nitrogenous bases – adenine (A), guanine (G), cytosine (C), or thymine (T). The two chains are joined to each other by hydrogen bonding between bases. The two purine bases (adenine and guanine) always bond with the pyrimidine bases (thymine and cytosine), and the pairing is quite specific: adenine with thymine and guanine with cytosine. The two chains are therefore complementary. The sequence of bases along the chain makes up a code – the genetic code – that determines the precise sequence of amino acids in proteins.

DNA is the hereditary material of all organisms with the exception of RNA viruses. *See also* RNA.

Döbereiner's triads /**doh**-ber-ÿ-nerz/ Triads of chemically similar elements in which the central member, when placed in order of increasing relative atomic mass, has a relative atomic mass approximately equal to the average of the outer two. Other chemical and physical properties of the central member also lie between those of the first and last members of the triad. The German chemist Johann Wolfgang

Döbereiner (1780–1849) noted this relationship in 1817; the triads are now recognized as consecutive members of a group of the periodic table; e.g. Ca, Sr, and Ba and Cl, Br, and I.

dodecanoic acid /doh-dek-ă-**noh**-ik/ (**lauric acid**; $CH_3(CH_2)_{10}COOH$) A white crystalline carboxylic acid, used as a plasticizer and for making detergents and soaps. Its glycerides occur naturally in coconut and palm oils.

dolomite /**dol**-ŏ-mÿt/ (**pearl spar**) A mineral, $CaCO_3.MgCO_3$, used as an ore of magnesium and for the lining of open-hearth steel furnaces and Bessemer converters.

donor **1.** The atom, ion, or molecule that provides the pair of electrons in forming a covalent bond.
2. The impurity atoms used in doping semiconductors.

doping The incorporation of impurities within the crystal lattice of a solid so as to alter its physical properties. For instance, silicon, when doped with boron, becomes semiconducting.

double bond A covalent bond between two atoms that includes two pairs of electrons, one pair being the single bond equivalent (the sigma pair) and the other forming an additional bond, the pi bond (π bond). It is conventionally represented by two lines, for example $H_2C=O$. *See* multiple bond; orbital.

double decomposition (**metathesis**) A chemical technique for making an insoluble salt by mixing two soluble salts. The ions from the two reactants 'change partners' to form a new soluble compound and the insoluble compound, which is precipitated. For example, mixing solutions of potassium iodide and lead nitrate forms a solution of potassium nitrate and a yellow precipitate of lead iodide.

double helix *See* DNA.

double salt When equivalent quantities of certain salts are mixed in aqueous solution and the solution evaporated, a salt may form, e.g. $FeSO_4.(NH_4)_2SO_4.6H_2O$. In aqueous solution the salt behaves as a mixture of the two individuals. These salts are called double salts to distinguish them from complex salts, which yield complex ions in solution.

Downs process An electrolytic process for making chlorine from fused sodium chloride. The anode is a central piece of graphite, surrounded by a cylindrical steel cathode. A grid and dome between the two keeps the products separate.

Dow process /dow/ An industrial method whereby magnesium is extracted from seawater via the precipitation of $Mg(OH)_2$ by $Ca(OH)_2$, followed by solvation of the precipitated hydroxide by hydrochloric acid. The process is named for the American industrial chemist Herbert Henry Dow (1866–1930), who founded the Dow Chemical Company (1897).

dry cell A voltaic cell in which the electrolyte is in the form of a jelly or paste. The Leclanché dry cell is extensively used for flashlights and other portable applications. *See also* Leclanché cell.

dryers Devices used in chemical processes to remove a liquid from a solid by evaporation. Drying equipment is classified by the method of transferring heat to a wet solid. This can be by direct contact between hot gases and the solid (direct dryers), heat transfer by conduction through a retaining metallic wall (indirect dryers), or infrared rays (infrared dryers).

dry ice Solid carbon dioxide, used as a refrigerant.

drying oil A natural oil, such as linseed oil, that hardens in air. Such oils contain unsaturated fatty acids, which polymerize on oxidation.

dubnium /**dub**-nee-ŭm/ A radioactive synthetic element made by bombarding ^{249}Cf

nuclei with ^{15}N atoms. It was first reported by workers at Dubna, a town near Moscow. Symbol: Db; p.n. 105; most stable isotope ^{262}Db (half-life 34 s).

Dulong and Petit's law /doo-**long** ănd pĕ-**teez** / The molar thermal capacity of a solid element is approximately equal to $3R$, where R is the gas constant (25 J K^{-1} mol^{-1}). The law applies only to elements with simple crystal structures at normal temperatures. At lower temperatures the molar heat capacity falls with decreasing temperature (it is proportional to T^3). Molar thermal capacity was formerly called *atomic heat* – the product of the atomic weight (relative atomic mass) and the specific thermal capacity. The law is named for the French physicists Pierre-Louis Dulong (1785–1838) and Alexis-Thérèse Petit (1791–1820).

Dumas' method /doo-**mahz**/ **1.** A method for determining the vapor densities of volatile liquids. The method utilizes a Dumas bulb; i.e. a glass bulb with a narrow entrance tube. The bulb is weighed 'empty' (i.e. full of air) then the sample is introduced and the bulb immersed in a heating bath so that the sample boils and expels all the air. When the surplus vapor has been expelled, the bulb is sealed off, cooled, dried, and weighed. The tip of the tube is then broken under water so that the water completely fills the tube and the whole weighed again. This enables the volume of the bulb to be calculated from the known density of water, and knowing the density of air one can compute the vapor density of the sample. The method is named for the French chemist Jean Baptiste André Dumas (1800–84). *See also* Hofmann's method; Victor Meyer's method.

2. A method of finding the amount of nitrogen in an organic compound by heating the compound with copper oxide to convert the nitrogen into nitrogen oxides. These are reduced by passing them over hot copper and the volume of nitrogen collected is measured.

Duralumin /dewr-**al**-yŭ-min/ (*Trademark*) A strong lightweight aluminum alloy containing 3–4% copper, with small amounts of magnesium, manganese, and sometimes silicon. It is widely used in aircraft bodies.

dye A coloring material for fabric, leather, etc. Most dyes are now synthetic organic compounds (the first such was the dye *mauve* synthesized from aniline in 1856 by W. H. Perkin). Dyes are often unsaturated organic compounds containing conjugated double bonds – the bond system responsible for the color is called the *chromophore*. *See also* azo compound.

dynamite A high explosive made by absorbing nitroglycerine into an earthy material such as diatomite (kieselguhr). Solid sticks of dynamite are much safer to handle than the highly sensitive liquid nitroglycerine.

dyne /dȳn/ Symbol: dyn The former unit of force used in the c.g.s. system. It is equal to 10^{-5} N.

dysprosium /dis-**proh**-see-ŭm, -shee-ŭm/ A soft malleable silvery element of the lanthanoid series of metals. It occurs in association with other lanthanoids. One of its few uses is as a neutron absorber in nuclear reactors; it is also a constituent of certain magnetic alloys.
Symbol: Dy; m.p. 1412°C; b.p. 22562°C; r.d. 8.55 (20°C); p.n. 66; r.a.m. 162.50.

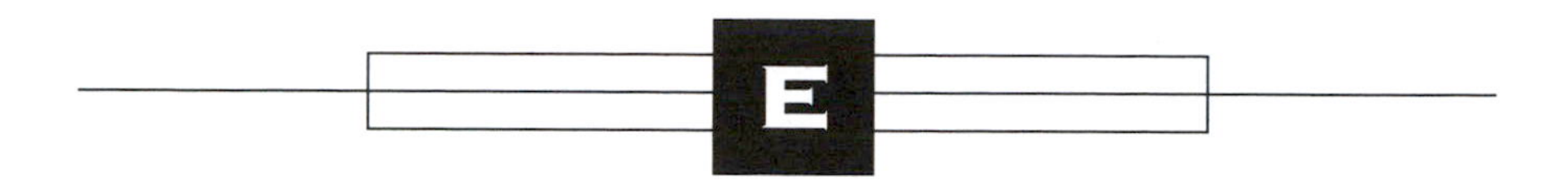

ebulioscopic constant /i-bul-ee-ŏ-**skop**-ik/ *See* elevation of boiling point.

ebullition /eb-ŭ-**lish**-ŏn/ The boiling or bubbling of a liquid.

eclipsed conformation *See* conformation.

Edison cell *See* nickel–iron accumulator.

edta /ee-dee-tee-**ay**/ (**ethylenediamine tetraacetic acid**) A compound with the formula

$$(HOOCCH_2)_2N(CH_2)_2N(CH_2COOH)_2$$

It is used in forming chelates of transition metals. *See* chelate.

effervescence /ef-er-**vess**-ĕns/ The evolution of gas in the form of bubbles in a liquid.

efficiency Symbol: η A measure used for processes of energy transfer; the ratio of the useful energy produced by a system or device to the energy input. For a reversible heat engine, the efficiency is given by

$$\eta = (T_1 - T_2)/T_1$$

T_1 being the source temperature and T_2 the 'sink' temperature.

efflorescence /ef-flŏ-**ress**-ĕns/ The process in which a crystalline hydrated solid loses water of crystallization to the air. A powdery deposit is gradually formed.

einsteinium /ÿn-**stÿ**-nee-ŭm/ A radioactive transuranic element of the actinoid series, not found naturally on Earth. It can be produced in milligram quantities by bombarding ^{239}Pu with neutrons to give ^{253}Es (half-life 20.47 days). Several other short-lived isotopes have been synthesized.

Symbol: Es; m.p. 860 ± 30°C; p.n. 99; most stable isotope ^{254}Es (half-life 276 days). The element is named for the German–Swiss–American theoretical physicist Albert Einstein (1879–1955).

elastomer /i-las-**tom**-ĕ-ter/ An elastic substance, e.g. a natural or synthetic rubber.

electrochemical equivalent /i-lek-trŏ-**kem**-ă-kăl/ Symbol: *z* The mass of an element released from a solution of its ion when a current of one ampere flows for one second during electrolysis.

electrochemical series (**electromotive series**) A series giving the activities of metals for reactions that involve ions in solution. In decreasing order of activity, the series is

K, Na, Ca, Mg, Al, Zn, Fe
Pb, H, Cu, Hg, Ag, Pt, Au

Any member of the series will displace ions of a lower member from solution. For example, zinc metal will displace Cu^{2+} ions:

$$Zn(s) + Cu^{2+}(aq) \rightarrow Zn^{2+}(aq) + Cu(s)$$

Zinc has a greater tendency than copper to form positive ions in solution. Similarly, metals above hydrogen displace hydrogen from acids:

$$Zn + 2HCl \rightarrow ZnCl_2 + H_2$$

The series is based on electrode potentials, which measure the tendency to form positive ions. The series is one of increasing electrode potential for half cells of the type $M^{n+}|M$. Thus, copper ($E^{\ominus}$ for $Cu^{2+}|Cu = +0.34$ V) is lower than zinc ($E^{\ominus}$ for $Zn^{2+}|Zn = -0.76$V). The hydrogen half cell has a value $E^{\ominus} = 0$.

electrochemistry /i-lek-trŏ-**kem**-iss-tree/ The study of the formation and behavior of ions in solutions. It includes electrolysis and the generation of electricity by chemical reactions in cells.

electrochromatography /i-lek-troh-kroh-mă-**tog**-ră-fee/ *See* electrophoresis.

electrocyclic reaction /i-lek-troh-**sÿ**-klik/ *See* pericyclic reaction.

electrode Any part of an electrical device or system that emits or collects electrons or other charge carriers. An electrode may also be used to deflect charged particles by the action of the electrostatic field that it produces. *See also* half cell.

electrodeposition /i-lek-troh-dep-ŏ-**zish**-ŏn/ (**electroplating**) The process of depositing a layer of solid (metal) on an electrode by electrolysis. Positive ions in solution gain electrons at the cathode and are deposited as atoms. Copper, for instance, can be deposited on a metal cathode from an acidified copper sulfate solution.

electrode potential Symbol: E A measure of the tendency of an element to form ions in solution. For example, a metal in a solution containing M^+ ions may dissolve in the solution as M^+ ions; the metal then has an excess of electrons and the solution an excess of positive ions – thus, the metal becomes negative with respect to the solution. Alternatively, the positive ions may gain electrons from the metal and be deposited as metal atoms. In this case, the metal becomes positively charged with respect to the solution. In either case, a potential difference is developed between solid and solution, and an equilibrium state will be reached at which further reaction is prevented. The equilibrium value of this potential difference would give an indication of the tendency to form aqueous ions.

It is not, however, possible to measure this for an isolated half cell – any measurement requires a circuit, which sets up another half cell in the solution. Therefore, electrode potentials (or *reduction potentials*) are defined by comparison with a hydrogen half cell, which is connected to the half cell under investigation by a salt bridge. The e.m.f. of the full cell can then be measured.

In referring to a given half cell the more reduced form is written on the right for a half-cell reaction. For the half cell $Cu^{2+}|Cu$, the half-cell reaction is a reduction:

$$Cu^{2+}(aq) + 2e \rightarrow Cu$$

The cell formed in comparison with a hydrogen electrode is:

$$Pt(s)H_2(g)|H^+(aq)|Cu^{2+}(aq)|Cu$$

The e.m.f. of this cell is +0.34 volt measured under standard conditions. Thus, the standard electrode potential (symbol: $E^{\ominus}$) is +0.34 V for the half cell $Cu^{2+}|Cu$. The standard conditions are 1.0 molar solutions of all ionic species, standard pressure, and a temperature of 298 K.

Half cells can also be formed by a solution of two different ions (e.g. Fe^{2+} and Fe^{3+}). In such cases, a platinum electrode is used under standard conditions.

electrodialysis /i-lek-troh-dÿ-**al**-ă-sis/ (*pl.* **electrodialyses**) A method of removing ions from water by selective flow through membranes under the influence of an electric field. In a simple arrangement a cell is divided into three compartments by two semipermeable membranes, one permeable to positive ions and the other permeable to negative ions. Electrodes are placed in the outer compartments of the cell – the positive electrode next to the membrane that allows negative ions to pass. In this arrangement, positive and negative ions pass through the membranes in leaving deionized water in the center compartment of the cell. In practice, an array of alternating membranes is used. Electrodialysis is used for making drinking water in areas in which brackish ground water is available.

electrolysis /i-lek-**trol**-ă-sis/ The production of chemical change by passing electric charge through certain conducting liquids (electrolytes). The current is conducted by migration of ions – positive ones (cations) to the cathode (negative electrode), and negative ones (anions) to the anode (positive electrode). Reactions take place at the electrodes by transfer of electrons to or from them.

In the electrolysis of water (containing a small amount of acid to make it conduct adequately) hydrogen gas is given off at the cathode and oxygen is evolved at the anode. At the cathode the reaction is:

$$H^+ + e^- \rightarrow H$$
$$2H \rightarrow H_2$$

At the anode:

$$OH^- \rightarrow e^- + OH$$
$$2OH \rightarrow H_2O + O$$
$$2O \rightarrow O_2$$

In certain cases the electrode material may dissolve. For instance, in the electrolysis of copper(II) sulfate solution with copper electrodes, copper atoms of the anode dissolve as copper ions

$$Cu \rightarrow 2e^- + Cu^{2+}$$

See Faraday's laws.

electrolyte /i-**lek**-trŏ-lÿt/ A liquid containing positive and negative ions that conducts electricity by the flow of those charges. Electrolytes can be solutions of acids or metal salts ('ionic compounds'), usually in water. Alternatively they may be molten ionic compounds – again the ions can move freely through the substance. Liquid metals (in which conduction is by free electrons rather than ions) are not classified as electrolytes. *See also* electrolysis.

electrolytic /i-lek-trŏ-**lit**-ik/ Relating to the behavior or reactions of ions in solution.

electrolytic cell *See* cell; electrolysis.

electrolytic corrosion Corrosion by electrochemical reaction. *See* rusting.

electrolytic refining A method of purifying metals by electrolysis. Copper is purified by making the impure metal the anode in an electrolytic cell with an acidified copper sulfate electrolyte. The cathode is a thin strip of pure copper. The copper at the anode dissolves (as Cu^{2+} ions) and pure copper is deposited on the cathode. In this particular process, gold and silver are obtained as by-products, deposited as an *anode sludge* on the bottom of the cell.

electromagnetic radiation Energy propagated by vibrating electric and magnetic fields. Electromagnetic radiation forms a whole electromagnetic spectrum, depending on frequency and ranging from high-frequency radio waves to low-frequency gamma rays. It can be thought of as waves (*electromagnetic waves*) or as streams of photons. The frequency and wavelength are related by

$$\lambda\nu = c$$

where c is the speed of light. The energy carried depends on the frequency.

electromotive series *See* electrochemical series.

electron /i-**lek**-tron/ An elementary particle of negative charge ($-1.602\,192 \times 10^{-19}$C) and rest mass $9.109\,558 \times 10^{-31}$ kg. Electrons are present in all atoms.

electron affinity Symbol: A The energy released when an atom (or molecule or group) gains an electron in the gas phase to form a negative ion. It is thus the energy of:

$$A + e^- \rightarrow A^-$$

ELECTROMAGNETIC SPECTRUM
(note: the figures are only approximate)

Radiation	*Wavelength (m)*	*Frequency (Hz)*
gamma radiation	-10^{-12}	$10^{19}-$
x-rays	$10^{-12}-10^{-9}$	$10^{17}-10^{20}$
ultraviolet radiation	$10^{-9}-10^{-7}$	$10^{15}-10^{18}$
visible radiation	$10^{-7}-10^{-6}$	$10^{14}-10^{15}$
infrared radiation	$10^{-6}-10^{-4}$	$10^{12}-10^{14}$
microwaves	$10^{-4}-1$	$10^{9}-10^{13}$
radio waves	$1-$	-10^{9}

A positive value of *A* (often in electronvolts) indicates that heat is given out. Often the molar enthalpy is given for this process of electron attachment (ΔH). Here the units are joules per mole (J mol^{-1}), and, by the usual convention, a negative value indicates that energy is released.

electron-deficient compounds Compounds in which the number of electrons available for bonding is insufficient for the bonds to consist of conventional two-electron covalent bonds. Diborane, B_2H_6, is an example in which each boron atom has two terminal hydrogen atoms bound by conventional electron-pair bonds and in addition the molecule has two hydrogen atoms bridging the boron atoms (B–H–B). In each bridge there are only two electrons for the bonding orbital. *See also* multicenter bond.

electron diffraction A technique used to determine the structure of substances, principally the shapes of molecules in the gaseous phase. A beam of electrons directed through a gas at low pressure produces a series of concentric rings on a photographic plate. The dimensions of these rings are related to the interatomic distances in the molecules. *See also* x-ray diffraction.

electron donor *See* reduction.

electronegative /i-lek-troh-**neg**-ă-tiv/ Describing an atom or molecule that attracts electrons, forming negative ions. Examples of electronegative elements include the halogens (chlorine etc.), which readily form negative ions (F^-, Cl^-, etc.). *See also* electronegativity.

electronegativity /i-lek-troh-neg-ă-**tiv**-ă-tee/ A measure of the tendency of an atom in a molecule to attract electrons to itself. Elements to the right-hand side of the periodic table are strongly electronegative (values from 2.5 to 4); those on the left-hand side have low electronegativities (0.8–1.5) and are sometimes called electropositive elements. Different electronegativities of atoms in the same molecule give rise to polar bonds and sometimes to polar molecules.

As the concept of electronegativity is not precisely defined it cannot be precisely measured and several electronegativity scales exist. Although the actual values differ the scales are in good relative agreement. *See also* electron affinity; ionization potential.

electronic energy levels *See* atom; energy levels.

electronic transition The demotion or promotion of an electron between electronic energy levels in an atom or molecule.

electron pair Two electrons in one orbital with opposing spins (*spin paired*), such as the electrons in a covalent bond or lone pair.

electron spin *See* atom.

electron spin resonance (ESR) A similar technique to nuclear magnetic resonance, but applied to unpaired electrons in a molecule (rather than to the nuclei). It is a powerful method of studying free radicals and transition-metal complexes.

electron-transfer reaction A chemical reaction that involves the transfer, addition, or removal of electrons. Many electron-transfer reactions involve complexes of transition metals. The rates of such reactions vary enormously and can be explained in terms of the way in which molecules of the solvent, which start off solvating the reactants, rearrange so as to solvate the products.

electronvolt /i-lek-tron-**vohlt**/ Symbol: eV A unit of energy equal to 1.602 191 7 × 10^{-19} joule. It is defined as the energy required to move an electron charge across a potential difference of one volt. It is often used to measure the kinetic energies of elementary particles or ions, or the ionization potentials of molecules.

electrophile /i-**lek**-trŏ-fÿl, -fil/ An electron-deficient ion or molecule that takes part in an organic reaction. The electrophile can be either a positive ion (H^+, NO_2^+) or a molecule that can accept an electron pair (SO_3, O_3). The electrophile attacks negatively charged areas of molecules, which usually arise from the presence in the molecule of an electronegative atom or group or of pi-bonds. *Compare* nucleophile.

electrophilic addition /i-lek-trŏ-**fil**-ik/ A reaction involving the addition of a small molecule to an unsaturated organic compound, across the atoms held by a double or triple bond. The reaction is initiated by the attack of an electrophile on the electron-rich area of the molecule. The mechanism of electrophilic addition is thought to be ionic, as in the addition of HBr to ethene:

$$H_2C{:}CH_2 + H^+Br^- \rightarrow H_3CCH_2^+ + Br^- \rightarrow H_3CCH_2Br$$

In the case of higher alkenes (more than two carbon atoms) several isomeric products are possible. The particular isomer produced depends on the stability of the alternative intermediates and this is summarized empirically by Markovnikoff's rule. *See also* addition reaction.

electrophilic substitution A reaction involving substitution of an atom or group of atoms in an organic compound with an electrophile as the attacking substituent. Electrophilic substitution is very common in aromatic compounds, in which electrophiles are substituted onto the ring. An example is the nitration of benzene:

$$C_6H_6 + NO_2^+ \rightarrow C_6H_5NO_2 + H^+$$

The nitronium ion (NO_2^+) is formed by mixing concentrated nitric and sulfuric acids:

$$HNO_3 + H_2SO_4 \rightarrow H_2NO_3^+ + HSO_4^-$$

$$H_2NO_3^+ \rightarrow NO_2^+ + H_2O$$

The accepted mechanism for a simple electrophilic substitution on benzene involves an intermediate of the form $C_6H_5HNO_2^+$. *See also* substitution reaction.

electrophoresis /i-lek-troh-fŏ-**ree**-sis/ The use of an electric field (between two electrodes) to cause charged particles of a colloid to move through a solution. The technique is used to separate and identify colloidal substances such as carbohydrates, proteins, and nucleic acids. Various experimental arrangements are used. One simple technique uses a strip of adsorbent paper soaked in a buffer solution with electrodes placed at two points on the paper. This technique is sometimes called *electrochromatography*. In *gel electrophoresis*, used to separate DNA fragments, the medium is a layer of gel.

electroplating /i-**lek**-trŏ-play-ting/ The process of coating a solid surface with a layer of metal by means of electrolysis (i.e. by electrodeposition).

electropositive /i-lek-troh-**poz**-ă-tiv/ Describing an atom or molecule that tends to lose electrons, forming positive ions. Examples of electropositive elements include the alkali metals (lithium, sodium, etc.), which readily form positive ions (Li^+, Na^+, etc.).

electrovalent bond /i-lek-trŏ-**vay**-lĕnt/ (**ionic bond**) A binding force between the ions in compounds in which the ions are formed by complete transfer of electrons from one element to another element or radical. For example, Na + Cl becomes Na^+ + Cl^-. The electrovalent bond arises from the excess of the net attractive force between the ions of opposite charge over the net repulsive force between ions of like charge. The magnitude of electrovalent interactions is of the order 10^2–10^3 kJ mol^{-1} and electrovalent compounds are generally solids with rigid lattices of closely packed ions. The strengths of electrovalent bonds vary with the reciprocal of the inter-ionic distances and are discussed in terms of lattice energies.

electrum /i-**lek**-trŭm/ Originally, a naturally occurring alloy of gold and silver (up to 45% silver) that resembles pure gold in appearance. A synthetic alloy of copper, nickel, and zinc is also known as electrum.

element /el-ĕ-mĕnt/ A substance that cannot be chemically decomposed into more simple substances. The atoms of an element all have the same proton number (and thus the same number of electrons, which determines the chemical activity).

At present there are 112 reported chemical elements, although research is continuing all the time to synthesize new ones. The elements from hydrogen (p.n. 1) to uranium (92) all occur naturally, with the exception of technetium (43), which is produced artificially by particle bombardment. Technetium and elements with proton numbers higher than 84 (polonium) are radioactive. Radioactive isotopes also exist for other elements, either naturally in small amounts or synthesized by particle bombardment. The elements with proton number higher than 92 are the *transuranic elements*. Neptunium (93) and plutonium (94) both occur naturally in small quantities in uranium ores, but the transuranics are all synthesized. Thus, neptunium and plutonium are made by neutron bombardment of uranium nuclei. Other transuranics are made by high-energy collision processes between nuclei. The higher proton number elements have been detected only in very small quantities – in some cases, only a few atoms have been produced.

There has been some controversy about the naming of the higher synthetic elements because of disputes about their discovery. A long-standing problem concerned element-104 (rutherfordium) which was formerly also known by its Russian name of *kurchatovium*. More recent confusion has been caused by differences between names suggested by the International Union of Pure and Applied Chemistry (IUPAC) and the names suggested by the American Chemical Union (ACU).

The original IUPAC names (1994) were:

mendelevium (Md, 101)
nobelium (No, 102)
lawrencium (Lr, 103)
dubnium (Db, 104)
joliotium (Jl, 105)
rutherfordium (Rf, 106)
bohrium (Bh, 107)
hahnium (Hn, 108)
meitnerium (Mt, 109)

The ACU names were:

mendelevium (Md, 101)
nobelium (No, 102)
lawrencium (Lr, 103)
rutherfordium (Rf, 104)
hahnium (Ha, 105)
seaborgium (Sg, 106)
nielsbohrium (Ns, 107)
hassium (Hs, 108)
meitnerium (Mt, 109)

A compromise list of names was adopted by IUPAC in 1997:

mendelevium (Md, 101)
nobelium (No, 102)
lawrencium (Lr, 103)
rutherfordium (Rf, 104)
dubnium (Db, 105)
seaborgium (Sg, 106)
bohrium (Bh, 107)
hassium (Hs, 108)
meitnerium (Mt, 109)

This is the list used in this dictionary. Dubnium is named after Dubna in Russia, where much work has been done on synthetic elements. Hassium is the Latin name for the German state of Hesse, where the German research group is located

In addition to the above elements, seven other elements have been reported with proton numbers 110, 111, 112, 113, 114, 115, and 116. Element 110 has been named darmstadtium for the town in Germany where it was first discovered. Element 111 has been named roentgenium (for Roentgen). So far the others have not been named and have their temporary systematic IUPAC names (*see* unnil-).

elevation of boiling point A colligative property of solutions in which the boiling point of a solution is raised relative to that of the pure solvent. The elevation is directly proportional to the number of solute molecules introduced rather than to any specific aspect of the solute composition. The proportionality constant, k_B, is called the boiling-point elevation constant or sometimes the *ebulioscopic constant*. The relationship is

$$\Delta t = k_B C_M$$

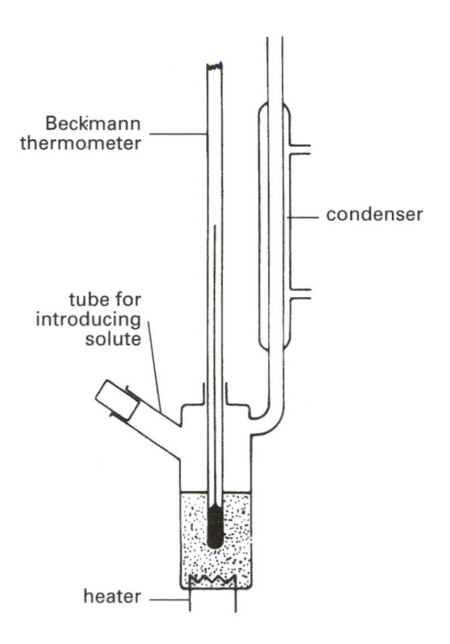

Apparatus for measuring elevation of boiling point

where Δt is the rise in boiling point and C_M is the molal concentration; the units of k_B are kelvins kilograms moles^{-1} (K kg mol^{-1}). The property permits the measurement of relative molecular mass of involatile solutes. An accurately weighed amount of pure solvent is boiled until the temperature is steady, a known weight of solute of known molecular mass is quickly introduced, the boiling continued, and the elevation measured using a Beckmann thermometer. The process is repeated several times and the average value of k_B obtained by plotting Δt against C_M. The whole process is then repeated with the unknown material and its relative molecular mass obtained using the value of k_B previously obtained.

There are several disadvantages with this method and it is therefore used largely for demonstration purposes. The main problem is that the exact amount of solvent remaining in the liquid phase is unknown and varies with the rate of boiling. The theoretical explanation of the effect is identical to that for the lowering of vapor pressure; the boiling points are those temperatures at which the vapor pressure equals the atmospheric pressure.

elimination reaction A reaction involving the removal of a small molecule, e.g. water or hydrogen chloride, from an organic molecule to give an unsaturated compound. An example is the elimination of a water molecule from an alcohol to produce an alkene.

An elimination reaction is often in competition with a substitution reaction and the predominant product will depend on the reaction conditions. The reaction of bromoethane with sodium hydroxide could yield either ethene (by elimination of HBr) or ethanol (by substitution of the Br with OH). The former product predominates if the reaction is carried out in an alcoholic solution and the latter if the solution is aqueous.

Elinvar /el-**in**-var/ (*Trademark*) An alloy of chromium, iron, and nickel. It is used for making hairsprings for clocks and watches because its elasticity does not vary with temperature.

elution /i-**loo**-shŏn/ The removal of an adsorbed substance in a chromatography column or ion-exchange column using a solvent (*eluent*), giving a solution called the *eluate*. The chromatography column can selectively adsorb one or more components from the mixture. To ensure efficient recovery of these components graded elution is used. The eluent is changed in a regular manner starting with a non-polar solvent and gradually replacing it by a more polar one. This will wash the strongly polar components from the column. *See* column chromatography.

emery A mineral form of corundum (aluminum oxide) with some iron oxide and silica impurities. It is extremely hard and is used as an abrasive and polishing material.

emission spectrum *See* spectrum.

empirical formula The formula of a compound showing the simplest ratio of the atoms present. The empirical formula is the formula obtained by experimental analysis of a compound and it can be related to a molecular formula only if the molecular weight is known. For example P_2O_5 is the empirical formula of phosphorus(V) oxide although its molecular formula is P_4O_{10}. *Compare* molecular formula; structural formula.

emulsion A colloid in which a liquid phase (small droplets with a diameter range 10^{-5}–10^{-7} cm) is dispersed or suspended in a liquid medium. Emulsions are classed as lyophobic (solvent-repelling and generally unstable) or lyophilic (solvent-attracting and generally stable).

enantiomer /en-**an**-tee-ŏ-mer/ (**enantiomorph**) A compound whose structure is not superimposable on its mirror image: one of any pair of optical isomers. *See also* isomerism; optical activity.

enantiotropy /en-an-tee-**ot**-rŏ-pee/ The existence of different stable allotropes of an element at different temperatures; sulfur, for example, exhibits enantiotropy. The phase diagram for an enantiotropic element has a point at which all the allotropes can coexist in a stable equilibrium. At temperatures above or below this point, one of the allotropes will be more stable than the other(s). *See also* allotropy; monotropy.

enclosure compound *See* clathrate.

endothermic /en-doh-**th'er**-mik/ Describing a process in which heat is absorbed (i.e. heat flows from outside the system, or the temperature falls). The dissolving of a salt in water, for instance, is often an endothermic process. *Compare* exothermic.

end point *See* equivalence point; volumetric analysis.

energy Symbol: *W* A property of a system; a measure of its capacity to do work. Energy and work have the same unit: the joule (J). It is convenient to divide energy into *kinetic energy* (energy of motion) and *potential energy* ('stored' energy). Names are given to many different forms of energy (chemical, electrical, nuclear, etc.); the only real difference lies in the system under discussion. For example, chemical energy is the kinetic and potential energies of electrons in a chemical compound.

energy level One of the discrete energies that an atom or molecule, for instance, can have according to quantum theory. Thus in an atom there are certain definite orbits that the electrons can be in, corresponding to definite *electronic energy levels* of the atom. Similarly, a vibrating or rotating molecule can have discrete vibrational and rotational energy levels.

energy profile A diagram that traces the changes in the energy of a system during the course of a reaction. Energy profiles are obtained by plotting the potential energy of the reacting particles against the reaction coordinate. To obtain the reaction coordinate the energy of the total interacting system is plotted against position for the molecules. The *reaction coordinate* is the pathway for which the energy is a minimum.

The energy profile for the hydrogenation of an alkene with and without a catalyst has certain interesting features: the activation energy without the catalyst is very much larger than that with the catalyst and any point along the horizontal axis represents the changes that the colliding particles experience as atoms are separated in bond breaking and as atoms are brought together in bond formation.

enol /**ee**-nol, -nohl/ An organic compound containing the C:CH(OH) group; i.e. one in which a hydroxyl group is attached to the carbon of a double bond. *See* keto-enol tautomerism.

enthalpy /en-**thal**-pee, **en**-thal-/ Symbol: *H* The sum of the internal energy (*U*) and the product of pressure (*p*) and volume (*V*) of a system:

$$H = U + pV$$

In a chemical reaction carried out at constant pressure, the change in enthalpy measured is the internal energy change plus the work done by the volume change:

$$\Delta H = \Delta U + p\Delta V$$

entropy /**en**-trŏ-pee/ (*pl.* **entropies**) Symbol: *S* In any system that undergoes a reversible change, the change of entropy is defined as the heat absorbed divided by the thermodynamic temperature:

$$\delta S = \delta Q/T$$

A given system is said to have a certain entropy, although absolute entropies are seldom used: it is change in entropy that is important. The entropy of a system measures the availability of energy to do work.

In any real (irreversible) change in a closed system the entropy always increases. Although the total energy of the system has not changed (first law of thermodynamics) the available energy is less – a consequence of the second law of thermodynamics.

The concept of entropy has been widened to take in the general idea of disorder – the higher the entropy, the more disordered the system. For instance, a chemical reaction involving polymerization may well have a decrease in entropy because there is a change to a more ordered system. The 'thermal' definition of entropy is a special case of this idea of disorder – here the entropy measures how the energy transferred is distributed among the particles of matter. *See also* Boltzmann formula.

enzyme /**en**-zÿm/ A protein, with a relative molecular mass between 10^5 and 10^7, that catalyzes a specific biochemical reaction. An enzyme mediates the conversion of one substance (the *substrate*) to another (the *product*) by combining with the substrate to form an intermediate complex. The enzyme molecule is very much larger than the substrate molecule. Enzymes differ from inorganic catalysts in the following ways:

1. they have a high degree of specificity – enzymes can even distinguish between enantiomorphs;
2. an increase in temperature can cause a decrease in the rate of reaction as a result of the enzyme being denatured;
3. they are destroyed by too great a change in pH;
4. they are inactivated by low concentrations of heavy-metal ions.

epimerism /i-**pim**-ĕ-riz-ăm/ A form of isomerism exhibited by carbohydrates in which the isomers (*epimers*) differ in the positions of –OH groups. The α- and β-forms of glucose are epimers. *See* sugar.

epoxide /i-**poks**-ÿd/ A type of organic compound containing a three-membered ring in which two carbon atoms are each bonded to the same oxygen atom.

epoxyethane /i-poks-ee-**eth**-ayn/ (**ethylene oxide**; C_2H_4O) A colorless gaseous ether. Epoxyethane has a cyclic structure with a three-membered ring. It is made by oxidation of ethene over a silver catalyst. The ring is strained and the compound is consequently highly reactive. It polymerizes to epoxy polymers (resins). Hydrolysis produces 1,2-ethanediol.

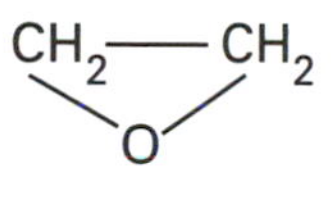

Epoxyethane

epoxy resin /i-**poks**-ee/ *See* epoxyethane.

Epsom salt /**ep**-sŏm/ *See* magnesium sulfate.

equation *See* chemical equation.

equation of state An equation that interrelates the pressure, temperature, and volume of a system, such as a gas. The ideal gas equation (*see* gas laws) and the van der Waals equation are examples.

equilibrium /ee-kwă-**lib**-ree-ŭm, ek-wă-/ (*pl.* **equilibriums** or **equilibria**) In a reversible chemical reaction:

$$A + B \rightleftharpoons C + D$$

The reactants are forming the products:

$$A + B \rightarrow C + D$$

which also react to give the original reactants:

$$C + D \rightarrow A + B$$

The concentrations of A, B, C, and D change with time until a state is reached at which both reactions are taking place at the same rate. The concentrations (or pressures) of the components are then constant – the system is said to be in a state of chemical equilibrium. Note that the equilibrium is a dynamic one; the reactions still take place but at equal rates. The relative proportions of the components determine the 'position' of the equilibrium, which may be *displaced* by changing the conditions (e.g. temperature or pressure).

equilibrium constant In a chemical equilibrium of the type

$$xA + yB \rightleftharpoons zC + wD$$

The expression:

$$[A]^x[B]^y/[C]^z[D]^w$$

where the square brackets indicate concentrations, is a constant (K_c) when the system is at equilibrium. K_c is the equilibrium constant of the given reaction; its units depend on the stoichiometry of the reaction. For gas reactions, pressures are often used instead of concentration. The equilibrium constant is then K_p, where $K_p = K_c^n$. Here n is the number of moles of product minus the number of moles of reactant; for instance, in

$$3H_2 + N_2 \rightleftharpoons 2NH_3$$

n is $2 - (1 + 3) = -2$.

equipartition of energy /ee-kwă-par-**tish**-ŏn, ek-wă-/ The principle that the total energy of a molecule is, on average, equally distributed among the available DEGREES OF FREEDOM. It is only approximately true in most cases.

equivalence point The point in a titration at which the reactants have been added in equivalent proportions, so that there is no excess of either. It differs slightly from the *end point*, which is the observed point of complete reaction, because of the effect of the indicator, errors, etc.

equivalent proportions, law of (**law of reciprocal proportions**) When two chemical elements both form compounds with a third element, a compound of the first two elements contains them in the relative proportions they have in compounds with the third element. For example, the mass ratio of carbon to hydrogen in methane (CH_4) is 12:4; the ratio of oxygen to hydrogen in water (H_2O) is 16:2. In carbon monoxide (CO), the ratio of carbon to oxygen is 12:16.

equivalent weight A measure of 'combining power' formerly used in calculations for chemical reactions. The equivalent weight of an element is the number of grams that could combine with or displace one gram of hydrogen (or 8 grams of oxygen or 35.5 grams of chlorine). It is the relative atomic mass (atomic weight) divided by the valence. For a compound the equivalent weight depends on the reaction considered. An acid, for instance, in acid–base reactions has an equivalent weight equal to its molecular weight divided by the number of acidic hydrogen atoms.

erbium /**er**-bee-ŭm/ A soft malleable silvery element of the lanthanoid series of metals. It occurs in association with other lanthanoids. Erbium has uses in the metallurgical and nuclear industries and in making glass for absorbing infrared radiation.

Symbol: Er; m.p. 1529°C; b.p. 2863°C; r.d. 9.066 (25°C); p.n. 68; r.a.m. 167.26.

erg A former unit of energy used in the c.g.s. system. It is equal to 10^{-7} joule.

Erlenmeyer flask /**er**-lĕn-mÿ-er/ A conical glass laboratory flask with a narrow neck. It is named for the German chemist Richard August Carl Emil Erlenmeyer (1825–1909).

ESR *See* electron spin resonance.

essential amino acid *See* amino acid.

essential oil Any pleasant-smelling volatile oil obtained from various plants, widely used in making flavorings and perfumes. Most consist of terpenes and they are obtained by steam distillation or solvent extraction.

ester A type of organic compound formed by reaction between an acid and an alcohol. If the acid is a carboxylic acid, the formula is RCOOR′, where R and R′ are organic groups. Esters of low-molecular weight acids and alcohols are volatile fragrant compounds. Esters of some long-chain carboxylic acids occur naturally in fats and oils. Esters can be prepared by direct reaction of acids and alcohols, with a dehydrating agent to take out the water and displace the equilibrium. Reaction of alcohols with acyl halides also gives esters. *See also* carboxylic acid; saponification; glyceride.

esterification /es-te-ri-fă-**kay**-shŏn/ The reaction of an acid with an alcohol to form an ester and water. The reaction is an equilibrium of the form:

$$CH_3COOH + C_2H_5OH \rightleftharpoons CH_3COOC_2H_5 + H_2O$$

In the preparation of esters the ester is distilled off or the water is removed to increase the yield. The reverse reaction is hydrolysis.

ethanal /**eth**-ă-nal/ (**acetaldehyde;** CH_3CHO) A water-soluble liquid aldehyde used as a starting material in the manufacture of several other compounds. Ethanal can be prepared by the oxidation of ethanol. It is manufactured by the catalytic oxidation of ethyne with oxygen using copper(II) chloride and palladium(II) chloride as catalysts. The mixture of gases is bubbled through an aqueous solution of the catalysts; the reaction involves formation of an intermediate organometallic complex with Pd^{2+} ions. With dilute acids ethanal polymerizes to *ethanal trimer* ($C_3O_3H_3(CH_3)_3$, formerly called *paraldehyde*), which is a sleep-inducing drug. Below 0°C *ethanal tetramer* is formed ($C_4O_4H_4(CH_3)_4$, formerly called *metaldehyde*), which is used as a slug poison and a fuel in small portable stoves.

ethane /**eth**-ayn/ (C_2H_6) A gaseous alkane obtained either from the gaseous fraction of crude oil or by the 'cracking' of heavier fractions. Ethane is the second member of the homologous series of alkanes.

ethanedioic acid /eth-ayn-dÿ-**oh**-ik/ (**oxalic acid;** $(COOH)_2$) A white crystalline organic acid that occurs naturally in rhubarb, sorrel, and other plants of the genus *Oxalis*. It is slightly soluble in water, highly toxic, and used in dyeing and as a chemical reagent.

ethane-1,2-diol (**ethylene glycol; glycol;** $CH_2(OH)CH_2(OH)$) A syrupy organic liquid commonly used as antifreeze and as a starting material in the manufacture of Dacron. The compound is manufactured from ethene by oxidation over suitable catalysts to form epoxyethane, with subsequent hydrolysis to the diol.

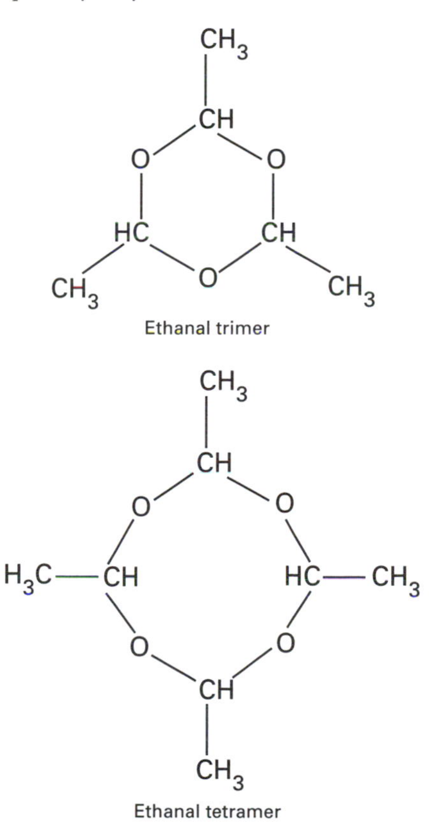

Ethanal trimer

Ethanal tetramer

ethanoate /eth-ă-**noh**-ayt/ (**acetate;** CH_3COO^-) A salt or ester of ETHANOIC ACID (acetic acid).

ethanoic acid /eth-ă-**noh**-ik/ (**acetic acid;** CH_3COOH) A colorless viscous liquid organic acid with a pungent odor (it is the acid in vinegar). Below 16.7°C it solidifies to a glassy solid (*glacial ethanoic acid*). It is made by the oxidation of ethanol or butane, or by the continued fermentation of beer or wine. It is made into ethenyl ethanoate (vinyl acetate) for making polymers. Cellulose ethanoate (acetate) is made from ethanoic anhydride. *See* cellulose acetate.

ethanol /**eth**-ă-nol, -nohl/ (**ethyl alcohol; alcohol;** C_2H_5OH) A colorless volatile liquid alcohol. Ethanol occurs in intoxicating drinks, in which it is produced by fermentation of a sugar:

$$C_6H_{12}O_6 \rightarrow 2C_2H_5OH + 2CO_2$$

Yeast is used to cause the reaction. At about 15% alcohol concentration (by volume) the reaction stops because the yeast is killed. Higher concentrations of alcohol are produced by distillation.

Apart from its use in drinks, alcohol is used as a solvent and to form ethanal. Formerly, the main source was by fermentation of molasses, but now catalytic hydration of ethene is used to manufacture industrial ethanol. *See also* methylated spirits.

ethanoyl chloride /eth-ă-**noh**-ăl/ (**acetyl chloride;** CH_3COCl) A liquid acyl chloride used as an acetylating agent.

ethanoyl group (**acetyl group**) The group RCO–.

ethene /**eth**-een/ (**ethylene;** C_2H_4) A gaseous alkene. Ethene is not normally present in the gaseous fraction of crude oil but can be obtained from heavier fractions by catalytic cracking. This is the principal industrial source. The compound is important as a starting material in the organic-chemicals industry (e.g. in the manufacture of ethanol) and as the starting material for the production of polyethene. Ethene is the first member of the homologous series of alkenes.

ether /**ee**-th'er/ A type of organic compound containing the group –O–. Simple ethers have the formula R–O–R′, where R and R′ are alkyl or aryl groups, which may or may not be the same. They are either gases or very volatile liquids and are very flammable. The commonest example is ethoxyethane (diethylether, $C_2H_5OC_2H_5$) used formerly as an anesthetic. Ethers now find application as solvents. They are prepared in the laboratory by the dehydration of alcohols with concentrated sulfuric acid. An excess of alcohol is used to ensure that only one molecule of water is removed from each pair of alcohol molecules. They are generally unreactive, but the C–O bond can be cleaved by reaction with HI or PCl_5.

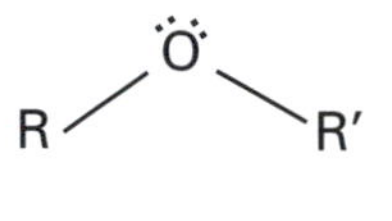

Ether

ethoxyethane /i-thoks-ee-**eth**-ayn/ (**ether; diethylether;** $C_2H_5OC_2H_5$) A colorless volatile liquid. Ether is well known for its characteristic smell and anesthetic properties, also for its extreme flammability. It still finds some application as an anesthetic when more modern materials are unsuitable; it is also an excellent solvent. Its manufacture is an extension of the laboratory synthesis: ethanol vapor is passed into a mixture of excess ethanol and concentrated sulfuric acid at 140°C:

$$C_2H_5OH + H_2SO_4 \rightarrow C_2H_5.O.SO_2.OH + H_2O$$

$$C_2H_5O.SO_2.OH + C_2H_5OH \rightarrow C_2H_5OC_2H_5 + H_2SO_4$$

ethyl acetate /**eth**-ăl/ *See* ethyl ethanoate.

ethyl alcohol *See* ethanol.

ethylamine /**eth**-ă-lă-meen, -lam-een/ (**aminoethane;** $C_2H_5NH_2$) A colorless liquid amine. It can be prepared from chloroethane heated with concentrated aqueous ammonia:

$$C_2H_5Cl + NH_3 \rightarrow C_2H_5NH_2 + HCl$$

It is used in manufacturing certain dyes.

ethyl bromide *See* bromoethane.

ethyl carbamate *See* urethane.

ethyl chloride *See* chloroethane.

ethyl dibromide /dȳ-**broh**-mȳd/ *See* dibromoethane.

ethylene /**eth**-ă-leen/ *See* ethene.

ethylenediamine /eth-ă-leen-**dȳ**-ă-meen/ An organic compound, $H_2NCH_2CH_2NH_2$. It is important in inorganic chemistry because it may function as a bidentantate ligand, coordinating to a metal ion by the lone pairs on the two nitrogen atoms. In the names of complexes it is given the abbreviation *en*.

ethylenediamine tetraacetic acid /eth-ă-leen-**dȳ**-ă-meen tet-ră-**see**-tik/ *See* edta.

ethylene glycol *See* ethane-1,2-diol.

ethylene oxide *See* epoxyethane.

ethylene tetrachloride /tet-ră-**klor**-ȳd, -id, -**kloh**-rȳd, -rid/ *See* tetrachloroethane.

ethyl ethanoate (**ethyl acetate**; $C_2H_5OOCCH_3$) An ester formed from ethanol and ethanoic acid. It is a fragrant liquid used as a solvent for plastics and in flavoring and perfumery.

ethyl iodide *See* iodoethane.

ethyne /**eth**-ȳn/ (**acetylene**; C_2H_2) A gaseous alkyne. Traditionally ethyne has found use in oxy-acetylene welding torches, since its combustion with oxygen produces a flame of very high temperature. It is also important in the organic chemicals industry for the production of chloroethene (vinyl chloride), which is the starting material for the production of polyvinyl chloride (PVC), and for the production of other vinyl compounds. Until recently, ethyne was manufactured by the synthesis and subsequent hydrolysis of calcium dicarbide, a very expensive procedure. Modern methods increasingly employ the cracking of alkanes.

ethynide /**eth**-ă-nȳd, -nid/ *See* carbide.

eudiometer /yoo-dee-**om**-ĕ-ter/ An apparatus for the volumetric analysis of gases.

europium /yû-**roh**-pee-ŭm/ A silvery element of the lanthanoid series of metals. It occurs in association with other lanthanoids. Its main use is in a mixture of europium and yttrium oxides widely employed as the red phosphor in television screens. The metal is used in some superconducting alloys.

Symbol: Eu; m.p. 822°C; b.p. 1597°C; r.d. 5.23 (25°C); p.n. 63; r.a.m. 151.965.

eutectic /yoo-**tek**-tik/ A mixture of two substances in such proportions that no other mixture of the same substances has a lower freezing point. If a solution containing the proportions of the eutectic is cooled no solid is deposited until the freezing point of the eutectic is reached, when two solid phases are simultaneously deposited in the same proportions as the mixture; these two solid phases form the eutectic. If one of the substances is water the eutectic can also be referred to as a *cryohydrate*.

evaporation **1.** A change of state from liquid to gas (or vapor). Evaporation can take place at any temperature, the rate increasing with temperature. Some molecules in the liquid have enough energy to escape into the gas phase (if they are near the surface and moving in the right direction). Because these are the molecules with higher kinetic energies, evaporation results in a cooling of the liquid.

2. A change from solid to vapor, especially occurring at high temperatures close to the melting point of the solid. Thin films of metal can be evaporated onto a surface in this way.

exa- Symbol: E A prefix denoting 10^{18}.

excitation /eks-ÿ-**tay**-shŏn/ The process of producing an excited state of an atom, molecule, etc.

excitation energy The energy required to change an atom, molecule, etc. from one quantum state to a state with a higher energy. The excitation energy (sometimes called *excitation potential*) is the difference between two energy levels of the system.

excited state A state of an atom, molecule, or other system, with an energy greater than that of the ground state. *Compare* ground state.

exclusion principle The principle, enunciated by Pauli in 1925, that no two electrons in an atom can have an identical set of quantum numbers.

exothermic /eks-oh-**th'er**-mik/ Denoting a chemical reaction in which heat is evolved (i.e. heat flows from the system or the temperature rises). Combustion is an example of an exothermic process. *Compare* endothermic.

explosive A substance or mixture which can rapidly decompose upon detonation producing large amounts of heat and gases. The three most important classes of explosives are:

1. *Propellants* which burn steadily, and are used as rocket fuels.
2. *Initiators* which are very sensitive and are used in small amounts to detonate less sensitive explosives.
3. *High explosives* which need an initiator, but are very powerful.

E–Z convention A convention for the description of a molecule showing *cis–trans* isomerism. In a molecule ABC=CDE, where A, B, D, and E are different groups, a sequence rule (*see* CIP system) is applied to the pair A and B to find which has priority and similarly to the pair C and D. If the two groups of higher priority are on the same side of the bond then the isomer is designated *Z* (from German *zusammen*, together). If they are on opposite sides the isomer is designated *E* (German *entgegen*, opposite).

F

face-centered cubic crystal (f.c.c.) A crystal structure in which the unit cell has atoms, ions, or molecules at each corner and at the center of each face of a cube (also called *cubic close-packed*). It has a coordination number of 12. The structure is close-packed and made up of layers of atoms in which each atom is surrounded by six others arranged hexagonally. Copper and aluminum have face-centered cubic structures. The layers are packed one on top of the other, with the second layer fitting into the holes of the first layer. The third layer is in a different set of holes formed by the second layer. If the layers are designated A, B, C, the packing is arranged

ABCABC

in contrast to hexagonal close packing, in which the layers are arranged

ABABAB.

Fahrenheit scale A TEMPERATURE SCALE in which the ice temperature is taken as 32° and the steam temperature is taken as 212° (both at standard pressure). The scale is not used for scientific purposes. To convert between degrees Fahrenheit (F) and degrees Celsius (C) the formula

$$C/5 = (F - 32)/9$$

is used. The scale is named for the German physicist Daniel Fahrenheit (1686–1736).

Fajan's rules /**faj**-ănz/ Rules that deal with variations in the degree of covalent character in ionic compounds in terms of polarization effects. Decrease in size and increase in charge for cations is regarded as making them more polarizing and for anions an increase in both charge and size makes them more polarizable. Thus covalent character is said to increase with increasing polarizing power of the cation and/or increasing polarizability of the anion (originally rules 1 and 2). Fajan's third rule states that covalent character is greater for cations with a non-rare gas configuration than for those with a complete octet, charge and size being approximately equal. For example, Cu^+ is more polarizing than Na^+ (approximately the same size) because the d-electrons around Cu^+ do not shield the nucleus as effectively as the complete octet in Na^+. The rules are named for the Polish–American physical chemist Kasimir Fajans (1887–1975).

farad /**fa**-răd, -rad/ Symbol: F The SI unit of capacitance. When the plates of a capacitor are charged by one coulomb and there is a potential difference of one volt between them, then the capacitor has a capacitance of one farad. One farad = 1 coulomb $volt^{-1}$ (CV^{-1}). The unit is named for the British physicist and chemist Michael Faraday (1791–1867).

faraday /**fa**-ră-day/ Symbol: F A unit of electric charge equal to the charge required to discharge one mole of a singly-charged ion. One faraday is $9.648\,670 \times 10^4$ coulombs. *See* Faraday's laws.

Faraday constant *See* Faraday's laws.

Faraday's laws (of electrolysis) Two laws resulting from the work of Michael Faraday on electrolysis:

1. The amount of chemical change produced is proportional to the electric charge passed.
2. The amount of chemical change produced by a given charge depends on the ion concerned.

More strictly it is proportional to the relative ionic mass of the ion, and the charge on it. Thus the charge Q needed to deposit *m* grams of an ion of relative ionic mass, *M*, carrying a charge *Z* is given by:

$$Q = FmZ/M$$

F, the *Faraday constant*, has a value of one faraday, i.e. 9.648 670 × 10^4 coulombs.

fat *See* glyceride.

fatty acid *See* carboxylic acid.

f-block elements The block of elements in the periodic table that have two outer *s*-electrons and an incomplete penultimate subshell of *f*-electrons. The f-block consists of the lanthanoids (cerium to lutetium) and the actinoids (thorium to lawrencium).

f.c.c. *See* face-centered cubic crystal.

Fehling's solution /fay-lingz/ A solution used to test for the aldehyde group (–CHO). It is a freshly made mixture of copper(II) sulfate solution with alkaline potassium sodium 2,3-dihydroxybutanedioate (tartrate). The aldehyde, when heated with the mixture, is oxidized to a carboxylic acid, and a red precipitate of copper(I) oxide and copper metal is produced. The tartrate is present to complex with the original copper(II) ions to prevent precipitation of copper(II) hydroxide. The solution is named for the German organic chemist Hermann Christian von Fehling (1812–85).

feldspar (**felspar**) A member of a large group of abundant crystalline igneous rocks, mainly silicates of aluminum with calcium, potassium, and sodium. The alkali feldspars have mainly potassium with a little sodium and no calcium; the plagioclase feldspars vary from sodium-only to calcium-only, with little or no potassium. Feldspars are used in making ceramics, enamels, and glazes.

femto- Symbol: f A prefix denoting 10^{-15}. For example, 1 femtometer (fm) = 10^{-15} meter (m).

femtochemistry /fem-toh-**kem**-iss-tree/ The study of atomic and molecular processes, particularly chemical reactions, that take place on timescales of about a femtosecond (10^{-15}s). Femtochemistry became possible with the construction of lasers that could be pulsed in femtoseconds. Femtochemistry can be regarded as very high speed 'photography' of chemical reactions. Typically, a short pulse of radiation can initiate a change, with subsequent pulses at short intervals used to investigate the intermediates spectroscopically. The technique has enabled entities that exist only for a short time, such as activated complexes, to be studied. The detailed mechanisms of many chemical reactions have been investigated in this way.

fermentation A chemical reaction produced by microorganisms (molds, bacteria, or yeasts). A common example is the formation of ethanol from sugars:

$$C_6H_{12}O_6 \rightarrow 2C_2H_5OH + 2CO_2$$

fermi /fer-mee/ A unit of length equal to 10^{-15} meter. It was formerly used in atomic and nuclear physics. The unit is named for the Italian–American physicist Enrico Fermi (1901–54).

fermium /fer-mee-ŭm/ A radioactive transuranic element of the actinoid series, not found naturally on Earth. It is produced in very small quantities by bombarding ^{239}Pu with neutrons to give ^{253}Fm (half-life 3 days). Several other short-lived isotopes have been synthesized.

Symbol: Fm; p.n. 100; most stable isotope ^{257}Fm (half-life 100.5 days).

ferric chloride /fe-rik/ *See* iron(III) chloride.

ferric oxide *See* iron(III) oxide.

ferric sulfate *See* iron(III) sulfate.

ferricyanide *See* cyanoferrate.

ferrite /fe-rÿt/ A nonconducting magnetic ceramic material, general formula MO.Fe_2O_4 (where M is a divalent transition metal such as cobalt, nickel, or zinc). Ferrites are used to make powerful magnets in high-frequency electronic devices such as radar sets.

ferrocene /**fe**-roh-seen/ ($Fe(C_5H_5)_2$) An orange crystalline solid. It is an example of a sandwich compound, in which an iron(II) ion is coordinated to two cyclopentadienyl ions. The bonding involves overlap of d orbitals on the iron with the pi electrons in the cyclopentadienyl ring. The compound melts at 173°C, is soluble in organic solvents, and can undergo substitution reactions on the rings. It can be oxidized to the blue cation $(C_5H_5)_2Fe^+$. The systematic name is di-π-cyclopentadienyl iron(II).

ferrocyanide /fe-roh-**sÿ**-ă-nÿd/ *See* cyanoferrate.

ferromagnetism /fe-roh-**mag**-nĕ-tiz-ăm/ *See* magnetism.

ferrosoferric oxide /fe-ros-oh-**fe**-rik/ *See* triiron tetroxide.

ferrous chloride /**fe**-rŭs/ *See* iron(II) chloride.

ferrous oxide *See* iron(II) oxide.

ferrous sulfate *See* iron(II) sulfate.

fertilizer A substance added to soil to increase its fertility. Artificial fertilizers generally consist of a mixture of chemicals designed to supply nitrogen (N), as well as some phosphorus (P) and potassium (K), known as PKN fertilizers. Typical chemicals include ammonium nitrate, ammonium phosphate, ammonium sulfate, and potassium carbonate.

Fick's law /fiks/ A law describing the diffusion that occurs when solutions of different concentrations come into contact, with molecules moving from regions of higher concentration to regions of lower concentration. Fick's law states that the rate of diffusion dn/dt, called the *diffusive flux* and denoted J, across an area A is given by: $dn/dt = J = -DA\partial c/\partial x$, where D is a constant called the *diffusion constant*, $\partial c/\partial x$ is the concentration gradient of the solute, and dn/dt is the amount of solute crossing the area A per unit time. D is constant for a specific solute and solvent at a specific temperature. Fick's law was formulated by the German physiologist Adolf Eugen Fick (1829–1901) in 1855.

filler A solid material used to modify the physical properties or reduce the cost of synthetic compounds, such as rubbers, plastics, paints, and resins. Slate powder, glass fiber, mica, and cotton are all used as fillers.

filter *See* filtration.

filter pump A type of vacuum pump in which a jet of water forced through a nozzle carries air molecules out of the system. Filter pumps cannot produce pressures below the vapor pressure of water. They are used in the laboratory for vacuum filtration, distillation, and similar techniques requiring a low-grade vacuum.

filtrate *See* filtration.

filtration The process of removing suspended particles from a fluid by passing or forcing the fluid through a porous material (the *filter*). The fluid that passes through the filter is the *filtrate*. In laboratory filtration, filter paper or sintered glass is commonly used.

fine organic chemicals Carbon compounds, such as pesticides, dyes, and drugs, produced only in small quantities. Their main requirement is that they must have a high degree of purity, often higher than 95%. They are manufactured for special purposes, e.g. for use in spectroscopy, pharmacology, and electronics.

fine structure Closely spaced lines seen at high resolution in a spectral line or band. Fine structure may be caused by vibration of the molecules or by electron spin. *Hyperfine structure*, seen at very high resolution, is caused by the atomic nucleus affecting the possible energy levels of the atom.

fireclay A refractory clay containing high proportions of alumina (aluminum oxide) and silica (silicon(IV) oxide), used for making firebricks and furnace linings.

firedamp Methane occurring in coal mines. Explosions can occur if quantities of methane accumulate. The air left after an explosion is called *blackdamp*. *Afterdamp* is the poisonous carbon monoxide formed.

first-order reaction A reaction in which the rate of reaction is proportional to the concentration of one of the reacting substances. The concentration of the reacting substance is raised to the power one; i.e. rate = k[A]. For example, the decomposition of hydrogen peroxide is a first-order reaction,

$$\text{rate} = k[H_2O_2]$$

Similarly the rate of decay of radioactive material is a first-order reaction,

$$\text{rate} = k[\text{radioactive material}]$$

For a first-order reaction, the time for a definite fraction of the reactant to be consumed is independent of the original concentration. The units of k, the rate constant, are s^{-1}.

Fischer projection /fish-er/ A way of representing the three-dimensional structure of a molecule in two dimensions. The molecule is drawn using vertical and horizontal lines. Horizontal lines represent bonds that come out of the paper. Vertical lines represent bonds that go into the paper (or are in the plane of the paper). Named for Emil Fischer, the convention was formerly used for representing the absolute configuration of sugars.

Fischer–Tropsch process /fish-er tropsh/ The catalytic hydrogenation of carbon monoxide in the ratio of 2:1 hydrogen to carbon monoxide at 200°C to produce hydrocarbons, especially motor fuel. The carbon monoxide is derived from *water gas*, which uses coal or natural gas as a source of carbon, and the catalyst is finely divided nickel. The process was extensively used in Germany during World War II and is still used in countries in which crude oil is not an easily available source of motor fuel. The process is named for the German chemists Franz Fischer (1877–1948) and Hans Tropsch (1889–1935).

Fittig reaction *See* Wurtz reaction.

fixation of nitrogen *See* nitrogen fixation.

fixing, photographic A stage in the development of exposed photographic film or paper in which unexposed silver halides in the emulsion are dissolved away. The fixing bath contains a solution of ammonium or sodium thiosulfate(IV), and is employed after the initial development (chemical reduction) of the latent silver image.

flame-ionization detector *See* gas chromatography.

flame test A preliminary test in qualitative analysis in which a small sample of a chemical is introduced into a nonluminous Bunsen-burner flame on a clean platinum wire. The flame vaporizes part of the sample and excites some of the atoms, which emit light at wavelengths characteristic of the metallic elements in the sample. Thus the observation of certain colors in the flame indicates the presence of these elements. The same principles are applied in the modern instrumental method of spectrographic analysis.

FLAME TEST COLORS

Element	*Flame color*
barium	green
calcium	brick red
lithium	crimson
potassium	pale lilac
sodium	yellow
strontium	red

flare stack A chimney at the top of which unwanted gases are burnt in an oil refinery or other chemical plant.

flash photolysis A technique for investigating free radicals in gases. The gas is held at low pressure in a long glass or quartz tube, and an absorption spectrum taken using a beam of light passing down the tube. The gas can be subjected to a very brief intense flash of light from a lamp outside the tube, producing free radicals, which are identified by their spectra. Measurements of the intensity of spectral lines can be made with time using an oscilloscope, and the kinetics of very fast reactions can thus be investigated.

flash point The lowest temperature at which sufficient vapor is given off by a flammable liquid to ignite in the presence of a spark.

flocculation /flok-yŭ-**lay**-shŏn/ (**coagulation**) The combining of the particles of a finely divided precipitate, such as a colloid, into larger particles or clumps that sink and are easier to filter off.

flocculent /**flok**-yŭ-lănt/ Describing a precipitate that has aggregated in wooly masses.

flotation *See* froth flotation.

fluid A state of matter that is not a solid – that is, a liquid or a gas. All fluids can flow, and the resistance to flow is the viscosity.

fluidization /floo-id-ă-**zay**-shŏn/ The suspension of a finely-divided solid in an upward-flowing liquid or gas. This suspension mimics many properties of liquids, such as allowing objects to 'float' in it. Fluidized beds so constructed are important industrially.

fluorescein /floo-ŏ-**ress**-ee-in/ A fluorescent dye used as an absorption indicator.

fluorescence /floo-ŏ-**ress**-ĕns/ The absorption of energy by atoms, molecules, etc., followed by immediate emission of electromagnetic radiation as the particles make transitions to lower energy states. *Compare* phosphorescence.

fluoridation /floo-ŏ-ră-**day**-shŏn/ The introduction of small quantities of fluoride compounds into the water supply as a public-health measure to reduce the incidence of tooth decay in children.

fluoride /**floo**-ŏ-rÿd, -rid/ *See* halide.

fluorination /floo-ŏ-ră-**nay**-shŏn/ *See* halogenation.

fluorine /**floo**-ŏ-reen, -rin/ A slightly greenish-yellow highly reactive gaseous element belonging to the halogens (group 17 of the periodic table, formerly VIIA). It occurs notably as fluorite (CaF_2) and cryolite (Na_3AlF_3) but traces are also widely distributed with other minerals. It is slightly more abundant than chlorine, accounting for about 0.065% of the Earth's crust. The high reactivity of the element delayed its isolation. Fluorine is now prepared by electrolysis of molten KF/HF electrolytes, using copper or steel apparatus. Its preparation by chemical methods is impossible.

Fluorine compounds are used in the steel industry, glass and enamels, uranium processing, aluminum production, and in a range of fine organic chemicals. Fluorine is the most reactive and most electronegative element known; in fact it reacts directly with almost all the elements, including the rare gas xenon. As the most powerful oxidizing agent known, it has the pronounced characteristic of bringing out the higher oxidation states when combined with other elements. Ionic fluorides contain the small F^- ion, which is very similar to the O^{2-} ion. With the more electronegative elements, fluorine forms an extensive range of compounds, e.g. SF_6, NF_3, and PF_5, in which the bonding is essentially covalent. A vast number of organic compounds are known in which hydrogen may be replaced by fluorine and in which the C–F bond is characteristically very stable. Examples are CF_2=CF_2, $C_6H_4F_2$, and CF_3COOH; the bonding is again essentially covalent.

Fluorine reacts explosively with hydrogen, even in the dark, to form hydrogen fluoride, HF, which polymerizes as a result of hydrogen bonding and has a boiling point very much *higher* than that of HCl

(HF b.p. 19°C; HCl b.p. –85°C). Unlike the other halogens, fluorine does not form higher oxides or oxyacids; oxygen difluoride in fact reacts with water to give hydrogen fluoride.

Fluorine is strongly electronegative and exhibits a strong electron withdrawing effect on adjacent bonds, thus CF_3COOH is a strong acid (whereas CH_3COOH is not). Although the coordination number seldom exceeds one there are cases in which fluorine bridging occurs, e.g. $(SbF_5)_n$. The element is sufficiently reactive to combine directly with the rare gas xenon at 400°C to form XeF_2 and XeF_4. At 600°C and high pressure it will even form XeF_6.

Fluorine and hydrogen fluoride are extremely dangerous and should only be used in purpose-built apparatus; gloves and face shields should be used when working with hydrofluoric acid and accidental exposure should be treated as a hospital emergency.

Symbol: F; b.p. –188.14°C; m.p. –219.62°C; d. 1.696 kg m^{-3} (0°C); p.n. 9; r.a.m. 18.99840.32.

fluorite /**floo**-ŏ-rÿt/ (**fluorspar**) A mineral form of calcium fluoride (CaF_2), used in making some cements and types of glass. *See also* calcium-fluoride structure.

fluorite structure *See* calcium-fluoride structure.

fluorocarbon /floo-ŏ-rŏ-**kar**-bŏn/ A compound derived from a hydrocarbon by replacing hydrogen atoms with fluorine atoms. Fluorocarbons are unreactive and most are stable up to high temperatures. They have a variety of uses – in aerosol propellants, oils and greases, and synthetic polymers such as PTFE. *See also* chlorofluorocarbon.

fluorspar /**floo**-er-spar/ *See* fluorite.

flux **1.** A substance used to keep metal surfaces free of oxide in soldering. *See* solder.
2. A substance used in smelting metals to react with silicates and other impurities and form a low-melting slag.

fluxional molecule A molecule in which the constituent atoms change their relative positions so quickly at room temperature that the normal concept of structure is inadequate; i.e. no specific structure exists for longer than about 10^{-2} second and the relative positions become indistinguishable. For example ClF_3 at –60°C has a distinct 'T' shape but at room temperature the fluorine atoms are visualized as moving rapidly over the surface of the chlorine atom in a state of exchange and are effectively identical.

foam A dispersion of bubbles of gas in a liquid, usually stabilized by a SURFACTANT. Solid foams, such as expanded polystyrene or foam rubber, are made by allowing liquid foams to set.

formaldehyde /for-**mal**-dĕ-hÿd/ *See* methanal.

formalin /**for**-mă-lin/ *See* methanal.

formate /**for**-mayt/ *See* methanoate.

formic acid /**for**-mik/ *See* methanoic acid.

formula (*pl.* **formulas** or **formulae**) A representation of a chemical compound using symbols for the atoms and subscript numbers to show the numbers of atoms present. *See* empirical formula; general formula; molecular formula; structural formula.

formyl group /**for**-măl/ The group HCO–.

fossil fuel A mineral fuel that forms underground from the remains of living organisms. Fossil fuels include coal, natural gas, peat, and petroleum.

fraction A mixture of liquids with similar boiling points collected by fractional distillation.

fractional crystallization Crystallization of one component from a mixture in solution. When two or more substances are present in a liquid (or in solution), on cooling to a lower temperature one substance will preferentially form crystals, leaving the other substance in the liquid (or dissolved) state. Fractional crystallization can thus be used to purify or separate substances if the correct conditions are known.

fractional distillation (**fractionation**) A distillation carried out with partial reflux, using a long vertical column (fractionating column). It utilizes the fact that the vapor phase above a liquid mixture is generally richer in the more volatile component. If the region in which refluxing occurs is sufficiently long, fractionation permits the complete separation of two or more volatile liquids. Fractionation is the fundamental process for producing petroleum from crude oil.

Unlike normal reflux, the fractionating column may be insulated to reduce heat loss, and special designs are used to maximize the liquid-vapour interface.

fractionation /frak-shŏ-**nay**-shŏn/ *See* fractional distillation.

francium /**fran**-see-ŭm/ A radioactive element of the alkali-metal group. It is found on Earth only as a short-lived product of radioactive decay, occurring in uranium ores in minute quantities. A large number of radioisotopes of francium are known.

Symbol: Fr; m.p. 27°C; b.p. 677°C; p.n. 87; most stable isotope ^{223}Fr (half-life 21.8 minutes).

Frasch process /frash/ *See* sulfur.

free electron An electron that can move from one atom or molecule to another under the influence of an applied electric field. Movement of free electrons in a conductor constitutes an electric current. Free electrons are also involved in metallic bonds (*see* metals).

free energy A measure of the ability of a system to do useful work. *See* Gibbs function; Helmholtz function.

free radical An atom or group of atoms with a single unpaired electron. Free radicals are produced by breaking a covalent bond; for example:

$$CH_3Cl \rightarrow CH_3\bullet + Cl\bullet$$

They are often formed in light-induced reactions. Free radicals are extremely reactive and can be stabilized and isolated only under special conditions.

free-radical substitution *See* substitution reaction.

freezing The process by which a liquid is converted into a solid by cooling; the reverse of melting.

freezing mixtures Two or more substances mixed together to produce a low temperature. A mixture of sodium chloride and ice in water (–20°C) is a common example.

freezing point The temperature at which a liquid is in equilibrium with its solid phase at standard pressure and below which the liquid freezes or solidifies. This temperature is always the same for a particular liquid and is numerically equal to the melting point of the solid.

freezing point constant *See* depression of freezing point.

French chalk *See* talc.

Frenkel defect /**frenk**-ĕl/ *See* defect.

Freons /**free**-onz/ (*Trademark*) Various chlorofluorocarbons (CFCs) and fluorocarbons used as refrigerants. *See* chlorofluorocarbon; fluorocarbon.

Friedel–Crafts reaction /**free**-dĕl **kraft**/ A method for the substitution of an alkyl or acyl group onto a benzene ring. The aromatic hydrocarbon (e.g. benzene) is refluxed with a haloalkane or acyl halide using aluminum chloride catalyst. The product is an alkylarene, e.g.:

$$C_6H_6 + CH_3I \rightarrow C_6H_5CH_3 + HI$$

or a ketone, for example:

$$C_6H_6 + CH_3COCl \rightarrow C_6H_5COCH_3 + HCl$$

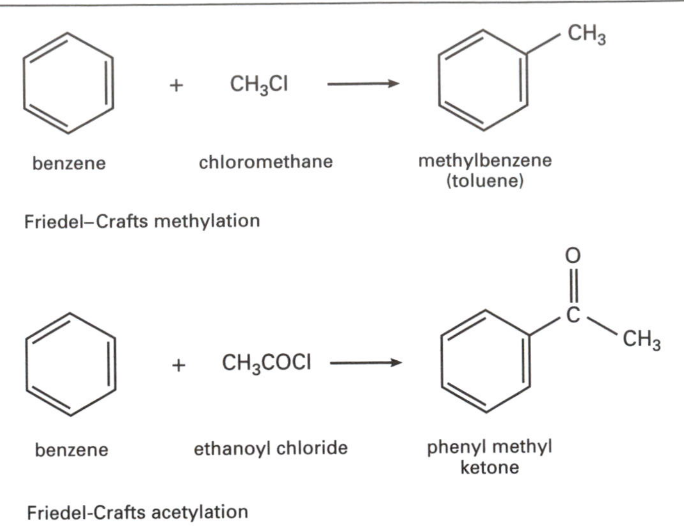

Friedel–Crafts reactions

The aluminum chloride accepts a lone pair of electrons on the chlorine, polarizing the chloroalkane to produce positive charge on the carbon. Electrophilic substitution then occurs. The reaction is named for the French chemist Charles Friedel (1832–99) and the American chemist James Crafts (1839–1917).

frontier orbital Either of two orbitals in a molecule: the highest occupied molecular orbital (the *HOMO*) or the lowest unoccupied molecular orbital (the *LUMO*). The HOMO is the orbital with the highest energy level occupied at absolute zero temperature. The LUMO is the lowest-energy unoccupied orbital at absolute zero. For a particular molecule the nature of these two orbitals is very important in determining the chemical properties. *Frontier-orbital theory*, which considers the symmetry of these orbitals, has been very successful in explaining such reactions as the DIELS–ALDER REACTION. *See also* Woodward–Hoffmann rules.

froth flotation An industrial technique for separating the required parts of ores from unwanted gangue. The mineral is pulverized and mixed with water, to which a frothing agent is added. The mixture is aerated and particles of one constituent are carried to the surface by bubbles of air, which adhere preferentially to one of the constituents. Various additives are also used to modify the surface properties of the particles (e.g. to increase the adherence of air bubbles).

fructose /**fruk**-tohs, **frûk**-/ (**fruit sugar**; $C_6H_{12}O_6$) A sugar found in fruit juices, honey, and cane sugar. It is a ketohexose, existing in a pyranose form when free. In combination (e.g. in sucrose) it exists in the furanose form.

fruit sugar *See* fructose.

fuel cell A type of cell in which fuel is converted directly into electricity. In one form, hydrogen gas and oxygen gas are fed to the surfaces of two porous nickel electrodes immersed in potassium hydroxide solution. The oxygen reacts to form hydroxyl (OH^-) ions, which it releases into the solution, leaving a positive charge on the electrode. The hydrogen reacts with the OH^- ions in the solution to form water, giving up electrons to leave a negative charge on the other electrode. Large fuel cells can generate tens of amperes. Usually the e.m.f.

is about 0.9 volt and the efficiency around 60%.

fullerene /**fûl**-ĕ-reen/ *See* buckminsterfullerene.

fullerite /**fûl**-ĕ-rÿt/ *See* buckminsterfullerene.

fuller's earth A natural clay used as an absorbent and industrial catalyst.

fumaric acid /fyoo-**ma**-rik/ *See* butenedioic acid.

fuming sulfuric acid *See* oleum.

functional group A group of atoms in a compound that is responsible for the characteristic reactions of the type of compound. Examples are:

alcohol $-OH$
aldehyde $-CHO$
amine $-NH_2$
ketone $=CO$.
carboxylic acid $-CO.OH$
acyl halide $-CO.X$ (X = halogen)
nitro compound $-NO_2$
sulfonic acid $-SO_2.OH$
nitrile $-CN$
diazonium salt $-N_2^+$
diazo compound $-N{=}N-$

fundamental units The units of length, mass, and time that form the basis of most systems of units. In SI, the fundamental units are the meter, the kilogram, and the second. *See also* base unit.

furan /**fyoo**-ran, fyoo-**ran**/ (**furfuran**; C_4H_4O) A heterocyclic liquid organic compound. Its five-membered ring contains four carbon atoms and one oxygen atom. The structure is characteristic of some monosaccharide sugars (furanoses).

furanose /**fyoo**-ră-nohs/ A SUGAR that has a five-membered ring (four carbon atoms and one oxygen atom).

furfuran /**fer**-fŭ-răn, fer-fŭ-**ran**/ *See* furan.

fused Describing a solid that has been melted and solidified into a single mass. Fused silica, for example, is produced by melting sand.

fused ring *See* ring.

fusel oil /**fyoo**-zĕl, -sĕl/ A mixture of high-molecular-weight alcohols together with some esters and fatty acids, formed from alcoholic fermentation and obtained during distillation. It is used as a source of higher alcohols.

fusion Another word for melting.

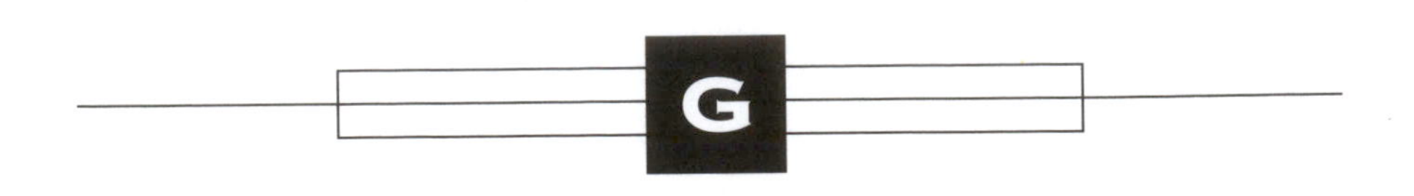

gadolinium /gad-ŏ-**lin**-ee-ŭm/ A ductile malleable silvery element of the lanthanoid series of metals. It occurs in association with other lanthanoids. Gadolinium is used in alloys, magnets, and in the electronics industry.

Symbol: Gd; m.p. 1313°C; b.p. 3266°C; r.d. 7.9 (25°C); p.n. 64; r.a.m. 157.25.

galactose /gă-**lak**-tohs/ ($C_6H_{12}O_6$) A SUGAR found in lactose and many polysaccharides. It is an aldohexose, isomeric with glucose.

galena /gă-**lee**-nă/ (**lead glance**) A mineral form of lead(II) sulfide. It is the principal ore of lead.

gallium /**gal**-ee-ŭm/ A soft silvery low-melting metallic element belonging to group 3 (formerly IIIA) of the periodic table. It is found in minute quantities in several ores, including zinc blende (ZnS) and bauxite ($Al_2O_3.H_2O$). Gallium is used in low-melting alloys, high-temperature thermometers, and as a doping impurity in semiconductors. Gallium arsenide is a semiconductor used in light-emitting diodes and in microwave apparatus.

Symbol: Ga; m.p. 29.78°C; b.p. 2403°C; r.d. 5.907 (solid at 20°C), 6.114 (liquid); p.n. 31; r.a.m. 69.723.

galvanic cell /gal-**van**-ik/ *See* cell.

galvanizing /**gal**-vă-nȳ-zing/ A process for coating steel with zinc by dipping in a bath of molten zinc, or by electrodeposition. The zinc protects the steel from corrosion.

gamma radiation A form of electromagnetic radiation emitted by changes in the nuclei of atoms. Gamma waves have very high frequency (short wavelength). Gamma radiation shows particle properties much more often than wave properties. The energy of a gamma photon, given by

$$W = h\nu$$

(where h is the Planck constant), can be very high. A gamma photon of 10^{24} Hz has an energy of 6.6×10^{-10} J. *See also* electromagnetic radiation.

gamma rays Streams of gamma radiation.

gangue /gang/ The rock or other undesirable material occurring in an ore.

gas The state of matter in which forces of attraction between the particles of a substance are small. The particles have freedom of movement and gases, therefore, have no fixed shape or volume. The atoms and molecules of a gas are in a continual state of motion and are continually colliding with each other and with the walls of the containing vessel. These collisions with the walls create the pressure of a gas.

gas chromatography A technique widely used for the separation and analysis of mixtures. Gas chromatography employs a column packed with either a solid stationary phase (*gas-solid chromatography* or *GSC*) or a solid coated with a non-volatile liquid (*gas-liquid chromatography* or *GLC*). The whole column is placed in a thermostatically controlled heating jacket. A volatile sample is introduced into the column using a syringe, and an unreactive carrier gas, such as nitrogen, passed through it. The components of the sample will be carried along in this mobile phase. However, some of the components will cling more readily to the stationary phase than others, either because they become attached to the solid surface or because they dissolve in the liquid. The time taken for different components to pass through the column is characteristic and can be used to identify them. The emergent sample is passed through a detector, which registers the presence of the different components in the carrier gas.

Two types of detector are in common use: the *katharometer*, which measures changes in thermal conductivity, and the flame-ionization detector, which turns the volatile components into ions and registers the change in electrical conductivity.

gas constant (**universal gas constant**) Symbol: R The universal constant 8.314 34 J mol^{-1} K^{-1} appearing in the equation of state for an ideal gas. *See* gas laws.

gaseous diffusion separation A technique for the separation of gases that relies on their slightly different atomic masses. For example, the isotopes of uranium (^{235}U and ^{238}U) can be separated by first preparing gaseous uranium hexafluoride from uranium ore. If this is then made to pass through a series of fine pores, the molecules containing the lighter isotope of uranium will pass through more quickly. As the gases pass through successive 'diaphragms', the proportion of lighter molecules increases and a separation of over 99% is attainable. This method has also been applied to the separation of the isotopes of hydrogen.

gas equation *See* gas laws.

gas laws Laws relating the temperature, pressure, and volume of a fixed mass of gas. The main gas laws are Boyle's law and Charles' law. The laws are not obeyed exactly by any real gas, but many common gases obey them under certain conditions, particularly at high temperatures and low pressures. A gas that would obey the laws over all pressures and temperatures is a *perfect* or *ideal gas*.

Boyle's and Charles' laws can be combined into an equation of state for ideal gases:

$$pV_m = RT$$

where V_m is the molar volume and R the molar gas constant. For n moles of gas

$$pV = nRT$$

All real gases deviate to some extent from the gas laws, which are applicable only to idealized systems of particles of negligible volume with no intermolecular forces. There are several modified equations of state that give a better description of the behavior of real gases, the best known being the van der Waals equation.

gas–liquid chromatography *See* gas chromatography.

gasohol /**gas**-ŏ-hôl/ Alcohol (ethanol) obtained by the industrial fermentation of sugar for use as a motor fuel. It has been produced on a large scale in some countries.

gas oil One of the main fractions obtained from PETROLEUM by distillation, used as a fuel for diesel engines. *See* diesel fuel.

gasoline /**gas**-ŏ-leen, gas-ŏ-**leen**/ *See* petroleum.

gas-solid chromatography *See* gas chromatography.

Gatterman–Koch reaction A reaction for substituting a formyl (methanoyl) group (HCO–) onto a benzene ring of an aromatic hydrocarbon. It is used in the industrial production of benzaldehyde from benzene:

$$C_6H_6 \rightarrow C_6H_5CHO$$

The aromatic hydrocarbon is mixed with a Lewis acid, such as aluminum chloride, and a mixture of carbon monoxide and the hydrogen chloride is passed through. The first stage is the production of an $H–C≡O^+$ ion:

$$HCl + CO + AlCl_3 \rightarrow HCO^+ + AlCl_4^-$$

Copper (I) chloride (CuCl) is also added as a catalyst. The HCO^+ ion acts as an electrophile in electrophilic substitution on the benzene ring.

A variation of the reaction in which the HCO– group is substituted onto the benzene ring of a phenol uses hydrogen cyanide rather than carbon monoxide. Typically, a mixture of zinc cyanide and hydrochloric acid is used, to give zinc chloride (which acts as a Lewis acid) and hydrogen cyanide. The electrophile in this case is protonated hydrogen cyanide:

$$HCN + HCl + ZnCl_2 \rightarrow HCNH^+ + ZnCl_3^-$$

The phenol is first substituted to give an imine:

$$C_6H_5OH \rightarrow HOC_6H_4CH{=}NH$$

This then hydrolyzes to the aromatic aldehyde:

$$HOC_6H_4CH{=}NH_2 \rightarrow HOC_6H_4CHO$$

Similar reactions using alkyl cyanides (nitriles) rather than hydrogen cyanide give aromatic ketones. This type of reaction, in which a cyanide is used to produce an aldehyde (or ketone) is often called the *Gatterman reaction.* The Gatterman–Koch reaction was reported by the German chemist Ludwig Gatterman (1860–1920) in 1897 (with J. C. Koch). The use of hydrogen cyanide was reported by Gatterman in 1907.

Gatterman reaction **1.** *See* Sandmeyer reaction.
2. *See* Gatterman–Koch reaction.

gauche conformation /gohsh/ *See* conformation.

gauss /gows/ Symbol: G The unit of magnetic flux density in the c.g.s. system. It is equal to 10^{-4} tesla.

Gay-Lussac's law **1.** Gases react in volumes that are in simple ratios to each other and to the products if they are gases (all volumes measured at the same temperature and pressure).
2. *See* Charles' law.

gel /jel/ A lyophilic colloid that is normally stable but may be induced to coagulate partially under certain conditions (e.g. lowering the temperature). This produces a pseudo-solid or easily deformable jelly-like mass, called a gel, in which intertwining particles enclose the whole dispersing medium. Gels may be further subdivided into elastic gels (e.g. gelatin) and rigid gels (e.g. silica gel).

gelatin /jel-ă-tin/ (**gelatine**) A pale yellow protein obtained from the bones, hides, and skins of animals, which forms a colloidal jelly when dissolved in hot water. It is used in jellies and other foods, to make capsules for various medicinal drugs, as an adhesive and sizing medium, and in photographic emulsions.

gel electrophoresis *see* electrophoresis.

gel filtration *See* molecular sieve.

gem positions Positions in a molecule on the same atom. For example, 1,1-dichloroethane (CH_3CHCl_2) is a gem dihalide.

general formula A representation of the chemical formula common to a group of compounds. For example, C_nH_{2n+2} is the general formula for an alkane, whose members form a homologous series. *See also* empirical formula; homologous series; molecular formula; structural formula.

geochemistry /jee-oh-**kem**-iss-tree/ The study of the chemistry of the Earth. The subject includes the chemical composition of the Earth, the abundance of the elements and their isotopes, the atomic structure of minerals, and chemical processes that occur at the high temperatures and pressures inside the Earth.

geometrical isomerism *See* isomerism.

germanium /jer-**may**-nee-ŭm/ A hard brittle gray metalloid element belonging to group 14 (formerly IVA) of the periodic table. It is found in sulfide ores such as argyrodite ($4Ag_2S.GeS_2$) and in zinc ores and coal. Most germanium is recovered during zinc or copper refining as a by-product. Germanium was extensively used in early semiconductor devices but has now been largely superseded by silicon. It is used as an alloying agent, catalyst, phosphor, and in infrared equipment.

Symbol: Ge; m.p. 937.45°C; b.p. 2830°C; r.d. 5.323 (20°C); p.n. 32; r.a.m. 72.61.

germanium(IV) oxide (GeO_2) A compound made by strongly heating germanium in air or by hydrolyzing germanium(IV) chloride. It occurs in two forms. One is slightly soluble in water forming an acidic solution; the other is insoluble in water. Germanium oxide dissolves in alkalis to form the anion GeO_3^{2-}.

German silver *See* nickel-silver.

Gibbs function (**Gibbs free energy**) Symbol: *G* A thermodynamic function defined by

$$G = H - TS$$

where *H* is the enthalpy, *T* the thermodynamic temperature, and *S* the entropy. It is useful for specifying the conditions of chemical equilibrium for reactions for constant temperature and pressure (*G* is a minimum). The function is named for the American mathematician and theoretical physicist Josiah Williard Gibbs (1839–1903). *See also* free energy.

giga- Symbol: G A prefix denoting 10^9. For example, 1 gigahertz (GHz) = 10^9 hertz (Hz).

glacial ethanoic (acetic) acid Pure water-free ethanoic acid.

glacial ethanoic acid (**glacial acetic acid**) Pure water-free ethanoic acid.

glass A hard transparent material made by heating calcium oxide (lime), sodium carbonate, and sand (silicon(IV) oxide). This produces a calcium silicate – the normal type of glass, called *soda glass*. Special types of glass can be obtained by incorporating boron oxide in the glass (borosilicate glass, used for laboratory apparatus) or by including metals other than calcium, e.g. lead or barium.

Glass is an amorphous substance, in the sense that there is no long-range ordering of the atoms on a lattice. It can be regarded as a supercooled liquid, which has not crystallized. Solids with similar noncrystalline structures are also called *glasses*

Glauber's salt /**glow**-ber, **glaw**-/ *See* sodium sulfate.

GLC Gas-liquid chromatography. *See* gas chromatography.

glove box A sealed box with gloves fitted to ports in one side and having a transparent top, used for safety reasons or to perform experiments in an inert or sterile atmosphere.

gluconic acid /gloo-**kon**-ik/ (**dextronic acid** $CH_2OH(CHOH)_4COOH$) A soluble crystalline organic acid made by the oxidation of glucose (using specific molds). It is used in paint strippers.

glucosan A POLYSACCHARIDE that is formed of glucose units. Cellulose and starch are examples.

glucose /**gloo**-kohs/ (**dextrose; grape sugar**; $C_6H_{12}O_6$) A monosaccharide occurring widely in nature as D-glucose. It occurs as glucose units in sucrose, starch, and cellulose. It is important to metabolism because it participates in energy-storage and energy-release systems. *See also* sugar.

glucoside /**gloo**-kŏ-sÿd/ *See* glycoside.

glue An adhesive, of which there are various types. Aqueous solutions of starch and ethyl cellulose are used as pastes for sticking paper; traditional wood glue is made by boiling animal bones (*see* gelatin); quick-drying adhesives are made by dissolving rubber or a synthetic polymer in a volatile solvent; and some polymers, such as epoxy resins and polyvinyl acetate (PVA), are themselves used as glues.

glutamic acid /gloo-**tam**-ik/ *See* amino acids.

glutamine /**gloo**-tă-meen, -min/ *See* amino acids.

glyceride /**gliss**-ĕ-rÿd, -rid/ An ester formed between glycerol (propane-1,2,3-triol) and one or more carboxylic acids. Glycerol has three alcohol groups, and if all three groups have formed esters, the compound is a *triglyceride*. Naturally occurring fats and oils are triglycerides of long-chain carboxylic acids (hence the name 'fatty acid'). The main carboxylic acids occurring in fats and oils are:
1. octadecanoic acid (stearic acid), a saturated acid $CH_3(CH_2)_{16}COOH$.
2. hexadecanoic acid (palmitic acid), a saturated acid $CH_3(CH_2)_{14}COOH$.
3. *cis*-9-octadecenoic acid (oleic acid), an unsaturated acid.
   $CH_3(CH_2)_7CH{:}CH(CH_2)_7COOH$

glycerine /**gliss**-ĕ-rin, -reen, gliss-ĕ-**reen**/ (**glycerin**) *See* propane-1,2,3-triol.

glycerol /**gliss**-ĕ-rol, -rohl/ *See* propane-1,2,3-triol.

glyceryl trinitrate /**gliss**-ĕ-răl trÿ-**nÿ**-trayt/ *See* nitroglycerine.

glycine /**glÿ**-seen, glÿ-**seen**/ *See* amino acids.

glycogen /**glÿ**-kŏ-jĕn/ (**animal starch**) A polysaccharide that is the main carbohydrate store of animals. It is composed of many glucose units linked in a similar way to starch. Glycogen is readily hydrolyzed in a stepwise manner to glucose itself. It is stored largely in the liver and in muscle but is found widely distributed in the body.

glycol /**glÿ**-kol. -kohl/ *See* diol.

glycoside /**glÿ**-kŏ-sÿd/ A compound made by replacing a hydroxyl group of a monosaccharide sugar with an alcoholic, phenolic, or other group. The bond is a *glycosidic link*. If the sugar is glucose, the compound is a glucoside. Glycosides occur in plants .

glycosidic link /glÿ-kŏ-**sid**-ik/ *See* glycoside.

gold A transition metal that occurs native. It is unreactive and is very ductile and malleable. Gold is used in jewelry, often alloyed with copper, and in electronics and colored glass. Pure gold is 24 carat; 9 carat indicates that 9 parts in 24 consist of gold.

Symbol: Au; m.p. 1064.43°C; b.p. 2807°C; r.d. 19.320 (20°C); p.n. 79; r.a.m. 196.96654.

gold(III) chloride (**gold trichloride; auric chloride**; $AuCl_3$) A compound prepared by dissolving gold in aqua regia. The bright yellow crystals (chloroauric acid) produced on evaporation are heated to form dark red crystals of gold(III) chloride. The chloride decomposes easily (at 175°C) to give gold(I) chloride and chlorine; at higher temperatures it decomposes to give gold and chlorine. Gold(III) chloride is used in photography. It exists as a dimer, Au_2Cl_6.

Goldschmidt process /**gohld**-shmit/ A process for extracting certain metals from their oxides by reduction with aluminum. It is named for the German chemist Johann (Hans) Wilhelm Goldschmidt (1861–1923). *See* thermite.

gold trichloride /trÿ-**klor**-ÿd, -**kloh**-rÿd/ *See* gold(III) chloride.

graft copolymer *See* polymerization.

Graham's law (of diffusion) The principle that gases diffuse at a rate that is inversely proportional to the square root of their density. Light molecules diffuse faster than heavy molecules. The principle is used in the separation of isotopes. The law is named for the Scottish chemist Thomas Graham (1805–69).

grain A crystal in a metal that has been prevented from attaining its regular geometrical form.

gram Symbol: g A unit of mass defined as 10^{-3} kilogram.

gram-atom *See* mole.

gram-equivalent The equivalent weight of a substance in grams.

gram-molecule *See* mole.

granulation A process for enlarging particles to improve the flow properties of solid reactants and products in industrial chemical processes. The larger a particle, and the freer from fine materials in a solid, the more easily it will flow. Dry granulation produces pellets from dry materials, which are crushed into the desired size. Wet granulation involves the addition of a liquid to the material, and the resulting paste is extruded and dried before cutting into the required size.

grape sugar *See* dextrose.

graphite An allotrope of CARBON. Graphite is a good conductor of heat and electricity. The atoms are arranged in layers which cleave easily and graphite is used as a solid lubricant.

gravimetric analysis /grav-ă-**met**-rik/ A method of quantitative analysis in which the final analytical measurement is made by weighing. There are many variations in the method but in essence they all consist of:

1. taking an accurately weighed sample into solution;
2. precipitation as a known compound by a quantitative reaction;
3. digestion and coagulation procedures.
4. filtration and washing;
5. drying and weighing as a pure compound.

Filtration is a key element in the method and a variety of special filter papers and sinter-glass filters are available.

gray Symbol: Gy The SI unit of absorbed energy dose per unit mass resulting from the passage of ionizing radiation through living tissue. One gray is an energy absorption of one joule per kilogram of mass. The unit is named for the British physician Harold Gray (1905–65).

Grignard reagent /gree-**nyar**/ A type of organometallic compound with the general formula RMgX, where R is an alkyl or aryl group and X is a halogen (e.g. CH_3MgCl). Grignard reagents are prepared by reacting the haloalkane or haloaryl compound with magnesium in dry ether:

$$CH_3Cl + Mg \rightarrow CH_3MgCl$$

Grignard reagents probably have the form $R_2Mg.MgCl_2$. They are used extensively in organic chemistry. With methanal a primary alcohol is produced:

$$RMgX + HCHO \rightarrow RCH_2OH + Mg(OH)X$$

Other aldehydes give secondary alcohols:

$$RMgX + R'CHO \rightarrow RR'CHOH + Mg(OH)X$$

Alcohols and carboxylic acids give hydrocarbons:

$$RMgX + R'OH \rightarrow RR' + Mg(OH)X$$

Water also gives a hydrocarbon:

$$RMgX + H_2O \rightarrow RH + Mg(OH)X$$

Solid carbon dioxide in acid solution gives a carboxylic acid:

$$RMgX + CO_2 + H_2O \rightarrow RCOOH + Mg(OH)X$$

ground state The lowest energy state of an atom, molecule, or other system. *Compare* excited state.

group 1. In the PERIODIC TABLE, a series of chemically similar elements that have similar electronic configurations. A group is thus a column of the periodic table. For example, the alkali metals, all of which have outer s^1 configurations, belong to group 1. 2. (FUNCTIONAL GROUP) In organic chemistry, an arrangement of atoms that bestows a particular type of property on a molecule and enables it to be placed in a particular class, e.g. the aldehyde group –CHO.

group 0 elements *See* rare gases.

group 1 elements *See* alkali metals.

group 2 elements *See* alkaline-earth metals.

group 3-12 elements *See* transition elements.

group 13 elements A group of elements in the periodic table consisting of the elements boron (B), aluminum (Al), gallium (Ga), indium (In), and thallium (T1). These elements were formerly classified as group III elements and belonged to the IIIA subgroup. The IIIB subgroup consisted of the elements scandium (Sc), yttrium (Y), lanthanum (La), and actinium (Ac). The group 13 elements all have three outer electrons (s^2p^1) with no partly filled inner shells. The group is the first group of elements in the p-block. As with all the groups there is an increase in metallic character as the group is descended.

The boron atom is both small and has a high ionization potential, so bonds to boron are largely covalent with only small degrees of polarization. In fact boron is classed as a metalloid. It has volatile hydrides and a weakly acidic oxide. As the group is descended the ionization potentials decrease and the atomic radii increase, leading to increasingly polar interactions and the formation of distinct M^{3+} ions. The increase in metallic character is clearly illustrated by the hydroxides: boric acid $B(OH)_3$ is acidic; aluminum and gallium hydroxides are amphoteric, dissolving in acids to give Al^{3+} and Ga^{3+} and in bases to give aluminates and gallates; the hydroxides of indium and thallium are distinctly basic. As the elements get heavier the bond energies with other elements become generally smaller. The monovalent state (removal of the p electron only) becomes progressively stable. For example gallium(I) is known in the gas phase and in complex systems sometimes known as '$GaCl_2$'. Monovalent indium halides are known, for example InX, and thallium(I) is distinctly more stable than thallium(III).

group 14 elements A group of elements in the periodic table consisting of the elements carbon (C), silicon (Si), germanium (Ge), tin (Sn), and lead (Pb). These elements were formerly classified as group IV elements and belonged to the IVA subgroup. The IVB subgroup consisted of the elements titanium (Ti), zirconium (Zr), and hafnium (Hf), which are transition metals. The group 14 elements all have electronic structures with four outer electrons (s^2p^2) and no partly filled inner shells. The group shows the common trend towards metallic character with the heavier elements; thus carbon is a typical non-metal, silicon and germanium are metalloids, and tin and lead are characteristically metallic. As observed with other groups the first member of the group is quite different from the rest. The carbon atom is smaller and has a higher ionization potential, both favouring predominance of covalence, but additional factors are:

1. The widespread nature of extensive catenation in carbon compounds.
2. The possibility of strong overlap of p orbitals or *double bonding.*
3. The unavailability of d orbitals in carbon compounds.

Significant differences are:

1. Both oxides of carbon are gaseous, CO and CO_2; the other elements form solid oxides, e.g. SiO_2, Pb_3O_4.
2. The heavier elements have no compounds analogous to aromatic compounds.
3. The heavier elements have no compounds analogous to ethene or ketones.
4. The heavier elements readily expand their coordination number, e.g. SiF_6^{2-} and $SnCl_6^{2-}$.

group 15 elements A group of elements in the periodic table consisting of the elements nitrogen (N), phosphorus (P), arsenic (As), antimony (Sb), and bismuth (Bi). These elements were formerly classified as group V elements and belonged to the VA subgroup. The VB subgroup consisted of the elements vanadium (V), niobium (Nb), and tantalum (Ta), which are transition elements.

The main-group elements all have electronic configurations corresponding to outer s^2p^3 with no vacancies in inner levels. The ionization potentials are high and the lighter members of the group, N and P, are distinctly electronegative and non-metallic in character. Arsenic and antimony are metalloids and bismuth is weakly metallic (Bi_2O_3 dissolves in acids to give $Bi(OH)_3$).

Nitrogen is dissimilar from the other members of the group in that it forms stable multiple bonds, it is limited to coordination number 4, e.g. NH_4^+, it is sufficiently electronegative to form hydrogen bonds and the nitride ion N^{3-}, and its oxides are irregular with the rest of the group.

The group displays a remarkable number of allotropes of its members showing the trend from non-metallic forms through to metallic forms. Thus nitrogen has only the diatomic form; phosphorus has a highly reactive form P_1 (brown), and a tetrahedral form P_4 (white), forms based on broken tetrahedra P_n (red and violet), and a hexagonal layer-type lattice (black). Arsenic and antimony have the As_4 and Sb_4 forms, which are less stable than the layer type in this case; bismuth has the layer-lattice form only.

group 16 elements A group of elements of the periodic table sometimes called the *chalcogens*, consisting of the elements oxygen (O), sulfur (S), selenium (Se), tellurium (Te), and polonium (Po). These elements were formerly classified as group VI elements and belonged to the VIA subgroup. The VIB subgroup consisted of the elements chromium (Cr), molybdenum (Mo), and tungsten (W), which are transition elements. The electronic configurations of the group 6 elements are all s^2p^4 with no vacant inner orbitals. These configurations are all just two electrons short of a rare gas structure, consequently they have high electron affinities and are almost entirely non-metallic in character.

The group shows the normal property of a trend towards metallic character as it is descended. Selenium, tellurium and polonium have 'metallic' allotropes and polonium has generally metalloid-type properties where they have been studied (Po is very rare). All the elements combine with a large number of other elements, both metallic and non-metallic, but in contrast to compounds of the halogens they are more generally insoluble in water, and even where soluble they do not ionize readily.

Oxygen behaves like the first members of other groups in having great differences from the rest of the group. For example, oxygen forms *hydrogen bonds* whereas sulfur and the others do not; oxygen cannot expand its outer shell and positive oxidation states are uncommon for oxygen (OF_2, O_2F_2, $O_2^+PtF_6^-$). Oxygen is paramagnetic. Excluding oxygen, group 16 shows similar trends with increasing size to those in group 15. For example decreasing stability of H_2X, greater tendency to form complexes such as $SeBr_6^{2-}$, decreasing stability of high formal positive oxidation states, and marginal metallic properties (e.g. Po forms a hydroxide).

group 17 elements A group of elements in the periodic table consisting of the elements fluorine (F), chlorine (Cl), bromine (Br), iodine (I), and astatine (At). Formerly they belonged to group VII and were classified in the VIIA subgroup. The VIIB subgroup consisted of the elements manganese (Mn), technetium (Te), and rhenium (Re), which are transition elements. *See* halogens.

group 18 elements *See* rare gases.

group theory A branch of mathematics used to analyze symmetry in a systematic way. A group consists of a set of elements A, B, C, etc. and a rule for forming the product of these elements, such that certain conditions are satisfied:
(1) Every product AB of two elements A and B is also an element of the set.
(2) Any three elements A, B, C, must satisfy $A(BC) = (AB)C$.
(3) There is an element of the set, denoted I, which is the *identity element*, i.e. for all A in the set $AI = IA = A$.
(4) For each element A of the set there is an *inverse*, denoted A^{-1}, which also belongs to the set which satisfies the relation $AA^{-1} = A^{-1} = A = I$.

In the case of symmetry operations the product is the successive performance of these operations. If a group has a finite number of elements it is said to be a *discrete group*. The *point groups*, which arise from the rotations and reflections of bodies such as isolated molecules are discrete groups.

GSC Gas–solid chromatography. *See* gas chromatography.

guanidine /**gwan**-ă-deen, -din, **gwah**-nă-/ (**iminourea**; $HN:C(NH_2)_2$) A strongly basic crystalline organic compound which can be nitrated to make a powerful explosive. It is also used in making dyestuffs, medicines and polymer resins.

guanine /gwah-neen, -nin/ A nitrogenous base found in DNA and RNA. Guanine has a purine ring structure.

guanosine /**gwah**-nŏ-seen, -sin/ (**guanine nucleoside**) A NUCLEOSIDE present in DNA and RNA and consisting of guanine linked to D-ribose via a β-glycosidic bond.

gun cotton *See* cellulose trinitrate.

gunmetal *See* bronze.

gunpowder A powdered mixture of sulfur, charcoal, and potassium nitrate, used as an explosive.

gypsum /**jip**-sŭm/ *See* calcium sulfate.

gyromagnetic ratio /jÿ-roh-mag-**net**-ik/ The ratio of the magnetic moment of a system to the angular momentum of that system. The gyromagnetic ratio is a useful quantity in the theory of electrons and atomic nuclei.

H

Haber process /**hay**-ber/ An important industrial process for the manufacture of ammonia, which is used for fertilizers and for making nitric acid. The reaction is the equilibrium:

$$N_2 + 3H_2 \rightleftharpoons 3NH_3$$

The nitrogen used is obtained by fractionation of liquid air and the hydrogen by the oxidation of hydrocarbons (from natural gas). The nitrogen and hydrogen are purified and mixed in the correct proportions. The equilibrium amount of ammonia is favoured by low temperatures, but in practice the reaction would never reach equilibrium at normal temperatures. An optimum temperature of about 450°C is therefore used. High pressure also favors the reaction and a pressure of about 250 atmospheres is used. The catalyst is iron with small amounts of potassium and aluminum oxides present. The yield is about 15%. Ammonia is condensed out of the gas mixture at –50°C. The process is named for the German physical chemist Fritz Haber (1868–1934).

habit *See* crystal habit.

hafnium /**haf**-nee-ŭm/ A transition metal found in zirconium ores. Hafnium is difficult to work and can burn in air. It is used in control rods for nuclear reactors and in certain specialized alloys and ceramics.

Symbol: Hf; m.p. 2230°C; b.p. 5197°C; r.d. 13.31 (20°C); p.n. 72; r.a.m. 178.49.

hahnium /**hah**-nee-ŭm/ Symbol: Ha or Hn A name formerly suggested for element-105 (dubnium) and also for element-108 (hassium). *See* element.

half cell An electrode in contact with a solution of ions. In general there will be an e.m.f. set up between electrode and solution by transfer of electrons to or from the electrode. The e.m.f. of a half cell cannot be measured directly since setting up a circuit results in the formation of another half cell. *See* electrode potential.

half-life (**half-life period**; Symbol: $t_{1/2}$) The time taken for half the nuclei of a sample of a radioactive nuclide to decay. The half-life of a nuclide is a measure of its stability. (Stable nuclei can be thought of as having infinitely long half-lives.) If N_0 is the original number of nuclei, the number remaining at the end of one half-life is $N_0/2$, at the end of two half-lives is $N_0/4$, etc.

half-sandwich compound *See* sandwich compound.

halide /**hal**-ÿd, **hay**-lÿd/ A compound containing a halogen. The inorganic halides of electropositive elements contain ions of the type Na^+Cl^- (sodium chloride) or K^+Br^- (potassium bromide). Transition metal halides often have some covalent bonding. Non-metal halides are covalent compounds, which are usually volatile. Examples are tetrachloromethane (CCl_4) and silicon tetrachloride ($SiCl_4$). Halides are named as bromides, chlorides, fluorides, or iodides.

halite /**hal**-ÿt, **hay**-lÿt/ (**rock salt**) A naturally occurring mineral form of sodium chloride, NaCl. It forms colorless or white crystals when pure, but is often colored by impurities.

haloalkane /hal-oh-**al**-kayn/ (**alkyl halide**) A type of organic compound in which one or more hydrogen atoms of an alkane have been replaced by halogen atoms. Haloalkanes can be made by direct reaction of the alkane with a halogen. Other methods are:

1. Reaction of an alcohol with the halogen acid (e.g. from NaBr + H_2SO_4) or with phosphorus halides (red phosphorus and iodine can be used):
   $$ROH + HBr \rightarrow RBr + H_2O$$
   $$ROH + PCl_5 \rightarrow RCl + POCl_3 + HCl$$
2. Addition of an acid to an alkene:
   $$RCH{:}CH_2 + HBr \rightarrow RCH_2CH_2Br$$

The haloalkanes are much more reactive than the alkanes, and are useful starting compounds for preparing a wide range of organic chemicals. In particular, they undergo nucleophilic substitutions in which the halogen atom is replaced by some other group (iodine compounds are the most reactive). Some reactions of haloalkanes are:

1. Refluxing with aqueous potassium hydroxide to give an alcohol:
   $$RI + OH^- \rightarrow ROH + I^-$$
2. Refluxing with potassium cyanide in alcoholic solution to give a nitrile:
   $$RI + CN^- \rightarrow RCN + I^-$$
3. Refluxing with an alkoxide to give an ether:
   $$RI + {}^-OR' \rightarrow ROR' + I^-$$
4. Reaction with alcoholic ammonia solution (100°C in a sealed tube) to give an amine:
   $$RI + NH_3 \rightarrow RNH_2 + HI$$
5. Boiling with alcoholic potassium hydroxide, to eliminate an acid and produce an alkene:
   $$RCH_2CH_2I + KOH \rightarrow KI + H_2O + RCH{:}CH_2$$

See also Fittig reaction; Grignard reagent; Wurtz reaction.

halobutyl /hal -oh-**byoo**-t'l/ *See* butyl rubber.

halocarbons /hal-oh-**kar**-bŏnz/ Chemical compounds that contain carbon atoms bound to halogen atoms and (sometimes) hydrogen atoms. The halocarbons include haloalkanes such as tetrachloromethane (CCl_4) and the haloforms ($CHCl_3$, $CHBr_3$, etc.). There are various types of halocarbon that are useful but are also significant pollutants. For example, the *chlorofluorocarbons* (CFCs) contain carbon, fluorine, and chlorine. They are useful as refrigerants, aerosol propellants, and in making rigid plastic foams. However, they are also thought to damage the ozone layer and an international agreement exists to phase out their use. Similar compounds are the *hydrochlorofluorocarbons* (HCFCs), which contain hydrogen as well as chlorine and fluorine, and the *hydrofluorocarbons* (HFCs), which contain hydrogen and fluorine.

The *halons* are a class of halocarbons that contain bromine as well as hydrogen and other halogens. Their main use is in fire extinguishers. They are, however, significantly more active than CFCs in their effect on the ozone layer. The halocarbons are also thought to contribute to global warming.

haloform /**hal**-ŏ-form/ Any of the four compounds CHX_3, where X is a halogen atom (F, fluoroform; Cl, chloroform; Br, bromoform; I, iodoform). The systematic names are fluoromethane, chloromethane, bromomethane, and iodomethane. *See also* haloform reaction.

haloform reaction A reaction of a methyl ketone with NaOX, where X is Cl, Br, or I, to give a haloform. With sodium chlorate(I), for example:

$$RCOCH_3 + 3NaOCl \rightarrow RCOCCl_3 + 3NaOH$$

$$RCOCl_3 + NaOH \rightarrow NaOCOR + CHCl_3$$

The reaction can be used to make carboxylic acids (from the NaOCOR), and is especially useful when R is an aromatic group because the starting ketone, $RCOCH_3$, can be produced by Friedel–Crafts acetylation. *See also* triiodomethane.

halogenating agent /**hal**-lŏ-jĕ-nayt-ing/ A compound used to introduce halogen atoms into a molecule. Examples are phosphorus trichloride (PCl_3) and aluminum trichloride ($AlCl_3$).

halogenation /hal-ŏ-jĕ-**nay**-shŏn/ A reaction in which a halogen atom is introduced into a molecule. Halogenations are specified as *chlorinations*, *brominations*, *fluorinations*, etc., according to the element involved. There are several methods.

1. Direct reaction with the element using high temperature or ultraviolet radiation:
$$CH_4 + Cl_2 \rightarrow CH_3Cl + HCl$$
2. Addition to a double bond:
$$H_2C{:}CH_2 + HCl \rightarrow C_2H_5Cl$$
3. Reaction of a hydroxyl group with a halogenating agent, such as PCl_3:
$$C_2H_5OH \rightarrow C_2H_5Cl + OH^-$$
4. In aromatic compounds direct substitution can occur using aluminum chloride as a catalyst:
$$2C_6H_6 + Cl_2 \rightarrow 2C_6H_5Cl$$
5. Alternatively in aromatic compounds, the chlorine can be introduced by reacting the diazonium ion with copper(I) chloride:
$$C_6H_5N_2^+ + Cl^- \rightarrow C_6H_5Cl + N_2$$

halogens /hal-ŏ-jĕnz/ A group of elements (group 17 of the periodic table) consisting of fluorine, chlorine, bromine, iodine, and the short-lived element astatine. The halogens all have outer valence shells that are one electron short of a rare-gas configuration. Because of this, the halogens are characterized by high electron affinities and high electronegativities, fluorine being the most electronegative element known. The high electronegativities of the halogens favor the formation of both uninegative ions, X^-, particularly combined with electropositive elements (e.g. NaCl, KBr, CsI, CaF_2), and the single covalent bond –X with elements of moderate to high electronegativity (e.g. HCl, SiF_4, SF_4, $BrCH_3$, Cl_2O). The halogens all form diatomic molecules, X_2, which are characterized by their high reactivities with a wide range of other elements and compounds. As a group, the elements increase in both size and polarizability as the proton number increases and there is an attendant decrease in electronegativity and reactivity. This means that there is no chemistry of positive species apart from a few cationic iodine compounds. The heavier halogens Cl, Br, and I, all form oxo-species with formal positive oxidation numbers +1 and +5, chlorine and iodine also forming +3 and +7 species (e.g. HOCl, $HBrO_3$, I_2O_5, HIO_4).

The elements decrease in oxidizing power in the order $F_2>Cl_2>Br_2>I_2$ and the ions X^- may be arranged in order of increasing reducing power $F^-<Cl^-<Br^-<I^-$. Thus any halogen will displace the elements below it from their salts in solution, for example
$$Cl_2 + 2Br^- \rightarrow Br_2 + 2Cl^-$$
A wide range of organic halides is formed in which the C–F bond is characteristically resistant to chemical attack; the C–Cl bond is also fairly stable, particularly in aryl compounds but the alkyl halogen compounds become increasingly susceptible to nucleophilic attack and generally more reactive.

halon /hal-on/ *See* halocarbon.

halothane /hal-ŏ-thayn/ ($CHBrClCF_3$) A colorless nonflammable liquid halocarbon used as a general anesthetic. The systematic name is 1-chloro-1-bromo-2,2,2-trifluoroethane.

hammer mill A device used in the chemical industry for crushing and grinding solid materials at high speeds to a specified size. The impact between the particles, grinding plates, and grinding hammers pulverizes the particles. Hammer mills can be used for a greater variety of soft material than other types of grinding equipment. *Compare* ball mill.

hardening (of oils) The conversion of liquid plant oils into a more solid form by hydrogenation using a nickel catalyst. Hardening of liquid oils is used in producing margerine. In vegetable oils the fatty acids present (as glycerides) contain double bonds (i.e. they are unsaturated). The hydrogenation process increases the amount of unsaturated material, increasing the melting point. The process still leaves some unsaturated fatty acids and, for this reason, it is claimed that margarines are healthier than animal fats (e.g. butter), which contain a much higher proportion of saturated fats. This is because the unsaturated fats are less likely to lead to cholesterol build-up in the body, and consequent risk of coronary heart disease. The hydrogenation process may, however, also affect the nature of the double bonds. Natural unsaturated fatty acids mostly have a *cis* configuration about the double bonds. In the hydrogenation process, a proportion of these are converted into fatty acids with a *trans* configuration. Glycerides of these are known as *trans-fats*. It has been claimed that there is also a link between trans-fats and coronary heart disease. *See also* Sabatier–Senderens process.

hardness (of water) Hard water is water that will not readily form a lather with SOAP owing to the presence of dissolved calcium, iron, and magnesium compounds. Such compounds can react with soap to produce insoluble salts, which collect as solid scum. The effectiveness of the cleansing solution is thus reduced. Hardness of water is of two types, TEMPORARY HARDNESS and PERMANENT HARDNESS. Only temporary hardness can be removed by boiling the water.

Hardness of water is usually expressed by assuming that all the hardness is due to dissolved calcium carbonate, which is present as ions. It can be estimated by titration with a standard soap solution or with edta. *See also* water softening.

hard vacuum *See* vacuum.

hard water *See* hardness.

hassium /**hass**-ee-ŭm/ A transactinide element that is formed artificially.

Symbol: Hs; p.n. 108; most stable isotope ^{265}Hs (half-life 2×10^{-3}s).

HCFC Hydrochlorofluorocarbon. *See* halocarbon.

heat Energy transferred as a result of a temperature difference. The term is often loosely used to mean internal energy (i.e. the total kinetic and potential energy of the particles). It is common in chemistry to define such quantities as *heat of combustion*, *heat of neutralization*, etc. These are in fact molar enthalpies for the change, given the symbol $\Delta H_M^{\ominus}$. The superscript symbol denotes standard conditions, while the subscript M indicates that the enthalpy change is for one mole. The unit is usually the kilojoule per mole (kJ mol^{-1}). By convention, ΔH is negative for an exothermic reaction. Molar enthalpy changes stated for chemical reactions are changes for standard conditions, which are defined as 298 K (25°C) and 101 325 Pa (1 atmosphere). Thus, the standard molar enthalpy of reaction is the enthalpy change for reaction of substances under these conditions producing reactants under the same conditions. The substances involved must be in their normal equilibrium physical states under these conditions (e.g. carbon as graphite, water as the liquid, etc.). Note that the measured enthalpy change will not usually be the standard change. In addition, it is common to specify the entity involved. For instance $\Delta H_f^{\ominus}(H_2O)$ is the standard molar enthalpy of formation for one mole of H_2O species.

heat engine (**thermodynamic engine**) A device for converting heat energy into work. Heat engines operate by transferring energy from a high-temperature source to a low-temperature sink. The theoretical operation of heat engines is useful in the theory of thermodynamics. *See* Carnot cycle.

heat exchangers Devices that enable the heat from a hot fluid to be transferred to a cool fluid without allowing them to come into contact. The normal arrangement is for one of the fluids to flow in a coiled tube through a jacket containing the second fluid. Both the cooling and heating effect may be of benefit in conserving the energy used in a chemical plant and in controlling the process.

heat of atomization The energy required in dissociating one mole of a substance into atoms. *See* heat.

heat of combustion The energy liberated when one mole of a substance burns in excess oxygen. *See* heat.

heat of crystallization The energy liberated when one mole of a substance crystallizes from a saturated solution of this substance.

heat of dissociation The energy required to dissociate one mole of a substance into its constituent elements.

heat of formation The energy change when one mole of a substance is formed from its elements. *See* heat.

heat of neutralization The energy liberated when one mole of an acid or base is neutralized.

heat of reaction The energy change when molar amounts of given substances react completely. *See* heat.

heat of solution The energy change when one mole of a substance is dissolved in a given solvent to infinite dilution (in practice, to form a dilute solution).

heavy hydrogen *See* deuterium.

heavy-metal pollution *See* pollution.

heavy water Deuterium oxide, D_2O.

hecto- Symbol: h A prefix denoting 10^2. For example, 1 hectometer (hm) = 10^2 meters (m).

Heisenberg's uncertainty principle /**hȳ**-zĕn-bergz/ The impossibility of making simultaneous measurements of both the position and the momentum of a subatomic particle (e.g. an electron) with unlimited accuracy. The uncertainty arises because, in order to detect the particle, radiation has to be 'bounced' off it, and this process itself disrupts the particle's position. Heisenberg's uncertainty principle is not a consequence of 'experimental error'. It represents a fundamental limit to objective scientific observation, and arises from the wave–particle duality of particles and radiation. In one direction, the uncertainty in position Δx and momentum Δp are related by $\Delta x \Delta p \sim h/4\pi$, where h is the Planck constant. The uncertainty principle is named for the German physicist Werner Karl Heisenberg (1901–76).

helicate /**hel**-ă-kayt/ *See* supramolecular chemistry.

helium /**hee**-lee-ŭm/ A colorless monatomic gas; the first member of the rare gases (group 18 of the periodic table). Helium has the electronic configuration $1s^2$ and consists of a nucleus of two protons and two neutrons (equivalent to an α-particle) with two extra-nuclear electrons. It has an extremely high ionization potential and is completely resistant to chemical attack of any sort. The gas accounts for only 5.2×10^{-4}% of the atmosphere; up to 7% occurs in some natural gas deposits. Helium is the second most abundant element in the universe, the primary process on the Sun being nuclear fusion of hydrogen to give helium.

Helium is recovered commercially from natural gas in both the USA and countries of the former USSR and it also forms part of ammonia plant tail gas if natural gas is used as a feedstock. Its applications are in fields in which inertness is required and where the cheaper alternatives, such as nitrogen, are too reactive; for example, high-temperature metallurgy, powder technology, and as a coolant in nuclear reactors. Helium is also favoured over nitrogen for diluting oxygen for deep-sea diving (lower solubility in blood) and as a pressurizer for liquefied gas fuels in rockets (total inertness). It is also used as an ideal gas for balloons (no fire risk) and for low-temperature physics research.

Helium is unusual in that it is the only known substance for which there is no triple point (i.e., no combination of pressure and temperature at which all three phases can co-exist). This is because the interatomic forces, which normally participate in the formation of solids, are so weak that they are of the same order as the zero-point energy. At 2.2 K helium undergoes a transition from liquid helium I to liquid helium II, the latter being a true liquid but exhibiting superconductivity and an immeasurably low viscosity (*superfluidity*).

The low viscosity allows the liquid to spread in layers a few atoms thick, described by some as 'flowing uphill'.

Helium also has an isotope. ^{3}He is formed in nuclear reactions and by decay of tritium. This also undergoes a phase change at temperatures close to absolute zero.

Symbol: He; m.p. 0.95 K (pressure); b.p. 4.216 K; d. 0.1785 kg m^{-3} (0°C); p.n. 2; r.a.m. 4.002602.

Hell–Volhard–Zelinsky reaction /hell voh-**lard** zem-**lin**-skee/ A method for the preparation of halogenated carboxylic acids using free halogen in the presence of a phosphorus halide. The halogenation occurs at the carbon atom adjacent to the –COOH group. With Br_2 and PBr_3:

$$RCH_2COOH \rightarrow RCHBrCOOH \rightarrow RCBr_2COOH$$

The reaction is named for the German chemists Carl Magnus von Hell (1849–1926) and Jacob Volhard (1834–1910) and for the Russian chemist Nicolai (Nikolay Dmitrievich) Zelinsky (1861–1953).

Helmholtz function (**Helmholtz free energy**) Symbol: F A thermodynamic function defined by

$$F = U - TS$$

where U is the internal energy, T the thermodynamic temperature, and S the entropy. It is a measure of the ability of a system to do useful work in an isothermal process. The function is named for the German physiologist and theoretical physicist Herman Ludwig Ferdinand von Helmholtz (1821–94). *See also* free energy.

hematite /**hem**-ă-tÿt, **hee**-mă-/ A mineral form of iron(III) oxide. It is the principal ore of iron.

heme /heem/ (**haeme**) An iron-containing porphyrin that is the prosthetic group in hemoglobin, myoglobin, and some cytochromes.

hemiacetal /hem-ee-**ass**-ĕ-tal/ *See* acetal.

hemihydrate /hem-ee-**hÿ**-drayt/ A crystalline compound with one molecule of water of crystallization per two molecules of compound (e.g. $2CaSO_4.H_2O$).

hemiketal /hem-ee-**kee**-t'l/ *See* ketal.

hemimorphite /hem-ee-**mor**-fÿt/ *See* calamine.

hemoglobin /hee-mŏ-**gloh**-bin, hem-ŏ-, **hee**-mŏ-gloh-bin/ The pigment of the red blood cells that is responsible for the transport of oxygen from the lungs to the tissues. It consists of a basic protein, globin, linked with four heme groups. Heme is a complex compound containing an iron atom.

The most important property of hemoglobin is its ability to combine reversibly with one molecule of oxygen per iron atom to form *oxyhemoglobin*, which has a bright red color. The iron is present in the divalent state (iron(II)) and this remains unchanged with the binding of oxygen. There are variations in the polypeptide chains, giving rise to different types of hemoglobins in different species. The binding of oxygen depends on the oxygen partial pressure; high pressure favors formation of oxyhemoglobin and low pressure favors release of oxygen.

henry /**hen**-ree/ Symbol: H The SI unit of inductance, equal to the inductance of a closed circuit that has a magnetic flux of one weber per ampere of current in the circuit. 1 H = 1 Wb A^{-1}. The unit is named for the American physicist Joseph Henry (1797–1878).

Henry's law The concentration (C) of a gas in solution is proportional to the partial pressure (p) of that gas in equilibrium with the solution, i.e. $p = kC$, where k is a proportionality constant.

The relationship is similar in form to that for Raoult's law, which deals with ideal solutions.

A consequence of Henry's law is that the 'volume solubility' of a gas is independent of pressure.

heparin /**hep**-ă-rin/ A POLYSACCHARIDE that inhibits the formation of thrombin from prothrombin and thereby prevents the clotting of blood. It is used in medicine as an anticoagulant.

heptahydrate /hep-ta-**hÿ**-drayt/ A crystalline hydrated compound containing one molecule of compound per seven molecules of water of crystallization.

heptane /**hep**-tayn/ (C_7H_{16}) A colorless liquid alkane obtained from petroleum refining. It is used as a solvent.

heptavalent /hep-tă-**vay**-lĕnt, hep-**tav**-ă-/ (**septavalent**) Having a valence of seven.

hertz /herts/ Symbol: Hz The SI unit of frequency, defined as one cycle per second (s^{-1}). Note that the hertz is used for regularly repeated processes, such as vibration or wave motion. The unit is named for the German physicist Heinrich Rudolph Hertz (1857–94).

Hess's law A derivative of the first law of thermodynamics. It states that the total heat change for a given chemical reaction involving alternative series of steps is independent of the route taken and is sometimes known as the *law of constant heat summation*. The law is named for the Swiss–Russian chemist Germain Henri Hess (1802–50).

hetero atom /**het**-ĕ-roh/ *See* heterocyclic compound.

heterocyclic compound /het-ĕ-rŏ-**sÿ**-klik, -**sik**-lik/ A compound that has a ring containing more than one type of atom. Commonly, heterocyclic compounds are organic compounds with at least one atom in the ring that is not a carbon atom. Pyridine and glucose are examples. The noncarbon atom is called a *hetero atom*. *Compare* homocyclic compound.

heterogeneous /het-ĕ-rŏ-**jee**-nee-ŭs/ Relating to more than one phase. A heterogeneous mixture, for instance, contains two or more distinct phases.

heterolytic fission /het-ĕ-rŏ-**lit**-ik/ (**heterolysis**) The breaking of a covalent bond so that both electrons of the bond remain with one fragment. A positive ion and a negative ion are produced:

$$RX \rightarrow R^+ + X^-$$

Compare homolytic fission.

heteropolymer *See* polymerization.

hexacyanoferrate /heks-ă-sÿ-an-oh-**fe**-rayt/ *See* cyanoferrate.

hexadecanoate /heks-ă-dek-ă-**noh**-ayt/ (**palmitate**) A salt or ester of hexadecanoic acid.

hexadecanoic acid /heks-ă-dek-ă-**noh**-ik/ (**palmitic acid**) A crystalline carboxylic acid:

$$CH_3(CH_2)_{14}COOH$$

It is present as GLYCERIDES in fats and oils.

hexagonal close-packed crystal /heks-**ag**-ŏ-năl/ A crystal structure in which layers of close-packed atoms are stacked in an ABABAB arrangement. The second layer B fits into the holes of the first layer A, with the third layer over the first. The coordination number is 12. Zinc and magnesium have hexagonal close-packed structures.

The unit cell is based on a hexagon with atoms occupying corner and mid positions. *See* face-centered cubic crystal.

hexagonal crystal *See* crystal system.

hexamethylene diamine /heks-ă-**meth**-ă-leen **dÿ**-ă-meen/ *See* 1,6-diaminohexane.

hexamethylenetetramine /heks-ă-meth-ă-leen-**tet**-ră-meen, -min/ *See* hexamine.

hexamine /**heks**-ă-meen, -min/ (**hexamethylenetetramine**; $C_6H_{12}N_4$) A white crystalline organic compound made by condensing methanal with ammonia. It is used as a fuel for camping stoves, in vulcanizing rubber, and as a urinary disinfectant. Hexamine can be nitrated to make the high explosive cyclonite.

hexane /**heks**-ayn/ (C_6H_{14}) A liquid alkane obtained from the light fraction of crude oil. The principal use of hexane is in gasoline and as a solvent.

hexanedioic acid /heks-ayn-dȳ-**oh**-ik/ (**adipic acid**; $HOOC(CH_2)_4COOH$) A colorless crystalline organic dicarboxylic acid that occurs in rosin. It is used in the manufacture of NYLON.

hexanoate /heks-ă-**noh**-ayt/ A salt or ester of hexanoic acid.

hexanoic acid /heks-ă-**noh**-ik/ (caproic acid, $CH_3(CH_2)_4COOH$) An oily carboxylic acid found (as glycerides) in cow's milk and some vegetable oils.

hexose /**heks**-ohs/ A SUGAR that has six carbon atoms in its molecules.

hexyl group /**heks**-ăl/ The group $C_5H_{11}CH_2$–, having a straight chain of carbon atoms.

HFC Hydrofluorocarbon. *See* halocarbon.

histidine /**his**-tă-deen, -din/ *See* amino acids.

Hofmann degradation /**hoff**-măn/ A method of preparing primary amines from acid amides. The amide is refluxed with aqueous sodium hydroxide and bromine:

$$RCONH_2 + NaOH + Br_2 \rightarrow$$
$$RCONHBr + NaBr + H_2O$$
$$RCONHBr + OH^- \rightarrow RCON^-Br + H_2O$$
$$RCON^-Br \rightarrow R\text{–}N = C = O + Br^-$$
$$RNCO + 2OH^- \rightarrow RNH_2 + CO_2^{2-}$$

The reaction is a 'degradation' in that a carbon atom is removed from the amide chain. The degradation is named for the German organic chemist August Wilhelm von Hofman (1818–92).

Hofmann's method A rather outmoded method for determining the vapor density of volatile liquids. A known weight of sample is introduced into a mercury barometer tube, which is surrounded by a heating jacket. The volume of vapor can thus be read off directly, the temperature is known, and the pressure is obtained by taking the atmospheric pressure minus the mercury height in the barometer (with corrections for the density of mercury at higher temperatures). The method's only advantage is that it may be used for samples that decompose at their normal boiling point. *See also* Dumas' method; Victor Meyer's method.

holmium /**hohl**-mee-ŭm/ A soft malleable silvery element of the lanthanoid series of metals. It occurs in association with other lanthanoids. It has few applications.

Symbol: Ho; m.p. 1474°C; b.p. 2695°C; r.d. 8.795 (25°C); p.n. 67; r.a.m. 164.93032.

HOMO /**hoh**-moh/ *See* frontier orbital.

homocyclic compound /hoh-mŏ-**sȳ**-klik, hom-ŏ-/ A compound containing a ring made up of the same atoms. Benzene is an example of a homocyclic compound. *Compare* heterocyclic compound.

homogeneous /hoh-mŏ-**jee**-nee-ŭs, hom-ŏ-/ Relating to a single phase. A homogeneous mixture, for instance, consists of only one phase.

homologous series A group of organic compounds possessing the same functional group and having a regular structural pattern so that each member of the series differs from the next one by a fixed number of atoms. The members of a homologous series can be represented by a general formula. For example, the homologous series of alkane alcohols CH_3OH, C_2H_5OH, C_3H_7OH, ..., has a general formula $C_nH_{2n+1}OH$. Each member differs by CH_2 from the next. Any two successive members of a series are called *homologs*.

homologs /**hom**-ŏ-lôgz/ *See* homologous series.

homolytic fission /hoh-mŏ-**lit**-ik, hom-ŏ-/ (**homolysis**) The breaking of a covalent bond so that one electron from the bond is left on each fragment. Two free radicals result:

$$RR' \rightarrow R\bullet + R'\bullet$$

Compare heterolytic fission.

homopolymer /hoh-mŏ-**pol**-i-mer, hom-ŏ-/ *See* polymerization.

hornblende /**horn**-blend/ A dark-colored rock-forming mineral consisting of silicates of calcium, iron, and magnesium. It is a major component of granite and other igneous and metamorphic rocks.

host–guest chemistry A branch of SUPRAMOLECULAR CHEMISTRY in which a molecular structure acts as a 'host' to hold an ion or molecule (the 'guest'). The guest may be coordinated to the host or may be trapped by its structure. For example, *calixarenes* are compounds with cup-shaped molecules that may accept guest molecules. *See also* crown ethers.

HPLC High-performance liquid chromatography; a sensitive analytical technique, that is similar to gas–liquid chromatography but using a liquid carrier. The carrier is specifically choosen for the particular substance to be detected.

Hückel rule /**hyoo**-kĕl/ *See* aromatic compound.

humectant /hyoo-**mek**-tănt/ A hygroscopic substance used to maintain moisture levels. Glycerol, mannitol, and sorbitol are commonly used in foodstuffs, tobacco, etc.

Humphreys series /**hum**-freez/ *See* hydrogen atom spectrum.

Hund's rule /hŭnts/ A rule that states that the electronic configuration in degenerate orbitals will have the minimum number of paired electrons. The rule is named for the German physicist Friedrich Hund (1896–1997).

hybrid orbital *See* orbital.

hydrate /**hȳ**-drayt/ A compound coordinated with water molecules. When water is bound up in a compound it is known as the *water of crystallization.*

hydration /hȳ-**dray**-shŏn/ The solvation of such species as ions in water.

hydraulic cement *See* cement.

hydrazine /**hȳ**-dră-zeen/ (N_2H_4) A colorless liquid that can be prepared by the oxidation of ammonia with sodium chlorate(I) or by the gas phase reaction of ammonia with chlorine. Hydrazine is a weak base, forming salts (e.g. $N_2H_4.HCl$) with strong acids. It is a powerful reducing agent, reducing salts of the noble metals to the metal. Anhydrous hydrazine ignites spontaneously in oxygen and reacts violently with oxidizing agents. The aqueous solution, *hydrazine hydrate*, has been used as a fuel for jet engines and for rockets. *See also* hydrazone.

hydrazoic acid /hȳ-dră-**zoh**-ik/ (**hydrogen azide; azoimide**; HN_3) A colorless liquid with a nauseating smell. It is highly poisonous and explodes in the presence of oxygen and oxidizing agents. It can be made by distilling a mixture of sodium azide (NaN_3) and a dilute acid. It is usually used as an aqueous solution. The salts of hydrazoic acid (*azides*), especially lead azide ($Pb(N_3)_2$), are used in detonators because of their ability to explode when given a mechanical shock.

hydrazone /**hȳ**-dră-zohn/ A type of organic compound containing the $C{:}NNH_2$ group, formed by the reaction between an aldehyde or ketone and hydrazine (N_2H_4). Derivatives of hydrazine are often used to produce crystalline products, which have sharp melting points that can be used to characterize the original aldehyde or ketone. Phenylhydrazine ($C_6H_5NH.NH_2$), for instance, produces *phenylhydrazones*.

hydride /**hȳ**-drȳd/ A compound of hydrogen. Ionic hydrides are formed with highly electropositive elements and contain the H^- ion (hydride ion). Non-metals form covalent hydrides, as in methane (CH_4) or silane (SiH_4). The boron hydrides are electron-deficient covalent compounds. Many transition metals absorb hydrogen to form interstitial hydrides.

hydrobromic acid /hÿ-droh-**broh**-mik/ (HBr) A colorless liquid produced by adding hydrogen bromide to water. It shows the typical properties of a strong acid and it is a strong reducing agent. A convenient way of producing hydrobromic acid is to bubble hydrogen sulfide through bromine water. Although it is not as strong as hydrochloric acid it dissociates extensively in water and is a good proton donor.

hydrocarbon /hÿ-droh-**kar**-bŏn/ Any compound containing only the elements carbon and hydrogen. Examples are the alkanes, alkenes, alkynes, and aromatics such as benzene and naphthalene.

hydrochloric acid /hÿ-droh-**klor**-ik, -**kloh**-rik/ (HCl) A colorless fuming liquid made by adding hydrogen chloride to water:

$$HCl(g) + H_2O1. \rightarrow H_3O^+(aq) + Cl^-(aq)$$

Dissociation into ions is extensive and hydrochloric acid shows the typical properties of a strong acid. It reacts with carbonates to give carbon dioxide and yields hydrogen when reacted with all but the most unreactive metals. Hydrochloric acid is used in the manufacture of dyes, drugs, and photographic materials. It is also used to pickle metals, i.e. clean the surface prior to electroplating. Hydrochloric acid donates protons with ease and is the strongest of the hydrohalic acids. The concentrated acid is oxidized to chlorine by such agents as potassium manganate(VII) and manganese(IV) oxide.

hydrochlorofluorocarbon /hÿ-droh-klor-ŏ-floo-ŏ-rŏ-**kar**-bŏn, -kloh-roh-/ *See* halocarbon.

hydrocyanic acid /hÿ-droh-sÿ-**an**-ik/ (**prussic acid**; HCN) A highly poisonous weak acid formed when hydrogen cyanide gas dissolves in water. Its salts are cyanides. Hydrogen cyanide is used in making acrylic plastics.

hydrofluoric acid /hÿ-droh-floo-**or**-ik, -**oh**-rik/ (HF) A colorless liquid produced by dissolving hydrogen fluoride in water. It is a weak acid, but will dissolve most silicates and hence can be used to etch glass. As the interatomic distance in HF is relatively small, the H–F bond energy is very high and hydrogen fluoride is not a good proton donor. It does, however, form hydrogen bonds. *See* hydrogen fluoride. – unable to donate protons easily – hydrofluoric acid is a weak acid. *See* bond energy.

hydrofluorocarbon /hÿ-droh-floo-ŏ-rŏ-**kar**-bŏn/ *See* halocarbon.

hydrogen /**hÿ**-drŏ-jĕn/ A colorless gaseous element. Hydrogen has some similarities to both the alkali metals (group 1) and the halogens (group 17), but is not normally classified in any particular group of the periodic table. It is the most abundant element in the Universe and the ninth most abundant element in the Earth's crust and atmosphere (by mass). It occurs principally in the form of water and petroleum products; traces of molecular hydrogen are found in some natural gases and in the upper atmosphere.

The gas may be prepared in the laboratory by the reaction of dilute hydrochloric acid with a metal that lies above hydrogen in the electromotive series, magnesium and zinc being commonly used:

$$Zn + 2HCl \rightarrow H_2 + ZnCl_2$$

Reactions of the amphoteric metals zinc and aluminum with dilute aqueous alkali, to form zincates or aluminates, are also a convenient source of hydrogen. Electrolysis of dilute mineral acids may be used to obtain hydrogen but care must be taken to avoid mixing with the oxygen released at the anode as this leads to explosive mixtures. Industrially, hydrogen is obtained as a by-product of the electrolytic cells used in the production of sodium hydroxide (reaction of the Na/Hg amalgam with water), or by the water-gas route in which steam is decomposed by hot coke.

The main use of hydrogen is as a chemical feedstock for the manufacture of ammonia and of a range of organic compounds. Small-scale uses include reducing atmospheres for metallurgy, hardening edible oils, and pharmaceutical manufacture. Hydrogen also has a potential future use as a fuel.

Hydrogen occupies a unique position among the elements as hydrogen atoms are the simplest of all atoms. The hydrogen atom consists of a proton (positive charge) with one extranuclear electron ($1s^1$). The chemistry of hydrogen depends on one of three processes:

1. Loss of the electron to form H^+.
2. Gain of an electron to form H^-.
3. Sharing of electrons by covalent bond formation as in H_2 or HCl.

The hydrogen atom with the 1s electron removed would have an extremely small ionic radius and the positive hydrogen ion occurs only in association with other species as in H^+NH_3 or H^+FH. The ion commonly written H^+ in solution is in fact the solvated proton (hydroxonium or hydronium) H_3O^+, which is formed by ionization of acids:

$$H_2O + HCl \rightarrow H_3O^+ + Cl^-$$

It is believed that the lifetime of any one H_3O^+ ion is extremely short as the protons appear to undergo very rapid exchange between water molecules.

Hydrogen also forms a number of compounds in which it is regarded as gaining an electron and becoming H^- (the hydride ion). It can form ionic hydrides only with the most electropositive elements (groups 1 and 2). The ionic nature of these compounds is indicated by the fact that the melts are good conductors and that electrolysis liberates hydrogen at the anode. These hydrides are prepared by heating the element in a stream of hydrogen:

$$H_2 + 2M \rightarrow 2MH$$

Examples are CH_4, NH_3, and HCl. These compounds are generally low-boiling. General methods of preparation are:

1. The hydrolysis of the appropriate '-ide' compound; e.g. silicides give silane, nitrides give ammonia, sulfides give H_2S.
2. Reduction of a chloride; for example:

$$SiCl_4 + LiAlH_4 \rightarrow SiH_4 + LiCl + AlCl_3$$

A third class of hydride is that of metallic hydrides formed between hydrogen and many transition metals. Palladium in particular is renowned for its ability to absorb hydrogen as PdH_x, where x takes values up to 1.8. Titanium and zirconium behave similarly but the exact nature of the compounds formed and of the bond type remains uncertain. There are changes in the magnetic properties of the metal (indicating some electron interaction) and in the lattice dimensions but discrete phases of the type MH or MH_2 are not isolated. The class is sometimes referred to as the *interstitial hydrides*.

Atomic nuclei possess the property of 'spin' and for diatomic molecules there exists the possibility of having the spins of adjacent nuclei aligned (*ortho*) or opposed (*para*). Because of the small mass of hydrogen, these forms are more important in hydrogen molecules than in other diatomic molecules. The two forms are in equilibrium with parahydrogen dominant at low temperatures, rising to 75% orthohydrogen at room temperatures. Although chemically identical the melting point and boiling point of the para form are both about 0.1° lower than the 3:1 equilibrium mixture.

Natural hydrogen in molecular or combined forms contains about one part in 2000 of deuterium, D, an isotope of hydrogen that contains one proton and one neutron in its nucleus. Although the effect of isotopes on chemical properties is normally small, in the case of hydrogen the difference in mass number leads to a lowering of some reaction rates known as the 'deuterium isotope effect'. Hydrogen also exhibits two less common forms of bonding. Boron hydrides form a wide variety of compounds in which the hydrogen acts as a bridging species involving 'three-centre two-electron' bonds. Such species are said to be 'electron deficient' as they do not have sufficient electrons for conventional two-electron covalent bonds. The second, less common, form is that of the coordinated hydrides in which the H^- ion acts as a ligand bound to a transition metal atom.

Symbol: H; m.p. 14.01 K; b.p. 20.28 K; d. 0.089 88 kg m^{-3} (0°C); p.n. 1; r.a.m. 1.0079.

hydrogenation /hÿ-droj-ĕ-**nay**-shŏn/ The reaction of a compound with hydrogen. An example is the hydrogenation of nitrogen to form ammonia in the Haber process. In organic chemistry, hydrogenation refers to the addition of hydrogen to multiple bonds, usually with the aid of a catalyst. Unsaturated natural liquid vegetable oils can be hydrogenated to form saturated semisolid fats – a reaction used in making types of margarine. *See* Bergius process; hardening.

hydrogen atom spectrum The spectrum of the hydrogen atom is characterized by several series of sharp spectral lines described by simple laws. The general law for these series of lines is:

$$1/\lambda = R\,(1/n^2_1 - 1/n^2_2),$$

where λ is the wavelength associated with a spectral line, R is the RYDBERG CONSTANT and n_1 and n_2 are integers, with $n_2 \geq n_1$.

The first of these series to be discovered was the BALMER SERIES in which $n_1 = 2$, $n_2 = 3,4,5,\ldots$ This series is in the visible region and was discovered by the Swiss mathematician and physicist Johann Jakob Balmer (1825–1898) in 1885. The series in which $n_1 = 1$ is the LYMAN SERIES which lies in the ultraviolet region. This series was discovered by the American physicist Theodore Lyman (1874–1954). The Lyman series is a conspicuous feature of the spectrum of the Sun.

In the PASCHEN SERIES ($n_1 = 3$), the *Brackett series* ($n_1 = 4$), the *Pfund series* ($n_1 = 5$) and the *Humphreys series* ($n_1 = 6$) the spectral lines occur in the infrared region.

The explanation for these regular series lies in the existence of discrete, quantized energy levels. In 1913 Niels Bohr was able to derive the formula for these series in terms of the ad hoc quantum assumptions of the BOHR THEORY. In the mid-1920s the formula was derived in a deductive way from quantum mechanics.

hydrogen azide *See* hydrazoic acid.

hydrogen bond (H-bonding) An intermolecular bond between molecules in which hydrogen is bound to a strongly electronegative element. Bond polarization by the electronegative element X leads to a positive charge on hydrogen $X^{\delta-}–H^{\delta+}$; this hydrogen can then interact directly with electronegative elements of adjacent molecules. The hydrogen bond is represented as a dotted line:

$$X^{\delta-} - H^{\delta+} \ldots\ldots X^{\delta-} - H^{\delta+} \ldots$$

The length of a hydrogen bond is characteristically 0.15–0.2 nm. Hydrogen bonding may lead to the formation of dimers (for example, in carboxylic acids) and is used to explain the anomalously high boiling points of H_2O and HF.

hydrogen bromide (HBr) A colorless sharp-smelling gas that is very soluble in water. It is produced by direct combination of hydrogen and bromine in the presence of a platinum catalyst or by the reaction of phosphorus tribromide with water. It dissolves in water to give hydrobromic acid. Hydrogen bromide is rather inactive chemically. It will not conduct electricity in the liquid state, indicating that it is a molecular compound.

hydrogencarbonate /hÿ-drŏ-jĕn-**kar**-bŏ-nayt/ (**bicarbonate**) A salt containing the ion $^-HCO_3$.

hydrogen chloride (HCl) A colorless gas that has a strong irritating odor and fumes strongly in moist air. It is prepared by the action of concentrated sulfuric acid on sodium chloride. The gas is made industrially by burning a stream of hydrogen in chlorine. It is not particularly reactive but will form dense white clouds of ammonium chloride when mixed with ammonia. It is very soluble in water and ionizes almost completely to give HYDROCHLORIC ACID. It will also dissolve in many nonaqueous solvents, including toluene, but will not ionize and the resultant solution shows no acidic properties. Hydrogen chloride is used in the manufacture of organic chlorine compounds, such as polyvinyl chloride (PVC).

Unlike the other hydrogen halides hydrogen chloride will not dissociate on heating, indicating a strong H–Cl bond.

hydrogen cyanide *See* hydrocyanic acid.

hydrogen electrode A type of half cell based on hydrogen, and assigned zero ELECTRODE POTENTIAL, so that other elements may be compared with it. It is also called the standard hydrogen half cell. The hydrogen is bubbled over a platinum electrode, coated in 'platinum black', in a 1 M acid solution. Hydrogen is adsorbed on the platinum black, which has a high surface area, enabling the equilibrium

$$H(g) \rightleftharpoons H^+(aq) + e^-$$

to be set up. The platinum is inert and has no tendency to form platinum ions in solution.

hydrogen fluoride (HF) A colorless liquid produced by the reaction of concentrated sulfuric acid on calcium fluoride:

$$CaF_2(s) + H_2SO_4(aq) \rightarrow CaSO_4(aq) + 2HF(l)$$

It produces toxic corrosive fumes and dissolves readily in water to give hydrofluoric acid.

Hydrogen fluoride is atypical of the hydrogen halides as the individual H–F units are associated into much larger units, forming zigzag chains and rings. This is caused by hydrogen bonds that form between the hydrogen and the highly electronegative fluoride ions. Hydrogen fluoride is used extensively as a catalyst in the petroleum industry. *See* hydrofluoric acid.

hydrogen ion A positively charged hydrogen atom, H^+, i.e. a proton. Hydrogen ions are produced by all ACIDS in water, in which they are hydrated to *hydroxonium* (*hydronium*) *ions*, H_3O^+. *See* pH.

hydrogen molecule ion The simplest type of molecule. It consists of two hydrogen nuclei and one electron. If the nuclei are regarded as being fixed the SCHRÖDINGER EQUATION for the hydrogen molecule ion can be solved exactly. This enables theories and approximation techniques concerned with chemical bonding to be tested quantitatively.

hydrogen peroxide (H_2O_2) A colorless syrupy liquid, usually used in solution in water. Although it is stable when pure, on contact with bases such as manganese(IV) oxide it gives off oxygen, the manganese(IV) oxide acting as a catalyst:

$$2H_2O_2 \rightarrow 2H_2O + O_2$$

Hydrogen peroxide can act as an oxidizing agent, converting iron(II) ions to iron(III) ions, or as a reducing agent with potassium manganate(VII). It is used as a bleach and in rocket fuel. The strength of solutions is usually given as *volume strength* – the volume of oxygen (dm^3) at STP given by decomposition of 1 dm^3 of the solution.

hydrogensulfate /hȳ-drŏ-jĕn-**sul**-fayt/ (**bisulfate**; HSO_4^-) An acidic salt or ester of sulfuric acid (H_2SO_4), in which only one of the acid's hydrogen atoms has been replaced by a metal or organic radical. An example is sodium hydrogensulfate, $NaHSO_4$.

hydrogen sulfide (**sulfuretted hydrogen**; H_2S) A colorless very poisonous gas with an odor of bad eggs. Hydrogen sulfide is prepared by reacting hydrochloric acid with iron(II) sulfide. It is tested for by mixing with lead nitrate, with which it gives a black precipitate. Its aqueous solution is weakly acidic. Hydrogen sulfide reduces iron(III) chloride to iron(II) chloride, forming hydrochloric acid and a yellow precipitate of sulfur. Hydrogen sulfide precipitates insoluble sulfides, and is used in qualitative analysis. It burns with a blue flame in oxygen to form sulfur(IV) oxide and water. Natural gas contains some hydrogen sulfide, which is removed before supply to the consumer.

hydrogensulfite /hȳ-drŏ-jĕn-**sul**-fȳt/ (**bisulfite**; HSO_3^-) An acidic salt or ester of sulfurous acid (H_2SO_3), in which only one of the acid's hydrogen atoms has been replaced by a metal or organic radical. An example is sodium hydrogensulfite, $NaHSO_3$.

hydrohalic /hȳ-droh-**hal**-ik/ Describing acids formed by *hydrogen halides*, e.g. HF, HCl, HBr, when dissolved in water.

hydroiodic acid /hÿ-droh-ÿ-**od**-ik/ *See* iodine.

hydrolysis /hÿ-**drol**-ă-sis/ A reaction between a compound and water. Some examples are:
Salts of weak acids
$$Na_2CO_3 + 2H_2O \rightarrow 2NaOH + H_2CO_3$$
Esters
$$CH_3COOC_2H_5 + H_2O \rightleftharpoons CH_3COOH + C_2H_5OH$$
Certain inorganic halides
$$SiCl_4 + 4H_2O \rightarrow Si(OH)_4 + 4HCl$$

hydron /**hÿ**-dron/ The positive ion H^+. The name is used when the isotope is not relevant, i.e. a hydron could be a proton, deuteron, or triton.

hydronium ion /hÿ-**droh**-nee-ŭm/ *See* acid; hydrogen ion.

hydrophilic /hÿ-drŏ-**fil**-ik/ Water attracting. *See* lyophilic.

hydrophobic /hÿ-drŏ-**foh**-bik/ Water repelling. *See* lyophobic.

hydroquinone /hÿ-droh-kwi-**nohn**, hÿ-droh-**kwin**-ohn/ *See* benzene-1,4-diol.

hydrosol /**hÿ**-drŏ-sol, -sohl/ A colloid in aqueous solution.

hydroxide /hÿ-**droks**-ÿd, -id/ A compound containing the ion OH^- or the group –OH.

hydroxonium ion /hÿ-droks-**oh**-nee-ŭm/ *See* hydrogen ion.

hydroxybenzene /hÿ-droks-ee-**ben**-zeen, -ben-**zeen**/ *See* phenol.

hydroxybenzoate /hÿ-droks-ee-**ben**-zoh-ayt/ *See* salicylate.

hydroxybenzoic acid /hÿ-droks-ee-ben-**zoh**-ik/ *See* salicylic acid.

hydroxyl group /hÿ-drox-ăl/ A group (–OH) containing hydrogen and oxygen, characteristic of alcohols and phenols, and some hydroxides. It should not be confused with the hydroxide ion (OH^-).

2-hydroxypropanoic acid /hÿ-droks-ee-proh-pă-**noh**-ok/ (**lactic acid**) A colorless liquid carboxylic acid:
$$CH_3CH(OH)COOH$$
See optical activity.

hygroscopic /hÿ-grŏ-**skop**-ik/ Describing a substance that absorbs moisture from the atmosphere. *See also* deliquescent.

hyperconjugation /hÿ-per-kon-jŭ-**gay**-shŏn/ The interaction of sigma bonds with pi bonds in a compound. It is sometimes described in terms of resonance structures of the type:
$$C_6H_5CH_3 \rightarrow C_6H_5CH_2^-H^+$$
to explain the interaction of the methyl group with the pi electrons of the benzene ring in methylbenzene (toluene).

hyperfine structure /**hÿ**-per-fÿn/ *See* fine structure.

hypertonic solution A solution that has a higher osmotic pressure than some other solution. *Compare* hypotonic solution.

hypo /**hÿ**-poh/ An old name for SODIUM THIOSULFATE(IV), especially when used in photography. It was formerly known as sodium hyposulfite.

hypobromous acid /hÿ-pŏ-**broh**-mŭs/ *See* bromic(I) acid.

hypochlorous acid /hÿ-pŏ-**klor**-ŭs, -**kloh**-rŭs/ *See* chloric(I) acid.

hypophosphorous acid /hÿ-pŏ-**fos**-fŏ-rŭs/ *See* phosphinic acid.

hyposulfuric acid /hÿ-pŏ-sul-**fyoor**-ik/ *See* dithionic acid.

hyposulfurous acid /hÿ-per-sul-**fyoor**-ŭs/ *See* dithionous acid.

hypotonic solution A solution that has a lower osmotic pressure than some other solution. *Compare* hypertonic solution.

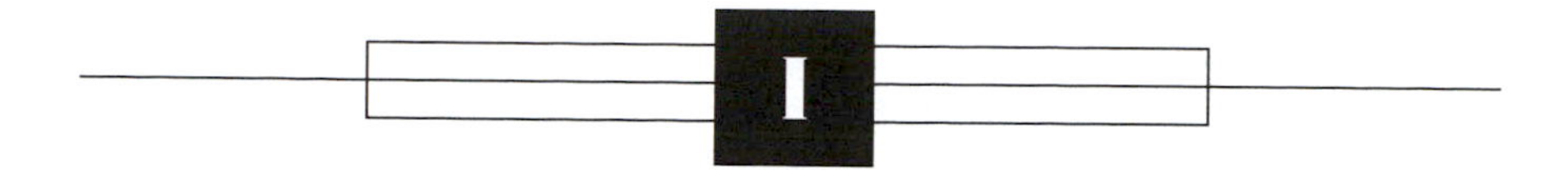

Iceland spar A pure transparent mineral form of calcite (calcium carbonate, $CaCO_3$), noted for its property of birefringence (double refraction).

ideal gas (**perfect gas**) *See* gas laws; kinetic theory.

ideal-gas scale *See* absolute temperature.

ideal solution A hypothetical solution that obeys RAOULT'S LAW.

ignis fatuus /**ig**-nis **fach**-û-ŭs/ (**will-o'-the-wisp**) A light sometimes seen over marshy ground. It is caused by methane produced by rotting vegetation, which is ignited by the presence of small amounts of spontaneously flammable phosphine (PH_3).

ignition temperature The temperature to which a substance must be heated before it will continue to burn (usually in air).

imide /**im**-ÿd, -id/ An organic compound containing the group –CO.NH.CO–, i.e. a –NH group attached to two carbonyl groups. Simple imides have the general formula R^1.CO.NH.CO.R^2, where R^1 and R^2 are alkyl or aryl groups. The group is known as the *imido group*, and it can form part of a ring in cyclic imides.

imido group /i-**mee**-doh, **im**-i-doh/ *See* imide.

imine /i-**meen**, **im**-in/ An organic compound containing the group C=N–, in which there is a double bond between the carbon and the nitrogen. A general formula for imines is $R^1R^2C{=}N{-}R^3$, where R^1, R^2, and R^3 are hydrocarbon groups or hydrogen. They can be made by the reaction of aldehydes and ketones with primary amines. For example, propanone (acetone; CH_3COCH_3) with ethylamine ($C_2H_6NH_2$):

$$CH_3COCH_3 + C_2H_6NH_2 \rightarrow (CH_3)_2C{=}N{-}C_2H_6 + H_2O$$

The reaction is acid-catalyzed. When ammonia is used the imine contains the C=N–H group:

$$CH_3COCH_3 + NH_3 \rightarrow (CH_3)_2C{=}N{-}H + H_2O$$

Most imines are unstable and can be detected only in solution unless R^1, R^2, or R^3 are aryl groups. Intermediates in which an imine adds a proton to form a positive ion (e.g. $R^1R^2C{=}NR^3H^+$) are known as *iminium ions*.

iminium ion *See* imine.

imino group /i-**mee**-noh, **im**-i-noh/ *See* imine.

iminourea /i-mee-noh-yû-**ree**-ă, im-ă-noh- / *See* guanidine.

immiscible /i-**miss**-ă-băl/ Describing two or more liquids that will not mix, such as oil and water. After being shaken together they form separate layers.

indene /**in**-deen/ (C_9H_8) A colorless flammable hydrocarbon. It has a benzene ring fused to a five-membered ring.

indicator A compound that reversibly changes color depending on the pH of the solution in which it is dissolved. The visual observation of this change is therefore a guide to the pH of the solution and it follows that careful choice of indicators permits a wide range of end points to be detected in acid–base titrations.

Redox titrations require either specific indicators, which detect one of the components of the reaction (e.g. starch for iodine, potassium thiocyanate for Fe^{3+}) or true redox indicators in which the transition potential of the indicator between oxidized and reduced forms is important. The transition potential of a redox indicator is analogous to the transition pH in acid–base systems.

Complexometric titrations require indicators that complex with metal ions and change color between the free state and the complex state. *See also* absorption indicator.

indigo ($C_{16}H_{10}N_2O_2$) A blue organic dye that occurs (as a glucoside) in plants of the genus *Indigofera*. It is a derivative of indole, and is now made synthetically.

indium /**in**-dee-ŭm/ A soft silvery metallic element belonging to group 13 of the periodic table. It is found in minute quantities, primarily in zinc ores and is used in alloys, in several electronic devices, and in electroplating.

Symbol: In; m.p. 155.17°C; b.p. 2080°C; r.d. 7.31 (25°C); p.n. 49; r.a.m. 114.818.

indole /**in**-dohl, -dol/ (**benzpyrrole**; C_8H_7N) A colorless solid organic compound, whose molecules consist of a benzene ring fused to a pyrrole ring. It occurs in coal tar and various plants, and is the basis of indigo and of several plant hormones.

inductive effect /in-**duk**-tiv/ The effect in which substituent atoms or groups in an organic compound can attract (–I) or push away electrons (+I), forming polar bonds.

inert gases *See* rare gases.

infrared /in-fră-**red**/ (**IR**) ELECTROMAGNETIC RADIATION with longer wavelengths than visible radiation. The wavelength range is approximately 0.7 μm to 1 mm. Many materials transparent to visible light are opaque to infrared, including glass. Rock salt, quartz, germanium, or polyethene prisms and lenses are suitable for use with infrared. Infrared radiation is produced by movement of charges on the molecular scale; i.e. by vibrational or rotational motion of molecules. Of particular importance in chemistry is the absorption spectrum of compounds in the infrared region. Certain bonds between pairs of atoms (C–C, C=C, C=O, etc.) have characteristic vibrational frequencies, which correspond to bands in the infrared spectrum. Infrared spectra are thus used in finding the structures of new organic compounds. They are also used to 'fingerprint' and thus identify known compounds. At shorter wavelengths, infrared absorption corresponds to transitions between rotational energy levels, and can be used to find the dimensions of molecules (by their moment of inertia).

inhibitor A substance that slows down the rate of reaction. Hydrogen sulfide, hydrogen cyanide, mercury salts, and arsenic compounds readily inhibit heterogeneous catalysts by adsorption. For example, arsenic compounds inhibit platinum catalysts in the oxidation of sulfur(IV) oxide to sulfur(VI) oxide. Inhibitors must not be confused with negative catalysts; inhibitors do not change the pathway of a reaction.

inner Describing a ring compound that is formed, or could be regarded as formed, by one part of a molecule reacting with another. For example, a LACTAM is an inner amide and a LACTONE is an inner ester.

inorganic chemistry The branch of chemistry concerned with elements other than carbon and with the preparation, properties, and reactions of their compounds. Certain simple carbon compounds are treated in inorganic chemistry, including the oxides, carbon disulfide, the halides, hydrogen cyanide, and salts, such as the cyanides, cyanates, carbonates, and hydrogencarbonates.

insertion reaction A reaction in which an atom or group is inserted between two other groups. *See* carbene.

insoluble /in-**sol**-yŭ-băl/ Describing a compound that has a very low solubility (in a specified solvent).

instrumentation /in-strŭ-men-**tay**-shŏn/ The measurement of the conditions and the control of processes within a chemical plant. The instruments can be classified into three groups: those for current information using mercury thermometers, weighing scales, and pressure gauges; those for recording viscosity, fluid flow, pressure, and temperature; and those instruments that control and maintain the desired conditions including pH and the flow of materials.

intercallation compound /in-terk-ă-**lay**-shŏn/ A compound that has a structure based on layers and in which there are layers of a different character interleaved in the basic structural units. For example, the micas phlogopite ($KMg_3(OH)_2Si_3AlO_{10}$) and muscovite ($KAl_2(OH)_2Si_3AlO_{10}$) are formed by interleaving K^+ ions replacing a quarter of the silicon layers in talc and pyrophyllite respectively. *See also* lamellar compound.

intermediate 1. A compound that requires further chemical treatment to produce a finished industrial product such as a dye or pharmaceutical chemical.
2. A transient chemical entity in a complex reaction. *See also* precursor.

intermediate bond A form of covalent bond that also has an ionic or electrovalent character. *See* polar bond.

intermediate bonding A form of covalent bond that also has an ionic or electrovalent character. *See* polar bond.

intermolecular forces /in-ter-mŏ-**lek**-yŭ-ler/ Forces of attraction between molecules rather than forces within the molecule (chemical bonding). If these intermolecular forces are weak the material will be gaseous and as their strength progressively increases materials become progressively liquids and solids. The intermolecular forces are divided into H-bonding forces and VAN DER WAALS FORCES, and the major component is the ELECTROSTATIC INTERACTION OF DIPOLES. *See* hydrogen bond.

internal energy Symbol: *U* The energy of a system that is the total of the kinetic and potential energies of its constituent particles (e.g. atoms and molecules). If the temperature of a substance is raised, by transferring energy to it, the internal energy increases (the particles move faster). Similarly, work done on or by a system results in an increase or decrease in the internal energy. The relationship between heat, work, and internal energy is given by the first law of thermodynamics. Sometimes the internal energy of a system is loosely spoken of as 'heat' or 'heat energy'. Strictly, this is incorrect; heat is the transfer of energy as a result of a temperature difference.

internal resistance Resistance of a source of electricity. In the case of a cell, when a current is supplied, the potential difference between the terminals is lower than the e.m.f. The difference (i.e. e.m.f. – p.d.) is proportional to the current supplied. The internal resistance (r) is given by:

$$r = (E - V)/I$$

where E is the e.m.f., V the potential difference between the terminals, and I the current.

interstitial /in-ter-**stish**-ăl/ *See* defect.

interstitial compound A crystalline compound in which atoms of a non-metal (e.g. carbon, hydrogen, or boron) occupy interstitial positions in the crystal lattice of a metal (usually a transition metal). Interstitial compounds are often non-stoichiometric. Their physical properties are often similar to those of metals; e.g. they have a metallic luster and are electrical conductors.

inversion /in-**ver**-shŏn/ A change from one optical isomer to the other. *See* Walden inversion.

invert sugar *See* sucrose.

iodic acid /ÿ-**od**-ik/ *See* iodic(V) acid.

iodic(V) acid (**iodic acid**; HIO_3) A colorless deliquescent crystalline solid produced by the reaction of concentrated nitric acid with iodine. Iodic(V) acid is a strong oxidizing agent. It will liberate iodine from solutions containing iodide ions and it reacts vigorously with organic materials, often producing flames. It dissociates extensively in water and hence is a strong acid.

iodic(VII) acid (**periodic acid**; H_5IO_6) A white crystalline solid made by low-temperature electrolysis of concentrated iodic(V) acid. It exists in a number of forms, the most common of which is *paraiodic(VII) acid.* Iodic(VII) acid is a powerful oxidizing agent. It is a weak acid, which – by cautious heating under vacuum – can be converted to *dimesoiodic(VII) acid* ($H_4I_2O_9$), and *metaiodic(VII) acid* (HIO_4). All three will oxidize manganese to manganate(VII) and this reaction can be used to determine small amounts of manganese in steel.

iodide /ÿ-ŏ-dÿd, -did/ *See* halide.

iodine /ÿ-ŏ-dÿn, -deen/ A dark-violet volatile solid element belonging to the halogens (group 17 of the periodic table). It occurs in seawater and is concentrated by various marine organisms in the form of iodides. Significant deposits also occur in the form of iodates. The element is conveniently prepared by the oxidation of iodides in acid solution (using MnO_2). Industrial methods similarly use oxidation of iodides or reduction of iodates to iodides by sulfur(IV) oxide (sulfur dioxide) followed by oxidation, depending on the source of the raw materials. Iodine and its compounds are used in chemical synthesis, photography, pharmaceuticals, and dyestuffs manufacture.

Iodine has the lowest electronegativity of the stable halogens and consequently is the least reactive. It combines only slowly with hydrogen to form *hydroiodic acid*, HI. Iodine also combines directly with many electropositive elements, but does so much more slowly than does bromine or chlorine. Because of the larger size of the iodine ion and the consequent low lattice energies, the iodides are generally more soluble than related bromides or chlorides. As with the other halides, iodides of Ag(I), Cu(I), Hg(I), and Pb(II) are insoluble unless complexing ions are present.

Iodine also forms a range of covalent iodides with the metalloids and non-metallic elements (this includes a vast range of organic iodides) but these are generally less thermodynamically stable and are more readily hydrolyzed than chlorine or bromine analogs.

Four oxides are known of which iodine(V) oxide, I_2O_5, is the most important. The other oxides, I_2O_4, I_4O_9, and I_2O_7 are much less stable and of uncertain structure. Like chlorine (but *not* bromine) iodine forms oxo-species based on IO^-, IO_2^-, IO_3^-, and IO_4^-. The chemistry of the other oxo-species in solution is complex. Elemental iodine reacts with alkalis in a similar way to bromine and chlorine.

The ionization potential of iodine is sufficiently low for it to form a number of compounds in which it is electropositive. It forms I^+ cations, for example, by reaction of solid silver nitrate with iodine solution, and such cations are sufficiently electrophilic to substitute aromatic compounds such as phenol. Iodine also exhibits the interesting property of forming solutions that are violet colored in non-donor type solvents, such as tetrachloromethane, but in donor solvents, such as ethanol or dioxan, there is a strong iodine–solvent interaction, which gives the solution a deep brown color. Even though iodine solutions were commonly used as antiseptic agents, the element is classified as toxic, and care should be taken to avoid eye intrusions or excessive skin contact.

Symbol: I; m.p. 113.5°C; b.p. 184°C; r.d. 4.93 (20°C); p.n. 53; r.a.m. 126.90447.

iodine monochloride /mon-ŏ-**klor**-ÿd, -id, -**kloh**-rÿd, -rid/ (ICl) A dark red liquid made by passing chlorine over iodine. It has properties similar to those of its constituent halogens. Iodine monochloride is used as a nonaqueous solvent and as an iodating agent in organic reactions.

iodine number (**iodine value**) A number that gives a measure of the number of unsaturated bonds in an organic compound such as a fat or oil. It is found by measuring the amount of iodine in grams taken up by 100 grams of the compound.

iodine(V) oxide (**iodine pentoxide**; I_2O_5) A white crystalline solid made by heating iodic(V) acid to a temperature of 200°C. It is a very strong oxidizing agent.

Iodine oxide is the acid anhydride of iodic(V) acid, which is reformed when water is added to the oxide. Its main use is in titration work, measuring traces of carbon monoxide in the air.

iodine pentoxide /pen-**toks**-ÿd/ *See* iodine(V) oxide.

iodine trichloride /trÿ-**klor**-ÿd, -**kloh**-rÿd/ *See* diiodine hexachloride.

iodoethane /ÿ-od-oh-**eth**-ayn/ (**ethyl iodide**; C_2H_5I) A colorless liquid alkyl halide made by reaction of ethanol with iodine in the presence of red phosphorus.

iodoform /ÿ-**od**-ŏ-form/ *See* triiodomethane.

iodoform reaction *See* triiodomethane.

iodomethane /ÿ-od-oh-**meth**-ayn/ (**methyl iodide**; CH_3I) A liquid alkyl halide made by reaction of methanol with iodine in the presence of red phosphorus.

ion /**ÿ**-on, -ŏn/ An atom or molecule that has a negative or positive charge as a result of losing or gaining one or more electrons. *See* electrolysis; ionization.

ion exchange A process that takes place in certain insoluble materials, which contain ions capable of exchanging with ions in the surrounding medium. Zeolites, the first ion exchange materials, were used for water softening. These have largely been replaced by synthetic resins made of an inert backbone material, such as polyphenylethene, to which ionic groups are weakly attached. If the ions exchanged are positive, the resin is a cationic resin. An anionic resin exchanges negative ions. When all available ions have been exchanged (e.g. sodium ions replacing calcium ions) the material can be regenerated by passing concentrated solutions (e.g. sodium chloride) through it. The calcium ions are then replaced by sodium ions. Ion-exchange techniques are used for a vast range of purification and analytical purposes.

For example, solutions with ions that can be exchanged for OH^- and H^+ can be estimated by titrating the resulting solution with an acid or a base.

ionic bond /ÿ-**on**-ik/ *See* electrovalent bond.

ionic crystal A crystal composed of ions of two or more elements. The positive and negative ions are arranged in definite patterns and are held together by electrostatic attraction. Common examples are sodium chloride and cesium chloride.

ionic product The product of concentrations:

$$K_W = [H^+][OH^-]$$

in water as a result of a small amount of self-ionization.

ionic radius A measure of the effective radius of an ion in a compound. For an isolated ion, the concept is not very meaningful, since the ion is a nucleus surrounded by an 'electron cloud'. Values of ionic radii can be assigned, however, based on the distances between ions in crystals.

Different methods exist for determining ionic radii and often different values are quoted for the same ion. The two main methods are those of Goldschmidt and of Pauling. The Goldschmidt radii are determined by substituting data from one compound to another, to produce a set of ionic radii for different ions. The Pauling radii are assigned by a more theoretical treatment for apportioning the distance between an anion and a cation.

ionic strength For an ionic solution a quantity can be introduced which emphasizes the charges of the ions present:

$$I = \frac{1}{2}\Sigma_i m_i z^2_i$$

where *m* is the molality and *z* the ionic charge. The summation is continued over all the different ions in the solution, *i*.

ionization /ÿ-ŏ-ni-**zay**-shŏn/ The process of producing ions. There are several ways in which ions may be formed from atoms or molecules. In certain chemical reactions ionization occurs by transfer of electrons; for example, sodium atoms and chlorine atoms react to form sodium chloride, which consists of sodium ions (Na^+) and chloride ions (Cl^-). Certain molecules can ionize in solution; acids, for example, form hydrogen ions as in the reaction

$$H_2SO_4 \rightarrow 2H^+ + SO_4^{2-}$$

The 'driving force' for ionization in a solution is solvation of the ions by molecules of the solvent. H^+, for example, is solvated as a hydroxonium (hydronium) ion, H_3O^+.

Ions can also be produced by ionizing radiation; i.e. by the impact of particles or photons with sufficient energy to break up molecules or detach electrons from atoms: $A \rightarrow A^+ + e^-$. Negative ions can be formed by capture of electrons by atoms or molecules: $A + e^- \rightarrow A^-$.

ionization energy *See* ionization potential.

ionization potential (IP; Symbol: *I*) The energy required to remove an electron from an atom (or molecule or group) in the gas phase, i.e. the energy required for the process:

$$M \rightarrow M^+ + e^-$$

It gives a measure of the ability of metals to form positive ions. The second ionization potential is the energy required to remove two electrons and form a doubly charged ion:

$$M \rightarrow M^{2+} + e^-$$

Ionization potentials stated in this way are positive; often they are given in electronvolts. *Ionization energy* is the energy required to ionize one mole of the substance, and is usually stated in kilojoules per mole (kJ mol^{-1}).

In chemistry, the terms 'second', 'third', etc., ionization potentials are usually used for the formation of doubly, triply, etc., charged ions. However, in spectroscopy and physics, they are often used with a different meaning. The second ionization potential is the energy to remove the second least strongly bound electron in forming a singly charge ion. For lithium ($ls^2 2s^1$) it would refer to removal of a 1s electron to produce an excited ion with the configuration $1s^1 2s^1$. Note also that ionization potentials are now stated as energies. Originally they were the potential through which an electron had to be accelerated to cause ionization by electron impact:

$$M + e^- \rightarrow M^{2+} + 2e^-$$

They were thus stated in volts.

ionizing radiation Radiation of sufficiently high energy to cause IONIZATION. It may be short-wavelength electromagnetic radiation (ultraviolet, x-rays, or gamma rays) or streams of particles.

ion pair A positive ion and a negative ion in close proximity in solution, held by the attractive force between their charges. *See* Debye–Hückel theory; electrolysis.

IP *See* ionization potential.

IR *See* infrared.

iridium /i-**rid**-ee-ŭm/ A white transition metal that is highly resistant to corrosion. It is used in electrical contacts, in spark plugs, and in jewelry.

Symbol: Ir; m.p. 2410°C; b.p. 4130°C; r.d. 22.56 (17°C); p.n. 77; r.a.m. 192.217.

iron /ÿ-ern/ A transition element occurring in many ores, especially the oxides (haematite and magnetite) and carbonate. It is extracted in a blast furnace using coke, limestone, and hot air. The coke and air form carbon monoxide, which then reduces the iron ore to iron. The limestone removes acidic impurities and forms a layer of slag above the molten iron at the base of the furnace. Steel is formed by reducing the carbon content to between 0.1 and 1.5%. Iron is used as a catalyst in the Haber process for ammonia production. It corrodes in air and moisture to hydrated iron(III) oxide (rust).

The most stable oxidation state is +3, which is yellow, but +2 (green) and +6 (easily reduced) also exist. Solutions of iron(II) ions give a green precipitate with sodium hydroxide solution, whereas iron(III) ions give a brown precipitate. The concentration of iron(II) ions can be estimated in acid solution by titration with standard potassium manganate(VII) solution. Iron(III) ions must first be reduced to iron(II) ions with sulfur(IV) oxide (sulfur dioxide).

Symbol: Fe; m.p. 1535°C; b.p. 2750°C; r.d. 7.874 (20°C); p.n. 26; r.a.m. 55.845.

iron(II) chloride (**ferrous chloride**; $FeCl_2$) A compound prepared by passing dry hydrogen chloride gas over heated iron. White feathery anhydrous crystals are produced. Hydrated iron(II) chloride ($FeCl_2.6H_2O$) is prepared by reacting excess iron with dilute or concentrated hydrochloric acid. Green crystals of the hexahydrate are obtained on crystallization from solution. Iron(II) chloride is readily soluble in water, producing an acidic solution due to salt hydrolysis.

iron(III) chloride (**ferric chloride**; $FeCl_3$) A compound prepared in the anhydrous state as dark red crystals by passing dry chlorine over heated iron. The product sublimes and is collected in a cooled receiver. The hydrated salt ($FeCl_3.6H_2O$) is prepared by adding excess iron(III) oxide to concentrated hydrochloric acid. On crystallization yellow-brown crystals of the hexahydrate are formed. Iron(III) chloride is very soluble in water and undergoes salt hydrolysis. At temperatures below 400°C the anhydrous salt exists as a dimer, Fe_2Cl_6.

iron(II) oxide (**ferrous oxide**; FeO) A black powder formed by the careful reduction of iron(III) oxide using either carbon monoxide or hydrogen. It can also be prepared by heating iron(II) oxalate in the absence of air. It is only really stable at high temperatures and disproportionates slowly on cooling to give iron(III) oxide and iron. Iron(II) oxide can be reduced by heating in a stream of hydrogen. When exposed to air, it is oxidized to iron(III) oxide. It is a basic oxide, dissolving readily in dilute acids to form iron(II) salt solutions. If heated to a high temperature in an inert atmosphere, iron(II) oxide disproportionates to give iron and triiron tetroxide.

iron(III) oxide (**ferric oxide**; Fe_2O_3) A rusty-brown solid prepared by the action of heat on iron(III) hydroxide or iron(II) sulfate. It occurs in nature as the mineral hematite. Industrially it is obtained by roasting iron pyrites. Iron(III) oxide dissolves in dilute acids to produce solutions of iron(III) salts. It is stable at red heat, decomposes around 1300°C to give triiron tetroxide, and can be reduced to iron by hydrogen at 1000°C. Iron(III) oxide is not ionic in character but has a structure similar to that of aluminum(III) oxide.

iron(II) sulfate (**ferrous sulfate; green vitriol**; $FeSO_4.7H_2O$) A compound that occurs in nature as the mineral melanterite (or copperas). It is made industrially from iron pyrites. In the laboratory iron(II) sulfate is prepared by dissolving excess iron in dilute sulfuric acid. On crystallization, green crystals of the heptahydrate are obtained. Careful heating of the hydrated salt yields anhydrous iron(II) sulfate; on further heating the sulfate decomposes to give iron(III) oxide, sulfur(IV) oxide, and sulfur(VI) oxide. The hydrated crystals oxidize easily on exposure to air owing to the formation of basic iron(III) sulfate. A freshly prepared solution of iron(II) sulfate absorbs nitrogen(II) oxide (brown-ring test).

Iron(II) sulfate crystals are isomorphous with the sulfates of zinc, magnesium, nickel, and cobalt.

iron(III) sulfate (**ferric sulfate**; $Fe_2(SO_4)_3$) A compound prepared in the hydrated state by the oxidation of iron(II) sulfate dissolved in dilute sulfuric acid, using an oxidizing agent, such as hydrogen peroxide or concentrated nitric acid. On crystallization, the solution deposits a white mass of the nonahydrate ($Fe_2(SO_4)_3.9H_2O$). The anhydrous salt can be prepared by gently heating the hydrated salt. Iron(III) sulfate decomposes on heating to give iron(III) oxide and sulfur(VI) oxide. It forms alums with the sulfates of the alkali metals.

irreversible change /i-ri-**ver**-să-băl/ *See* reversible change.

irreversible reaction A reaction in which conversion to products is complete; i.e. there is little or no back reaction.

isobars /**ÿ**-sŏ-barz/ **1.** Two or more nuclides that have the same nucleon numbers but different proton numbers.
2. Lines joining points of equal pressure.

isocyanide /ÿ-sŏ-**sÿ**-ă-nÿd/ *See* isonitrile.

isocyanide test (**carbylamine reaction**) A test for the primary amine group. The suspected primary amine is warmed with trichloromethane in an alcoholic solution of potassium hydroxide. The resulting isocyanide (RNC) has a characteristic smell of bad onions (and is very toxic):

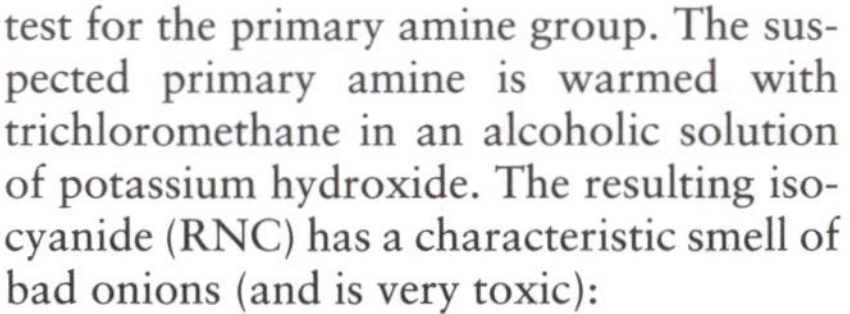

$$CHCl_3 + 3KOH + RNH_2 \rightarrow RNC + 3KCl + 3H_2O$$

isoelectronic /ÿ-soh-i-lek-**tron**-ik/ Describing compounds that have the same number of electrons. For example, carbon monoxide (CO) and nitrogen (N_2) are isoelectronic.

isoleucine /ÿ-sŏ-**loo**-seen, -**loo**-sin/ *See* amino acids.

isomer /**ÿ**-sŏ-mer/ *See* isomerism.

isomerism /ÿ-**som**-ĕ-riz-ăm/ The existence of two or more chemical compounds with the same molecular formulae but different structural formulae or different spatial

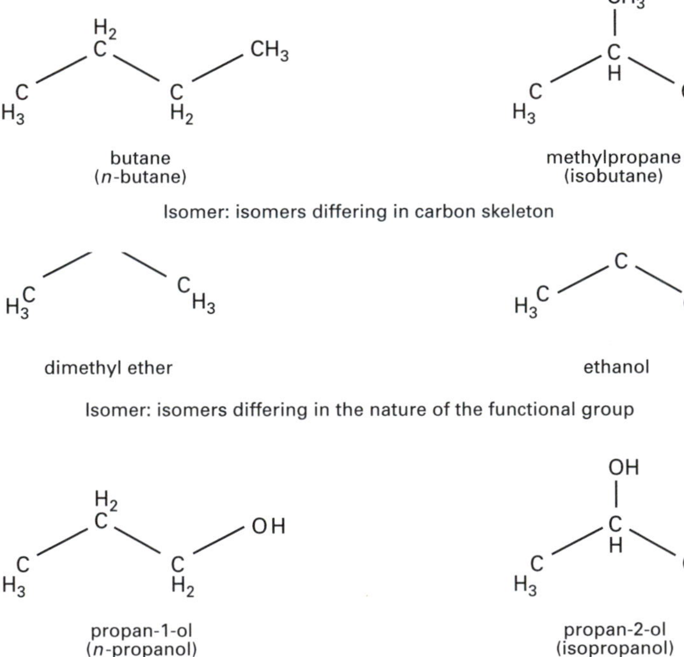

Isomer: isomers differing in carbon skeleton

Isomer: isomers differing in the nature of the functional group

Isomer: isomers differing in the position of a functional group

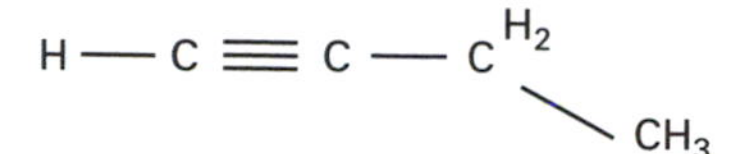

propan-1-yne

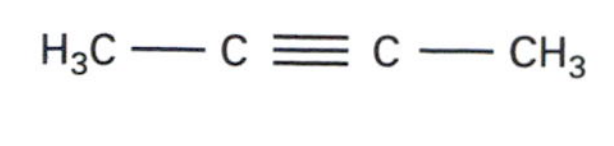

propan-2-yne

Isomer: isomers differing in the position of a multiple bond

arrangements of atoms. The different forms are known as *isomers*. For example, the compound C_4H_{10} may be butane ($CH_3CH_2CH_2CH_3$, with a straight chain of carbon atoms) or 2-methyl propane ($CH_3CH(CH_3)CH_3$, with a branched chain).

Structural isomerism is the type of isomerism in which the structural formulae of the compounds differ. There are two main forms. In one the isomers are different types of compound. An example of this is the compounds ethanol (C_2H_5OH) and methoxymethane (dimethyl ether, CH_3O-CH_3), both having the molecular formula C_2H_6O but quite different functional groups. In the other type of structural iso-

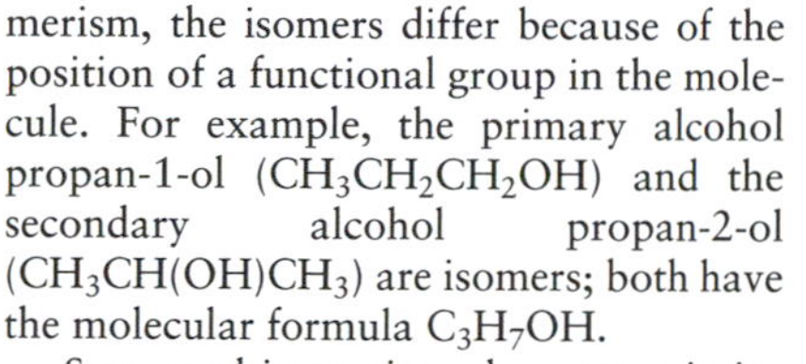

merism, the isomers differ because of the position of a functional group in the molecule. For example, the primary alcohol propan-1-ol ($CH_3CH_2CH_2OH$) and the secondary alcohol propan-2-ol ($CH_3CH(OH)CH_3$) are isomers; both have the molecular formula C_3H_7OH.

Structural isomerism also occurs in inorganic chemistry. A particular case is found in complexes in which a ligand may coordinate to a metal ion in two ways. For example, NO_2 can coordinate through N (the *nitro ligand*) or through O (the *nitrido ligand*). Such ligands are said to be *ambidentate*. Complexes that differ only in the way in which the ligand coordinates are said to show *linkage isomerism*.

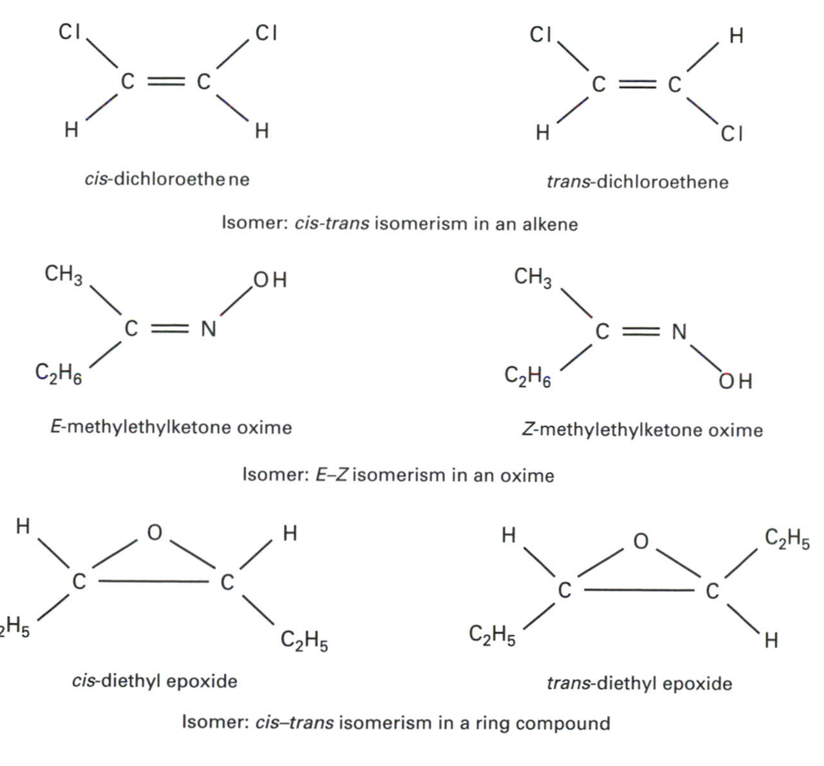

cis-dichloroethene

trans-dichloroethene

Isomer: *cis-trans* isomerism in an alkene

E-methylethylketone oxime

Z-methylethylketone oxime

Isomer: *E–Z* isomerism in an oxime

cis-diethyl epoxide

trans-diethyl epoxide

Isomer: *cis–trans* isomerism in a ring compound

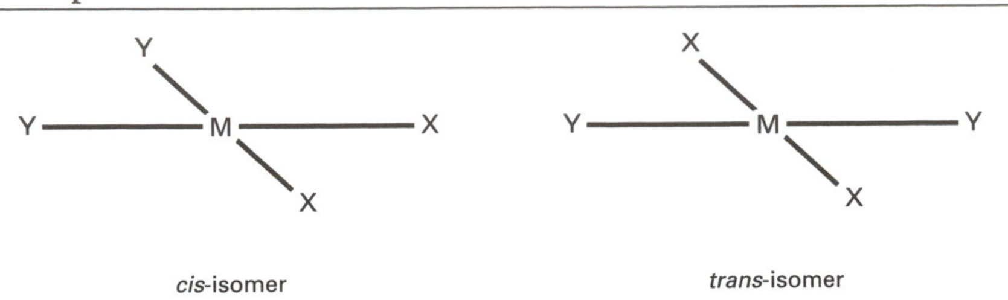

cis-isomer

trans-isomer

Isomerism

Stereoisomerism occurs when two compounds with the same molecular formulae and the same groups differ only in the arrangement of the groups in space. There are two types of stereoisomerism.

Cis–trans (or *syn–anti*) *isomerism* occurs when there is restricted rotation about a bond between two atoms (e.g. a double bond or a bond in a ring). Groups attached to each atom may be on the same side of the bond (the *cis-* or *syn-isomer*) or opposite sides (the *trans-* or *anti-isomer*). *Cis-trans* isomerism also occurs in square-planar complexes of the type MX_2Y_2, where M is a metal ion and X and Y are different ligands. If the X ligands are adjacent the isomer is a *cis*-isomer; if they are opposite, it is a *trans*-isomer. This type

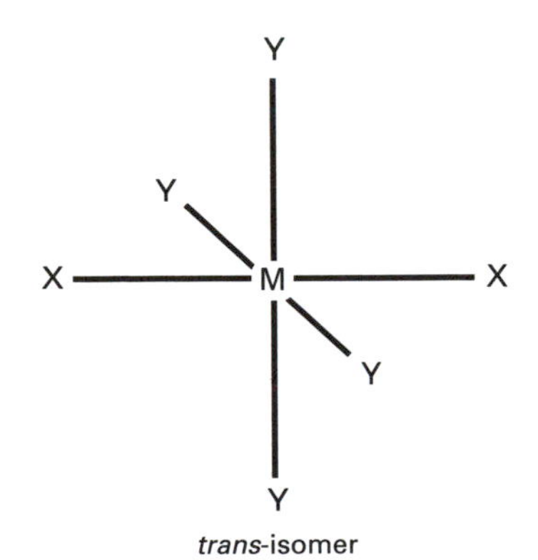

trans-isomer

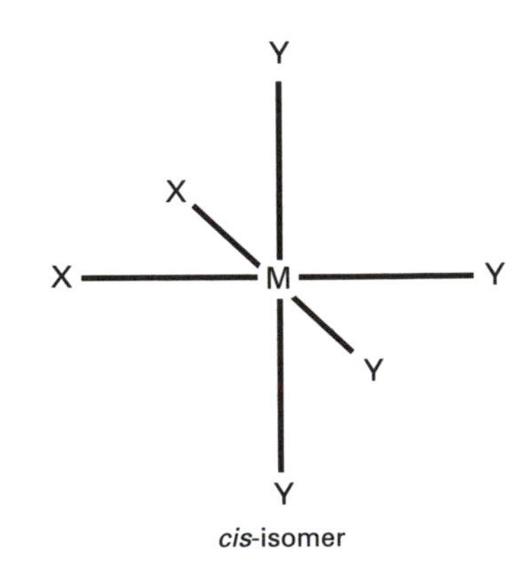

cis-isomer

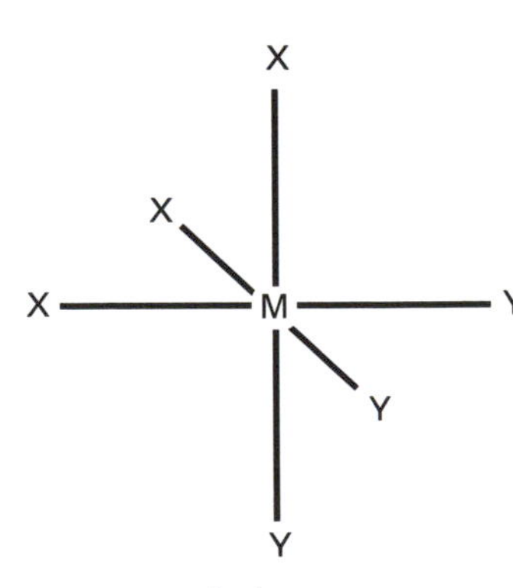

fac-isomer

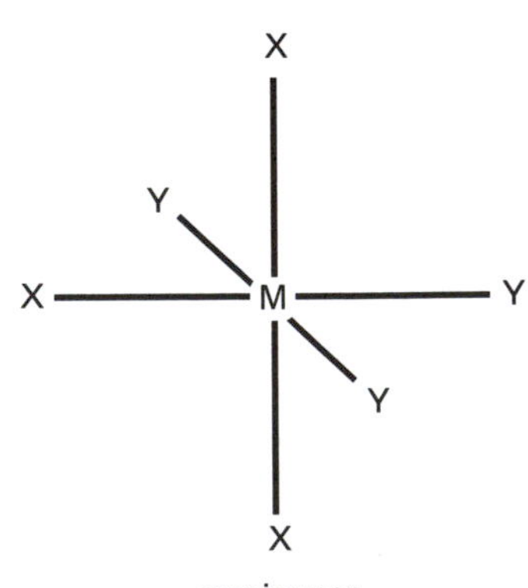

mer-isomer

Isomerism

of isomerism can also occur in octahedral complexes of the type MX_2Y_4. *Cis–trans* isomerism was formerly called *geometrical isomerism*.

Octahedral complexes of the type MX_3Y_3 show a different type of stereoisomerism. If the three X ligands are in a plane that includes the M ion (with the three Y ligands in a plane at right angles), then the structure is called the *mer-isomer* ('mer' stands for meridional). If the three X (and Y) ligands are all on a face of the octahedron, the structure is the *fac-isomer* ('fac' stands for facial). *See also E–Z* convention

Optical isomerism occurs when the compound has no plane of symmetry and can exist in left- and right-handed forms that are mirror images of each other. Such molecules always contain a CHIRAL ELEMENT. Molecules that are mirror images of each other are more properly called *enantiomers*. Stereoisomers that are not mirror images are called *diastereoisomers*. For example, ANOMERS are diastereoisomers. *See also* optical activity.

isomorphism /ÿ-sŏ-**mor**-fiz-ăm/ The existence of compounds with the same crystal structure.

isomorphism, law of *See* Mitscherlich's law.

isonitrile /ÿ-sŏ-**nÿ**-trăl, -tril, -trÿl/ (**isocyanide**) An organic compound of the formula R–NC.

isoprene /**ÿ**-sŏ-preen/ *See* methylbuta-1,3-diene.

isopropanol /ÿ-sŏ-**proh**-pă-nol, -nohl/ Propan-2-ol. *See* propanol.

isotactic polymer *See* polymerization.

isotherm /**ÿ**-sŏ-therm/ A line on a chart or graph joining points of equal temperature. *See also* isothermal change.

isothermal change /ÿ-sŏ-**therm**-ăl/ A process that takes place at a constant temperature. Throughout an isothermal process, the system is in thermal equilibrium with its surroundings. For example, a cylinder of gas in contact with a constant-temperature box may be compressed slowly by a piston. The work done appears as energy, which flows into the reservoir to keep the gas at the same temperature. Isothermal changes are contrasted with *adiabatic changes*, in which no energy enters or leaves the system, and the temperature changes. In practice no process is perfectly isothermal and none is perfectly adiabatic, although some can approximate in behavior to one of these ideals.

isotones /**ÿ**-sŏ-tohnz/ Two or more nuclides that have the same neutron numbers but different proton numbers.

isotonic /ÿ-sŏ-**tonn**-ik/ Describing solutions that have the same osmotic pressure.

isotopes /**ÿ**-sŏ-tohps/ Two or more species of the same element differing in their mass numbers because of differing numbers of neutrons in their nuclei. The nuclei must have the same number of protons (an element is characterized by its proton number). Isotopes of the same element have very similar properties because they have the same electron configuration, but differ slightly in their physical properties. An unstable isotope is termed a *radioactive isotope* or *radioisotope*

Isotopes of elements are useful in chemistry for studies of the mechanisms of chemical reactions. A standard technique is to *label* one of the atoms in a molecule by using an isotope of the element. It is then possible to trace the way in which this atom behaves throughout the course of the reaction. For example, in the esterification reaction:

$$ROH + R'COOH \rightleftharpoons H_2O + R'COOR$$

it is possible to find which bonds are broken by using a labeled oxygen atom. If the reaction is performed using ^{18}O in the alcohol it is found that this nuclide appears in the ester, showing that the C–OH bond of the acid is broken in the reaction. In labeling, radioisotopes are detected by counters; stable isotopes can also be used, and detected by a mass spectrum.

Isotopes are also used in kinetic studies. For example, if the bond between two atoms X–Y is broken in the rate-determining step, and Y is replaced by a heavier isotope of the element, Y*, then the reaction rate will be slightly lower with the Y* present. This *kinetic isotope effect* is particularly noticeable with hydrogen and deuterium compounds, because of the large relative difference in mass.

isotope separation /ÿ-sŏ-tohp/ The separation of different isotopes of a chemical element, making use of differences in the physical properties of these isotopes. For small-scale isotope separation a MASS SPECTROMETER is frequently used. For large-scale isotope separation the methods used include diffusion in the gas phase (used for uranium using the gas uranium hexafluoride), distillation (which was used to produce heavy water), electrolysis (also used for heavy water), and the use of centrifuges. Laser separation may also be employed, with excitation of one isotope occurring, followed by its separation using an electromagnetic field.

isotopic mass /ÿ-sŏ-**top**-ik/ (**isotopic weight**) The mass number of a given isotope of an element.

isotopic number The difference between the number of neutrons in an atom and the number of protons.

isotopic weight *See* isotopic mass.

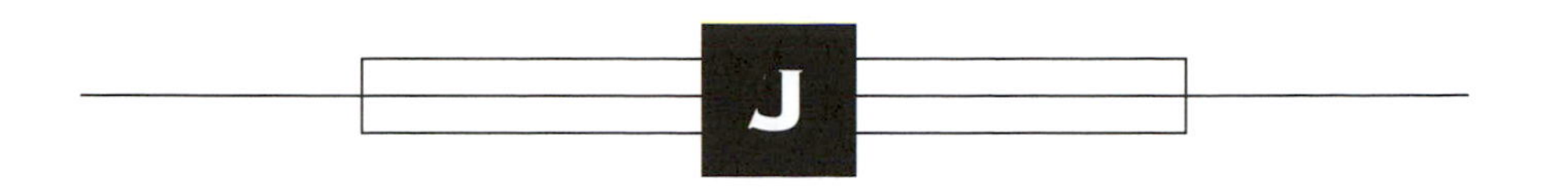

jeweler's rouge Iron(III) oxide (Fe_2O_3), used as a mild abrasive.

joliotium /zhoh-lee-**oh**-shee-ŭm/ Symbol: Jl A name formerly suggested for element-105 (dubnium). *See* element.

joule /jool/ Symbol: J The SI unit of energy and work, equal to the work done when the point of application of a force of one newton moves one meter in the direction of action of the force. 1 J = 1 N m. The joule is the unit of all forms of energy. The unit is named for the British physicist James Prescott Joule (1818–89).

kainite /**kay**-ă-nÿt, **kÿ**-nÿt/ A mineral form of hydrated crystalline magnesium sulfate ($MgSO_4$) containing also some potassium chloride. It is used as a fertilizer and source of potassium compounds.

kaolin /**kay**-ŏ-lin/ *See* china clay.

kaolinite /**kay**-ŏ-li-nÿt/ A hydrated aluminosilicate mineral, $Al_2(OH)_4Si_2O_5$, the major constituent of kaolin. It is formed by the weathering of FELDSPARS from granites.

katal /**kay**-t'l/ Symbol: kat The SI derived unit of catalytic activity, equal to the amount of substance in moles that can be catalyzed per second. 1 kat = 1 mol s^{-1}.

katharometer /kath-ă-**rom**-ĕ-ter/ *See* gas chromatography.

Kekulé structure /**kay**-koo-lay/ A structure of BENZENE in which there is a hexagonal ring with alternate double and single bonds. Two possible Kekulé structures contribute to the RESONANCE hybrid of benzene. The structure is named for the German chemist Friedrich Augus Kekulé von Stradonitz (1829–96).

kelvin Symbol: K The SI base unit of thermodynamic temperature. It is defined as the fraction 1/273.16 of the thermodynamic temperature of the triple point of water. Zero kelvin (0 K) is absolute zero. One kelvin is the same as one degree on the Celsius scale of temperature. The unit is named for the British theoretical and experimental physicist William Thompson, Baron Kelvin (1824–1907).

keratin /**ke**-ră-tin/ Any of a class of fibrous proteins found in hair, feathers, horn, and hooves. The keratins contain polypeptide chains linked by disulfide bonds between cystein amino acids.

kerosene /**ke**-rŏ-seen, ke-rŏ-**seen**/ *See* petroleum.

ketal /**kee**-t'l/ A type of organic compound formed by addition of an alcohol to a ketone. Addition of one molecule gives a *hemiketal*; two molecules give a full ketal. *See* acetal.

ketene /**kee**-teen/ (**keten**) A member of a group of organic compounds, general formula $R_2C{:}CO$, where R is hydrogen or an organic radical. The simplest member is the colorless gas ketene, $CH_2{:}CO$; the other ketenes are colored because of the presence of the double bonds. Ketenes are unstable and react with other unsaturated compounds to give cyclic compounds.

keto–enol tautomerism /**kee**-toh **ee**-nol/ An equilibrium between two isomers in which one isomer is a ketone (the *keto form*) and the other an enol. The equilibrium is by transfer of a hydrogen atom between the oxygen atom of the carbonyl group and an adjacent carbon atom.

keto form /**kee**-toh/ *See* keto–enol tautomerism.

ketohexose /kee-toh-**heks**-ohs/ A ketose SUGAR with six carbon atoms.

ketone /**kee**-tohn/ A type of organic compound with the general formula RCOR, having two alkyl or aryl groups bound to a carbonyl group. They are made by oxidizing secondary alcohols (just as ALDEHYDES are made from primary alcohols). Simple

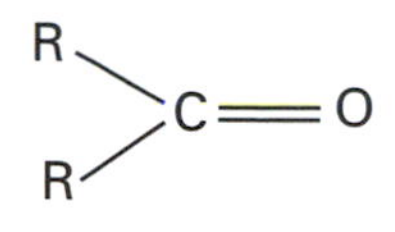

Ketone

examples are propanone (acetone, CH_3COCH_3) and butanone (methyl ethyl ketone, $CH_3COC_2H_5$).

The chemical reactions of ketones are similar in many ways to those of aldehydes. The carbonyl group is polarized, with positive charge on the carbon and negative charge on the oxygen. Thus nucleophilic addition can occur at the carbonyl group. Ketones thus:

1. Undergo addition reactions with hydrogen cyanide and hydrogensulfite ions.
2. Undergo condensation reactions with hydroxylamine, hydrazine, and their derivatives.
3. Are reduced to (secondary) alcohols. They are not, however, easily oxidized. Strong oxidizing agents give a mixture of carboxylic acids. They do not react with Fehling's solution or Tollen's reagent, and do not easily polymerize.

ketopentose /kee-toh-**pen**-tohs/ A ketose SUGAR with five carbon atoms.

ketose /**kee**-tohs/ A SUGAR containing a keto- (=CO) or potential keto- group.

kieselguhr /**kee**-zĕl-goor/ *See* diatomite.

killed spirits Zinc(II) chloride solution used as a flux for solder, so called because it is made by adding zinc to hydrochloric acid ('spirits of salt').

kilo- Symbol: k A prefix denoting 10^3. For example, 1 kilometer (km) = 10^3 meters (m).

kilocalorie /**kil**-ŏ-kal-ŏ-ree/ *See* calorie.

kilogram /**kil**-ŏ-gram/ (**kilogramme**; Symbol: kg) The SI base unit of mass, equal to the mass of the international prototype of the kilogram, which is a piece of platinum–iridium kept at Sèvres in France.

kilogramme /**kil**-ŏ-gram/ An alternative spelling of *kilogram*.

kilowatt-hour /**kil**-ŏ-wot/ Symbol: kWh A unit of energy, usually electrical, equal to the energy transferred by one kilowatt of power in one hour. It has a value of 3.6×10^6 joules.

kinetic energy /ki-**net**-ik/ *See* energy.

kinetic isotope effect *See* isotopes.

kinetics /ki-**net**-iks/ A branch of physical chemistry concerned with the study of rates of chemical reactions and the effect of physical conditions that influence the rate of reaction, e.g. temperature, light, concentration, etc. The measurement of these rates under different conditions gives information on the mechanism of the reaction; i.e. on the sequence of processes by which reactants are converted into products.

kinetic theory A theory explaining physical properties in terms of the motion of particles. The kinetic theory of gases assumes that the molecules or atoms of a gas are in continuous random motion and the pressure (*p*) exerted on the walls of a con-

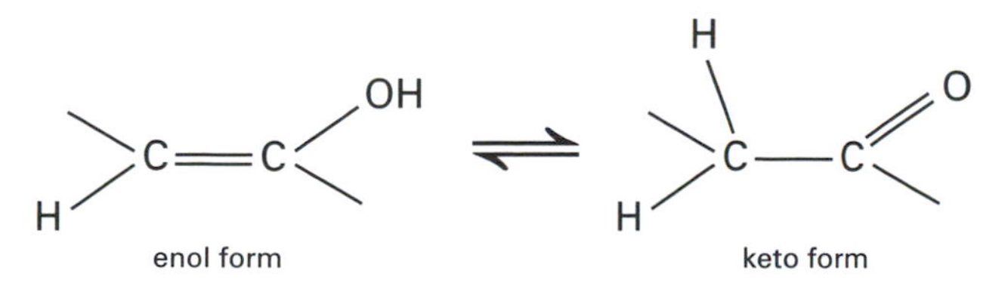

Keto-enol tautomerism

taining vessel arises from the bombardment by these fast moving particles. When the temperature is raised the speeds increase; so consequently does the pressure. If more particles are introduced or the volume is reduced there are more particles to bombard unit area of the walls and the pressure also increases. When a particle collides with the wall it experiences a rate of change of momentum, which is proportional to the force exerted. For a large number of particles this provides a steady pressure on the wall.

Additional assumptions are:

1. The particles behave as if they are hard smooth perfectly elastic points.
2. They do not exert any appreciable force on each other except during collisions.
3. The volume occupied by the particles themselves is a negligible fraction of the volume of the gas.
4. The duration of each collision is negligible compared with the time between collisions.

By considering the change in momentum on impact with the walls it can be shown that

$$p = \rho c^{2/3}$$

where ρ is the density of the gas and c is the root-mean-square speed of the molecules. The mean-square speed of the molecules is proportional to the absolute temperature:

$$Nmc^2 = RT$$

See also degrees of freedom.

Kipps apparatus /kips/ An apparatus for the production of a gas from the reaction of a liquid on a solid. It consists of three globes, the upper globe being connected via a wide tube to the lower globe. The upper globe is the liquid reservoir. The middle globe contains the solid and also has a tap at which the gas may be drawn off. When the gas is drawn off the liquid rises from the lower globe to enter the middle globe and reacts with the solid, thereby releasing more gas. When turned off the gas released forces the liquid back down into the lower globe and up into the reservoir, thus stopping the reaction. The apparatus is named for the Dutch chemist Petrus Jacobus Kipps (1808–64).

Kjeldahl's method /**kyel**-dahlz/ A method used for the determination of nitrogen in organic compounds. The nitrogenous substance is converted to ammonium sulfate by boiling with concentrated sulfuric acid (often with a catalyst such as $CuSO_4$) in a specially designed long-necked *Kjeldahl flask*. The mixture is then made alkaline and the ammonia distilled off into standard acid for measurement by titration. The method is named for the Danish chemist Johan Gustav Christoffer Thorsager Kjeldahl (1849–1900).

Kolbe electrolysis /**kol**-bee/ The electrolysis of sodium salts of carboxylic acids to prepare alkanes. The alkane is produced at the anode after discharge of the carboxylate anion and decomposition of the radical:

$$RCOO^- \rightarrow RCOO\bullet + e^-$$
$$RCOO\bullet \rightarrow R\bullet + CO_2$$
$$\rightarrow R-R + CO_2$$

and

$$2R\bullet \rightarrow R-R$$

As the reaction is a coupling reaction, only alkanes with an even number of carbon atoms in the chain can be prepared in this way. The process is named for the German chemist Adolph Wilhelm Hermann Kolbe (1818–84).

Kroll process /krol/ A method of obtaining certain metals by reducing the metal chloride with magnesium. Titanium can be obtained in this way:

$$TiCl_4 + 2Mg \rightarrow 2MgCl_2 + Ti$$

The process is named for the Luxembourg metallurgist William Justin Kroll (1889–1873).

krypton /**krip**-ton/ A colorless odorless monatomic element of the rare-gas group, known to form unstable compounds with fluorine. It occurs in minute quantities (0.001% by volume) in air. Krypton is used in fluorescent lights. Symbol: Kr; m.p. –156.55°C; b.p. –152.3°C; d. 3.749 (0°C) kg m^{-3}; p.n. 36; r.a.m. 83.80.

kurchatovium /ker-chă-**toh**-vee-ŭm/ Symbol: Ku A former name for element-104 (rutherfordium). *See* element.

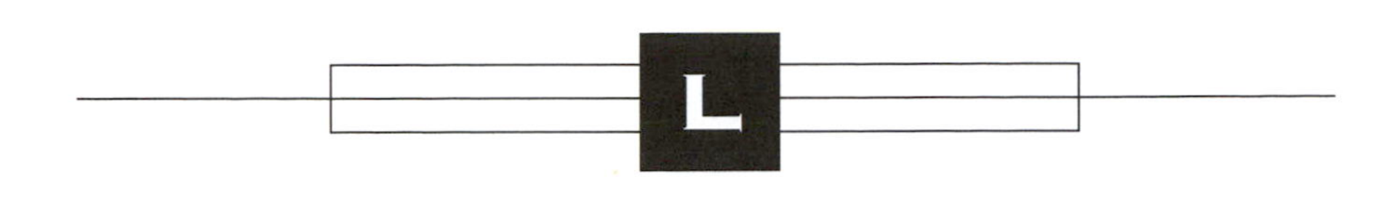

label A stable or radioactive nuclide used to investigate some process, such as a chemical reaction. *See* isotopes.

labile /lay-băl/ Describing a complex with ligands that can easily be replaced by more strongly bonded ligands.

lactam /lak-tam/ A type of organic compound containing the –NH.CO– group as part of a ring in the molecule. Lactams can be regarded as formed from a straight-chain compound that has an amine group ($-NH_2$) at one end of the molecule and a carboxylic acid group (–COOH) at the other; i.e. from an amino acid. The reaction of the amine group with the carboxylic acid group, with elimination of water, leads to the cyclic lactam, which is thus an internal amide. Lactams may exist in an alternate tautomeric form in which the hydrogen atom has migrated from the N onto the O of the carbonyl group. This, the *lactim* form, contains the group –N=C(OH)–.

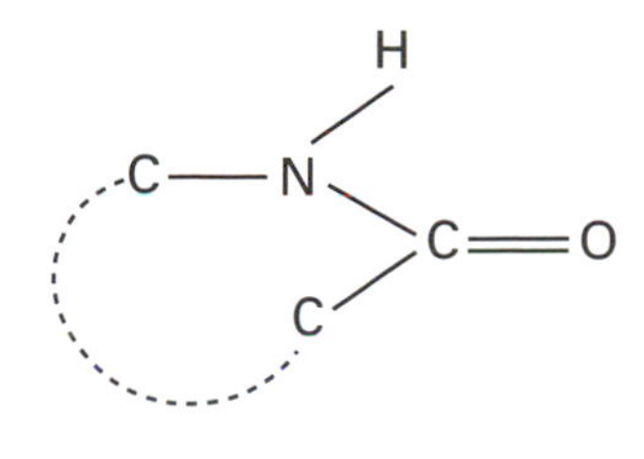

Lactam

lactate /lak-tayt/ A salt or ester of lactic acid.

lactic acid /lak-tik/ *See* 2-hydroxypropanoic acid.

lactim /lak-tim/ *See* lactam.

lactone /lak-tohn/ A type of organic compound containing the group –O.CO– as part of a ring in the molecule. A lactone can be regarded as formed from a compound with an alcohol (–OH) group on one end of the chain and a carboxylic acid (–COOH) group on the other. The lactone then results from reaction of the –OH group with the –COOH group; i.e. it is an internal ester.

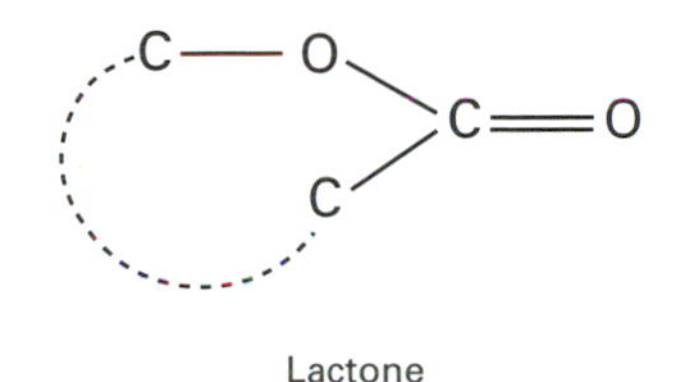

Lactone

lactose /lak-tohs/ (milk sugar; $C_{12}H_{22}O_{11}$) A sugar found in milk. It is a disaccharide composed of glucose and galactose units.

Ladenburg benzene /lan-dĕn-berg/ A structure once suggested for BENZENE in which the six carbon atoms are at the corners of a triangular prism with each carbon atom bonded to a hydrogen atom. It is named for the German chemist Albert Ladenburg (1842–1911). The actual compound with this structure, known as *prismane*, was synthesized in 1973.

lake A pigment made by combining an organic dyestuff with an inorganic compound (e.g. aluminum oxide).

lamellar compound /lă-mel-er/ A compound with a crystal structure composed of thin plates or layers. Silicates form many compounds with distinct layers. Typical examples are talc ($Mg_3(OH)_2Si_4O_{10}$) and pyrophyllite ($Al_2(OH)_2Si_4O_{10}$). *See also* intercallation compound.

lamp black A black pigment; a finely divided form of carbon formed by incomplete combustion of an organic compound.

Langmuir isotherm /**lang**-myoor/ An equation used to describe the adsorption of a gas onto a plane surface at a fixed temperature. It can be written in the form:

$$\theta = bp/(1 + bp),$$

where θ is the fraction of the surface covered by the gas, p is the gas pressure, and b is a constant known as the *adsorption coefficient*, which is the equilibrium constant for the adsorption process. The equation was derived by the American chemist Irving Langmuir (1881–1957) in 1916 using the kinetic theory of gases.

lanolin /**lan**-ŏ-lin/ A yellowish viscous substance obtained from wool fat. It contains cholesterol and terpene compounds, and is used in cosmetics, in ointments, and in treating leather.

lanthanides /**lan**-thă-nÿdz, -nidz/ *See* lanthanoids.

lanthanoids /**lan**-thă-noidz/ (**lanthanides**; **lanthanons**) A group of elements whose electronic configurations display back filling of the 4f-level. There is a maximum of 14 electrons in an f-orbital. The element lanthanum itself has no f-electrons (La $[Xe]5d^16s^2$) and is thus strictly not a lanthanoid, but it is included by convention, thus giving a closely related series of 15 elements. Excluding lanthanum, the elements all have $4f^x6s^2$ configurations but gadolinium and lutecium have an additional $5d^1$ electron.

The characteristic oxidation state is M^{3+} and the great similarity in the size of the ions leads to a very close similarity of chemical properties and hence to great difficulties of separation using conventional methods. Chromatographic and solvent-extraction methods have been specially developed for the lanthanoids.

lanthanons /**lan**-thă-nŭs/ *See* lanthanoids.

lanthanum /**lan**-thă-nŭm/ A soft ductile malleable silvery metallic element that is the first member of the lanthanoid series. It is found associated with other lanthanoids in many minerals, including monazite and bastnaesite. Lanthanum is used in several alloys (especially for lighter flints), as a catalyst, and in making optical glass.

Symbol: La; m.p. 921°C; b.p. 3457°C; r.d. 6.145 (25°C); p.n. 57; r.a.m. 138.9055.

lapis lazuli /**lap**-is **laz**-yŭ-lÿ, -lee/ A deep blue mineral consisting of sodium aluminum sulfate with some sulfur, widely used for ornaments and as a semiprecious gemstone. It was the original source of the pigment ultramarine.

laser /**lay**-zer/ An acronym for Light Amplification by Stimulated Emission of Radiation. A laser device produces high-intensity, monochromatic, coherent beams of light. In the laser process the molecules of a sample (such as ruby doped with Cr^{3+} ions) are promoted to an excited state. As the sample is in a cavity between two reflective surfaces, when a molecule emits spontaneously, the photon so generated ricochets backwards and forwards. In this way other molecules are stimulated to emit photons of the same energy. If one of the reflective surfaces is partially transmitting this radiation can be tapped.

laser spectroscopy Spectroscopy that uses lasers as a radiation source. Laser spectroscopy is particularly important when an intense beam of monochromatic radiation is needed, as in the RAMAN EFFECT. Pulsed-laser techniques are also used (*see* femtochemistry).

latent heat /**lay**-tĕnt/ The heat evolved or absorbed when a substance changes its physical state, e.g. the latent heat of fusion is the heat absorbed when a substance changes from a solid to a liquid.

laterite /**lat**-ĕ-rÿt/ A red fine-grained type of clay formed in tropical climates by the weathering of igneous rocks. Its color comes from the presence of iron(III) hydroxide.

lattice A regular three-dimensional arrangement of points. A lattice is used to describe the positions of the particles (atoms, ions, or molecules) in a crystalline solid. The lattice structure can be examined by x-ray diffraction techniques.

lattice energy The energy released when ions of opposite charge are brought together from infinity to form one mole of a given crystal. The lattice energy is a measure of the stability of a solid ionic substance, with respect to ions in the gas. *See also* Born–Haber cycle.

laughing gas *See* dinitrogen oxide.

lauric acid /**lô**-rik/ *See* dodecanoic acid.

law of conservation of energy *See* conservation of energy; law of.

law of conservation of mass *See* conservation of mass; law of.

law of constant composition *See* constant proportions; law of.

law of constant heat summation *See* Hess's law.

law of constant proportions *See* constant proportions; law of.

law of definite proportions *See* definite proportions; law of.

law of equivalent proportions *See* equivalent proportions, law of.

law of isomorphism *See* Mitscherlich's law.

law of mass action *See* mass action; law of.

law of octaves *See* Newlands' law.

law of reciprocal proportions *See* equivalent proportions, law of.

lawrencium /lor-**en**-see-ŭm/ A radioactive transuranic element of the actinoid series, not found naturally on Earth. Several very short-lived isotopes have been synthesized by bombarding ^{252}Cf with boron nuclei or ^{249}Bk with ^{18}O nuclei.

Symbol: Lr; p.n. 103; most stable isotope ^{262}Lr (half-life 261 minutes).

laws of chemical combination *See* chemical combination; laws of.

lazurite /**laz**-yŭ-rÿt/ An uncommon typically azure blue mineral consisting of a silicate of sodium, calcium, and aluminum with some sulfur. It and its parent rock, *lapis lazuli*, are widely used for ornaments and as semiprecious gemstones. Lazurite was the original source of the pigment *ultramarine*.

leaching The washing out of a soluble material from an insoluble solid using a solvent. This is often carried out in batch tanks or by dispersing the crushed solid in a liquid.

lead /led/ A dense, dull, gray, soft metallic element; the fifth member of group 14 (formerly VA) of the periodic table and the end product of radioactive decay series. It occurs in small quantities in a wide variety of minerals but only a few are economically important. The most important one is galena (PbS), found in Australia, Mexico, the USA, and Canada. Other minerals are anglesite ($PbSO_4$), litharge (PbO), and cerussite ($PbCO_3$).

Galena is often associated with zinc ores and the smelting operations of both lead and zinc industries are closely integrated. The ore is concentrated by froth flotation and the concentrate roasted then reduced,

$$2PbS + 3O_2 \rightarrow 2PbO + 2SO_2$$
$$PbS + 2O_2 \rightarrow PbSO_4$$
$$2PbO + 2C \rightarrow 2Pb + 2CO$$
$$2PbO + PbS \rightarrow 3Pb + SO_2$$
$$PbSO_4 + 2C \rightarrow Pb + 2CO + SO_2$$

Silver is often recovered in economic quantities from crude lead.

The outer s^2p^2 configurations of tin and lead give rise to similar properties for the two metals. There is however a much greater predominance of the divalent state for lead. Both oxides are amphoteric, lead(II) oxide (PbO) leading to plumbites and lead(IV) oxide (PbO_2) to plumbates on dissolution in alkalis. Lead also forms mixed oxides Pb_2O_3 (a yellow solid, better written as Pb(II)Pb(IV)O_3) and Pb_3O_4 (red lead, Pb_2(II)Pb(IV)O_4). Both metals have low melting points and neither is attacked by dilute acids; they differ however in their reaction with concentrated nitric acid

$$3Pb + 8HNO_3 \rightarrow 3Pb(NO_3)_2 + 2NO + 4H_2O$$

Tin gives hydrated Sn(IV) oxide. Concentrated hydrochloric acid will not attack lead.

Lead, like tin, forms halides, PbX_2, with all halogens. $PbCl_4$ is the only known lead(IV) halide (all halides of SnX_4 are known). The great stability of the Pb(II) state leads to Pb(IV) compounds being powerful oxidizing agents.

Lead is used in accumulators (lead–acid), alloys, radiation shielding, and water and sound proofing. It is also used in the petrochemical, paint, and glass industries.

Symbol: Pb; m.p. 327.5°C; b.p. 1830°C; r.d. 11.35 (20°C); p.n. 82; r.a.m. 207.2.

lead(II) acetate *See* lead(II) ethanoate.

lead–acid accumulator A type of electrical accumulator used in vehicle batteries. It has two sets of plates: spongy lead plates connected in series to the negative terminal and lead(IV) oxide plates connected to the positive terminal. The material of the electrodes is held in a hard lead-alloy grid. The plates are interleaved. The electrolyte is dilute sulfuric acid.

The e.m.f. when fully charged is about 2.2 V. This falls to a steady 2 V when current is drawn. As the accumulator begins to run down, the e.m.f. falls further. During discharge the electrolyte becomes more dilute and its relative density falls. To recharge the accumulator, charge is passed through it in the opposite direction to the direction of current supply. This reverses the cell reactions and increases the relative density of the electrolyte (*c.* 1.25 for a fully charged accumulator).

The electrolyte contains hydrogen ions (H^+) and sulfate ions (SO_4^{2-}). During discharge, H^+ ions react with the lead(IV) oxide to give lead(II) oxide and water

$$PbO_2 + 2H^+ + 2e^- \rightarrow PbO + H_2O$$

This reaction takes electrons from the plate, causing the positive charge. There is a further reaction to yield soft lead sulfate:

$$PbO + SO_4^{2-} + 2H^+ \rightarrow PbSO_4 + H_2O + 2e^-$$

Electrons are released to the electrode, producing the negative charge. During charging the reactions are reversed:

$$PbSO_4 + 2e^- \rightarrow Pb + SO_4^{2-}$$

$$PbSO_4 + 2H_2O \rightarrow PbO_2 + 4H^+ + SO_4^{2-} + 2e^-$$

lead(II) carbonate ($PbCO_3$) A white poisonous powder that occurs naturally as the mineral cerussite. It forms rhombic crystals and can be precipitated by reacting a cold aqueous solution of a soluble lead salt (e.g. lead(II) nitrate) with ammonium carbonate.

lead(II) carbonate hydroxide (**white lead**; $2PbCO_3.Pb(OH)_2$) The most important basic lead carbonate. It can be manufactured electrolytically. It has been used as a pigment in white and colored paints but it has the considerable drawback that it is poisonous.

lead-chamber process A former process for the manufacture of sulfuric acid by oxidizing sulfur(IV) oxide with nitrogen monoxide. The reaction, which was carried out in large lead chambers, has now been replaced by the contact process.

lead dioxide /dÿ-**oks**-ÿd, -id/ *See* lead(IV) oxide.

lead(II) ethanoate (**lead(II) acetate**; $Pb(CH_3CO_2)_2$) A compound usually occurring as the hydrate $Pb(CH_3CO_2)_2.3H_2O$, forming monoclinic crystals. At

100°C it loses ethanoic acid and water, forming a basic lead(II) ethanoate. Its solubility in water is 50 g per 100 g of water at 25°C. Lead(II) ethanoate can be obtained as an anhydrous salt. Its chief asset is that it is one of the few common lead(II) salts that are soluble in water.

lead-free fuel Vehicle fuel that contains none of the anti-knock agent lead tetraethyl. *See* lead tetraethyl.

lead(IV) hydride *See* plumbane.

lead monoxide /mon-**oks**-ȳd, -id/ *See* lead(II) oxide.

lead(II) oxide (**lead monoxide**; PbO) A yellow crystalline powder formed by roasting molten lead in air. *Litharge*, the most common form, is obtained when lead(II) oxide is heated above its melting point. If prepared below its melting point, another form, called *massicot*, is obtained. Litharge is used in the rubber industry, in the manufacture of paints and varnishes, and in the manufacture of lead glazes for pottery.

lead(IV) oxide (**lead dioxide**; PbO_2) A dark-brown solid forming hexagonal crystals. On heating to 310°C it decomposes into lead(II) oxide and oxygen. It can be prepared electrolytically or by reacting lead(II) oxide with potassium chlorate. Lead(IV) oxide has been used in the manufacture of matches.

lead(II) sulfate ($PbSO_4$) A white crystalline solid that occurs naturally as the mineral anglesite. It is almost insoluble in water and can be precipitated by reacting an aqueous solution containing sulfate ions with a solution of a soluble lead(II) salt (e.g. lead(II) ethanoate). It forms basic lead(II) sulfates when shaken together with lead(II) hydroxide and water. Basic lead(II) sulfates are useful as pigments.

lead(II) sulfide (PbS) A black solid that occurs naturally as the mineral galena. It can be prepared as a precipitate by reacting a solution of a soluble lead(II) salt with hydrogen sulfide or with a solution of a soluble sulfide. Lead(II) sulfide is useful as a rectifier in electrical components.

lead tetraethyl /tet-ră-**eth**-ăl/ (**tetraethyl lead**; $Pb(CH_2CH_3)_4$) A poisonous liquid that is insoluble in water but soluble in organic solvents. It is manufactured by the reaction of an alloy of sodium and lead with 1-chloroethane. The product is obtained by steam distillation. Lead tetraethyl is used as an additive in internal-combustion engine fuel to increase its octane number and thus prevent preignition (knocking).

leaving group The group leaving a molecule in a substitution or elimination reaction. The nature of the leaving group is often an important factor in the progress of the reaction.

Leblanc process /lă-**blahnk**/ An obsolete process for the manufacture of sodium carbonate. Sodium chloride is converted to sodium sulfate by heating with sulfuric acid. This sulfate is then roasted in a rotary furnace where it is first reduced to the sulfide using carbon, and then immediately converted to the carbonate by the action of limestone. Sodium carbonate solution is obtained by leaching with water and this is subsequently dried and calcined to obtain the solid. The process is named for the French physician Nicolas Leblanc (1742–1806).

Le Chatelier's principle /lă-sha-**tel**-yayz/ If a system is at equilibrium and a change is made in the conditions, the equilibrium adjusts so as to oppose the change. The principle can be applied to the effect of temperature and pressure on chemical reactions. A good example is the Haber process for synthesis of ammonia:

$$N_2 + 3H_2 \rightleftharpoons 2NH_3$$

The 'forward' reaction

$$N_2 + 3H_2 \rightarrow 2NH_3$$

is exothermic. Thus, reducing the temperature displaces the equilibrium towards production of NH_3 (as this tends to increase temperature). Increasing the pressure also favors formation of NH_3, because this leads to a reduction in the total number of molecules (and hence pressure). The process is named for the French chemist Henri Louis Le Chatelier (1850–1936).

Leclanché cell /lă-**klahn**-shay/ A primary voltaic cell consisting, in its 'wet' form, of a carbon-rod anode and a zinc cathode, with a 10–20% solution of ammonium chloride as electrolyte. Manganese(IV) oxide mixed with crushed carbon in a porous bag or pot surrounding the anode acts as a depolarizing agent. The dry form (*dry cell*) is widely used for flashlight batteries, transistor radios, etc. It has a mixture of ammonium chloride, zinc chloride, flour, and gum forming an electrolyte paste. Sometimes the dry cell is arranged in layers to form a rectangular battery, which has a longer life than the cylinder type. The cell is named for the French engineer and inventor Georges Leclanché (1839–82).

LEED (**low-energy electron diffraction**) A method of electron diffraction that is used to investigate surfaces. A surface is bombarded with a beam of low-energy electrons, with the diffracted electrons hitting a fluorescent screen or other detector. The electron beam has a low energy so that it interacts with the surface rather than the bulk of the material. The technique gives useful information about the structure of surfaces and processes occurring at surfaces.

leucine /**loo**-seen/ *See* amino acids.

levo-form /**lee**-vo-form/ *See* optical activity.

levorotatory /lee-voh-**roh**-tă-tor-ee. -toh-ree/ *See* optical activity.

Lewis acid /**loo**-is/ A substance that can accept an electron pair to form a coordinate bond; a *Lewis base* is an electron-pair donor. In this model, the neutralization reaction is seen as the acquisition of a stable octet by the acid, for example:

$$Cl_3B + :NH_3 \rightarrow Cl_3B:NH_3$$

Metal ions in coordination compounds are also electron-pair acceptors and therefore Lewis acids. The definition includes the 'traditional' Brønsted acids since H^+ is an electron acceptor, but in common usage the terms Lewis acid and Lewis base are reserved for systems without acidic hydrogen atoms. The acid is named for the American physical chemist Gilbert Newton Lewis (1875–1946).

Lewis base *See* Lewis acid.

Lewis octet theory *See* octet.

Lewis structure (**Lewis formula**) A two-dimensional depiction, by means of chemical element and electron dot symbols, of the possible structure of a molecule or ion, showing each atom in relation to its neighbors, the bonds that hold the atoms together, and the lone pairs in each atom's outer shell.

***l*-form** *See* optical activity.

L-form *See* optical activity.

Liebig condenser /**lee**-big/ A simple type of laboratory condenser. It consists of a straight glass tube, in which the vapor is condensed, with a surrounding glass jacket through which cooling water flows. The condenser is named for the German chemist Justus von Liebig (1803–73).

ligand /**lig**-ănd/ A molecule or ion that forms a coordinate bond to a metal atom or ion in a COMPLEX. *See also* chelate.

ligand-field theory *See* crystal-field theory.

light (**visible radiation**) A form of electromagnetic radiation able to be detected by the human eye. Its wavelength range is between about 400 nm (far red) and about 700 nm (far violet). The boundaries are not precise as individuals vary in their ability to detect extreme wavelengths; this ability also declines with age.

lignite /**lig**-nÿt/ (**brown coal**) The poorest grade of coal, containing up to 45% carbon and with a high moisture content.

ligroin /**lig**-roh-in/ A mixture of hydrocarbons obtained from PETROLEUM, used as a general solvent. It has a boiling range of 80 to 120°C.

lime *See* calcium oxide; calcium hydroxide.

limestone A natural form of calcium carbonate. It is used in making calcium compounds, carbon dioxide, and cement.

lime water A solution of calcium hydroxide in water. If carbon dioxide is bubbled through lime water a milky precipitate of calcium carbonate forms. Prolonged bubbling of carbon dioxide turns the solution clear again as a result of the formation of soluble calcium hydrogencarbonate ($Ca(HCO_3)_2$).

limonite /**lÿ**-mŏ-nÿt/ (**bog iron ore**) A dark-colored mineral that consists mainly of the hydroxide and oxides of iron, a major iron ore. It is also used as a pigment.

Linde process /**lin**-dĕ/ A method of liquefying gases by compression followed by expansion through a nozzle. The process is used for producing liquid oxygen and nitrogen by liquefying air and then fractionating it. It is named for the German engineer Karle von Linde (1842–1934).

line defect *See* defect.

line spectrum A SPECTRUM composed of a number of discrete lines corresponding to single wavelengths of emitted or absorbed radiation. Line spectra are produced by atoms or simple (monatomic) ions in gases. Each line corresponds to a change in electron orbit, with emission or absorption of radiation.

linkage isomerism *See* isomerism.

linoleic acid /lin-ŏ-**lee**-ik, lă-**noh**-lee-ik/ ($C_{17}H_{31}COOH$) An unsaturated carboxylic acid that occurs in LINSEED OIL and other plant oils. It is used in making paints and varnishes.

linolenic acid /lin-ŏ-**lee**-nik, -**len**-ik/ ($C_{17}H_{29}COOH$) A liquid unsaturated carboxylic acid that occurs in LINSEED OIL and other plant oils. It contains three double bonds.

linseed oil An oil extracted from the seeds of flax (linseed). It hardens on exposure to air (it is a drying oil) because it contains LINOLEIC ACID and LINOLENIC ACID, and is used in enamels, paints, putty, and varnishes.

lipid /**lip**-id/ A collective term used to describe a group of substances in cells characterized by their solubility in organic solvents such as ether and benzene, and their low solubility in water. The group is rather heterogeneous in terms of both function and structure. The simple lipids include *neutral lipids* or glycerides, which are esters of glycerol and fatty acids, and the waxes, which are esters of long-chain monohydric alcohols and fatty acids.

liquefied natural gas /**lik**-wĕ-fÿd/ (**LNG**) Liquid METHANE, obtained from NATURAL GAS and used as a fuel.

liquefied petroleum gas (**LPG**) A mixture of liquefied hydrocarbon gases (mainly PROPANE) extracted from PETROLEUM, used as a fuel for internal combustion engines and for heating.

liquid The state of matter in which the particles of a substance are loosely bound by intermolecular forces. The weakness of these forces permits movement of the particles and consequently liquids can change their shape within a fixed volume. The liquid state lacks the order of the solid state. Thus, amorphous materials, such as glass, in which the particles are disordered and can move relative to each other, can be classed as liquids.

liquid air A pale blue liquid. It is a mixture of liquid oxygen (boiling at −183°C) and liquid nitrogen (boiling at −196°C).

liquid crystal A substance that can flow like a viscous liquid but has a considerable amount of molecular order. There are three types of liquid crystal. In a *smectic crystal* there are layers of aligned molecules, with the long axes of the molecules being perpendicular to the layers. In a *cholesteric crystal* there are also layers of aligned molecules, but with the axes of the molecules being parallel to the planes of the layers. In a *nematic crystal* the molecules are all aligned in the same direction but not arranged in layers.

liter /**lee**-ter/ Symbol: l A unit of volume now defined as 10^{-3} meter^3; i.e. 1000 cm^3. The milliliter (ml) is thus the same as the cubic centimeter (cm^3). However, the name is not recommended for precise measurements since the liter was formerly defined as the volume of one kilogram of pure water at 4°C and standard pressure. On this definition, one liter is the same as 1000.028 cubic centimeters.

litharge /**lith**-arj/ *See* lead(II) oxide.

lithia /**lith**-ee-ă/ *See* lithium oxide.

lithia water *See* lithium hydrogencarbonate.

lithium /**lith**-ee-ŭm/ A light silvery moderately reactive metal; the first member of the alkali metals (group 1 of the periodic table). It occurs in a number of complex silicates, such as spodumene, lepidolite, and petalite, and a mixed phosphate, tryphilite. It is a rare element accounting for 0.0065% of the Earth's crust. Lithium ores are treated with concentrated sulfuric acid and lithium sulfate subsequently separated by crystallization. The element can be obtained by conversion to the chloride and electrolysis of the fused chloride or of solutions of LiCl in pyridine.

Lithium has the electronic configuration $1s^22s^1$ and is entirely monovalent in its chemistry. The lithium ion is, however, much smaller than the ions of the other alkali metals; consequently it is polarizing and a certain degree of covalence occurs in its bonds. Lithium also has the highest ionization potential of the alkali metals.

The element reacts with hydrogen to form lithium hydride, LiH, a colorless high-melting solid, which releases hydrogen at the anode during electrolysis (confirming the ionic nature Li^+H^-). The compound reacts with water to release hydrogen and is also frequently used as a reducing agent in organometallic synthesis. Lithium reacts with oxygen to give Li_2O (sodium gives the peroxide) and with nitrogen to form Li_3N on fairly gentle warming. The metal itself reacts only slowly with water, giving the hydroxide (LiOH) but lithium oxide reacts much more vigorously to give again the hydroxide; the nitride is hydrolyzed to ammonia. The metal reacts with halogens to form halides (LiX).

Apart from the fluoride the halides are readily soluble both in water and in oxygen-containing organic solvents. In this property lithium partly resembles magnesium which has a similar charge/size ratio. Compared to the other carbonates of group 1, lithium carbonate is thermally unstable decomposing to Li_2O and CO_2. This is because the small Li^+ ion leads to particularly high lattice energies favouring the formation of Li_2O.

Lithium also forms a wide range of alkyl- and aryl-compounds with organic compounds, which are particularly useful in organic synthesis. Lithium compounds impart a characteristic purple color to flames.

Symbol: Li; m.p. 180.54°C; b.p. 1347°C; r.d. 0.534 (20°C); p.n. 3; r.a.m. 6.941.

lithium aluminum hydride *See* lithium tetrahydridoaluminate(III).

lithium carbonate (Li_2CO_3) A white solid obtained by the addition of excess sodium carbonate solution to a solution of a lithium salt. It is soluble in the presence of excess carbon dioxide, forming lithium hydrogencarbonate. If heated (to 780°C) in a stream of hydrogen it undergoes thermal decomposition to give lithium oxide and carbon dioxide. Lithium carbonate forms monoclinic crystals and differs from the other group 1 carbonates in being only sparingly soluble in water. It has a close resemblance to the carbonates of magnesium and calcium – an example of a diagonal relationship in the periodic table.

lithium chloride (LiCl) A white solid produced by dissolving lithium carbonate or oxide in dilute hydrochloric acid and crystallizing the product. Below 19°C the dihydrate ($LiCl.2H_2O$) is obtained; at 19°C the chloride loses one molecule of water and becomes anhydrous at 93.5°C. Lithium chloride is used as a flux in welding aluminum. The compound forms cubic crystals; the anhydrous salt is isomorphous with sodium chloride. It dissolves in such organic solvents as alcohols, ketones, and esters. Lithium chloride is probably the most deliquescent substance known and forms hydrates with 1, 2, and 3 molecules of water.

lithium hydride (LiH) A white crystalline solid produced by direct combination of the elements at a temperature above 500°C. Electrolysis of the fused salt yields hydrogen at the anode. Lithium hydride reacts violently with water to give lithium hydroxide and hydrogen; the reaction is highly exothermic. The compound is used as a reducing agent and in the preparation of other hydrides. Lithium hydride has a cubic structure and is more stable than the group 1 hydrides.

lithium hydrogencarbonate ($LiHCO_3$) A compound known only in solution, produced by the action of carbon dioxide, water, and lithium carbonate. When a solution of the hydrogencarbonate is heated, carbon dioxide and lithium carbonate are produced. Solutions of the hydrogencarbonate are sold and used in medicine under the name of *lithia water*.

lithium hydroxide (LiOH) A white solid made industrially as the monohydrate ($LiOH.H_2O$) by reacting lime with a lithium ore or with a salt made from the ore. Lithium hydroxide has a closer resemblance to the group 2 hydroxides than to the group 1 hydroxides.

lithium oxide (**lithia**; Li_2O) A white solid produced by burning metallic lithium in air above its melting point. It can also be produced by the thermal decomposition of the carbonate or hydroxide. Lithium oxide reacts slowly with water forming a solution of lithium hydroxide; the reaction is exothermic. The compound has a calcium-fluoride structure.

lithium sulfate (Li_2SO_4) A white solid prepared by the addition of excess lithium oxide or carbonate to a solution of sulfuric acid. It is readily soluble in water from which it crystallizes as the monohydrate. In contrast to the other group 1 sulfates it does not form alums and it is not isomorphous with these other sulfates.

lithium tetrahydridoaluminate(III) /tet-ră-hȳ-dri-doh-ă-**loo**-mă-nit, -nayt/ (**lithium aluminum hydride**; $LiAlH_4$) A white solid produced by action of lithium hydride on aluminum chloride, the hydride being in excess. Lithium tetrahydridoaluminate reacts violently with water. It is a powerful reducing agent, reducing ketones and carboxylic acids to their corresponding alcohols. In inorganic chemistry it is used in the preparation of hydrides.

litmus /**lit**-mŭs/ A natural pigment that changes color when in contact with acids and alkalis; above a pH of 8.3 it is blue and below a pH of 4.5 it is red. Thus it gives a rough indication of the acidity or basicity of a solution; because of its rather broad range of color change it is not used for precise work. Litmus is used both in solution and as litmus paper.

lixiviation /lik-siv-ee-**ay**-shŏn/ The process of separating soluble components from a mixture by washing them out with water.

LNG *See* liquefied natural gas.

localized bond A bond in which the electrons contributing to the bond remain between the two atoms concerned, i.e. the bonding orbital is localized. The majority of bonds are of this type. *Compare* delocalized bond.

lodestone /**lohd**-stohn/ A naturally occurring magnetic oxide of iron, a form of MAGNETITE (Fe_3O_4). A piece of lodestone is a natural magnet and lodestones were used as magnetic compasses in ancient times.

lone pair A pair of valence electrons having opposite spin that are located together on one atom, i.e. are not shared as in a covalent bond. Lone pairs occupy similar positions in space to bond pairs and account for the shapes of molecules. A molecule with a lone pair can donate this to an electron acceptor, such as H^+ or a metal ion, to form coordinate bonds. *See* complex; Lewis base.

long period *See* period.

lowering of vapor pressure A colligative property of solutions in which the vapor pressure of a solvent is lowered as a solute is introduced. When both solvent and solute are volatile the effect of increasing the solute concentration is to lower the partial vapor pressure of each component. When the solute is a solid of negligible vapor pressure the lowering of the vapor pressure of the solution is directly proportional to the number of species introduced rather than to their nature and the proportionality constant is regarded as a general solvent property. Thus the introduction of the same number of moles of any solute causes the same lowering of vapor pressure, if dissociation does not occur. If the solute dissociates into two species on dissolution the effect is doubled. The kinetic model for the lowering of vapor pressure treats the solute molecules as occupying part of the surface of the liquid phase and thereby restricting the escape of solvent molecules. The effect can be used in the measurement of relative molecular masses, particularly for large molecules, such as polymers. *See also* Raoult's law.

Lowry–Brønsted theory /**low**-ree **bron**-sted/ *See* acid.

LPG *See* liquefied petroleum gas.

LSD *See* lysergic acid diethylamide.

lumen /**loo**-mĕn/ (*pl.* **lumens** or **lumina**) Symbol: lm The SI unit of luminous flux, equal to the luminous flux emitted by a point source of one candela in a solid angle of one steradian. 1 lm = 1 cd sr.

luminescence /loo-mă-**ness**-ĕns/ The emission of radiation from a substance in which the particles have absorbed energy and gone into excited states. They then return to lower energy states with the emission of electromagnetic radiation. If the luminescence persists after the source of excitation is removed the process is called *phosphorescence*: if not, it is called *fluorescence*.

lutetium /loo-**tee**-shee-ŭm/ A silvery element of the lanthanoid series of metals. It occurs in association with other lanthanoids. Lutetium is a very rare lanthanoid and has few uses.

Symbol: Lu; m.p. 1663°C; b.p. 3395°C; r.d. 9.84 (25°C); p.n. 71; r.a.m. 174.967.

lux /luks/ (*pl.* **lux**) Symbol: lx The SI unit of illumination, equal to the illumination produced by a luminous flux of one lumen falling on a surface of one square meter. 1 lx = 1 lm m^{-2}.

Lyman series /**lȳ**-măn/ *See* hydrogen atom spectrum. It is named for the American physicist Theodore Lyman (1874–1954).

lyophilic /lȳ-ŏ-**fil**-ik/ Solvent attracting. When the solvent is water, the word *hydrophilic* is often used. The terms are applied to:

1. Ions or groups on a molecule. In aqueous or other polar solutions ions or polar groups are lyophilic. For example, the $-COO^-$ group on a soap is the lyophilic (hydrophilic) part of the molecule.
2. The disperse phase in colloids. In lyophilic colloids the dispersed particles have an affinity for the solvent, and the

colloids are generally stable. *Compare* lyophobic.

lyophobic /lÿ-ŏ-**foh**-bik/ Solvent repelling. When the solvent is water, the word *hydrophobic* is used. The terms are applied to:

1. Ions or groups on a molecule. In aqueous or other polar solvents, the lyophobic group will be non-polar. For example, the hydrocarbon group on a soap molecule is the lyophobic (hydrophobic) part.
2. The disperse phase in colloids. In lyophobic colloids the dispersed particles are not solvated and the colloid is easily solvated. Gold and sulfur sols are examples. *Compare* lyophilic.

lysergic acid diethylamide /lÿ-**ser**-jik; dÿ-eth-ăl-**am**-ÿd/ (**LSD**) A synthetic organic compound that has physiological effects similar to those produced by alkaloids in certain fungi. Even small quantities, if ingested, produce hallucinations and extreme mental disturbances. The initials LSD come from the German form of the chemical's name, *Lysergic-Saure-Diathylamide*.

lysine /**lÿ**-seen, -sin/ *See* amino acids.

M

macromolecular crystal /mak-roh-mŏ-**lek**-yŭ-ler/ A crystal composed of atoms joined together by covalent bonds, which form giant three-dimensional or two-dimensional networks. Diamond and silicates are examples of macromolecular crystals.

macromolecule /mak-roh-**mol**-ĕ-kyool/ A large molecule; e.g. a natural or synthetic polymer.

magic acid *See* superacid.

magnesia /mag-**nee**-zhă, -shă/ *See* magnesium oxide.

magnesite /**mag**-nĕ-sÿt/ A mineral form of magnesium carbonate, $MgCO_3$.

magnesium /mag-**nee**-zee-ŭm, -seee-ŭm/ A light metallic element; the second member of group 2 of the periodic table (the alkaline earths). It has the electronic configuration of neon with two additional outer 3s electrons. The element accounts for 2.09% of the Earth's crust and is eighth in order of abundance. It occurs in a wide variety of minerals such as brucite ($Mg(OH)_2$), carnallite ($MgCl_2.KCl.6H_2O$), epsomite ($MgSO_4. 7H_2O$), magnesite ($MgCO_3$), and dolomite ($MgCO_3.CaCO_3$); and also as $MgCl_2$ in sea water. The metal is obtained by several routes, depending on the mineral used, but all leading to the chloride. This is followed by electrolysis of the fused chloride.

The element has a fairly low ionization potential and is an electropositive element. It reacts directly with oxygen, nitrogen, and sulfur on heating, to form the oxide, MgO, the nitride, Mg_3N_2, and the sulfide MgS.

Magnesium halides are also formed by direct reaction with the halogen. They show a slight deviation in the general behavior of halides by dissolving normally in water and only undergoing a significant amount of hydrolysis when solutions are evaporated:

$$MgX_2 + 2H_2O \rightarrow Mg(OH)_2 + 2HX$$

Magnesium hydroxide is considerably less soluble than the hydroxides of the heavier elements in group 2 and is a weaker base. It forms a soluble sulfate, chlorate, and nitrate. Like the heavier alkaline earths, magnesium forms an insoluble carbonate, but $MgCO_3$ is the least thermally stable of the group. Unlike calcium, it does not form a carbide.

Magnesium metal is industrially important as a major component in lightweight alloys (with aluminum and zinc). The metal surfaces develop an impervious oxide film which protects them from progressive deterioration.

Because of its ionic nature magnesium forms very few coordination compounds, Donor–acceptor species in aqueous solutions are short-lived, but magnesium bromide and iodide have sufficient acceptor properties to dissolve in donor solvents such as alcohols and ketones. Magnesium appears to be an essential element for life, occurring in chlorophyll.

Certain specialized organometallic compounds of magnesium are important in organic chemistry as synthetic reagents. Magnesium forms Grignard reagents, RMgX, and related di-alkyls, R_2Mg, by reaction of the metal directly with the alkyl bromide or iodide.

The element crystallizes with the close-packed hexagonal structure.

Symbol: Mg; m.p. 649°C; b.p. 1090°C; r.d. 1.738 (20°C); p.n. 12; r.a.m. 24.3050.

magnesium bicarbonate *See* magnesium hydrogencarbonate.

magnesium carbonate ($MgCO_3$) A white solid that occurs naturally as the mineral magnesite and in association with calcium carbonate in dolomite. Magnesium carbonate is sparingly soluble in water but reacts with dilute acids to give the salt, carbon dioxide, and water. Because of this it is used in medicine as a mild antacid. Magnesium carbonate also decomposes easily to give magnesium oxide, which is an important refractory material.

magnesium chloride ($MgCl_2$) A solid that exists in a variety of hydrated forms, most commonly as the hexahydrate, $MgCl_2.6H_2O$. When heated, this hydrated form is hydrolyzed by its water of crystallization, with the evolution of hydrogen chloride and the formation of the oxide:

$$MgCl_2 + H_2O \rightarrow MgO + 2HCl$$

The anhydrous chloride must therefore be prepared by evaporating an aqueous solution in an atmosphere of hydrogen chloride. It is used in the preparation of cotton fabrics.

magnesium hydrogencarbonate (**magnesium bicarbonate**; $Mg(HCO_3)_2$) The solid hydrogencarbonate is unknown at room temperature. The compound is produced in solution when water containing carbon dioxide dissolves magnesium carbonate:

$$MgCO_3 + H_2O + CO_2 \rightarrow Mg(HCO_3)_2$$

It is a cause of temporary hardness in water.

magnesium hydroxide ($Mg(OH)_2$) A white solid that dissolves sparingly in water to give an alkaline solution. It can be prepared by adding an alkali to a soluble magnesium compound. Magnesium hydroxide is used in medicine as an antacid.

magnesium oxide (**magnesia**; MgO) A white solid that occurs naturally as the mineral periclase. It can be prepared by heating magnesium in oxygen or by decomposition of magnesium hydroxide, carbonate, or nitrate. Magnesium oxide is weakly basic because of the attraction of the oxide ions for protons from water molecules:

$$O^{2-} + H_2O \rightarrow 2OH^-$$

Magnesium oxide melts at 2800°C and is widely used as a refractory lining for furnaces.

magnesium peroxide (MgO_2) A white insoluble solid prepared by reacting sodium or barium peroxide with a concentrated solution of a magnesium salt. Magnesium peroxide is used as a bleach for dyestuffs and silks.

magnesium sulfate ($MgSO_4$) A solid that occurs naturally in many salt deposits in combination with other minerals. One hydrated form, $MgSO_4.7H_2O$, is known as *Epsom salt* or epsomite. Epsom salt is made on a commercial scale by reacting magnesium carbonate with sulfuric acid. Magnesium sulfate is used as a purgative drug and as an antidote for barium and barbiturate poisoning. It is also used in the dyeing and sizing of textiles, in synthetic fiber production, and as a source of magnesium in fertilizers.

magnetic constant *See* magnetic permeability.

magnetic moment The ratio of the torque exerted on a magnet or a loop that carries an electrical current or on a moving electrical charge in a magnetic field to the field strength of that magnetic field. An electron has an *orbital magnetic moment* because of the magnetic field generated by its orbital motion round the nucleus and a *spin magnetic moment* because of its quantum mechanical spin. Atomic nuclei can also have magnetic moments because of their spin. The existence of electron and nuclear magnetic moments is made use of in various branches of spectroscopy and in magnetochemistry.

magnetic permeability symbol μ. A quantity that characterizes the response of a medium to a magnetic field. The *permeability of free space*, which is denoted μ_0 and is also called the *magnetic constant* has the value $4\pi \times 10^{-7}$ Hm^{-1} in SI units. The magnetic permeability of a medium relates the magnetic flux density **B** to the magnetic field strength **H** by the equation: $\mathbf{B} = \mu\mathbf{H}$. The *relative magnetic permeability* of a medium is defined by $\mu = \mu_r\mu_0$.

magnetic quantum number *See* atom.

magnetic separation A method of separating crushed mineral mixtures using the magnetic properties that a component may possess, such as ferromagnetism or diamagnetism.

magnetic susceptibility Symbol χ. A dimensionless quantity, related to the MAGNETIC PERMEABILITY of a medium, that characterizes the magnetic nature of that medium. The relation between the magnetic susceptibility and the relative magnetic permeability μ_r is given by $\chi = \mu_r^{-1}$. If the material is diamagnetic χ has a small negative value. If the material is paramagnetic χ has a small positive value. If the material is ferromagnetic χ has a large positive value.

magnetism /mag-nĕ-tiz-ăm/ The study of the nature and cause of magnetic force fields, and how different substances are affected by them. Magnetic fields are produced by moving charge – on a large scale (as with a current in a coil, forming an *electromagnet*), or on the small scale of the moving charges in the atoms. It is generally assumed that the Earth's magnetism and that of other planets, stars, and galaxies have the same cause.

Substances may be classified on the basis of how samples interact with fields. Different types of magnetic behavior result from the type of atom. *Diamagnetism*, which is common to all substances, is due to the orbital motion of electrons. *Paramagnetism* is due to electron spin, and a property of materials containing unpaired electrons. It is particularly important in transition-metal chemistry, in which the complexes often contain unpaired electrons. Magnetic measurements can give information about the bonding in these complexes. *Ferromagnetism*, the strongest effect, also involves electron spin and the alignment of magnetic moments in domains.

magnetite /mag-nĕ-tÿt/ A black mineral form of iron(II)–iron(III) oxide, Fe_2O_3, a major type of iron ore. It is also used as a flux in making ceramics. *See also* lodestone.

magnetochemistry /mag-nee-toh-**kem**-iss-tree/ The use of magnetic measurements to give information about bonding.

magneton /mag-nĕ-ton/ Symbol μ. A unit for measuring magnetic moments at the atomic and subatomic scales. The *Bohr magneton* μ_B is given by $\mu_B = eh/4\pi m_e$, where e and m are the charge and mass of an electron respectively, and h is the Planck constant. The Bohr magneton is the smallest unit the orbital magnetic moment of an electron in an atom can have. The *nuclear magneton* μ_N is given by replacing the mass of the electron by the mass of the proton and is therefore about 1840 times smaller than the Bohr magneton.

malachite /mal-ă-kÿt/ A green mineral consisting of copper(II) carbonate and hydroxide ($CuCO_3.Cu(OH)_2$). It is used as an ore and a pigment.

maleic acid /mă-**lee**-ik/ *See* butenedioic acid.

malic acid /mal-ik/ (**2-hydroxybutanedioic acid**; $HCOOCH_2CH(OH)COOH$) A colorless crystalline carboxylic acid that is found in unripe fruits. It tastes of apples and is used in food flavorings.

malonic acid /mă-**lon**-ik/ *See* propanedioic acid.

maltose /môl-tohs/ ($C_{12}H_{22}O_{11}$) A sugar found in germinating cereal seeds. It is a disaccharide composed of two glucose units. Maltose is an important intermediate in the enzyme hydrolysis of starch. It is further hydrolyzed to glucose.

manganate(VI) /**mang**-gă-nayt/ A salt containing the ion MnO_4^{2-}.

manganate(VII) (**permanganate**) A salt containing the ion MnO_4^-. Manganate(VII) salts are purple, and strong oxidizing agents. In basic solutions the dark green manganate(VI) ion is formed.

manganese /**mang**-gă-neez, -nees/ A transition metal occurring naturally as oxides, e.g. pyrolusite (MnO_2). Nodules found on the ocean floor are about 25% manganese. Its main use is in alloy steels made by adding pyrolusite to iron ore in an electric furnace. Manganese decomposes cold water and dilute acids to give hydrogen and reacts with oxygen and nitrogen when heated. The oxidation states are +7, +6, +4, and (the most stable) +2. Manganese(II) salts are pale pink and with alkali the solutions precipitate manganese(II) hydroxide, which rapidly oxidizes in air to brown manganese(III) oxide.

Symbol: Mn; m.p. 1244°C; b.p. 1962°C; r.d. 7.44 (20%C); p.n. 25; r.a.m. 54.93805.

manganese dioxide /dȳ-**oks**-ȳd, -id/ *See* manganese(IV) oxide.

manganese(II) oxide (**manganous oxide**; MnO) A green powder prepared by heating manganese(II) carbonate or oxalate in the absence of air. Alternatively it may be prepared by heating the higher manganese oxides in a stream of hydrogen. Manganese(II) oxide is a basic oxide and is almost insoluble in water. At high temperatures it is reduced by hydrogen to manganese. On exposure to air, manganese(II) oxide rapidly oxidizes. The compound has a crystal lattice similar to that of sodium(I) chloride.

manganese(III) oxide (**manganic oxide**; **manganese sesquioxide**; Mn_2O_3) A black powder obtained by igniting manganese(IV) oxide or a manganese(II) salt in air at 800°C. It reacts slowly with cold dilute acids to form manganese(III) salts. Manganese(III) oxide occurs in nature as braunite ($3Mn_2O_3.MnSiO_3$) and the monohydrate ($Mn_2O_3.H_2O$) occurs in the mineral manganite. The manganese(III) oxide crystal lattice contains Mn^{3+} and O^{2-} ions. With concentrated alkalis, it undergoes disproportionation to give manganese(II) and manganese(IV) ions.

manganese(IV) oxide (**manganese dioxide**; MnO_2) A black powder prepared by the action of heat on manganese(II) nitrate. A hydrated form occurs naturally as the mineral *pyrolusite*. Manganese(IV) oxide is insoluble in water. It is a powerful oxidizing agent: it reacts with hot concentrated hydrochloric acid to produce chlorine and with warm sulfuric acid to give oxygen. The dihydrate ($MnO_2.2H_2O$) is formed when potassium manganate(VII) is reduced in alkaline solution. It is used as a catalyst in the laboratory preparation of chlorine, as a depolarizer in electric dry cells, and in the glass industry. At 500–600°C, manganese(IV) oxide decomposes to give manganese(III) oxide and trimanganese tetroxide. It has a good electrical conductivity.

manganese sesquioxide /ses-kwee-**oks**-ȳd, -id/ *See* manganese(III) oxide.

manganic oxide /man-**gan**-ik/ *See* manganese(III) oxide.

manganin /**mang**-gă-nin/ An alloy of copper, manganese, and nickel. It has a high electrical resistivity and is used to make resistors.

manganous oxide /**mang**-gă-nŭs/ *See* manganese(II) oxide.

mannitol /**man**-ă-tol, -tohl/ ($HOCH_2(CHOH)_4CH_2OH$) A soluble hexahydric alcohol that occurs in many plants and fungi. It is used in medicines and as a sweetener (particularly in foods for diabetics). It is an isomer of sorbitol.

mannose /**man**-ohs/ ($C_6H_{12}O_6$) A simple sugar found in many polysaccharides. It is an aldohexose, isomeric with glucose.

manometer /mă-**nom**-ĕ-ter/ A device for measuring pressure. A simple type is a U-shaped glass tube containing mercury or other liquid. The pressure difference between the arms of the tube is indicated by the difference in heights of the liquid.

marble A dense mineral form of calcium carbonate ($CaCO_3$), formed from limestone.

Markovnikoff's rule /mar-**koff**-ni-koffs/ A rule that predicts the quantities of the products formed when an acid (HA) adds to the double bond in an alkene. If the alkene is not symmetrical two products may result; for instance $(CH_3)_2C{:}CH_2$ can yield either $(CH_3)_2HCCH_2A$ or $(CH_3)_2ACCH_3$. The rule states that the major product will be the one in which the hydrogen atom attaches itself to the carbon atom with the larger number of hydrogen atoms. In the example above, therefore, the major product is $(CH_3)_2ACCH_3$.

The Markovnikoff rule is explainable if the mechanism is ionic. The first step is addition of H^+ to one side of the double bond, forming a carbonium ion. The more stable form of carbonium ion will be the form in which the positive charge appears on the carbon atom with the largest number of alkyl groups – thus the hydrogen tends to attach itself to the other carbon. The positive charge is partially stabilized by the electron-releasing (inductive) effect of the alkyl groups.

Additions of this type do not always follow the Markovnikoff rule. Under certain conditions the reaction may involve the free radicals H• and A•, in which case the opposite (anti-Markovnikoff) effect occurs. The rule is named for the Russian chemist Vladimir Vasilyevich Markovnikoff (1838–1904).

marsh gas Methane that is produced in marshes by decomposing vegetation.

Marsh's test *See* arsine.

mass Symbol: m A measure of the quantity of matter in an object. Mass is determined in two ways: the *inertial mass* of a body determines its tendency to resist change in motion; the *gravitational mass* determines its gravitational attraction for other masses. The SI unit of mass is the kilogram.

mass action, law of At constant temperature, the rate of a chemical reaction is directly proportional to the active mass of the reactants, the active mass being taken as the concentration. In general, the rate of reaction decreases steadily as the reaction proceeds; a measure of the concentration of any of the reactants will give a measure of the rate of reaction.

For the reaction A + B → products, the law of mass action states that

$$\text{rate} = k[A][B]$$

where [A] represents the concentration of A in mol dm^{-3} and k is a constant dependent on the reaction. The term *active mass* can equal the concentration in mol dm^{-3} only if there is no interaction or interference between the reacting molecules. Many systems exhibit interactions and interference and consequently the concentration has to be multiplied by an activity coefficient in order to obtain the effective active mass. *See* activity coefficient.

mass-energy equation The equation $E = mc^2$, where E is the total energy (rest-mass energy + kinetic energy + potential energy) of a mass m, c being the speed of light. The equation is a consequence of Einstein's Special theory of relativity; mass is a form of energy and energy also has mass. Conversion of rest-mass energy into kinetic energy is the source of power in radioactive substances and the basis of nuclear-power generation.

massicot /**mass**-ă-kot/ *See* lead(II) oxide.

mass number *See* nucleon number.

mass spectrometer An instrument for producing ions in a gas and analyzing them according to their charge/mass ratio. The earliest experiments by J. J. Thomson used a stream of positive ions from a discharge tube, which were deflected by parallel electric and magnetic fields at right angles to the beam. Each type of ion formed a parabolic trace on a photographic plate (a *mass spectrograph*).

In modern instruments, the ions are produced by ionizing the gas with electrons. The positive ions are accelerated out of this ion source into a high-vacuum region. Here, the stream of ions is deflected and focused by a combination of electric and magnetic fields, which can be varied so that different types of ion fall on a detector. In this way, the ions can be analyzed according to their mass, giving a *mass spectrum* of the material. Mass spectrometers are used for accurate measurements of relative atomic mass and for analysis of isotope abundance. They can also be used to identify compounds and analyze mixtures. An organic compound bombarded with electrons forms a number of fragment ions. (Ethane (C_2H_6), for instance, might form CH^+, $C_2H_5^+$, CH_2^+, etc.) The relative proportions of different types of ions is used to find the structure of new compounds. The characteristic spectrum can also identify compounds by comparison with standard spectra.

mass spectrum *See* mass spectrometer.

matrix /may-triks, mat-riks/ (*pl.* **matrices** or **matrixes**) A continuous solid phase in which particles of a different solid phase are embedded.

matte /mat/ A mixture of iron and copper sulfides obtained at an intermediate stage in smelting copper ores.

maxwell /maks-wĕl/ Symbol: Mx A unit of magnetic flux used in the c.g.s. system. It is equal to 10^{-8} Wb. The unit is named for the British physicist James Clerk Maxwell (1831–79).

mechanism A step-by-step description of the events taking place in a chemical reaction. It is a theoretical framework accounting for the fate of bonding electrons and illustrates which bonds are broken and which are formed. For example, in the chlorination of methane to give chloromethane:

step 1

$$Cl{:}Cl \rightarrow 2Cl\bullet$$

step 2

$$Cl\bullet + CH_4 \rightarrow HCl + CH_3\bullet$$

step 3

$$CH_3\bullet + Cl{:}Cl \rightarrow CH_3Cl + Cl\bullet$$

mega- Symbol: M A prefix denoting 10^6. For example, 1 megahertz (MHz) = 10^6 hertz (Hz).

meitnerium /mÿt-**neer**-ee-ŭm/ A radioactive metallic element not found naturally on Earth. Only a few atoms of the element have ever been detected; it can be made by bombarding a bismuth target with iron nuclei. The isotope ^{266}Mt has a half-life of about 3.4×10^{-3}s.

Symbol: Mt; p.n. 109.

melamine /mel-ă-mÿn/ (**triaminotriazine;** $C_3N_3(NH_2)_3$) A white solid organic compound whose molecules consist of a six-membered heterocyclic ring of alternate carbon and nitrogen atoms with three amino groups attached to the carbons. Condensation polymerization with methanal or other aldehydes produces melamine resins, which are important thermosetting plastics.

melting (**fusion**) The process by which a solid is converted into a liquid by heat or pressure.

melting point The temperature at which a solid is in equilibrium with its liquid phase at standard pressure and above which the solid melts. This temperature is always the same for a particular solid. Ionically bonded solids generally have much higher melting points than those in which the forces are covalent or intermolecular.

membrane A thin pliable sheet of tissue or other material acting as a boundary. The membrane may be either natural (as in cells, skin, etc.) or synthetic modifications of natural materials (cellulose derivatives or rubbers). In many physicochemical studies membranes are supported on porous materials, such as porcelain, to provide mechanical strength. Membranes are generally permeable to some degree.

Membranes can be prepared to permit the passage of other molecules and micromolecular material. Because of permeability effects, concentration diffferences at a membrane give rise to a whole range of membrane-equilibrium studies, of which osmosis, dialysis, and ultrafiltration are examples. *See also* semipermeable membrane.

mendelevium /men-dĕ-**lee**-vee-ŭm/ A radioactive transuranic element of the actinoid series, not found naturally on Earth. Several short-lived isotopes have been synthesized.

Symbol: Md; p.n. 101; most stable isotope ^{258}Md (half-life 57 minutes).

Mendius reaction /**men**-dee-ŭs/ The reduction of the cyanide group to a primary amine group using sodium in alcohol:

$$RCN + 2H_2 \rightarrow RCH_2NH_2$$

It is a method of increasing the chain length of compounds in ascending a homologous series.

mer *See* polymer.

mercaptan /mer-**kap**-tăn/ *See* thiol.

mercuric chloride /mer-**kyoo**-rik/ *See* mercury(II) chloride.

mercuric oxide *See* mercury(II) oxide.

mercuric sulfide *See* mercury(II) sulfide.

mercurous chloride /mer-**kyoo**-rŭs/ *See* mercury(I) chloride.

mercurous oxide *See* mercury(I) oxide.

mercurous sulfide *See* mercury(I) sulfide.

mercury /**mer**-kyŭ-ree/ A transition metal that occurs naturally as cinnabar (mercury(II) sulfide); small drops of metallic mercury also occur in cinnabar and in some volcanic rocks. The vapor is very poisonous. Mercury is used in thermometers, special amalgams for dentistry, scientific apparatus, and in mercury cells. Mercury compounds are used as fungicides, timber preservatives, and detonators.

Symbol: Hg; m.p. −38.87°C; b.p. 356.58°C; r.d. 13.546 (20°C); p.n. 80; r.a.m. 200.59. *See* amalgam; zinc group.

mercury cell A voltaic or electrolytic cell in which one or both of the electrodes consists of mercury or an amalgam. Amalgam electrodes are used in the Daniell cell and the Weston cadmium cell. Flowing mercury electrodes are also used in electrolytic cells (*see* polarography). In the production of chlorine, sodium chloride solution is electrolyzed in a cell with carbon anodes and a flowing mercury cathode. At the cathode, sodium metal is formed, which forms an amalgam with the mercury.

mercury(I) chloride (**mercurous chloride; calomel**; Hg_2Cl_2) A white precipitate prepared by adding dilute hydrochloric acid to a mercury(I) salt solution or by subliming mercury(II) chloride with mercury. Mercury(I) chloride is sparingly soluble in water and is blackened by both ammonia gas, which it absorbs, and by alkalis. It was formerly used as a purgative.

mercury(II) chloride (**mercuric chloride; corrosive sublimate**; $HgCl_2$) A colorless crystalline compound prepared by direct combination of mercury with cold dry chlorine. Mercury(II) chloride is soluble in water; it also dissolves in concentrated hydrochloric acid because of the formation of complex ions, $HgCl_4^{2-}$ and $HgCl_3^-$. On heating, mercury(II) chloride sublimes forming a white translucent mass. It is extremely poisonous, but in dilute solution (1:1000) it is used as an antiseptic.

mercury(I) oxide (**mercurous oxide**; Hg_2O) The black precipitate formed on addition of sodium hydroxide solution to a solution of mercury(I) nitrate is thought by some to be mercury(I) oxide; others, who doubt its existence, think that the blackness of the precipitate is due to some free mercury. X-ray examination of this black compound has shown it to be an intimate mixture of mercury(II) oxide and mercury.

mercury(II) oxide (**mercuric oxide**; HgO) A poisonous compound formed as a yellow powder by the addition of sodium(I) hydroxide to a solution of mercury(II) nitrate or as a red solid by heating mercury at 350°C for a long time. The difference in color is simply one of particle size. If the oxide is strongly heated it decomposes to give mercury and oxygen.

mercury(I) sulfide (**mercurous sulfide**; Hg_2S) The brownish black precipitate formed when mercury is treated with cold concentrated sulfuric acid for a long time is thought to be mercury(I) sulfide. Alternatively it may be prepared by the action of hydrogen sulfide or an alkaline sulfide on a mercury(I) salt solution. As soon as the mercury(I) sulfide has been formed it disproportionates to give mercury(II) sulfide and mercury.

mercury(II) sulfide (**mercuric sulfide**; **vermilion**; HgS) A compound that occurs in nature as the minerals *cinnabar* (a red solid) and *metacinnabar* (a black solid). It is prepared as a black precipitate by the action of hydrogen sulfide on a soluble mercury(II) salt solution. On heating, mercury(II) sulfide sublimes and becomes red. It is insoluble in dilute hydrochloric and nitric acids but will dissolve in concentrated nitric acid and aqua regia. Mercury(II) sulfide is used as the pigment vermilion.

***meso*-form** /**mess**-oh-form/ *See* optical activity.

mesomerism /mes-**som**-ĕ-riz-ăm/ *See* resonance.

meta- 1. Designating a benzene compound with substituents in the 1,3 positions. The position on a benzene ring that is two carbon atoms away from a substituent is the meta position. This was used in the systematic naming of benzene derivatives. For example, meta-dinitrobenzene (or *m*-dinitrobenzene) is 1,3-dinitrobenzene.

2. Certain acids regarded as formed from an anhydride and water are named meta acids to distinguish them from the more hydrated ortho acids. For example, H_2SiO_3 (SiO_2 + H_2O) is metasilicic acid; H_4SiO_4 (SiO_2 + $2H_2O$) is orthosilicic acid.

See also ortho-; para-.

metabolism /mĕ-**tab**-ŏ-liz-ăm/ The biochemical reactions that take place in cells. The molecules taking part in these reactions are termed *metabolites*. Metabolic reactions characteristically occur in small steps that together make up a *metabolic pathway*. They involve the breaking down of molecules to provide energy (catabolism) and the building up of more complex molecules and structures from simpler molecules (anabolism).

metabolite /mĕ-**tab**-ŏ-lÿt/ A substance that takes part in a metabolic reaction, either as reactant or product. Some are synthesized within the organism itself, whereas others have to be taken in as food.

metaiodic(VII) acid /met-ă-ÿ-**od**-ik/ *See* iodic(VII) acid.

metal *See* metals.

metal carbonyl A coordination compound formed between a metal and carbonyl groups. Transition metals form many such compounds.

metaldehyde /mĕ-**tal**-dĕ-hÿd/ *See* ethanal.

metallic bond A bond formed between atoms of a metallic element in its zero oxidation state and in an array of similar atoms. The outer electrons of each atom are regarded as contributing to an 'electron gas', which occupies the whole crystal of the metal. It is the attraction of the positive atomic cores for the negative electron gas that provides the strength of the metallic bond.

Quantum mechanical treatment of the electron gas restricts its energy to a series of 'bands'. It is the behavior of electrons in these bands that gives rise to semiconductors.

metallic crystal A crystal formed by metal atoms in the solid state. Each atom contributes its valence (outer) electrons to a 'sea' of electrons, which are free to migrate through the solid, and the ions remaining are arranged in a lattice. The ability of electrons to move through the lattice accounts for the electrical and thermal conductivity of metals.

metallocene /mĕ-**tal**-ŏ-seen/ A sandwich compound in which a metal atom or ion is coordinated to two cyclopentadienyl ions. Ferrocene ($Fe(C_5H_5)_2$) is the commonest example.

metalloid /**met**-ă-loid/ Any of a class of chemical elements that are intermediate in properties between metals and non-metals. Examples are germanium, arsenic, and tellurium.

There is, in fact, no clear-cut distinction between metals and non-metals. In the periodic table, there is a change from metallic to non-metallic properties across the table, and an increase in metallic properties down a group. Consequently there is a diagonal around the center of the table (B, Si, As, Te) in which there is a borderline between metals and non-metals, and the metalloids are the borderline cases. Elements such as arsenic, germanium, and tellurium are semiconductors, but other elements are often said to be metalloids according to their chemical properties. Tin, for instance, forms salts with acids but also forms stannates with alkalis. Its oxide is amphoteric. Note also that tin has metallic (white tin) and non-metallic (gray tin) allotropes.

metalloporphyrin /mĕ-tal-oh-**por**-fă-rin/ *See* porphyrins.

metallurgy /**met**-ă-ler-jee/ The study of metals, especially methods of extracting metals from their ores and the formation and properties of alloys.

metals Any of a class of chemical elements with certain characteristic properties. In everyday usage, metals are elements (and alloys) such as iron, aluminum, and copper, which are lustrous malleable solids – usually good conductors of heat and electricity. Note that this is not a strict definition – some metals are poor conductors, mercury is a liquid, etc.

In chemistry, metals are distinguished by their chemical properties, and there are two main groups. Reactive metals, such as the alkali metals and alkaline-earth metals, are electropositive elements. They are high in the electromotive series and tend to form compounds by losing electrons to give positive ions. They have basic oxides and hydroxides. This typical metallic behavior decreases across the periodic table and increases down a group in the table.

The other type of metals are the transition elements, which are less reactive, have variable valences, and tend to form complexes. In the solid and liquid states metals have metallic bonds, formed by positive ions with free electrons. *See also* metalloid; non-metal.

metastable species /met-ă-**stay**-băl/ An excited state of an atom, ion, or molecule, that has a relatively long lifetime before reverting to the ground state. Metastable species are intermediates in some reactions.

metastable state A condition of a system or body in which it appears to be in stable equilibrium but, if disturbed, can settle into a lower energy state. For example, supercooled water is liquid below 0°C (at standard pressure). When a small crystal of ice or dust (for example) is introduced, rapid freezing occurs.

metathesis /mĕ-**tath**-ĕ-sis/ *See* double decomposition.

meter /**mee**-ter/ Symbol: m The SI base unit of length, defined as the distance traveled by light in vacuum in 1/(2.99 792 458 × 10^8) second. This definition was adopted in 1983 to replace the 1967 definition of a length equal to 1 650 763.73 wavelengths in vacuum corresponding to the transition between the levels $2p^{10}$ and $5d^5$ of the ^{86}Kr atom.

methanal /**meth**-ă-nal/ (**formaldehyde;** HCOH) A colorless gaseous aldehyde. It is manufactured by the oxidation of methanol (500°C and a silver catalyst):

$$2CH_3OH + O_2 \rightarrow 2HCOH + 2H_2O$$

The compound is used in the manufacture of urea–formaldehyde resins. A solution of methanal (40%) in water is called *formalin*. It is extensively used as a preservative for biological specimens.

If an aqueous solution of methanal is evaporated a polymer – *polymethanal* – is formed:

$$-O-CH_2-O-CH_2-O-CH_2-$$

This was formerly called *paraformaldehyde*. If methanal is distilled from acidic solutions a cyclic *methanal trimer* ($C_3O_3H_6$) is produced.

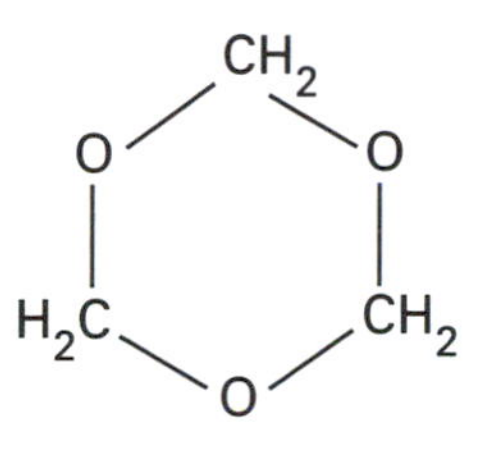

Methanal trimer

methane /**meth**-ayn/ (CH_4) A gaseous alkane. Natural gas is about 99% methane and this provides an important starting material for the organic-chemicals industry. Methane can be chlorinated directly to produce the more reactive chloromethanes, or it can be 'reformed' by partial oxidation or using steam to give mixtures of carbon oxides and hydrogen. Methane is the first member of the homologous series of alkanes.

methanide /**meth**-ă-nÿd/ *See* carbide.

methanoate /meth-ă-**noh**-ayt/ (**formate**) A salt or ester of methanoic acid.

methanoic acid /meth-ă-**noh**-ik/ (**formic acid;** HCOOH) A liquid carboxylic acid made by the action of sulfuric acid on sodium methanoate (NaOOCH). It is a strong reducing agent. Methanoic acid occurs naturally in ants and nettles.

methanol /**meth**-ă-nol, -nohl/ (**methyl alcohol; wood alcohol;** CH_3OH) A colorless liquid alcohol, which is used as a solvent and in the manufacture of methanal (formaldehyde) for the plastics and drugs industries. Methanol was originally produced from the distillation of wood. Now it is manufactured by the catalytic oxidation of methane from natural gas.

methionine /meth-**ÿ**-ŏ-neen, -nin/ *See* amino acids.

methoxy group /meth-**oks**-ee/ The group CH_3O-.

methyl acetate /**meth**-ăl/ *See* methyl ethanoate.

methyl alcohol *See* methanol.

methylamine /meth-ăl-ă-**meen**, -**am**-een/ (CH_3NH_2) A colorless flammable gas that smells like ammonia. It is the simplest primary amine, used for making herbicides and other organic chemicals.

methylaniline /meth-ăl-**an**-ă-lin, -lÿn/ *See* toluidine.

methylated spirits /**meth**-ă-lay-tid/ Ethanol to which is added methanol (about 9.5%), pyridine (about 0.5%), and a blue dye. The ethanol is denatured in this way so that it can be sold without excise duty for use as a fuel and solvent.

methylation /meth-ă-**lay**-shŏn/ A reaction in which a methyl group (CH_3-) is introduced in a compound. Friedel–Crafts reactions involving halo-methanes are examples of methylations.

methylbenzene /meth-ăl-**ben**-zeen, -been-**zeen**/ (**toluene;** $C_6H_5CH_3$) A colorless liquid hydrocarbon, similar to benzene both in structure and properties. As methylbenzene is much less toxic than benzene it is more widely used, especially as a solvent.

Large quantities are required for the manufacture of TNT (trinitrotoluene).

Methylbenzene is itself produced either from the fractional distillation of coal tar or from methylcyclohexane (a constituent of some crude oils):

$$C_6H_{11}CH_3 \rightarrow C_6H_5CH_3 + 3H_2$$

A catalyst of aluminum and molybdenum oxides is employed at high temperatures and pressures.

methyl bromide *See* bromomethane.

2-methylbuta-1,3-diene (**isoprene**; $CH_2{:}CH(CH_3)CH{:}CH_2$) A colorless unsaturated liquid hydrocarbon, which occurs in terpenes and natural rubber. It is used to make synthetic rubber.

methyl chloride *See* chloromethane.

methyl cyanide (**acetonitrile**; CH_3CN) A pleasant-smelling poisonous colorless liquid organic nitrile. Methyl cyanide is a polar solvent, and is widely used for dissolving both inorganic and organic compounds.

methylene /**meth**-ă-leen/ *See* carbene.

methylene group The group $:CH_2$.

methyl ethanoate (**methyl acetate**; CH_3COOCH_3) A colorless liquid ester with a fragrant odor. It is commonly used as a solvent.

methyl ethyl ketone *See* butanone.

methyl group The group CH_3–.

methyl iodide *See* iodomethane.

methyl methacrylate /meth-**ak**-ră-layt/ ($CH_2{:}CCH_3COOCH_3$) The methyl ester of methacrylic acid. The compound is used in the manufacture of a number of acrylic polymers, such as Plexiglas (polymethylmethacrylate).

methyl orange An acid–base indicator that is red in solutions below a pH of 3 and yellow above a pH of 4.4. As the transition range is clearly on the acid side, methyl orange is suitable for the titration of an acid with a moderately weak base, such as sodium carbonate.

methylphenols /meth-ăl-**fee**-nol, -nohl/ (**cresols**; $HOC_6H_4CH_3$) Compounds with both methyl and hydroxyl groups substituted onto the benzene ring. There are three isomers, with the methyl group in the 2–, 3–, and 4– positions, respectively. A mixture of the isomers can be obtained from coal tar. It is used as a germicide (known as Lysol).

methyl red An acid–base indicator that is red in solutions below a pH of 4.2 and yellow above a pH of 6.3. It is often used for the same types of titration as methyl orange but the transition range of methyl red is nearer neutral (pH7) than that of methyl orange. The two molecules are structurally similar.

methyl salicylate (**oil of wintergreen; methyl 2-hydroxybenzoate**; $C_8H_8O_3$) The methyl ester of SALICYLIC ACID, which occurs in certain plants. It is absorbed through the skin and used medicinally to relieve rheumatic symptoms. It is also used in perfumes and as a flavoring agent in various foods.

metric system A system of units based on the meter and the kilogram and using multiples and submultiples of 10. SI units, c.g.s. units, and m.k.s. units are all scientific metric systems of units.

metric ton *See* tonne.

mho /moh/ *See* siemens.

mica /**mÿ**-kă/ A member of an important group of aluminosilicate minerals that have a characteristic layered structure. The three main types are biotite, lepidolite, and muscovite, which differ in their content of other elements (such as potassium, magnesium, and iron). Mica flakes are used as electrical insulators, dielectrics, and small heat-proof windows.

micelle /mi-**sell**/ An aggregate of molecules in a COLLOID.

micro- Symbol: μ A prefix denoting 10^{-6}. For example, 1 micrometer (μm) = 10^{-6} meter (m).

micron /**mÿ**-kron/ Symbol: μ A unit of length equal to 10^{-6} meter.

microwaves /**mÿ**-kroh-wayvz/ A form of ELECTROMAGNETIC RADIATION, ranging in wavelength from about 1 mm (where it merges with infrared) to about 120 mm (bordering on radio waves). Microwaves are produced by various electronic devices including the klyston; they are often carried over short distances in tubes of rectangular section called *waveguides*.

Spectra in the microwave region can give information on the rotational energy levels of certain molecules.

migration **1.** The movement of an atom, group, or double bond from one position to another in a molecule.
2. The movement of ions in an electric field.

milk of lime *See* calcium hydroxide.

milk sugar *See* lactose.

milli- Symbol: m A prefix denoting 10^{-3}. For example, 1 millimeter (mm) = 10^{-3} meter (m).

millimeter of mercury *See* mmHg.

mineral A naturally occurring inorganic compound, usually crystalline and of definite chemical composition; a mineral's physical properties are also more or less constant. All rocks are composed of mixtures of individual minerals. *See* rock.

mineral acid An inorganic acid, especially an acid used commercially in large quantities. Examples are hydrochloric, nitric, and sulfuric acids.

mirror image A shape that is identical to another except that its structure is reversed as if viewed in a mirror. If an object is not symmetrical it cannot be superimposed on its mirror image. For example, the left hand is the mirror image of the right hand.

misch metal /mish/ A pyrophoric alloy of cerium and other lanthanoids, used in lighter flints.

miscible /**miss**-ă-băl/ Denoting combinations of substances that, when mixed, give rise to only one phase; i.e. substances that dissolve in each other. *See* solid solution; solution.

Mitscherlich's law /**mich**-er-lik/ (**law of isomorphism**) The law stating that substances that crystallize in isomorphous forms (i.e. have identical crystalline forms and form mixed crystals) have similar chemical compositions. The law can be used to indicate the formulae of compounds. For instance, the fact that chromium(III) oxide is isomorphous with Fe_2O_3 and Al_2O_3 implies that its formula is Cr_2O_3. The law is named for the German chemist Eilhard Mitscherlich (1794–1863).

mixed indicator A mixture of two or more indicators so as to decrease the pH range or heighten the color change, etc.

mixture Two or more substances forming a system in which there is no chemical bonding between the two. In homogeneous mixtures (e.g. solutions or mixtures of gases) the molecules of the substances are mixed, and there is only one phase. In heterogeneous mixtures (e.g. gunpowder or certain alloys) different phases can be distinguished. Mixtures differ from chemical compounds in that:

1. The chemical properties of the components of a mixture are the same as those of the pure substances.
2. The mixture can be separated by physical means (e.g. distillation or crystallization) or mechanically.
3. The proportions of the components can vary. Some mixtures (e.g. certain solutions) can only vary in proportions between definite limits.

m.k.s. system A system of units based on the meter, the kilogram, and the second. It formed the basis for SI units.

mmHg (**millimeter of mercury**) A former unit of pressure defined as the pressure that will support a column of mercury one millimeter high under specified conditions. It is equal to 133.322 4 Pa, and is almost identical to the torr.

mobile phase *See* chromatography.

moiety /**moi**-ĕ-tee/ A distinct part of a molecule; for example, the sugar moiety in a nucleoside.

molal concentration /**moh**-lăl/ *See* concentration.

molar /**moh**-ler/ **1.** Denoting a physical quantity divided by the amount of substance. In almost all cases the amount of substance will be in moles. For example, volume (V) divided by the number of moles (n) is molar volume $V_m = v/n$.
2. A *molar solution* contains one mole of solute per cubic decimeter of solvent.

molarity /moh-**la**-ră-tee/ A measure of the concentration of solutions based upon the number of molecules or ions present, rather than on the mass of solute, in any particular volume of solution. The molarity (M) is the number of moles of solute in one cubic decimeter (litre). Thus a 0.5M solution of hydrochloric acid contains 0.5 × (1 + 35.5)g HCl per dm^3 of solution.

mole Symbol: mol The SI base unit of amount of substance, defined as the amount of substance that contains as many elementary entities as there are atoms in 0.012 kilogram of ^{12}C. The elementary entities may be atoms, molecules, ions, electrons, photons, etc., and they must be specified. The amount of substance is proportional to the number of entities, the constant of proportionality being the Avogadro number. One mole contains $6.022\,045 \times 10^{23}$ entities. One mole of an element with relative atomic mass A has a mass of A grams (this mass was formerly called one *gram-atom* of the element).

molecular beam /mŏ-**lek**-yŭ-ler/ A beam of molecules (or atoms or ions) at low pressure such that the entities in the beam are moving in the same direction with few intermolecular collisions. Molecular beams are used in spectroscopy and in the study of intermolecular forces, chemical reactions, and surfaces.

molecular crystal /mŏ-**lek**-yŭ-ler/ A crystal in which molecules, as opposed to atoms, occupy lattice points. Examples include iodine and solid carbon dioxide (dry ice). Because the forces holding the molecules together are weak, molecular crystals have low melting points. When the molecules are small, the crystal structure approximates to a close-packed arrangement. *See* close packing.

molecular formula The formula of a compound showing the number and types of the atoms present in a molecule, but not the arrangement of the atoms. For example, C_2H_6O represents the molecular formula both of ethanol (C_2H_5OH) and methoxymethane (CH_3OCH_3). The molecular formula can be determined only if the molecular weight is known. *Compare* empirical formula; general formula; structural formula.

molecularity /mŏ-lek-yŭ-**la**-ră-tee/ The total number of reacting molecules in the individual steps of a chemical reaction. Thus, a unimolecular step has molecularity 1, a bimolecular step 2, etc. Molecularity is always an integer, whereas the order of a reaction need not necessarily be so. The molecularity of a reaction gives no information about the mechanism by which it takes place.

molecular orbital *See* orbital.

molecular sieve A substance through which molecules of a limited range of sizes can pass, enabling volatile mixtures to be separated. Zeolites and other metal aluminum silicates can be manufactured with pores of constant dimensions in their molecular structure. When a sample is passed through a column packed with granules of this material, some of the molecules enter these pores and become trapped. The remainder of the mixture passes through the interstices in the column. The trapped molecules can be recovered by heating. Molecular-sieve chromatography is widely used in chemistry and biochemistry laboratories. A modified form of molecular sieve is used in *gel filtration.* The sieve is a continuous gel made from a polysaccharide. In this case, molecules larger than the largest pore size are totally excluded from the column.

molecular spectrum The absorption or emission spectrum that is characteristic of a molecule. Molecular spectra are usually band spectra.

molecular weight *See* relative molecular mass.

molecule /**mol**-ĕ-kyool/ A particle formed by the combination of atoms in a whole-number ratio. A molecule of an element (combining atoms are the same, e.g. O_2) or of a compound (different combining atoms, e.g. HCl) retains the properties of that element or compound. Thus, any quantity of a compound is a collection of many identical molecules. Molecular sizes are characteristically 10^{-10} to 10^{-9} m.

Many molecules of natural products are so large that they are regarded as giant molecules (macromolecules); they may contain thousands of atoms and have complex structural formulae that require very advanced techniques to identify. *See also* formula; relative molecular mass.

mole fraction The number of moles of a given component in a mixture divided by the total number of moles present of all the components. The mole fraction of component A is

$$n_A/(n_A + n_B + n_C + \ldots)$$

where n_A is the number of moles of A, etc.

molybdenum /mŏ-**lib**-dĕ-nŭm/ A transition element that occurs naturally in molybdenite (MoS_2) and wulfenite ($PbMoO_4$). It is used in alloy steels, lamp bulbs, and catalysts. The compound ammonium molybdate, dissolved in nitric acid, is used as a test for phosphates(V). Molybdenum sulfide (MoS_2) is used in lubricants to enhance viscosity.

Symbol: Mo; m.p. 2620°C; b.p. 4610°C; r.d. 10.22 (20°C); p.n. 42; r.a.m. 95.94.

monatomic /mon-ă-**tom**-ik/ Denoting a molecule, radical, or ion consisting of only one atom. For example, helium is a monatomic gas and H• is a monatomic radical.

Mond process /mohnt/ *See* nickel.

monobasic acid /mon-oh-**bay**-sik/ An acid that has only one active proton, such as hydrochloric acid. *See also* dibasic acid.

monochlorobenzene /mon-ŏ-klor-oh-**ben**-zeen, -klor-oh-ben-**zeen**, -kloh-roh-/ *See* chlorobenzene.

monoclinic crystal /mon-ŏ-**klin**-ik/ *See* crystal system.

monohydrate /mon-ŏ-**hÿ**-drayt/ A salt that has a single molecule of WATER OF CRYSTALLIZATION, such as sodium carbonate monohydrate, $Na_2CO_3.H_2O$.

monohydric alcohol /mon-ŏ-**hÿ**-drik/ *See* alcohol.

monomer /**mon**-ŏ-mer/ The molecule, group, (or compound) from which a dimer, trimer, or POLYMER is formed.

monosaccharide /mon-ŏ-**sak**-ă-rÿd, -rid/ A sugar that cannot be hydrolyzed to simpler carbohydrates of smaller carbon content. Glucose and fructose are examples.

monosodium glutamate /mon-ŏ-**soh**-dee-ŭm **gloo**-tă-mayt/ (**MSG**) A white crystalline solid compound, made from soya-bean protein. It is a sodium salt of glutamic acid (*see* amino acid) used as a flavor enhancer, particularly in Chinese cuisine. Monosodium glutamate can cause an allergic reaction in people who are ultrasensitive to it.

monoterpene *See* terpene.

monotropy /mŏ-**not**-rŏ-pee/ The existence of a single allotrope of an element that is always more stable than the other(s) regardless of temperature; phosphorus, for example, exhibits monotropy. The phase diagram for a monotropic element shows that one allotrope always has a lower vapor pressure than the other(s) at all temperatures; this is the stable allotrope.

monovalent /mon-ŏ-**vay**-lĕnt, mŏ-**nov**-ă-/ (**univalent**) Having a valence of one.

mordant /**mor**-dănt/ An inorganic compound used to fix dye in cloth. The mordant (e.g. aluminum hydroxide or chromium salts) is precipitated in the fibers of the cloth, and the dye then absorbs in the particles.

Moseley's law Lines in the x-ray spectra of elements have frequencies that depend on the proton number of the element. For a set of elements, a graph of the square root of the frequency of x-ray emission against proton number is a straight line (for spectral lines corresponding to the same transition). The law is named for the British physicist Henry Gwyn Jeffreys Moseley (1887–1915).

mother liquor The solution remaining after the formation of crystals.

MSG *See* monosodium glutamate.

multicenter bond /mul-ti-**sen**-trik/ A two-electron bond formed by the overlap of orbitals from more than two atoms (usually 3). The bridging in diborane is believed to take place by overlap of an sp^3 hybrid orbital from each boron atom with the 1s orbital on the hydrogen atom. This multicenter bond is called a two-electron three-center bond. The molecule is electron-deficient. *See also* electron-deficient compounds.

multidecker sandwich compound *See* sandwich compound.

multidentate ligand /mul-ti-**den**-tayt/ A ligand that possesses at least two sites at which it can coordinate.

multiple bond A bond between two atoms involving more than one pair of electrons, e.g. a double bond or a triple bond. This additional bonding arises from overlap of atomic orbitals that are perpendicular to the inter-nuclear axis and gives rise to an increase in electron density above and below the inter-nuclear axis. Such bonds are called pi bonds. If the sigma bond axis is taken as the z-axis (i.e. the inter-nuclear axis) then overlap of orbitals along the x-axis gives rise to a pi bond in the xz plane. Similarly orbitals on the y-axis form a pi bond in the yz plane.

multiple proportions, law of Proposed by Dalton in 1804, the principle that when two elements A and B combine to form more than one compound, the weights of B that combine with a fixed weight of A are in small whole-number ratios. For example, in dinitrogen oxide, N_2O, nitrogen monoxide, NO, and dinitrogen tetroxide, N_2O_4, the amounts of nitrogen combined with a fixed weight of oxygen are in the ratio 4:2:1.

multiple-range indicator *See* universal indicator.

Mumetal /myoo-**met**-ăl/ (*Trademark*) A magnetic alloy containing about 75% nickel, the remainder being iron, copper, and chromium and sometimes molybdenum. It is easily magnetized and demagnetized, has a high permeability, and is used in transformer cores and electromechanical equipment.

muriate /**myoor**-ee-ayt/ An obsolete name for a chloride; e.g. muriate of potash (KCl).

mustard gas (**2,2′-dichlorodiethyl sulfide;** $((CH_2ClCH_2)_2S)$) A poisonous vesicant gas used as a war gas.

mutarotation /myoo-tă-roh-**tay**-shŏn/ A change in the optical rotation of a solution with time, as when one optical isomer disappears and another is formed.

myoglobin /mÿ-oh-**gloh**-bin/ A globular protein formed of a heme group and a single polypeptide chain. It occurs in muscle tissue where it acts as an oxygen store.

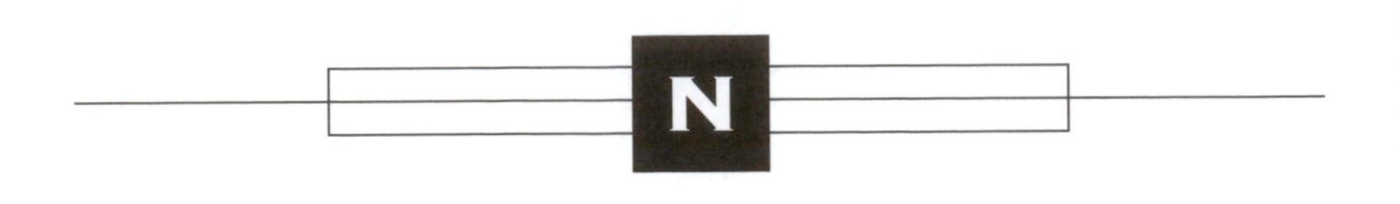

nano- Symbol: n A prefix denoting 10^{-9}. For example, 1 nanometer (nm) = 10^{-9} meter (m).

nanotechnology /nan-oh-tek-**nol**-ŏ-jee/ The technology of devices at the nanometer scale. In such small devices the quantum mechanical nature of electrons has to be taken into account fully. Instruments such as the ATOMIC FORCE MICROSCOPE which can identify and manipulate individual atoms are very useful in nanotechnology.

nanotubes /**nan**-ŏ-tewbz/ Tubular structures with diameters of a few nanometers (1nm = 10^{-9} m). Examples of nanotubes are the carbon structures known as 'bucky tubes', which have a structure similar to that of BUCKMINSTERFULLERENE. Interest has been shown in nanotubes as possible microscopic probes in experiments, as semiconductor materials, and as a component of composite materials. Nanotubes can also be produced by joining amino acids to give tubular polypeptide structures.

naphtha /**naf**-thă, **nap**-thă/ (**solvent naphtha**) An imprecise term for a mixture of hydrocarbons obtained from coal and petroleum. It has a boiling range of 70–160°C and is used as a solvent and as a raw material for making various other organic chemicals.

naphthalene /**naf**-thă-leen, **nap**-thă-leen/ ($C_{10}H_8$) A white crystalline solid with a distinctive smell of mothballs. Naphthalene is found in both the middle- and heavy-oil fractions of crude oil and is obtained by fractional crystallization. It is used in the manufacture of benzene-1,2-dicarboxylic anhydride (phthalic anhydride) and thence in the production of plastics and dyes.

The structure of naphthalene is 'benzene-like', having two six-membered rings fused together. The reactions are characteristic of AROMATIC COMPOUNDS.

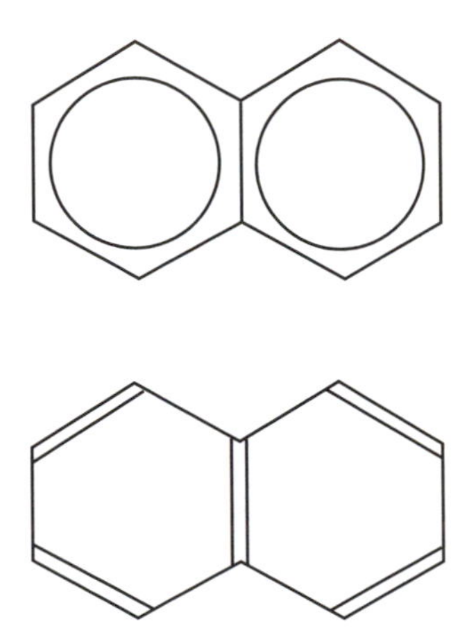

Representations of naphthalene

nascent hydrogen /**nay**-sĕnt/ A particularly reactive form of hydrogen, which is believed to exist briefly between its generation (e.g. by the action of dilute acid on magnesium) and its appearance as bubbles of normal hydrogen gas. It is thought that part of the free energy of the production reaction remains with the hydrogen molecules for a short time. Nascent hydrogen may be used to produce the hydrides of phosphorus, arsenic, and antimony, which are not readily formed from ordinary hydrogen.

natron /**nay**-tron/ A naturally occurring form of hydrated SODIUM CARBONATE

($Na_2CO_3.10H_2O$), found on the beds of dried-out soda lakes.

Natta process /**nat**-ă/ A method for the manufacture of isotactic polypropene using Ziegler catalysts. The process is named for the Italian chemist Giulio Natta (1903–79). *See also* addition polymerization; Ziegler process.

natural abundance *See* abundance.

natural gas Gas obtained from underground deposits and often associated with sources of petroleum. It contains a high proportion of methane and other volatile hydrocarbons.

neighboring-group participation An effect in an organic reaction in which groups close to the point at which reaction occurs affect the rate of reaction or stereochemistry of the products in some way.

nematic crystal /ni-**mat**-ik/ *See* liquid crystal.

neodymium /nee-ŏ-**dim**-ee-ŭm/ A toxic silvery element belonging to the lanthanoid series of metals. It occurs in association with other lanthanoids. Neodymium is used in various alloys, as a catalyst, in compound form in carbon-arc searchlights, etc., and in the glass industry.

Symbol: Nd; m.p. 1021°C; b.p. 3068°C; r.d. 7.0 (20°C); p.n. 60; r.a.m. 144.24.

neon /**nee**-on/ An inert colorless odorless monatomic element of the rare-gas group. Neon forms no compounds. It occurs in minute quantities (0.0018% by volume) in air and is obtained from liquid air. It is used in neon signs and lights, electrical equipment, and gas lasers.

Symbol: Ne; m.p. –248.67°C; b.p. –246.05°C; d. 0.9 kg m^{-3} (0°C); p.n. 10; r.a.m. 20.18.

neoprene /**nee**-ŏ-preen/ A type of synthetic rubber made by polymerization of 2-chlorobuta-1,2,-diene (H_2C:CHCCl:CH_2). It is more resistant to oil, solvents, and temperature than natural rubbers.

neptunium /nep-**tew**-nee-ŭm/ A toxic radioactive silvery element of the actinoid series of metals that was the first transuranic element to be synthesized (1940). Found on Earth only in minute quantities in uranium ores, it is obtained as a by-product from uranium fuel elements.

Symbol: Np; m.p. 640°C; b.p. 3902°C; r.d. 20.25 (20°C); p.n. 93; most stable isotope ^{237}Np (half-life 2.14×10^6 years).

neutralization The stoichiometric reaction of an acid and a base in volumetric analysis. The neutralization point or end point is detected with indicators.

neutral oxide An oxide that forms neither an acid nor a base on reaction with water. Examples include carbon monoxide (CO), dinitrogen oxide (N_2O), and water (H_2O). *See* acidic oxide; amphoteric; basic oxide.

neutron diffraction A method of structure determination that makes use of the quantum mechanical wave nature of neutrons. Thermal neutrons with average kinetic energies of about 0.025 eV have a wavelength of about 0.1 nanometer, making them suitable for investigating the structure of matter at the atomic level. Neutron diffraction is particularly useful for identifying the positions of hydrogen atoms. These are difficult to establish using x-rays since x-rays interact with electrons, and hence are scattered weakly by hydrogen atoms, which have only one electron. Protons scatter neutrons strongly, meaning that the positions of protons can readily be determined using neutron scattering. Neutron diffraction has also been used in investigations of the magnetic ordering in solids because there is an interaction between the magnetic moments of the neutrons and the magnetic moments of atoms in the sample.

neutron number Symbol: *N* The number of neutrons in the nucleus of an atom; i.e.

the nucleon number (*A*) minus the proton number (*Z*).

Newlands' law (**law of octaves**) The observation that, when the elements are arranged in order of increasing relative atomic mass (atomic weight), there is a similarity between members that are eight elements apart (inclusive). For example, lithium has similarities to sodium, beryllium to magnesium, and so on. Newlands discovered this relationship in 1863 before the announcement of Mendeléev's famous periodic law (1869). It is now recognized that Newlands' law arises from there being eight members in each short period. The law is named for the British chemist John Alexander Reina Newlands (1837–98).

newton Symbol: N The SI unit of force, equal to the force needed to accelerate one kilogram by one meter second^{-2}. 1 N = 1 kg m s^{-2}. The unit is named for the English physicist and mathematician Sir Isaac Newton (1642–1727).

Nichrome /**nÿ**-krohm/ (*Trademark*) Any of a group of nickel–chromium–iron alloys, containing 60–80% nickel and about 16% chromium; small amounts of other elements, such as carbon or silicon, may be added. They can withstand very high temperatures and their high electrical resistivity makes them suitable for use in heating elements.

nickel /**nik**-ăl/ A transition metal that occurs naturally as the sulfide and silicate. It is extracted by the *Mond process*, which involves reduction of nickel oxide using carbon monoxide followed by the formation and subsequent decomposition of volatile nickel carbonyl. Nickel is used as a catalyst in the hydrogenation of alkenes, e.g. margarine manufacture, and in coinage alloys. Its main oxidation state is +2 and these compounds are usually green.

Symbol: Ni; m.p. 1453°C; b.p. 2732°C; r.d. 8.902 (25°C); p.n. 28; r.a.m. 58.6934.

nickel carbonyl (**tetracarbonyl nickel(0)**; $Ni(CO)_4$) A colorless liquid with a highly toxic vapor. It is prepared by passing carbon monoxide over finely divided nickel at a temperature of approximately 60°C. The product is purified by fractional distillation. Nickel carbonyl decomposes on heating to give pure nickel. It is used as a catalyst and in the preparation of nickel (*Mond process*).

nickelic oxide *See* nickel(III) oxide.

nickel–iron accumulator (**Edison cell; Nife or NIFE cell**) A type of secondary cell in which the electrodes are formed by steel grids. The positive electrode is impregnated with a nickel–nickel hydride mixture and the negative electrode is impregnated with iron oxide. Potassium hydroxide solution forms the electrolyte. The nickel–iron cell is lighter and more durable than the lead accumulator, and it can work with higher currents. Its e.m.f. is approximately 1.3 volts. The cell is also named for the American physicist and inventor Thomas Alva Edison (1847–1931).

nickelous oxide *See* nickel(II) oxide.

nickel(II) oxide (**nickelous oxide**; NiO) A pale green powder formed by heating nickel(II) hydroxide, nitrate, or carbonate in the absence of air. It is a basic oxide dissolving in dilute acids to give green solutions of nickel(II) salts. Nickel(II) oxide may be reduced by carbon, carbon monoxide, or hydrogen.

nickel(III) oxide (**nickelic oxide**; Ni_2O_3) A black powder formed by the action of heat on nickel(II) oxide in air. It can also be prepared by igniting nickel(II) nitrate or carbonate. It exists as the dihydrate, $Ni_2O_3.2H_2O$.

nickel-silver (**German silver**) Any of a group of alloys containing nickel, copper, and zinc in various proportions (but no silver). They are white or silvery in color, can be highly polished, and have good corrosion-resistance. They are used, for example, in silver plating, chromium plating, and enameling.

nielsbohrium /neels-**bor**-ee-ŭm/ Symbol:

Ns A name formerly suggested for element-107 (bohrium). *See* element.

NIFE cell *See* nickel–iron accumulator.

ninhydrin /nin-**hÿ**-drin/ A colorless organic compound that gives a blue coloration with amino acids. It is used as a test for amino acids, in particular to show the positions of spots of amino acids in paper chromatography.

niobium /nÿ-**oh**-bee-ŭm/ A soft silvery transition element used in welding, special steels, and nuclear reactor work.

Symbol: Nb; m.p. 2468°C; b.p. 4742°C; r.d. 8.570 (20°C); p.n. 41; r.a.m. 92.90638.

niter /**nÿ**-ter/ *See* potassium nitrate.

nitrate /**nÿ**-trayt/ A salt or ester of nitric acid.

nitration /nÿ-**tray**-shŏn/ A reaction introducing the nitro ($-NO_2$) group into an organic compound. Nitration of aromatic compounds is usually carried out using a mixture of concentrated nitric and sulfuric acids, although the precise conditions differ from compound to compound. The attacking species is NO_2^+ (the nitryl cation), and the reaction is an example of electrophilic substitution.

nitric acid /**nÿ**-trik/ (HNO_3) A colorless fuming corrosive liquid that is a strong acid. Nitric acid can be made in a laboratory by the distillation of a mixture of an alkali metal nitrate and concentrated sulfuric acid. Commercially it is prepared by the catalytic oxidation of ammonia and is supplied as concentrated nitric acid, which contains 68% of the acid and is often colored yellow by dissolved oxides of nitrogen.

Nitric acid is a strong oxidizing agent. Most metals are converted to their nitrates with the evolution of oxides of nitrogen (the composition of the mixture of the oxides depends on the temperature and on the concentration of the nitric acid used). Some non-metals (e.g. sulfur and phosphorus) react to produce oxyacids. Organic substances (e.g. sawdust and ethanol) react violently, but the more stable aromatic compounds, such as benzene and toluene, can be converted to nitro compounds in controllable reactions.

nitric oxide *See* nitrogen monoxide.

nitride /**nÿ**-trÿd/ A compound of nitrogen with a more electropositive element such as magnesium or calcium.

nitriding /**nÿ**-trÿ-ding/ *See* case hardening.

nitrification /ny-tră-fă-**kay**-shŏn/ The action of nitrifying bacteria in converting ammonia and nitrites to nitrates. It is an important stage in the NITROGEN CYCLE. *See also* nitrogen fixation.

nitrile /**nÿ**-trăl, -tril, -trÿl/ (**cyanide**) A type of organic compound containing the –CN group. Nitriles are colorless liquids with pleasant smells. They can be prepared by either refluxing an organic halogen compound with an alcoholic solution of potassium cyanide, or by dehydration of an amide with phosphorus(V) oxide. Characteristic reactions are hydrolysis (to the carboxylic acid and ammonia) and reduction with hydrogen gas (to an amine).

nitrile rubber A copolymer of butadiene and propenonitrile (acrylonitrile) ($CH_2{=}CHCN$). It is an important rubber due to its resistance to oil and solvents.

nitrite /**nÿ**-trÿt/ A salt or ester of nitrous acid.

nitrobenzene /nÿ-troh-**ben**-zeen, -ben-**zeen**/ ($C_6H_5NO_2$) A yellow organic oil obtained by refluxing benzene with a mixture of concentrated nitric and sulfuric acids. The reaction is a typical electrophilic substitution on the benzene ring by the nitryl cation (NO_2^+).

nitrocellulose /nÿ-troh-**sell**-yŭ-lohs/ *See* cellulose trinitrate.

nitro compound /**nÿ**-troh/ A type of or-

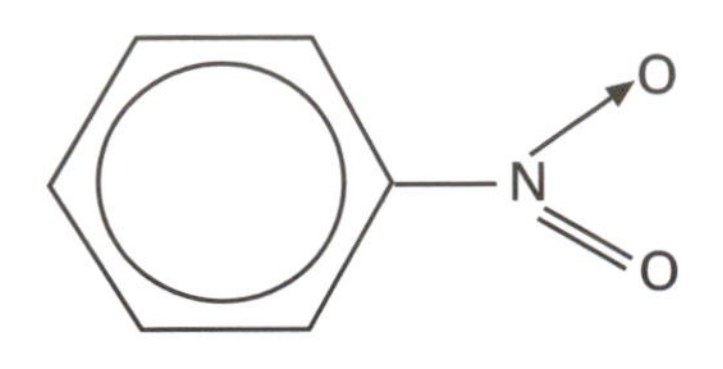

Nitro compound

ganic compound containing the nitro ($-NO_2$) group attached to an aromatic ring. Nitro compounds can be prepared by nitration using a mixture of concentrated nitric and sulfuric acids. They can be reduced to aromatic amines:

$$RNO_2 + 3H_2 \rightarrow RNH_2 + 2H_2O$$

They can also undergo further substitution on the benzene ring. The nitro group directs substituents into the 3 position.

nitrogen /nȳ-trŏ-jĕn/ The first element of group 15 of the periodic table; a very electronegative element existing in the uncombined state as gaseous diatomic N_2 molecules. The nitrogen atom has the electronic configuration $[He]2s^22p^3$. It is typically non-metallic and its bonding is primarily by polarized covalent bonds. With electropositive elements the nitride ion N^{3-} may be formed.

Nitrogen accounts for about 78% of the atmosphere (by volume) and it also occurs as sodium nitrate in various mineral deposits. It is separated for industrial use by the fractionation of liquid air. Pure nitrogen is prepared by thermal decomposition of azides;

$$2NaN_3 \rightarrow 2Na + 3N_2$$

The 'active nitrogen' gas obtained by passing an electric discharge through nitrogen is a mixture of ordinary nitrogen molecules and excited single nitrogen atoms.

Nitrogen is a diatomic molecule, which is effectively triple-bonded and has a high dissociation energy (940 kJ mol^{-1}). It is therefore inert and it only reacts readily with lithium and other highly electropositive elements. The direct combination of nitrogen and hydrogen occurs at elevated temperatures and pressures (400–600°C, 100 atmospheres) and is the basis of the industrially important Haber process for the manufacture of ammonia.

Nitrogen forms a number of oxides:

1. Dinitrogen monoxide (nitrous oxide, N_2O) – a neutral oxide.
2. Nitrogen monoxide (nitric oxide, NO) – a neutral oxide.
3. Nitrogen(III) oxide (nitrogen sesquioxide, N_2O_3) – an unstable oxide that is the anhydride of nitrous acid.
4. dioxide (NO_2), which gives a mixture of nitrous and nitric acids in water.
5. Dinitrogen tetroxide (N_2O_4), which is in equilibrium with nitrogen dioxide.
6. Dinitrogen pentoxide (N_2O_5) – the anhydride of nitric acid.
7. Nitrogen trioxide (NO_3).

Nitrogen also forms a number of binary halides such as NF_3, NCl_3, and halogen azides such as ClN_3. Except for NF_3 these are all very explosive, and even NF_3, which is reputedly stable, has been known to explode.

Transition metals form nitrides that are different from the ionic or covalent nitrides, the stoichiometries are, for example, ZrN, W_2N, Mn_4N. They are often called 'interstitial' nitrides. They are very hard and their formation is the basis of surface hardening by exposing hot metal to ammonia gas (*see* case hardening).

Nitrogen has two isotopes; ^{14}N, the common isotope, and ^{15}N (natural abundance 0.366%), which is used as a marker in mass spectrometric studies.

Symbol: N; m.p. −209.86°C; b.p. −195.8°C; d. 1.2506 kg m^{-3} (0°C); p.n. 7; r.a.m. 14.

nitrogen cycle The circulation of nitrogen and its compounds in the environment, one of the major natural cycles of an element. Nitrogen occurs mainly as a gas in the atmosphere and as nitrates in the soil. Denitrification of these nitrates converts them to nitrogen. Plants also take up nitrates and are eaten by animals; the plants and animals convert them to proteins, which after the death of the organisms appears as ammonia, which nitrification converts back to nitrates. Fixation of atmospheric nitrogen by bacteria or lightning also gives rise to nitrates. *See also* nitrification; nitrogen fixation.

nitrogen dioxide /dÿ-**oks**-ÿd, -id/ (NO_2) A brown gas produced by the dissociation of dinitrogen tetroxide (with which it is in equilibrium), the dissociation being complete at 140°C. Further heating causes dissociation to colorless nitrogen monoxide and oxygen:

$$2NO_2(g) = 2NO(g) + O_2(g)$$

Nitrogen dioxide can also be made by the action of heat on metal nitrates (not the nitrates of the alkali metals or some of the alkaline-earth metals).

nitrogen fixation A reaction in which atmospheric nitrogen is converted into nitrogen compounds. Nitrogen fixation occurs naturally by certain bacteria (in the soil and in the roots of leguminous plants). Small amounts of nitrogen(II) oxide (NO) are also formed in thunderstorms by direct combination of the elements in the electrical discharge. Fixation of nitrogen is important because nitrogen is an essential element for plant growth, and vast amounts of nitrogenous fertilizers are also required for agriculture. Former processes for fixing nitrogen were the Birkeland-Eyde process (reacting N_2 and O_2 in an electric arc – the equivalent of the thunderstorm in nature), and the cyanamide process (heating calcium dicarbide with nitrogen to give $CaCN_2$). Now the major method is the Haber process for making ammonia.

nitrogen monoxide /mon-**oks**-ÿd, -id/ (**nitric oxide**; NO) A colorless gas that is insoluble in water but dissolves in a solution containing iron(II) ions owing to the formation of the complex ion $(FeNO)^{2+}$: the nitrogen monoxide can be released by heating. Nitrogen monoxide is prepared by the action of nitric acid on copper turnings; the impure product can be purified by using a solution of iron(II) ions to absorb the product. Commercially, nitrogen monoxide is prepared by the catalytic oxidation of ammonia or by the direct union of nitrogen and oxygen in an electric arc. Nitrogen monoxide is the most heat-stable of the oxides of nitrogen, only decomposing above 1000°C. At ordinary temperatures it combines immediately with oxygen to give nitrogen dioxide:

$$2NO(g) + O_2(g) \rightarrow 2NO_2(g)$$

nitrogenous base /nÿ-**troj**-ĕ-nŭs/ *See* quaternary ammonium compound.

nitrogen oxides Any of the various compounds formed between nitrogen and oxygen. *See* dinitrogen oxide; nitrogen; nitrogen dioxide; nitrogen monoxide.

nitroglycerine /nÿ-troh-**gliss**-ĕ-rin, -reen/ (**glyceryl trinitrate**) A highly explosive substance used in dynamite. It is obtained by treating glycerol (1,2,3-trihydroxypropane) with a mixture of concentrated nitric and sulfuric acids. It is not a nitro compound, but a nitrate ester

$$CH_2(NO_3)CH(NO_3)CH_2(NO_3)$$

nitro group The group $-NO_2$; the functional group of nitro compounds.

nitro ligand *See* isomerism.

nitronium ion /nÿ-**troh**-nee-ŭm/ *See* nitryl ion.

nitrophenols /nÿ-troh-**fee**-nolz, -nohlz/ ($C_6H_4(OH)NO_2$) Organic compounds formed directly or indirectly by the nitration of phenol. Three isomeric forms are possible. The 2 and 4 isomers are produced by the direct nitration of phenol and can be separated by steam distillation, the 2 isomer being steam volatile. The 3 isomer is produced from nitrobenzene by formation of 1,3-dinitrobenzene, conversion to 3-nitrophenylamine, and thence by diazotization to 3-nitrophenol.

nitroso ligand *See* isomerism.

nitrous acid /**nÿ**-trŭs/ (HNO_2) A weak acid known only in solution, obtained by acidifying a solution of a nitrite. It readily decomposes on warming or shaking to nitrogen monoxide and nitric acid. The use of nitrous acid is very important in the dyestuffs industry in the diazo reaction: nitrous acid is liberated by acidifying a solution of a nitrite (usually sodium nitrite) in the presence of the compound to be diazotized. Nitrous acid and the nitrites are normally reducing agents but in certain circumstances they can behave as oxidizing agents, e.g. with sulfur dioxide and hydrogen sulfide.

nitrous oxide *See* dinitrogen oxide.

nitryl ion /**nÿ**-trăl/ (**nitronium ion**) The positive ion NO_2^+. *See* nitration.

NMR *See* nuclear magnetic resonance.

nobelium /noh-**bel**-ee-ŭm/ A radioactive transuranic element of the actinoid series, not found naturally on Earth. Several very short-lived isotopes have been synthesized.

Symbol: No; p.n. 102; most stable isotope ^{259}No (half-life 58 minutes).

noble gases *See* rare gases.

nonbenzenoid aromatic /non-**ben**-zĕ-noid/ *See* aromatic compound.

nonessential amino acid *See* amino acid.

nonionic detergent /non-ÿ-**on**-ik/ *See* detergents.

nonlocalized bond *See* delocalized bond.

nonmetal Any of a class of chemical elements. Non-metals lie in the top right-hand region of the periodic table. They are electronegative elements with a tendency to form covalent compounds or negative ions. They have acidic oxides and hydroxides. In the solid state, nonmetals are either covalent volatile crystals or macromolecular (giant-covalent) crystals. *See also* metal; metalloid.

nonpolar compound A compound that has molecules with no permanent dipole moment. Examples of non-polar compounds are hydrogen, tetrachloromethane, and carbon dioxide.

nonpolar solvent *See* solvent.

nonstoichiometric compound A chemical compound whose molecules contain fractional numbers of atoms. For example, titanium(IV) oxide in the form of the mineral rutile has the chemical formula $TiO_{1.8}$.

normality The number of gram equivalents per cubic decimeter of a given solution.

normal solution A solution that contains one gram equivalent weight per liter of solution. Values are designated by the symbol N, e.g. 0.2N, N/10, etc. Because there is not a clear definition of equivalent weight suitable for all reactions, a solution may have one value of normality for one reaction and another value in a different reaction. Because of this many workers prefer the molar solution notation.

NTP *See* STP.

nuclear magnetic resonance (NMR) A method of investigating nuclear spin. In an external magnetic field the nucleus can have certain quantized energy states, corresponding to certain orientations of the spin magnetic moment. Hydrogen nuclei, for instance, can have two energy states, and transitions between the two occur by absorption of radiofrequency radiation. In chemistry, this is the basis of a spectroscopic technique for investigating the structure of molecules. Radiofrequency radiation is fed to a sample and the magnetic field is changed slowly. Absorption of the radiation is detected when the difference between the nuclear levels corresponds to absorption of a quantum of radiation. This difference depends slightly on the electrons around the nucleus – i.e. the position of the atom in the molecule. Thus a different absorption frequency is seen for each type of hydrogen atom (this is an example of a chemical shift). In ethanol, for example, there are three frequencies, corresponding to hydrogen atoms on the CH_3, the CH_2, and the OH. The intensity of absorption also depends on the number of hydrogen atoms (3:2:1). NMR spectroscopy is a powerful method of finding the structures of organic compounds.

nuclear magneton *See* magneton.

nucleic acids Organic acids whose molecules consist of chains of alternating sugar and phosphate units, with nitrogenous bases attached to the sugar units. They occur in the cells of all organisms. In DNA the sugar is deoxyribose; in RNA it is ribose. *See* DNA; RNA.

nucleon number (**mass number**) Symbol: *A* The number of nucleons (protons plus neutrons) in an atomic nucleus.

nucleophile /**new**-klee-ŏ-fÿl/ An electron-rich ion or molecule that takes part in an organic reaction. The nucleophile can be a negative ion (Br^-, CN^-) or a molecule with a lone pair of electrons (NH_3, H_2O). The nucleophile attacks positively charged parts of molecules, which usually arise from the presence of an electronegative atom elsewhere in the molecule. *Compare* electrophile.

nucleophilic addition /new-klee-ŏ-**fil**-ik/ A class of reaction involving the addition of a small molecule to the double bond in an unsaturated organic compound. Since these reactions are initiated by nucleophiles, the unsaturated bond must contain an electronegative atom, which creates an electron-deficient area in the molecule. Nucleophilic addition is a characteristic reaction of aldehydes and ketones and is often followed by the subsequent elimination of a different small molecule, particularly water. The addition of hydrogen cyanide to propanone is an example:

$$CH_3COCH_3 + HCN \rightarrow CH_3C(CN)(OH)CH_3$$

See also condensation reaction.

nucleophilic reagent A compound that contains electron-rich groups of atoms that can donate electrons during a chemical reaction. They are generally oxidizing agents. *See* oxidation.

nucleophilic substitution A reaction involving the substitution of an atom or group of atoms in an organic compound with a nucleophile as the attacking substituent. Since nucleophiles are electron-rich species, nucleophilic substitution occurs in compounds in which a strongly electronegative atom or group leads to a dipolar bond. The electron-deficient center can then be attacked by the electron-rich nucleophile causing the electronegative atom or group to be displaced. In general terms:

$$R\text{–}Le + Nu^- \rightarrow R\text{–}Nu + Le^-$$

where Nu^- represents the incoming nucleophile and Le^- represents the leaving group.

There are two possible mechanisms for nucleophilic substitution. In a *S_N1 reaction* the molecule first forms a carbonium ion; for example:

$$RCH_2Cl \rightarrow RCH_2^+ + Cl^-$$

The nucleophile then attaches itself to this carbonium ion:

$$RCH_2^+ + OH^- \rightarrow RCH_2OH$$

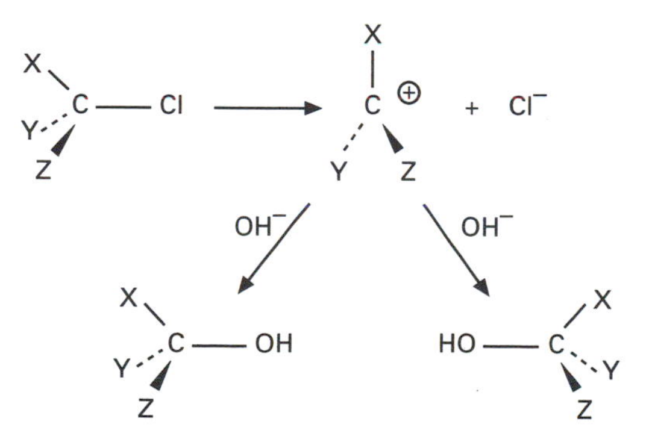

Nucleophilic substitution: the S_N1 reaction

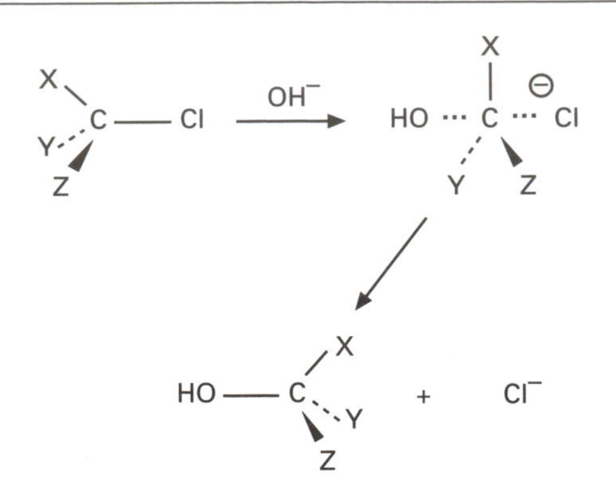

Nucleophilic substitution: the S_N2 reaction

In a *S_N2 reaction* the nucleophile approaches as the other group leaves, forming a transition state in which the carbon has five attached groups.

The preferred mechanism depends on several factors:

1. The stability of the intermediate in the S_N1 mechanism.
2. Steric factors affecting the formation of the transition state in the S_N2 mechanism.
3. The solvent in which the reaction occurs: polar solvents will stabilize polar intermediates and so favor the S_N1 mechanism.

In practice, however, both routes or some intermediate mechanism is thought to operate in many cases.

See also substitution reaction.

nucleoside /**new**-klee-ŏ-sÿd/ A molecule consisting of a purine or pyrimidine base linked to a sugar, either ribose or deoxyribose. ADENOSINE, CYTIDINE, GUANOSINE, THYMIDINE, and URIDINE are common nucleosides.

nucleotide /**new**-klee-ŏ-tÿd/ The compound formed by condensation of a nitrogenous base (a purine, pyrimidine, or pyridine) with a sugar (ribose or deoxyribose) and phosphoric acid. The nucleic acids are *polynucleotides* (consisting of chains of many linked nucleotides).

nucleus /**new**-klee-ŭs/ (*pl.* **nuclei** or **nucleuses**) The compact positively charged center of an atom made up of one or more nucleons (protons and neutrons) around which is a cloud of electrons. The density of nuclei is about 10^{15} kg m^{-3}. The number of protons in the nucleus defines the element, being its proton number (or atomic number). The nucleon number, or atomic mass number, is the sum of the protons and neutrons. The simplest nucleus is that of a hydrogen atom, ^{1}H, being simply one proton (mass 1.67×10^{-27} kg). The most massive naturally occurring nucleus is ^{238}U of 92 protons and 146 protons (mass 4×10^{-25} kg, radius 9.54×10^{-15} m). Only certain combinations of protons and neutrons form stable nuclei. Others undergo spontaneous decay.

A nucleus is depicted by a symbol indicating nucleon number (mass number), proton number (atomic number), and element name. For example, $^{23}_{11}Na$ represents a nucleus of sodium having 11 protons and mass 23, hence there are (23 – 11) = 12 neutrons.

nuclide /**new**-klÿd/ A nuclear species that is characterized by a given number of protons and neutrons; for example, ^{23}Na, ^{24}Na, and ^{24}Mg are all different nuclides. Thus:

$^{23}_{11}Na$ has 11 protons and 12 neutrons
$^{24}_{11}Na$ has 11 protons and 13 neutrons
$^{24}_{12}Mg$ has 12 protons and 12 neutrons

The term is applied to the nucleus and often also to the atom.

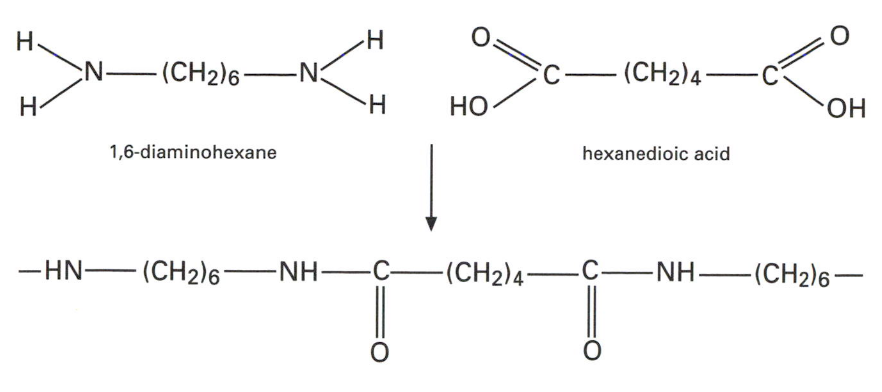

Formation of nylon

nylon A type of synthetic polymer linked by amide groups –NH.CO–. Nylon polymers can be made by copolymerization of a molecule containing two amine groups with one containing two carboxylic acid groups.

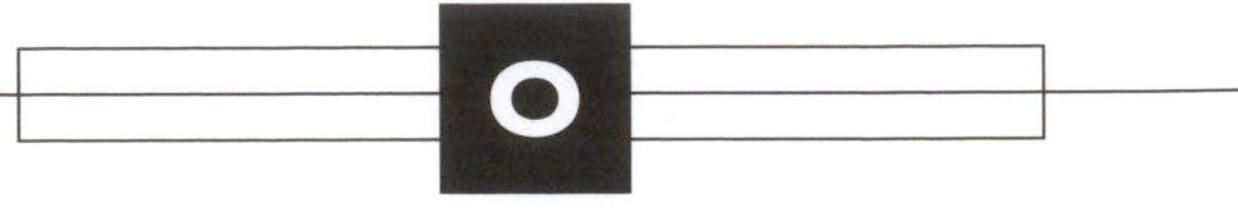

occlusion /ŏ-**kloo**-zhŏn/ **1.** The process in which small amounts of one substance are trapped in the crystals of another; for example, pockets of liquid occluded during crystallization from a solution.
2. Absorption of a gas by a solid; for example, the occlusion of hydrogen by palladium.

ocher /**oh**-ker/ A clay mineral containing iron(III) oxide (Fe_2O_3), used as a yellow, orange, or brown pigment.

octadecanoic acid /ok-tă-dek-ă-**noh**-ik/ (**stearic acid**; $CH_3(CH_2)_{16}COOH$) A solid carboxylic acid present in fats and oils as the glyceride.

octadecenoic acid /ok-tă-dek-ă-**noh**-ik/ *cis*-9-octadecenoic acid (or oleic acid) is a naturally occurring carboxylic acid present (as glycerides) in fats and oils:

$$CH_3(CH_2)_7CH{:}CH(CH_2)_7COOH$$

octahedral complex /ok-tă-**hee**-drăl/ *See* complex.

octahydrate /ok-tă-**hȳ**-drayt/ A crystalline compound containing eight molecules of water of crystallization per molecule of compound.

octane /**ok**-tayn/ (C_8H_{18}) A liquid alkane obtained from the light fraction of crude oil. Octane and its isomers are the principal constituents of gasoline, which is obtained as the refined light fraction from crude oil. *See also* octane rating.

octane rating (**octane number**) A rating for the performance of gasoline in internal-combustion engines. The octane rating measures the freedom from 'knocking' – i.e. preignition of the fuel in the engine. This depends on the relative proportions of branched-chain and straight-chain hydrocarbons present. High proportions of branched-chain alkanes are better in high-performance engines. In rating fuels, 2,2,4-trimethylpentane (isooctane) is given a value of 100 and heptane is given a value 0. The performance of a fuel is compared with a mixture of these hydrocarbons.

octavalent /ok-tă-**vay**-lĕnt/ Having a valence of eight.

octaves, law of *See* Newlands' law.

octet /ok-**tet**/ A stable shell of eight electrons in an atom. The completion of the octet gives rise to particular stability and this is the basis of the *Lewis octet theory* (named for the American physical chemist Gilbert Newton Lewis (1875–1946)), thus:

1. The rare gases have complete octets and are chemically inert
2. The bonding in small covalent molecules is frequently achieved by the central atom completing its octet by sharing electrons with other atoms, e.g. CH_4.
3. The ions formed by electropositive and electronegative elements are generally those with a complete octet, e.g. Na^+, Ca^{2+}, O^{2-}, Cl^-.

oersted Symbol: Oe A unit of magnetic field strength in the c.g.s. system. It is equal to $10^3/4\pi$ A m^{-1}. The unit is named for the Danish physicist Hans Christian Oersted (1777–1851).

ohm /ohm/ Symbol: Ω The SI unit of electrical resistance, equal to a resistance that passes a current of one ampere when there is an electric potential difference of one volt across it. 1 Ω = 1 V A^{-1}. Formerly, it was defined in terms of the resistance of a column of mercury under specified conditions. The unit is named for the German physicist Georg Simon Ohm (1787–1854).

oil Any of various viscous liquids. *See* glyceride; petroleum.

oil of vitriol /**vit**-ree-ŏl/ Sulfuric(VI) acid.

oil of wintergreen *See* methyl salicylate.

oil shale A sedimentary rock that includes in its structure 30–60% of organic matter, mainly in the form of bitumen. Heated in the absence of air it produces an oily substance resembling petroleum, which is rich in nitrogen and sulfur compounds.

oleate /**oh**-lee-ayt/ A salt or ester of oleic acid; i.e. an octadecenoate.

olefin /**oh**-lĕ-fin/ *See* alkene.

oleic acid /**oh**-lee-ik/ *See* octadecenoic acid.

oleum /**oh**-lee-ŭm/ (**pyrosulfuric acid**; $H_2S_2O_7$) A colorless fuming liquid formed by dissolving sulfur(VI) oxide in concentrated sulfuric acid. It gives sulfuric acid on diluting with water. *See* contact process.

oligomer /ŏ-**lig**-ŏ-mer/ A POLYMER formed from a relatively few monomer molecules.

oligosaccharide /ol-ă-goh-**sak**ă-rÿd, -rid/ A carbohydrate formed of a small number of monosaccharide units (up to around 20).

olivine /**ol**-ă-veen, ol-ă-**veen**/ A greenish rock-forming silicate mineral containing iron and magnesium. Its clear green form is the gemstone *peridot*.

one-pot synthesis A synthesis of a chemical compound in which the reactants form the required product in a single reaction mixture.

onium ion /**oh**-nee-ŭm/ An ion formed by addition of a proton (H^+) to a molecule. The hydronium ion (H_3O^+) and ammonium ion (NH_4^+) are examples.

oolite /**oh**-ŏ-lÿt/ A sedimentary rock, a type of limestone, that contains small spherical masses of calcium carbonate ($CaCO_3$).

opal /**oh**-păl/ A naturally occurring hydrated and noncrystalline form of silica (SiO_2), valued as a gemstone. Various layers within a gemstone opal can reflect and refract light to give a play of spectral colors. There are also milky white and black varieties.

open-hearth process A method of making steel by heating pig iron in an open-hearth furnace by burning a mixture of gas and air. Oxygen is injected through pipes to oxidize impurities and burn off carbon, and lime is used to form a slag.

optical activity /**op**-tă-kăl/ The ability of certain compounds to rotate the plane of polarization of plane-polarized light when the light is passed through them. Optical activity can be observed in crystals, gases, liquids, and solutions. The amount of rotation depends on the concentration of the active compound.

Optical activity is caused by the interaction of the varying electric field of the light with the electrons in the molecule. It occurs when the molecules are asymmetric – i.e. they have no plane of symmetry. Such molecules have a mirror image that cannot be superimposed on the original molecule. In organic compounds this usually means that they contain a carbon atom attached to four different groups, forming a chiral center. The two mirror-image forms of an asymmetric molecule are optical isomers. One isomer will rotate the polarized light in one sense and the other by the same amount in the opposite sense. Such isomers are described as *dextrorotatory* or *levorotatory*, according to whether they rotate the plane to the 'right' or 'left' respectively (rotation to the left is clockwise to an observer viewing the light coming toward the

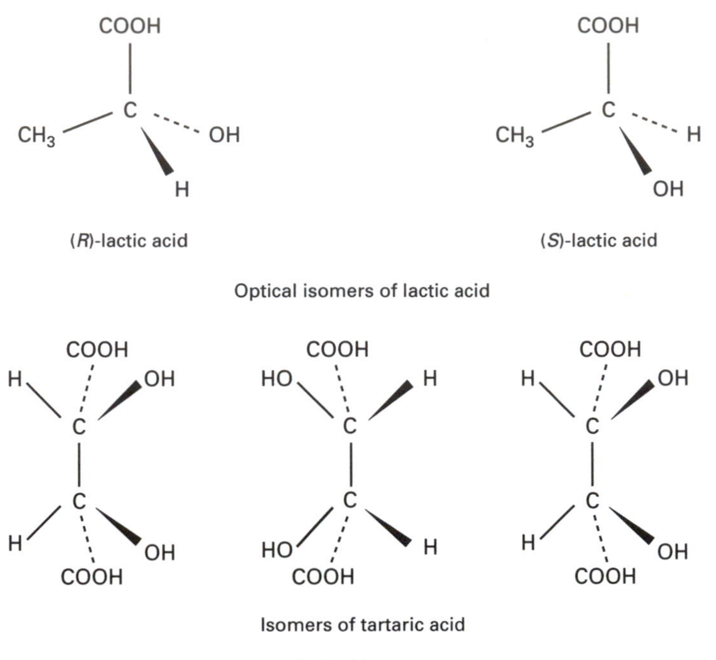

(*R*)-lactic acid (*S*)-lactic acid

Optical isomers of lactic acid

Isomers of tartaric acid

Optical isomers

observer). Dextrorotatory compounds are given the symbol *d* or (+) and levorotatory compounds *l* or (–). A mixture of the two isomers in equal amounts does not show optical activity. Such a mixture is sometimes called the (±) or *dl*-form, a *racemate*, or a *racemic mixture*

Optical isomers have identical physical properties (apart from optical activity) and cannot be separated by fractional crystallization or distillation. Their general chemical behavior is also the same, although they do differ in reactions involving other optical isomers. Many naturally occurring substances are optically active (only one optical isomer exists naturally) and biochemical reactions occur only with the natural isomer. For instance, the natural form of glucose is *d*-glucose and living organisms cannot metabolize the *l*-form.

The terms 'dextrorotatory' and 'levorotatory' refer to the effect on polarized light. A more common method of distinguishing two optical isomers is by their *D-form* or *L-form*. This convention (the *D–L convention*) refers to the absolute structure of the isomer according to specific rules (i.e. the *absolute* configuration of the molecule). Sugars are related to a particular configuration of glyceraldehyde (2,3-dihydroxypropanal). For alpha amino acids the *corn rule* is used: the structure of the acid $RC(NH_2)(COOH)$ H is drawn with H at the top; viewed from the top the groups spell CORN in a clockwise direction for all D-amino acids (i.e. the clockwise order is –*COOH*,*R*,NH_2). The opposite is true for L-amino acids. Note that this convention does not refer to optical activity: L-alanine is dextrorotatory by D-cystine is levorotatory.

An alternative is the *R–S convention* for showing configuration. There is an order of priority of attached groups based on the proton number of the attached atom:
I, Br, Cl, SO_3H, $OCOCH_3$, OCH_3, OH, NO_2, NH_2, $COOCH_3$, $CONH_2$, $COCH_3$, CHO, CH_2OH, C_6H_5, C_2H_5, CH_3, H

Hydrogen has the lowest priority. The chiral carbon is viewed such that the group of lowest priority is hidden behind it. If the other three groups are in descending prior-

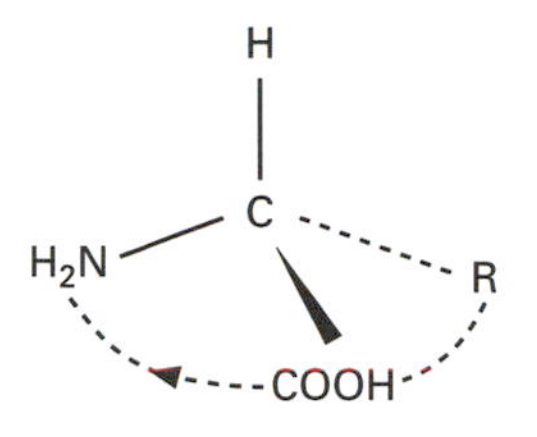

The corn rule for absolute configuration of alpha amino acids

R-configuration

S-configuration

The R/S system

Absolute configuration

ity in a clockwise direction, the compound is R-. If descending priority is anticlockwise it is S-.

The existence of a carbon atom bound to four different groups is not the strict condition for optical activity. The essential point is that the molecule should be asymmetric. Inorganic octahedral complexes, for example, can show optical isomerism. It is also possible for a molecule to contain asymmetric carbon atoms and still have a plane of symmetry. One structure of tartaric acid has two parts of the molecule that are mirror images, thus having a plane of symmetry. This (called the *meso-form*) is not optically active. *See also* resolution).

optical isomerism *See* isomerism; optical activity.

optical rotary dispersion (ORD) The phenomenon in which the amount of rotation of plane-polarized light by an optically active substance depends on the wavelength of the light. Plots of rotation against wavelength can be used to give information about the molecular structure of optically active compounds.

optical rotation Rotation of the plane of polarization of plane-polarized light by an optically active substance.

orbit The path of an electron as it moves around the nucleus in an atom.

orbital A region around an atomic nucleus in which there is a high probability of finding an electron. The modern picture of the atom according to quantum mechanics does not have electrons moving in fixed elliptical orbits. Instead, there is a finite probability that the electron will be found in any small region at all possible distances from the nucleus. In the hydrogen atom the probability is low near the nucleus, increases to a maximum, and falls off to infinity. It is useful to think of a region in space around the nucleus – in the case of hydrogen the region within a sphere – within which there is a high chance of finding the electron. Each of these, called an *atomic orbital*, corresponds to a sub-shell and can 'contain' a single electron or two electrons with opposite spins. Another way of visualizing an orbital is as a cloud of electron charge (the average distribution with time).

Similarly, in molecules the electrons move in the combined field of the nuclei and can be assigned to *molecular orbitals*. In considering bonding between atoms it is useful to treat molecular orbitals as formed by overlap of atomic orbitals.

It is possible to calculate the shapes and energies of atomic and molecular orbitals by quantum theory. The shapes of atomic orbitals depend on the orbital angular momentum (the sub-shell). For each shell there is one s orbital, three p orbitals, five d orbitals, etc. The s orbitals are spherical, the p orbitals each have two lobes; d orbitals have more complex shapes, typically with four lobes.

Molecular orbitals are formed by overlap of atomic orbitals, and again there are different types. If the orbital is completely symmetrical about an axis between the nuclei, it is a *sigma orbital*. This can occur, for instance, by overlap of two s orbitals, as in the hydrogen atom, or two p orbitals with their lobes along the axis. However, two p orbitals overlapping at right angles to the axis form a different type of molecular orbital – a *pi orbital* – with regions above and below the axis. Pi orbitals are also formed by overlap of d orbitals. Each molecular orbital can contain a pair of electrons, forming a sigma bond or pi bond. A double bond, for example the bond in ethene, is a combination of a sigma bond and a pi bond.

Hybrid orbitals are atomic orbitals formed by combinations of s, p, and d atomic orbitals, and are useful in describing the bonding in compounds. There are various types. In carbon, for instance, the electron configuration is $1s^2 2s^2 2p^2$. Carbon, in its outer (valence) shell, has one

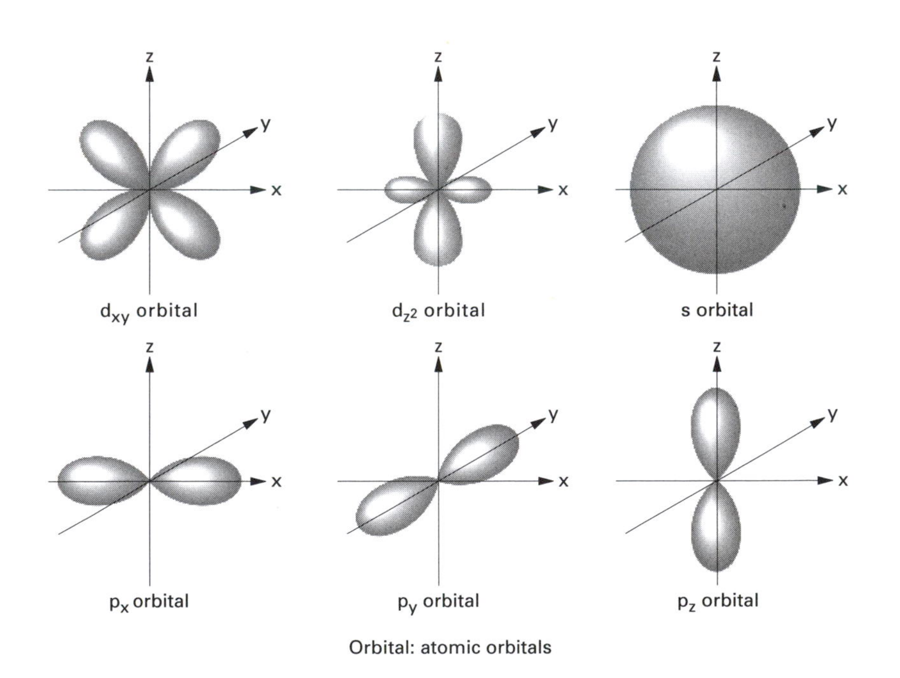

Orbital: atomic orbitals

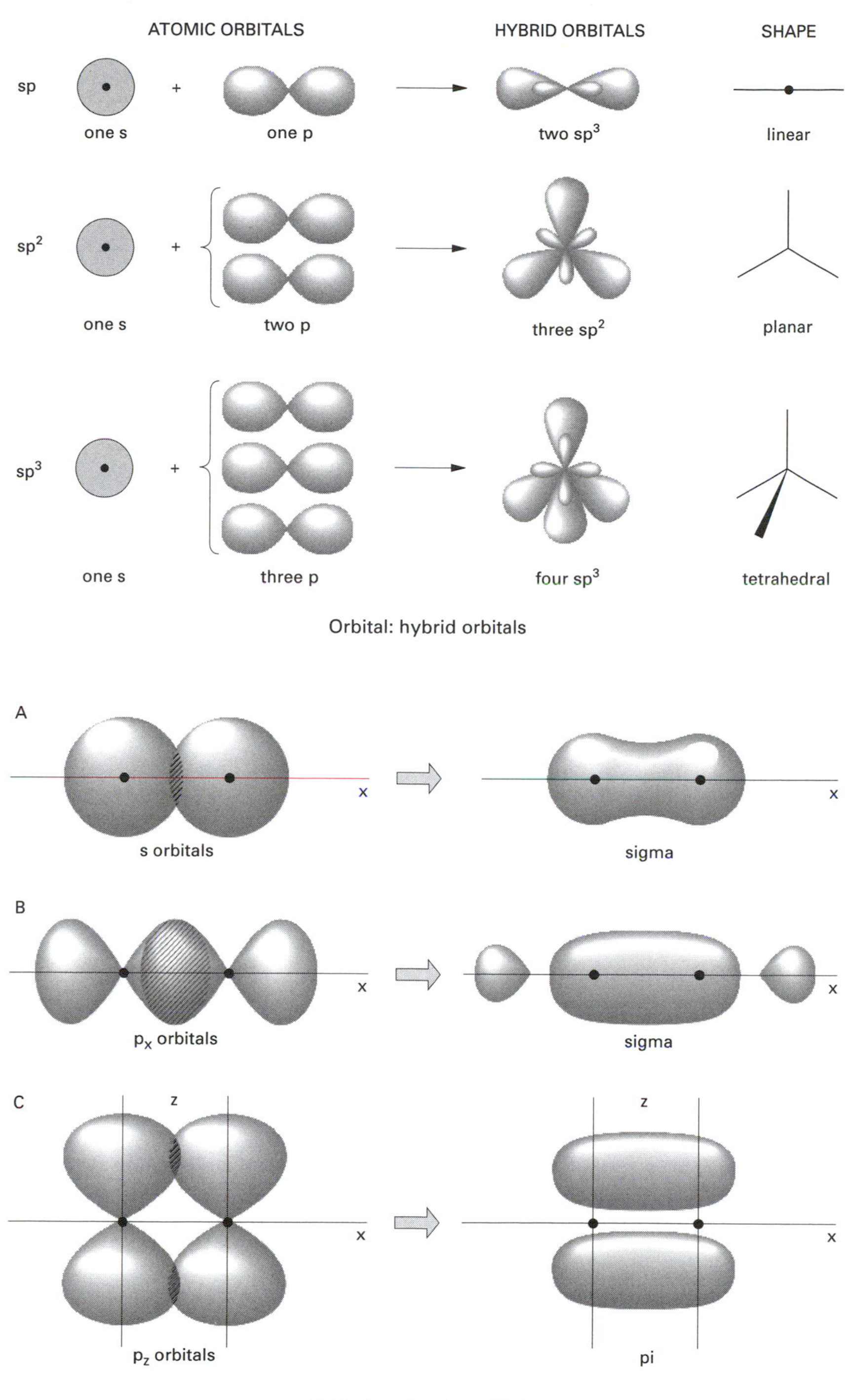

Orbital: hybrid orbitals

Orbital: molecular orbitals

filled s orbital, two filled p orbitals, and one 'empty' p orbital. These four orbitals may *hybridize* (sp^3 hybridization) to act as four equal orbitals arranged tetrahedrally, each with one electron. In methane, each hybrid orbital overlaps with a hydrogen s orbital to form a sigma bond. Alternatively, the s and two of the p orbitals may hybridize (sp^2 hybridization) and act as three orbitals in a plane at 120°. The remaining p orbital is at right angles to the plane, and can form pi bonds. Finally, sp hybridization may occur, giving two orbitals in a line. More complex types of hybridization, involving d orbitals, explain the geometries of inorganic complexes.

The combination of two atomic orbitals in fact produces two molecular orbitals. One – the *bonding orbital* – has a concentration of electron density between the nuclei, and thus tends to hold the atoms together. The other – the *antibonding orbital* – has slightly higher energy and tends to repel the atoms. If both atomic orbitals are filled, the two molecular orbitals are also filled and cancel each other out – there is no net bonding effect. If each atomic orbital has one electron, the pair occupies the lower energy bonding orbital, producing a net attraction.

ORD *See* optical rotary dispersion.

order The sum of the indices of the concentration terms in the expression that determines the rate of a chemical reaction. For example, in the expression

$$\text{rate} = k[A]^x[B]^y$$

x is called the order with respect to A, y the order with respect to B, and $(x + y)$ the order overall. The values of x and y are not necessarily equal to the coefficients of A and B in the molecular equation. Order is an experimentally determined quantity derived without reference to any equation or mechanism. Fractional orders do occur. For example, in the reaction

$$CH_3CHO \rightarrow CH_4 + CO$$

the rate is proportional to $[CH_3CHO]^{1.5}$ i.e. it is of order 1.5.

ore A mineral source of a chemical element.

ore dressing *See* beneficiation.

organic acid /or-**gan**-ik/ An organic compound that can release hydrogen ions (H^+) to a base, such as a CARBOXYLIC ACID or a PHENOL. *See* acid.

organic base An organic compound that can function as a base. Organic bases are typically amines that gain H^+ ions. *See* amine; base.

organic chemistry The chemistry of compounds of carbon. Originally the term *organic chemical* referred to chemical compounds present in living matter, but now it covers any carbon compound with the exception of certain simple ones, such as the carbon oxides, carbonates, cyanides, and cyanates. These are generally studied in inorganic chemistry. The vast numbers of synthetic and natural organic compounds exist because of the ability of carbon to form chains of atoms (catenation). Other elements are involved in organic compounds; principally hydrogen and oxygen but also nitrogen, halogens, sulfur, and phosphorus.

organometallic compound /or-gă-noh-mĕ-**tal**-ik/ An organic compound containing a carbon–metal bond. Tetraethyl lead, $(C_2H_5)_4Pb$, is a typical example of an organometallic compound, used as an additive in petrol.

ortho- **1.** Designating the form of a diatomic molecule in which both nuclei have the same spin direction; e.g. orthohydrogen, orthodeuterium.

2. Designating a benzene compound with substituents in the 1,2 positions. The position next to a substituent is the ortho position on the benzene ring. This was used in the systematic naming of benzene derivatives. For example, orthodinitrobenzene (or *o*-dinitrobenzene) is 1,2-dinitrobenzene.

3. Certain acids, regarded as formed from an anhydride and water, were named ortho acids to distinguish them from the less hydrated meta acids. For example, H_4SiO_4 (from SiO_2 + $2H_2O$) is orthosilicic acid;

H_2SiO_3 (SiO_2 + H_2O) is metasilicic acid. *See also* meta-; para-.

orthoboric acid /or-thŏ-**bô**-rik/ *See* boric acid.

orthohydrogen /or-thoh-**hȳ**-drŏ-jĕn/ *See* hydrogen.

orthophosphoric acid /or-thoh-fos-**fô**-rik/ *See* phosphoric(V) acid.

orthophosphorous acid /or-thoh-**fos**-fŏ-rŭs/ *See* phosphonic acid.

orthorhombic crystal /or-thŏ-**rom**-bik/ *See* crystal system.

osmiridium /oz-mă-**rid**-ee-ŭm/ A naturally occurring alloy of osmium and iridium, used to make the tips of pen nibs. *See* iridium; osmium.

osmium /**oz**-mee-ŭm/ A transition metal that is found associated with platinum. Osmium is the most dense of all metals. It has a characteristic smell resulting from the production of osmium(VIII) oxide (osmium tetroxide, OsO_4). The metal is used in catalysts and in alloys for pen nibs, pivots, and electrical contacts.

Symbol: Os; m.p. 3054°C; b.p. 5027°C; r.d. 22.59 (20°C); p.n. 76; r.a.m. 190.23.

osmium(IV) oxide (**osmium tetroxide**; OsO_4) A volatile yellow crystalline solid with a penetrating odor, used as an oxidizing agent and, in aqueous solution, as a catalyst for organic reactions.

osmium tetroxide /te-troks-ȳd/ *See* osmium(IV) oxide.

osmosis /oz-**moh**-sis/ Systems in which a solvent is separated from a solution by a semipermeable membrane approach equilibrium by solvent molecules on the solvent side of the membrane migrating through it to the solution side; this process is called osmosis and always leads to dilution of the solution. The phenomenon is quantified by measurement of the osmotic pressure. The process of osmosis is of fundamental importance in transport and control mechanisms in biological systems; for example, plant growth and general cell function.

Osmosis is a colligative property and its theoretical treatment is similar to that for the lowering of vapor pressure. The membrane can be regarded as equivalent to the liquid-vapour interface, i.e. one that permits free movement of solvent molecules but restricts the movement of solute molecules. The solute molecules occupy a certain area at the interface and therefore inhibit solvent egress from the solution. Just as the development of a vapor pressure in a closed system is necessary for liquid–vapour equilibrium, the development of an OSMOTIC PRESSURE on the solution side is necessary for equilibrium at the membrane.

osmotic pressure Symbol: π The pressure that must be exerted on a solution to prevent the passage of solvent molecules into it when the solvent and solution are separated by a semipermeable membrane. The osmotic pressure is therefore the pressure required to maintain equilibrium between the passage of solvent molecules through the membrane in either direction and thus prevent the process of osmosis proceeding. The osmotic pressure can be measured by placing the solution, contained in a small perforated thimble covered by a semipermeable membrane and fitted with a length of glass tubing, in a beaker of the pure solvent. Solvent molecules pass through the membrane, diluting the solution and thereby increasing the volume on the solution side and forcing the solution to rise up the glass tubing. The process continues until the pressure exerted by the solvent molecules on the membrane is balanced by the hydrostatic pressure of the solution in the tubing. A sample of the solution is then removed and its concentration determined. Osmosis is a colligative property; therefore the method can be applied to the determination of relative molecular masses, particularly for large molecules, such as proteins, but it is restricted by the difficulty of preparing good semipermeable membranes.

As the osmotic pressure is a colligative property it is directly proportional to the molar concentration of the solute if the temperature remains constant; thus π is proportional to the concentration n/V, where n is the number of moles of solute, and V the solvent volume. The osmotic pressure is also proportional to the absolute temperature. Combining these two proportionalities gives $\pi V = nCT$, which has the same form as the gas equation, $PV = nRT$, and experimental values of C are similar to those for R, the universal gas constant. This gives considerable support to the kinetic theory of colligative properties.

Ostwald's dilution law /**ost**-valts/ *See* dissociation constant.

oxalate /**oks**-ă-layt/ A salt or ester of ethanedioic acid (oxalic acid). *See* ethanedioic acid.

oxalic acid /oks-**al**-ik/ *See* ethanedioic acid.

oxidant /**oks**-ă-dănt/ An oxidizing agent. In rocket fuels, the oxidant is the substance that provides the oxygen for combustion (e.g. liquid oxygen or hydrogen peroxide).

oxidation /oks-ă-**day**-shŏn/ An atom, an ion, or a molecule is said to undergo oxidation or to be oxidized when it loses electrons. The process may be effected chemically, i.e. by reaction with an *oxidizing agent*, or electrically, in which case oxidation occurs at the anode. For example,

$$2Na + Cl_2 \rightarrow 2Na^+ + 2Cl^-$$

where chlorine is the oxidizing agent and sodium is oxidized, and

$$4CN^- + 2Cu^{2+} \rightarrow C_2N_2 + 2CuCN$$

where Cu^{2+} is the oxidizing agent and CN^- is oxidized.

The *oxidation state* of an atom is indicated by the number of electrons lost or effectively lost by the neutral atom, i.e. the oxidation number. The oxidation number of a negative ion is negative. The process of oxidation is the converse of reduction. *See also* redox.

oxidation–reduction *See* redox.

oxide /**oks**-ÿd/ A compound of oxygen and another element. Oxides can be made by direct combination of the elements or by heating a hydroxide, carbonate, or other salt. *See also* acidic oxide; basic oxide; neutral oxide; peroxide.

oxidizing acid /**oks**-ă-dÿ-zing/ An acid that acts as an oxidizing agent, e.g. sulfuric(VI) acid or nitric(V) acid. Because it is oxidizing, nitric(V) acid can dissolve metals below hydrogen in the electromotive series:

$$2HNO_3 + M \rightarrow MO + 2NO_2 + H_2O$$
$$MO + 2HNO_3 \rightarrow H_2O + M(NO_3)_2$$

oxidizing agent *See* oxidation.

oxime /**oks**-eem, -im/ A type of organic compound containing the C:NOH group, formed by reaction of an aldehyde or ketone with hydroxylamine (NH_2OH).

oxo process A method of manufacturing aldehydes by passing a mixture of carbon monoxide, hydrogen, and alkanes over a cobalt catalyst at high pressure (100 atmospheres and 150°C). The aldehydes can subsequently be reduced to alcohols, making the process a useful source of alcohols of high molecular weight.

2-oxopropanoic acid /oks-oh-proh-pă-**noh**-ik/ (**pyruvic acid**; $CH_3COCOOH$) A colorless liquid carboxylic acid containing a keto (–CO) group. It is important in metabolic reactions in the body.

oxyacid /oks-ee-**ass**-id/ An ACID in which the replaceable hydrogen atom is part of a hydroxyl group, including carboxylic acids, phenols and inorganic acids such as phosphoric(V) acid and sulfuric(VI) acid.

oxygen /**oks**-ă-jĕn/ A colorless odorless diatomic gas; the first member of group 16 of the periodic table. It has the electronic configuration $[He]2s^22p^4$ and its chemistry involves the acquisition of electrons to form either the di-negative ion, O^{2-} or two covalent bonds. In each case the oxygen atom attains the configuration of the rare gas neon.

Oxygen is the most plentiful element in the Earth's crust accounting for over 40% by weight. It is present in the atmosphere (20%) and is a constituent of the majority of minerals and rocks (e.g. sandstones, SiO_2, carbonates, $CaCO_3$, aluminosilicates) as well as the major constituent of the sea. Oxygen is an essential element for almost all living things. Elemental oxygen has two forms: the diatomic molecule O_2 and the less stable molecule trioxygen (ozone), O_3, which is formed by passing an electric discharge through oxygen gas. There are several convenient laboratory routes to oxygen:

1. Heat on unstable oxides.

$$2HgO \rightarrow 2Hg + O_2$$

2. Decomposition of peroxides.

$$2H_2O_2 \rightarrow 2H_2O + O_2$$

3. Heat on '-ate' compounds such as chlorates or manganate(VII).

$$2KClO_3 \rightarrow KCl + 3O_2$$

$$2KMnO_4 \rightarrow K_2MnO_4 + MnO_2 + O_2$$

Industrially oxygen is obtained by the fractionation of liquid air.

Oxygen reacts with elements from all groups of the periodic table except group 0 and even this unreactive group forms several compounds containing oxygen (XeO_3, XeO_4, $XeOF_2$, XeO_2^{2-}).

The nature of the oxides has long been a convenient guide to metallic and nonmetallic character. Thus, metals typically combine with oxygen to form oxides, which if soluble react with water to form *bases* (via OH^- ion); the more electropositive the metal the more vigorous the reaction. Most of the group 1 and group 2 metals behave like this forming oxides, O^{2-}, (e.g., Na_2O, CaO), peroxides, O_2^{2-}, (e.g., Na_2O_2, BaO_2), and superoxides, O_2^-, (e.g., RbO_2, CsO_2). Oxides of metallic elements are solids and x-ray studies show that discrete O^{2-} ions do exist in the solid state even though they do not exist in solution:

$$O^{2-} + H_2O \rightarrow 2OH^-$$

The non-metals typically form molecular oxides such as SO_2, ClO_2, and NO_2. When soluble in water they generally dissolve to form *acids*. The less electronegative elements among the non-metals generally form weaker acids, e.g. $B(OH)_3$ and H_2CO_3. The more electronegative elements form stronger acids. The higher the oxidation state of the central atom the stronger the acid ($HOCl < HClO_2 < HClO_4$). Nonmetal oxides may be gaseous (SO_2, CO_2), liquid (N_2O_4), or solid (P_2O_5). The weakly non-metallic elements such as silicon may require bases for dissolution and in some cases the oxides have extended polymeric structures, e.g., SiO_2. In between the acidic and basic oxides are the *amphoteric oxides*, which behave acidically to strong bases and basically towards strong acids. This behavior is associated with elements that are physically metallic but are chemically only weakly cationic, e.g., ZnO gives Zn^{2-} with acids and the zincates, $Zn(OH)_4^{2-}$, with bases. There is also a small class of *mixed oxides* that can be regarded as salts between a 'basic' and an 'acidic' oxide of the same metal: for example,

$$FeO + Fe_2O_3 \rightarrow Fe(FeO_2)_2 = Fe_3O_4$$

$$2PbO + PbO_2 \rightarrow Pb_3O_4$$

The neutral oxides are essentially insoluble non-metal oxides such as N_2O, NO, CO, F_2O, although extreme conditions can lead to some degree of acidic behavior in the case of CO, for example,

$$CO + OH^- \rightarrow HCOO^-$$

Oxygen occurs in three natural isotopic forms, ^{16}O (99.76%), ^{17}O (0.0374%), ^{18}O (0.2039%); the rarer isotopes are used in detailed studies of the behavior of oxygen-containing groups during reactions (tracer studies).

Oxygen is very limited in its catenation but an interesting range of O–O species is formed by the following: O_2^{2-} peroxide, O_2^- superoxide, O_2 oxygen gas, O_2^+ oxygenyl cation. The O–O bond gets both progressively shorter and stronger in the series. The peroxides are formed with Na, Ca, Sr, and Ba and are formally related to hydrogen peroxide (prepared by the action of acids on ionic peroxides). The superoxides of K, Rb, and Cs are prepared by burning the metal in air (the less electropositive metals, Na, Mg, and Zn form less stable superoxides). Oxygenyl cation, O_2^+, is formed by reaction of PtF_6 (which has a high electron affinity) with oxygen at room temperature. Other oxygenyl cation species can be made with group 15 fluorides, e.g. $O_2^+AsF_6^-$.

Symbol: O; m.p. –218.4°C; b.p. –182.962°C; d. 1.429 kg m^{-3} (0°C); p.n. 8; r.a.m. 15.9994.

ozone /**oh**-zohn/ (**trioxygen**; O_3) A poisonous, blue-colored allotrope of oxygen made by passing oxygen through a silent electric discharge. Ozone is unstable and decomposes to oxygen on warming. It is present in the upper layers of the atmosphere, where it screens the Earth from harmful short-wave ultraviolet radiation. There is concern that the ozone layer is possibly being depleted by the use of fluorocarbons and other compounds produced by industry.

Ozone is a strong oxidizing agent. Its concentration can be determined by reacting it with iodide ions and titrating the iodine formed with standard sodium thiosulfate(VI). It forms ozonides with alkenes. *See* ozonolysis.

ozonide /**oh**-zŏ-nÿd/ *See* ozonolysis.

ozonolysis /oh-zŏ-**nol**-ă-sis/ The addition of ozone (O_3) to alkenes and the subsequent hydrolysis of the ozonide into hydrogen peroxide and a mixture of carbonyl compounds. The carbonyl compounds can be separated and identified, which in turn, identifies the groups and locates the position of the double bond in the original alkene. Ozonolysis was formerly an important analytical technique.

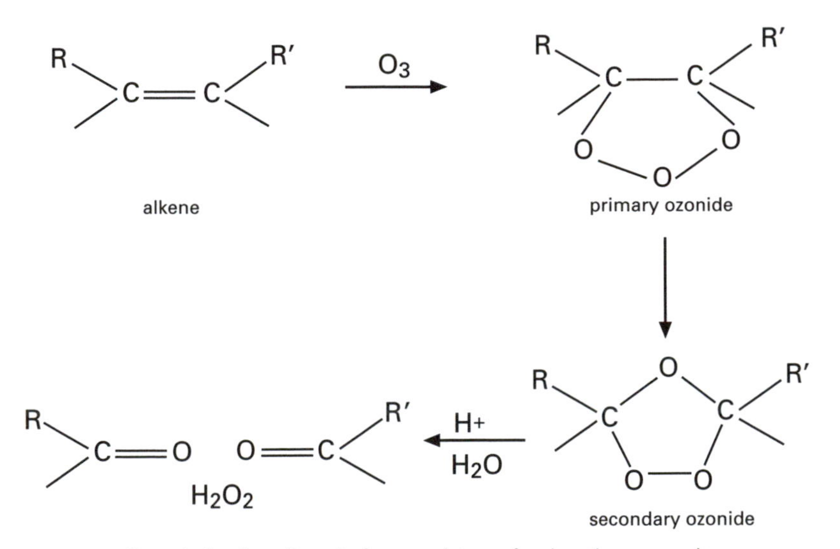

Ozonolysis of an alkene to form a mixture of carbonyl compounds

PAH *See* particulate matter.

paint A suspension of powdered pigment in a liquid vehicle, used for decorative and protective surface coatings. In oil-based paints, the vehicle contains solvents and a drying oil or synthetic resin. In water-based or emulsion paints, the vehicle is mostly water with an emulsified resin.

palladium /pă-**lay**-dee-ŭm/ A silvery white ductile transition metal occurring in platinum ores in Canada and as the native metal in Brazil. Most palladium is obtained as a by-product in the extraction of copper and zinc. It is used in electrical relays and as a catalyst in hydrogenation processes. Hydrogen will diffuse through a hot palladium barrier.

Symbol: Pd; m.p. 1552°C; b.p. 3140°C; r.d. 12.02 (20°C); p.n. 46; r.a.m. 106.42.

palmitic acid /pal-**mit**-ik/ *See* hexadecanoic acid.

paper chromatography A technique widely used for the analysis of mixtures. Paper chromatography usually employs a specially produced paper as the stationary phase. A base line is marked in pencil near the bottom of the paper and a small sample of the mixture is spotted onto it using a capillary tube. The paper is then placed vertically in a suitable solvent, which rises up to the base line and beyond by capillary action. The components within the sample mixture dissolve in this mobile phase and are carried up the paper. However, the paper holds a quantity of moisture and some components will have a greater tendency than others to dissolve in this moisture than in the mobile phase. In addition, some components will cling to the surface of the paper. Therefore, as the solvent moves through the paper, certain components will be left behind and separated from each other.

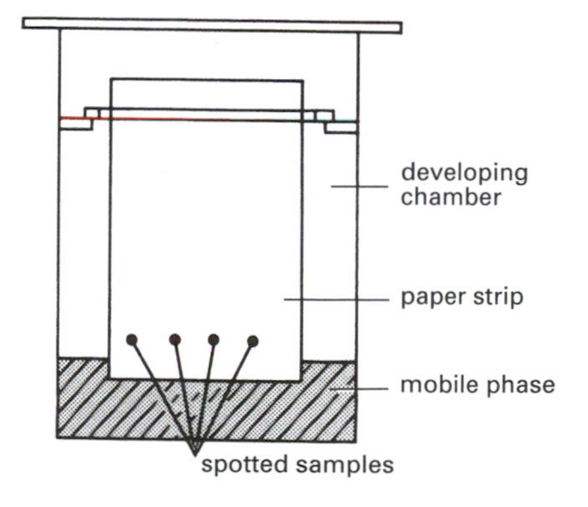

Apparatus for paper chromatography

When the solvent has almost reached the top of the paper, the paper is removed and quickly dried. The paper is developed to locate the positions of colorless fractions by spraying with a suitable chemical, e.g. ninhydrin, or by exposure to ultraviolet radiation. The components are identified by comparing the distance they have traveled up the paper with standard solutions that have been run simultaneously, or by computing an R_F value. A simplified version of paper chromatography uses a piece of filter paper. The sample is spotted at the center of the paper and solvent passed through it. Separation of the components of the mixture again takes place as the mobile phase spreads out on the paper.

Paper chromatography is an application of the partition law.

para- **1.** Designating the form of a diatomic molecule in which both nuclei have

opposite spin directions; e.g. parahydrogen, paradeuterium.
2. Designating a benzene compound with substituents in the 1,4 positions. The position on a benzene ring directly opposite a substituent is the para position. This was used in the systematic naming of benzene compounds. For example, para-dinitrobenzene (or *p*-dinitrobenzene) is 1,4-dinitrobenzene.
See also meta-; ortho-.

paraffin /**pa**-ră-fin/ **1.** *See* petroleum. **2.** *See* alkane.

paraffin oil *See* kerosene. *See also* petroleum.

paraffin wax A solid mixture of hydrocarbons, obtained from petroleum.

paraformaldehyde /pa-ră-for-**mal**-dĕ-hÿd/ *See* methanal.

parahydrogen /pa-ră-**hÿ**-drŏ-jĕn/ *See* hydrogen.

paraiodic (VII) acid /pa-ră-ÿ-**od**-ik/ *See* iodic (VII) acid.

paraldehyde /pă-**ral**-dĕ-hÿd/ *See* ethanal.

paramagnetism /pa-ră-**mag**-nĕ-tiz-ăm/ *See* magnetism.

partial ionic character The electrons of a covalent bond between atoms or groups with different electronegativities will be polarized towards the more electronegative constituent; the magnitude of this effect can be measured by the ionic character of the bond. When the effect is small the bond is referred to simply as a polar bond and is adequately treated using dipole moments; as the effect grows the theoretical treatment requires other contributions to ionic character. Examples of partial ionic character (from nuclear quadrupole resonance) are H–I, 21%; H–Cl, 40%; Li–Br, 90%;.

partial pressure In a mixture of gases, the contribution that one component makes to the total pressure. It is the pressure that the gas would have if it alone were present in the same volume. *See* Dalton's law.

particulate matter (**PM**) An airborne pollutant consisting of small particles of silicate, carbon, or large polyaromatic hydrocarbons (*PAHs*). These pollutants are often referred to as *particulates*. They are classified according to size; e.g. PM10 is particulate matter formed of particles less than 10 μm diameter.

particulates *See* particulate matter.

partition coefficient If a solute dissolves in two non-miscible liquids, the partition coefficient is the ratio of the concentration in one liquid to the concentration in the other liquid.

pascal /**pas**-kăl/ Symbol: Pa The SI unit of pressure, equal to a pressure of one newton per square meter ($1 \text{ Pa} = 1 \text{ N m}^{-2}$). The unit is named for the French mathematician, physicist, and religious philosopher Blaise Pascal (1623–62).

Paschen series /**pash**-ĕn/ A series of lines in the infrared spectrum emitted by excited hydrogen atoms. The lines correspond to the atomic electrons falling into the third lowest energy level and emitting energy as radiation. The wavelength (λ) of the radiation in the Paschen series is given by $1/\lambda = R(1/3^2 - 1/n^2)$ where n is an integer and R is the Rydberg constant. The series is named for the German physicist Louis Paschen (1865–1947). *See also* spectral series.

passive Describing a metal that is unreactive because of the formation of a thin protective layer. Iron, for example, does not dissolve in concentrated nitric(V) acid because of the formation of a thin oxide layer.

Pauli exclusion principle /**pow**-lee/ *See* exclusion principle.

p-block elements The elements of the main groups 13 (B to Tl), 14 (C to Pb), 15 (N to Bi), 16 (O to Po), 17 (F to At), and 18

(He to Rn). They are so called because they have outer electronic configurations that have occupied p levels.

PCB *See* polychlorinated biphenyl.

pearl A lustrous accretion of calcium carbonate (nacre, $CaCO_3$) that builds up on a particle of foreign matter inside the shells of oysters and some other mollusks.

pearl spar *See* dolomite.

peat A brown or black material formed by the partial decomposition of plant remains in marshy ground, the first stage in the formation of coal. It is used for making charcoal and compost, and, when dried, can be burned as a fuel.

PEC *See* photoelectrochemical cell.

pentahydrate /pen-tă-**hȳ**-drayt/ A crystalline compound that has five molecules of water of crystallization per molecule of compound.

pentane /**pen**-tayn/ (C_5H_{12}) A straight-chain alkane obtained by distillation of crude oil.

pentanoic acid /pen-tă-**noh**-ik/ (**valeric acid**; $CH_3(CH_2)_3COOH$) A colorless liquid carboxylic acid, used in making perfumes.

pentavalent /pen-tă-**vay**-lĕnt, pen-**tav**-ă-/ (**quinquevalent**) Having a valence of five.

pentose /**pen**-tohs/ A SUGAR that has five carbon atoms in its molecules.

pentyl group /**pen**-tăl/ (**amyl group**) The group

$$CH_3CH_2CH_2CH_2CH_2–.$$

peptide /**pep**-tÿd/ A compound formed by linkage of two or more amino-acid groups together.

percentage composition A way of expressing the composition of a chemical compound in terms of the percentage (by mass) of each of the elements that make it up. It is calculated by dividing the mass of each element (taking into account the number of atoms present) by the relative molecular mass of the whole molecule. For example, methane (CH_4) has a relative molecular mass of 16 and its percentage composition is 12/16 = 75% carbon and (4 × 1)/16 = 25% hydrogen.

perchloric acid /per-**klor**-ik, -**kloh**-rik/ *See* chloric(VII) acid.

perfect gas (**ideal gas**) *See* gas laws; kinetic theory.

pericyclic reaction /pe-ră-**sȳ**-klik/ A type of CONCERTED REACTION in which the transition state is cyclic and can be regarded as formed by movement of electrons in a circle, from one bond to an adjacent bond. It is usual to distinguish three types of pericyclic reaction:

Cycloadditions. In these reactions a conjugated diene adds to a double bond to form a ring. This involves the formation of two new sigma bonds. The DIELS–ALDER REACTION is a cycloaddition. The reverse reaction, in which a ring containing a double bond breaks to a diene and a compound containing a double bond, is also a pericyclic reaction.

Sigmatropic rearrangements. In these reactions a sigma bond breaks and another is formed. The COPE REARRANGEMENT is an example.

Electrocyclic reactions. In these reactions a ring is formed across the ends of a conjugated system of double bonds. The reverse reaction, in which a ring containing two double bonds breaks by movement of elec-

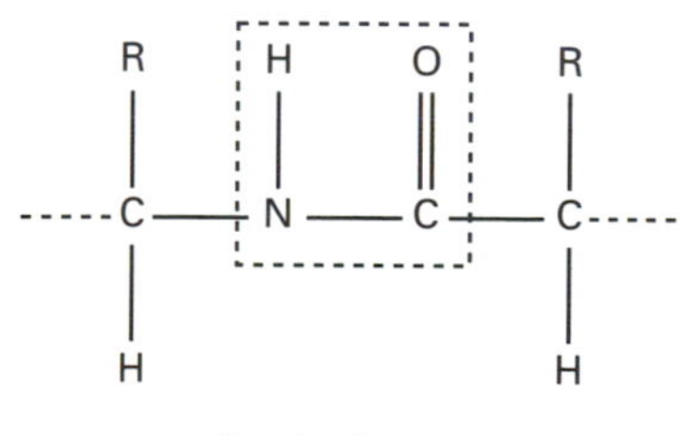

Peptide linkage

trons in a cycle, is also an electrocyclic reaction.

Pericyclic reactions are often treated using FRONTIER-ORBITAL theory or the WOODWARD–HOFFMANN RULES.

peridot /**pe**-ră-dot/ *See* olivine.

period One of the horizontal rows in the conventional periodic table. Each period represents the elements arising from progressive filling of the outer shell (i.e. the addition of one extra electron for each new element), the elements being arranged in order of ascending proton number. In a strict sense hydrogen and helium represent one period but convention refers to the elements lithium to neon (8 elements) as the first *short period* ($n = 2$), and the elements sodium to argon (8 elements) as the second *short period* ($n = 3$). With entry to the $n = 4$ level there is filling of the 4s, then back filling of the 3d, before the 4p are filled. Thus this set contains a total of 18 electrons (potassium to krypton) and is called a *long period*. The next set, rubidium to xenon, is similarly a long period. *See also* Aufbau principle.

periodic acid *See* iodic(VII) acid.

periodic law The law upon which the modern periodic table is based. Enunciated in 1869 by Mendeléev, this law stated that the properties of the elements are a periodic function of their atomic weights: if arranged in order of increasing atomic weight then elements having similar properties occur at fixed intervals. Certain exceptions or gaps in the table lead to the view that the nuclear charge is a more characteristic function, thus the modern statement of the periodic law is that the physical and chemical properties of elements are a periodic function of their proton number.

periodic table A table of the elements arranged in order of increasing proton number to show similarities in chemical behavior between elements. Horizontal rows of elements are called PERIODS. Across a period there is a general trend from metallic to non-metallic behavior. Vertical columns of related elements are called GROUPS. Down a group there is an increase in atomic size and in electropositive (metallic) behavior.

Originally the periodic table was arranged in eight groups with the alkali metals as group I, the halogens as group VII, and the rare gases as group 0. The transition elements were placed in a block in the middle of the table. Groups were split into two sub-groups. For example, group I contained the main-group elements, Li, Na, K, Rb, Cs, in subgroup IA and the subgroup IB elements, Cu, Ag, Au. The system was confusing because there was a difference in usage for subgroups and the current form of the table has 18 groups (*see* Appendix).

permanent gas A gas that cannot be liquefied by pressure alone – that is, a gas above its critical temperature. *See* critical temperature.

permanent hardness A type of water hardness caused by the presence of dissolved calcium, iron, and magnesium sulfates or chlorides. This form of hardness cannot be removed by boiling. *Compare* temporary hardness. *See also* water softening.

permanganate /per-**mang**-gă-nayt, -nit/ *See* manganate(VII).

Permutit (*Trademark*) A substance used to remove unwanted chemicals that have dissolved in water. It is a zeolite consisting of a complex chemical compound, sodium aluminum silicate. When hard water is passed over this material, calcium and magnesium ions exchange with sodium ions in the Permutit. This is a good example of ion exchange. Once all the available sodium ions have been used up, the Permutit can be regenerated by washing it with a saturated solution of sodium chloride. The excess of sodium ions now exchange with the calcium and magnesium ions in the Permutit.

peroxide /pĕ-**roks**-ȳd/ **1.** An oxide containing the $^{-}O–O^{-}$ ion.

2. A compound containing the -O–O– group.

peroxosulfuric(VI) acid /pĕ-roks-oh-sul-**fyoor**-ik/ (**Caro's acid**; H_2SO_5) A white crystalline solid formed by reacting hydrogen peroxide with sulfuric acid. It is a powerful oxidizing agent. The acid is also named for the German chemist Heinrich Caro (1834–1910).

perturbation theory An approximation technique in which a system is divided into two parts, with one part that can be solved exactly and a second part that makes the system too complicated to be solved exactly. For perturbation theory to be applicable it is necessary that the second part is a small correction to the first part. The effects of the small corrections are expressed as an infinite series, called a *perturbation series*, which modifies the results that could be calculated exactly. A classic example of perturbation theory is the modification of the orbits of planets round the Sun because of their gravitational interactions with other planets. In quantum mechanics an application of perturbation theory is the calculation of the effects of small electric or magnetic fields on atomic energy levels.

peta- Symbol: P A prefix denoting 10^{15}.

petrochemicals /pet-roh-**kem**-ă-kălz/ Chemicals that are obtained from crude oil or from natural gas.

petrol *See* petroleum.

petroleum A mixture of hydrocarbons formed originally from marine animals and plants, found beneath the ground trapped between layers of rock. It is obtained by drilling (also called *crude oil*). Different oilfields produce petroleum with differing compositions. The mixture is separated into fractions by fractional distillation in a vertical column. The main fractions are:
Diesel oil (*gas oil*) in the range 220–350°C, consisting mainly of C_{13}–C_{25} hydrocarbons. It is used in diesel engines.
Kerosene (*paraffin*) in the range 160–250°C, consisting mainly of C_{11} and C_{12} hydrocarbons. It is a fuel both for domestic heating and jet engines.
Gasoline (*petrol*) in the range 40–180°C, consisting mainly of C_5–C_{10} hydrocarbons. It is used as motor fuel and as a raw material for making other chemicals.
Refinery gas, consisting of C_1–C_4 gaseous hydrocarbons.

In addition lubricating oils and paraffin wax are obtained from the residue. The black material left is bitumen tar.

petroleum ether A flammable mixture of hydrocarbons, mainly pentane and hexane, used as a solvent. Note that it is not an ether.

pewter A tarnish-resistant alloy of tin and lead, usually containing 80–90% tin. Modern pewter is hardened by antimony and copper, which in the best grades replace the lead content. It is lighter than old pewter and having a bright or a soft satiny finish is often used for decorative and ceremonial purposes.

pH The logarithm to base 10 of the reciprocal of the hydrogen-ion concentration of a solution. In pure water at 25°C the concentration of hydrogen ions is 1.00×10^{-7} mol l^{-1}, thus the pH equals 7 at neutrality. An increase in acidity increases the value of $[H^+]$, decreasing the value of the pH below 7. An increase in the concentration of hydroxide ion $[OH^-]$ proportionately decreases $[H^+]$, therefore increasing the value of the pH above 7 in basic solutions. pH values can be obtained approximately by using indicators. More precise measurements use electrode systems.

phase /fayz/ One of the physically separable parts of a chemical system. For example, a mixture of ice (solid phase) and water (liquid phase) consists of two phases. A system consisting of only one phase is said to be homogeneous. A system consisting of more than one phase is said to be heterogeneous.

phase diagram A graphical representation of the state in which a substance will occur at a given pressure and temperature. The lines show the conditions under which more than one phase can coexist at equilibrium. For one-component systems (e.g. water) the point at which all three phases can coexist at equilibrium is called the triple point and is the point on the graph at which the pressure–temperature curves intersect.

phase equilibrium A state in which the proportions of the various phases in a chemical system are fixed. When two or more phases are present at fixed temperature and pressure a dynamic condition will be established in which individual particles will leave one phase and enter another; at equilibrium this will be balanced by an equal number of particles making the reverse change, leaving the total composition unchanged.

phase rule In a system at equilibrium the number of phases (*P*), number of components (C), and number of degrees of freedom (*F*) are related by the formula:

$$P + F = C + 2$$

phenol /**fee**-nol, -nohl/ **1.** (carbolic acid, hydroxybenzene, C_6H_5OH) A white crystalline solid used to make a variety of other organic compounds.

2. A type of organic compound in which at least one hydroxyl group is bound directly to one of the carbon atoms of an aromatic ring. Phenols do not show the behavior typical of alcohols. In particular they are more acidic because of the electron-withdrawing effect of the aromatic ring. The preparation of phenol itself is by fusing the sodium salt of the sulfonic acid with sodium hydroxide:

$$C_6H_5SO_2.ONa + 2NaOH \rightarrow C_6H_5ONa + Na_2SO_3 + H_2O$$

The phenol is then liberated by sulfuric acid:

$$2C_6H_5ONa + H_2SO_4 \rightarrow 2C_6H_5OH + Na_2SO_4$$

Reactions of phenol include:

1. Replacement of the hydroxyl group with a chlorine atom using phosphorus(V) chloride.
2. Reaction with acyl halides to form esters of carboxylic acids.
3. Reaction with haloalkanes under alkaline conditions to give mixed alkyl–aryl ethers.

In addition phenol can undergo further substitution on the benzene ring. The hydroxyl group directs other substituents into the 2- and 4-positions.

phenolphthalein /fee-nol-**thal**-een, -nohl-, **-fthal-**/ An acid–base indicator that is colorless in acid solutions and becomes red if the pH rises above the transition range of 8–9.6. It is used as the indicator in titrations for which the end point lies clearly on the basic side (pH > 7), e.g. oxalic acid or potassium hydrogentartrate against caustic soda.

phenoxy resin /fi-**noks**-ee/ A type of thermoplastic resin made by condensation of phenols.

phenylalanine /fen-ăl-**al**-ă-neen, fee-năl-/ *See* amino acids.

phenylamine /fen-al-**am**-een, fee-năl-/ *See* aniline.

phenylethene /fen-ăl-**eth**-een, fee-năl-/ (**styrene**; $C_6H_5CHCH_2$) A liquid hydrocarbon used as the starting material for the production of polystyrene. The manufacture of phenylethene is by dehydrogenation of ethyl benzene:

$$C_6H_5C_2H_5 \rightarrow C_6H_5CH{:}CH_2 + H_2$$

phenyl group /**fen**-ăl, **fee**-năl/ The group C_6H_5–, derived from benzene.

phenylhydrazone /fen-ăl-**hÿ**-dră-zohn, fee-năl-/ *See* hydrazone.

phenylmethanol /fen-ăl-**meth**-ă-nol, fee-năl-, -nohl/ (**benzyl alcohol**; $C_6H_5CH_2OH$) An aromatic primary alcohol. Phenylmethanol is synthesized by Cannizzaro's reaction, which involves the simultaneous oxidation and reduction of benzenecarbaldehyde (benzaldehyde) by refluxing in an aqueous solution of sodium hydroxide:

$$2C_6H_5CHO \rightarrow C_6H_5CH_2OH + C_6H_5COOH$$

Phenylmethanol undergoes the reactions characteristic of alcohols, especially those in which the formation of a stable carbonium ion as an intermediate ($C_6H_5CH_2^+$) enhances the reaction. Substitution onto the benzene ring is also possible; the $-CH_2OH$ group directs into the 2- or 4-position by the donation of electrons to the ring.

phenyl methyl ketone (**acetophenone**; $C_6H_5COCH_3$) A colorless sweet-smelling organic liquid, which solidifies below 20°C. It is used as a solvent for methyl and ethyl cellulose plastics.

3-phenylpropenoic acid /fen-ăl-proh-pĕ-**noh**-ik, fee-năl-/ (**cinnamic acid**; $C_6H_5CH{:}CHCOOH$) A white pleasant-smelling crystalline carboxylic acid. It occurs in amber but can be synthesized and is used in perfumes and flavorings.

Phillips process A method for the manufacture of high-density polyethene using a catalyst of chromium(III) oxide on a promoter of silica and alumina. The reaction conditions are 150°C and 30 atm pressure. *See also* Ziegler process.

Philosopher's stone *See* alchemy.

phlogiston theory /floh-**jis**-tŏn/ A former theory of combustion (disproved in the eighteenth century by Lavoisier). Flammable compounds were supposed to contain a substance (phlogiston), which was released when they burned, leaving ash.

phosgene /**fos**-jeen/ *See* carbonyl chloride.

phosphates /**fos**-fayts/ *See* phosphoric(V) acid.

phosphide /**fos**-fÿd, -fid/ A compound of phosphorus with a more electropositive element.

phosphine /**fos**-feen, -fin/ (**phosphorus(III) hydride**; PH_3) A colorless gas that is slightly soluble in water. It has a characteristic fishy smell. It can be made by reacting water and calcium phosphide or by the action of yellow phosphorus on a concentrated alkali. Phosphine usually ignites spontaneously in air because of contamination with diphosphine. It decomposes into its elements if heated to 450°C in the absence of oxygen and it burns in oxygen or air to yield phosphorus oxides. It reacts with solutions of metal salts to precipitate phosphides. Like its nitrogen analog ammonia it forms salts, called *phosphonium salts*. It also forms complex addition compounds with metal ions. As in ammonia, one or more of the hydrogen atoms can be replaced by alkyl groups.

phosphinic acid /fos-**fin**-ik/ (**hypophosphorous acid**; H_3PO_2) A white crystalline solid. It is a monobasic acid forming the anion $H_2PO_2^-$ in water. The sodium salt, and hence the acid, can be prepared by heating yellow phosphorus with sodium hydroxide solution. The free acid and its salts are powerful reducing agents.

phosphonic acid /fos-**fon**-ik/ (**phosphorous acid; orthophosphorous acid**; H_3PO_3) A colorless deliquescent solid that can be prepared by the action of water on phosphorus(III) oxide or phosphorus(III) chloride. It is a dibasic acid producing the anions $H_2PO_3^-$ and HPO_3^{2-} in water. The acid and its salts are slow reducing agents. On warming, phosphonic acid decomposes to phosphine and phosphoric(V) acid.

phosphonium ion /fos-**foh**-nee-ŭm/ The ion PH_4^+ derived from phosphine.

phosphonium salt *See* phosphine.

phosphor /**fos**-fer/ A substance that shows luminescence or phosphorescence.

phosphorescence /fos-fŏ-**ress**-ĕns/ **1.** The absorption of energy by atoms followed by emission of electromagnetic radiation. Phosphorescence is a type of luminescence, and is distinguished from fluorescence by the fact that the emitted radiation continues for some time after the source of excitation has been removed. In phosphorescence the excited atoms have relatively long lifetimes before they make transitions to lower energy states. However, there is no defined time distinguishing phosphorescence from fluorescence.

2. In general usage the term is applied to the emission of 'cold light' – light produced without a high temperature. The name comes from the fact that white phosphorus glows slightly in the dark as a result of a chemical reaction with oxygen. The light comes from excited atoms produced directly in the reaction – not from the heat produced. It is thus an example of *chemiluminescence*. There are also a number of biochemical examples termed bioluminescence; for example, phosphorescence is sometimes seen in the sea from marine organisms, or on rotting wood from certain fungi (known as 'fox fire').

phosphoric(V) acid /fos-**for**-ik, -**foh**-rik/ (**orthophosphoric acid**; H_3PO_4) A white solid that can be made by reacting phosphorus(V) oxide with water or by heating yellow phosphorus with nitric acid. The naturally occurring *phosphates* (orthophosphates, M_3PO_4) are salts of phosphoric(V) acid. An aqueous solution of phosphoric(V) acid has a sharp taste and has been used in the manufacture of soft drinks. At about 220°C phosphoric(V) acid is converted to *pyrophosphoric acid* ($H_4P_2O_7$). Pure pyrophosphoric acid in the form of colorless crystals is made by warming solid phosphoric(V) acid with phosphorus(III) chloride oxide. *Metaphosphoric acid* is obtained as a polymer, $(HPO_3)_n$, by heating phosphoric(V) acid to 320°C. Both meta- and pyrophosphoric acids change to phosphoric(V) acid in aqueous solution; the change will take place faster at higher temperatures. Phosphoric(V) acid is tri-basic, forming three series of salts with the anions $H_2PO_4^-$, HPO_4^{2-}, and PO_4^{3-}. These series of salts are acidic, neutral, and alkaline respectively. The alkali metal phosphates (except lithium phosphate) are soluble in water but other metal phosphates are insoluble. Phosphates are useful in water softening and as fertilizers.

phosphorous acid /**fos**-fŏ-rŭs, fos-**for**-ŭs, -**foh**-rŭs/ *See* phosphonic acid.

phosphorus /**fos**-fŏ-rŭs/ A reactive solid non-metallic element; the second element in group 15 of the periodic table. It has the electronic configuration $[Ne]3s^23p^3$ and is therefore formally similar to nitrogen. It is however very much more reactive than nitrogen and is never found in nature in the uncombined state. Phosphorus is widespread throughout the world; economic sources are phosphate rock ($Ca_3(PO_4)_2$) and the apatites, variously occurring as both fluoroapatite ($3Ca_3(PO_4)_2.CaF_2$) and as chloroapatite ($3Ca_3(PO_4)_2CaCl_2$). Guano formed from the skeletal phosphate of fish in sea-bird droppings is also an important source of phosphorus. The largest amounts of phosphorus compounds produced are used as fertilizers, with the detergents industry producing increasingly large tonnages of phosphates. Phosphorus is an essential constituent of living tissue and bones, and it plays a very important part in metabolic processes and muscle action.

The element is obtained industrially by the reduction of phosphate rock using sand (SiO_2) and coke (C) in an electric furnace. The phosphorus is distilled off as P_4 molecules and collected under water as white phosphorus.

$$2Ca_3(PO_4)_2 + 6SiO_2 + 10C \rightarrow P_4 + 6CaSiO_3 + 10CO$$

There are three common allotropes of phosphorus and several other modifications of these, some of which have indefinite structures. The common forms are:

1. white (or yellow) phosphorus (distillation of phosphorus; soluble in some organic solvents)
2. red phosphorus (heat on white phosphorus, insoluble in organic solvents)
3. black phosphorus (heat on white phosphorus under high pressures).

Phosphorus, being in group 15 is nonmetallic in character and has a sufficiently high ionization potential for the chemistry of bonds to phosphorus to be almost entirely covalent. Phosphorus compounds are formally P(III) and P(V) compounds. The element forms a hydride, PH_3 (phosphine), which is analogous to ammonia, but it is not formed by direct combination. Phosphorus(V) hydrides are not known.

When phosphorus is burned in air the product is the highly hygroscopic white solid P_4O_{10} (commonly called phosphorus pentoxide, P_2O_5); this is the P(V) oxide. In a limited supply of air the P(III) oxide, P_4O_6 (called phosphorus trioxide) contaminated with unreacted phosphorus is obtained. The oxides both react with water to form acids;

$P_4O_6 + 6H_2O \rightarrow$ $4H_3PO_3$ phosphor*ous* acid

$P_4O_{10} + 2H_2O \rightarrow$ $4HPO_3$ metaphosph*oric* acid

$P_4O_{10} + 6H_2O \rightarrow$ $4H_3PO_4$ orthophosph*oric* acid.

There are also many poly-phosphoric acids and salts with different ring sizes and different chain lengths. Pyrophosphoric acid is obtained by melting phosphoric acid, further dehydration above 200°C leads to metaphosphoric acids,

$$2H_3PO_4 \rightarrow H_4P_2O_7 + H_2O$$

$$H_4P_2O_7 \rightarrow 2HPO_3 + H_2O$$

Polyphosphates with molecular weights from 1000 to 10 000 are industrially important in detergent formulations and the lower members are used to complex metal ions (sequestration).

Phosphorus forms the P(III) halides, PX_3, with all the halogens, and P(V) halides, PX_5, with X = F, Cl, Br. The P(III) halides are formed by direct combination but because of hazards in the reaction of phosphorus and fluorine, PF_3 is prepared from PCl_3

The P(III) halide molecules are all pyramidal (as is PH_3) and are readily hydrolyzed. The P(V) halides are also readily hydrolyzed by water in two stages giving first the oxyhalide, then phosphoric acid,

$$PX_5 + H_2O \rightarrow POX_3 + 2HX$$

$$POX_3 + 3H_2O \rightarrow H_3PO_4 + 3HX$$

In the gas phase the trihalides are pyramidal and the pentahalides are trigonal bipyramids. However, in tetrachloromethane solution phosphorus pentachloride is a dimer with two chlorine bridges exhibiting the ready expansion of the coordination number to six. In the solid state, (PCl_5) is $[PCl_4]^+[PCl_6]^-$ (the first tetrahedral, the second octahedral) obtained by transfer of a chloride ion.

Phosphorus interacts directly with several metals and metalloids to give:

1. Ionic phosphides such as Na_3P, Ca_3P_2, Sr_3P_2.
2. Molecular phosphides such as P_4S_3 and P_4S_5.
3. Metallic phosphides such as Fe_2P.

Symbol: P; m.p. 44.1°C (white) 410°C (red under pressure); b.p. 280.5°C; r.d. 1.82 (white) 2.2 (red) 2.69 (black) (all at 20°C); p.n. 15; r.a.m. 30.973762.

phosphorus(III) bromide (**phosphorus tribromide**; PBr_3) A colorless liquid made by reacting phosphorus with bromine. It is readily hydrolyzed by water to phosphonic acid and hydrogen bromide. Phosphorus(III) bromide is important in organic chemistry, being used to replace a hydroxyl group with a bromine atom.

phosphorus(V) bromide (**phosphorus pentabromide**; PBr_5) A yellow crystalline solid that sublimes easily. It can be made by the reaction of bromine and phosphorus(III) bromide. Phosphorus(V) bromide is readily hydrolyzed by water to phosphoric(V) acid and hydrogen bromide. Its main use is in organic chemistry to replace a hydroxyl group with a bromine atom.

phosphorus(III) chloride (**phosphorus trichloride**; PCl_3) A colorless liquid formed from the reaction of phosphorus with chlorine. It is rapidly hydrolyzed by water to phosphonic acid and hydrogen chloride. Phosphorus(III) chloride is used in organic chemistry to replace a hydroxyl group with a chlorine atom.

phosphorus(V) chloride (**phosphorus pentachloride**; PCl_5) A white easily sublimed solid formed by the action of chlorine on phosphorus(III) chloride. It is hydrolyzed by water to phosphoric(V) acid and hydrogen chloride. Its main use is as a chlorinating agent in organic chemistry to replace a hydroxyl group with a chlorine atom.

phosphorus(III) chloride oxide (**phosphorus trichloride oxide; phosphorus oxychloride**; phosphoryl chloride, $POCl_3$) A colorless liquid that can be obtained by reacting phosphorus(III) chloride with oxygen or by distilling phosphorus(III) chloride with potassium chlorate. The reactions of phosphorus(III) chloride oxide are similar to those of phosphorus(III) chloride. The chlorine atoms can be replaced by alkyl groups using Grignard reagents or by alkoxo groups using alcohols. Water hydrolysis yields phosphoric(V) acid. Phosphorus(III) chloride oxide forms complexes with many metal ions.

phosphorus(III) hydride *See* phosphine.

phosphorus(III) oxide (**phosphorus trioxide**; P_2O_3) A white waxy solid with a characteristic smell of garlic; it usually exists as P_4O_6 molecules. It is soluble in organic solvents, such as benzene and carbon disulfide, and is made by burning phosphorus in a limited supply of air. Phosphorus(III) oxide oxidizes in air to phosphorus(V) oxide at room temperature and inflames above 70°C. It dissolves slowly in cold water or dilute alkalis to give phosphonic acid or one of its salts. Phosphorus(III) oxide reacts violently with hot water to form phosphine and phosphoric(V) acid.

phosphorus(V) oxide (**phosphorus pentoxide**; P_2O_5) A white powder that is soluble in organic solvents. It usually exists as P_4O_{10} molecules. Phosphorus(V) oxide can be prepared by burning phosphorus in a plentiful supply of oxygen. It readily combines with water to form phosphoric(V) acid and is therefore used as a drying agent for gases. It is a useful dehydrating agent because it is able to remove the elements of water from certain oxyacids and other compounds containing oxygen and hydrogen; for example, it reacts with nitric acid (HNO_3) to give nitrogen pentoxide (N_2O_5) on heating.

phosphorus oxychloride /oks-ă-**klor**-ÿd, -**kloh**-rÿd/ *See* phosphorus(III) chloride oxide.

phosphorus pentabromide /pen-tă-**broh**-mÿd/ *See* phosphorus(V) bromide.

phosphorus pentachloride /pen-tă-**klor**-ÿd, -**kloh**-rÿd/ *See* phosphorus(V) chloride.

phosphorus pentoxide /pen-**toks**-ÿd/ *See* phosphorus(V) oxide.

phosphorus tribromide /trÿ-**broh**-mÿd/ *See* phosphorus(III) bromide.

phosphorus trichloride /trÿ-**klor**-ÿd, -**kloh**-rÿd/ *See* phosphorus(III) chloride.

phosphorus trichloride oxide /trÿ-**klor**-ÿd, -**kloh**-rÿd/ *See* phosphorus(III) chloride oxide.

phosphorus trioxide /trÿ-**oks**-ÿd/ *See* phosphorus(III) oxide.

phosphoryl chloride /**fos**-fŏ-răl, fos-**for**-ăl, -**foh**-răl/ *See* phosphorus(III) chloride oxide.

phot /foht, fot/ A unit of illumination in the c.g.s. system, equal to an illumination of one lumen per square centimeter. It is equal to 10^4 lx.

photochemical reaction /foh-toh-**kem**-ă-kăl/ A reaction brought about by light; examples include the bleaching of colored material, the reduction of silver halides, and the photosynthesis of carbohydrates. Chemical changes only occur when the reacting atoms or molecules absorb photons of the appropriate energy. The amount of substance that reacts is proportional to the quantity of energy absorbed. For example, in the reaction between hydrogen and chlorine, it is not the concentrations of hydrogen or chlorine that dictate the rate of reaction but the intensity of the radiation.

photochemistry /foh-toh-**kem**-iss-tree/ The branch of chemistry dealing with reactions induced by light or ultraviolet radiation.

photoelectric effect The emission of electrons from a solid (or liquid) surface when it is irradiated with electromagnetic radiation. For most materials the photoelectric effect occurs with ultraviolet radiation or radiation of shorter wavelength; some show the effect with visible radiation.

In the photoelectric effect, the number of electrons emitted depends on the intensity of the radiation and not on its frequency. The kinetic energy of the electrons that are ejected depends on the frequency of the radiation. This was explained, by Einstein, by the idea that electromagnetic radiation consists of streams of photons. The *photon energy* is $h\nu$, where h is the Planck constant and ν the frequency of the radiation. To remove an electron from the solid a certain minimum energy must be supplied, known as the *work function*, ϕ. Thus, there is a certain minimum threshold frequency ν_0 for radiation to eject electrons: $h\nu_0 = \theta$. If the frequency is higher than this threshold the electrons are ejected. The maximum kinetic energy (W) of the electrons is given by *Einstein's equation*:

$$W = h\nu - \phi$$

The photoelectric effect also occurs with gases. *See* photoionization.

photoelectrochemical cell /foh-toh-i-lek-troh-**kem**-ă-kăl/ (PEC) A type of voltaic cell in which one of the electrodes is a light-sensitive semiconductor (such as gallium arsenide). A current is generated by light falling on the semiconductor, causing electrons to pass between the electrode and the electrolyte and produce chemical reactions in the cell.

photoelectron /foh-toh-i-**lek**-tron/ An electron ejected from a solid, liquid, or gas by the photoelectric effect or by photoionization.

photoelectron spectroscopy *See* photoionization.

photoemission /foh-toh-i-**mish**-ŏn/ The emission of photoelectrons by the photoelectric effect or by photoionization.

photoionization /foh-toh-ÿ-ŏ-ni-**zay**-shŏn/ The ionization of atoms or molecules by electromagnetic radiation. Photons absorbed by an atom may have sufficient photon energy to free an electron from its attraction by the nucleus. The process is

$$M + h\nu \rightarrow M^+ + e^-$$

As in the photoelectric effect, the radiation must have a certain minimum threshold frequency. The energy of the photoelectrons ejected is given by $W = h\nu - I$, where I is the ionization potential of the atom or molecule. Analysis of the energies of the emitted electrons gives information on the ionization potentials of the substance – a technique known as *photoelectron spectroscopy*.

photolysis /fŏ-**tol**-ă-sis/ A chemical reaction that is produced by electromagnetic radiation (light or ultraviolet radiation). Many photolytic reactions involve the formation of free radicals.

photosynthesis /foh-tŏ-**sin**-th'ĕ-sis/ The photochemical process green plants use to synthesize organic compounds from carbon dioxide and water.

phthalic acid /**thal**-ik, **fthal**-ik/ *See* benzene-1,2-dicarboxylic acid.

phthalic anhydride *See* benzene-1,2-dicarboxylic acid.

physical change A change to a substance that does not alter its chemical properties. Physical changes (e.g. melting, boiling, and dissolving) are comparatively easy to reverse.

physical chemistry The branch of chemistry concerned with physical properties of compounds and how these depend on the chemical bonding.

physisorption /fiz-ă-**sorp**-shŏn, **-zorp-**/ *See* adsorption.

piano-stool compound *See* sandwich compound.

pi bond *See* orbital.

pico- /**pÿ**-koh/ Symbol: p A prefix denoting 10^{-12}. For example, 1 picofarad (pF) = 10^{-12} farad (F).

pi complex A complex in which the metal ion is bound to a pi-electron system, as in FERROCENE and ZEISE'S SALT.

picrate /**pik**-rayt/ A salt of PICRIC ACID; metal picrates are explosive.

picric acid /**pik**-rik/ (**2,4,6-trinitrophenol**; $C_6H_2(NO_3)_3OH$) A yellow crystalline solid made by nitrating phenolsulfonic acid. It is used as a dye and as an explosive. With aromatic hydrocarbons picric acid forms characteristic charge-transfer complexes (misleadingly called *picrates*), used in analysis for identifying the hydrocarbon.

pig iron The crude iron produced from a blast furnace, containing carbon, silicon, and other impurities. The molten iron is run out of the furnace into channels (called 'sows'), which branch out into a number of offshoots (called 'pigs') in which the metal is allowed to cool.

pigment An insoluble, particulate coloring material.

pine-cone oil *See* turpentine.

pi orbital *See* orbital.

pipette /pi-**pet**/ A device used to transfer a known volume of solution from one container to another; in general, several samples of equal volume are transferred for individual analysis from one stock solution. Pipettes are of two types, bulb pipettes, which transfer a known and fixed volume, and graduated pipettes, which can transfer variable volumes. Pipettes were at one time universally mouth-operated but safety pipettes using a plunger or rubber bulb are now preferred.

pitchblende /**pich**-blend/ (**uraninite**) A black mineral consisting mostly of uranium(VI) oxide, UO_3. It is the chief ore of uranium and radium.

pK The logarithm to the base 10 of the reciprocal of an acid's dissociation constant:

$$\log_{10}(1/K_a)$$

Planck constant /plank/ Symbol: h A fundamental constant; the ratio of the energy (W) carried by a photon to its frequency (v). A basic relationship in the quantum theory of radiation is $W = hv$. The value of h is $6.626\,196 \times 10^{-34}$ J s. The Planck constant appears in many relationships in which some observable measurement is quantized (i.e. can take only specific discrete values rather than any of a range of values). The constant is named for the German physicist Max Planck (1858–1947).

plane polarization A type of polarization of electromagnetic radiation in which the vibrations take place entirely in one plane.

plasma /**plaz**-mă/ A mixture of ions and electrons, as in an electrical discharge. Sometimes, a plasma is described as a fourth state of matter.

plaster of Paris *See* calcium sulfate.

plasticizer /**plas**-tă-sÿ-zer/ A substance added to a synthetic resin to make it more flexible.

plastics A common term for synthetic polymers, especially when mixed with other substances such as plasticizers, fillers, preservatives, and colorants.

platinic chloride /plă-**tin**-ik/ *See* chloroplatinic acid.

platinum /**plat**-ă-nŭm/ A silvery-white malleable ductile transition metal. It occurs naturally in Australia and Canada, either free or in association with other platinum metals. It is resistant to oxidation and is not attacked by acids (except aqua regia) or alkalis. Platinum is used as a catalyst for ammonia oxidation (to make nitric acid) and in catalytic converters. It is also used in jewelry.

Symbol: Pt; m.p. 1772°C; b.p. 3830 ± 100°C; r.d. 21.45 (20°C); p.n. 78; r.a.m. 195.08.

platinum black A finely divided black form of platinum produced, as a coating, by evaporating platinum onto a surface in an inert atmosphere. Platinum-black coatings are used as pure absorbent electrode coatings in experiments on electric cells. They are also used, like carbon-black coatings, to improve the ability of a surface to absorb radiation.

platinum(II) chloride ($PtCl_2$) A gray-brown powder prepared by the partial decomposition of platinum(IV) chloride. It may also be prepared by passing chlorine over heated platinum. Platinum(II) chloride is insoluble in water but dissolves in concentrated hydrochloric acid to form a complex acid.

platinum(IV) chloride ($PtCl_4$) A reddish hygroscopic solid prepared by the action of heat on chloroplatinic acid. Crystals of the pentahydrate, $PtCl_4.5H_2O$, are formed; anhydrous platinum(IV) chloride is prepared by treating the hydrated crystals with concentrated sulfuric acid. The chloride dissolves in water to produce a strongly acidic solution. Platinum(IV) chloride forms a series of hydrates having 1, 2, 4, and 5 molecules of water; the tetrahydrate is the most stable. The strongly acidic solution produced on dissolving in water probably contains $H_2[PtCl_4(OH)_2]$.

platinum–iridium An alloy of platinum containing up to 30% iridium. Hardness and resistance to chemical attack increase as the iridium content is increased. It is used in jewelry, electrical contacts, and hypodermic needles.

platinum metals The group of transition metals ruthenium (Ru), osmium (Os), rhodium (Rh), iridium (Ir), palladium (Pd), and platinum (Pt). They are sometimes classed together as they have related properties and compounds.

Plexiglas /**pleks**-ă-glass/ (*Trademark*) A widely-used acrylic resin, polymethylmethacrylate.

plumbane /**plum**-bayn/ (**lead(IV) hydride**; PbH_4) A colorless unstable gas that can be obtained by the action of acids on a mixture of magnesium and lead pellets.

plumbic /**plum**-bik/ Designating a lead(IV) compound.

plumbous /**plum**-bŭs/ Designating a lead(II) compound.

plutonium /ploo-**toh**-nee-ŭm/ A radioactive silvery element of the actinoid series of metals. It is a transuranic element found on Earth only in minute quantities in uranium ores but readily obtained, as ^{239}Pu, by neutron bombardment of natural uranium. The readily fissionable ^{239}Pu is a major nuclear fuel and nuclear explosive. Plutonium is highly toxic because of its radioactivity; in the body it accumulates in bone.

Symbol: Pu; m.p. 641°C; b.p. 3232°C; r.d. 19.84 (25°C); p.n. 94; most stable isotope ^{244}Pu (half-life 8.2×10^7 years).

PM *See* particulate matter.

point defect *See* defect.

poison **1.** A substance that destroys catalyst activity.
2. Any substance that endangers biological activity, whether by physical or chemical means.

polar Describing a compound with molecules that have a permanent dipole moment. Hydrogen chloride and water are examples of polar compounds.

polar bond A covalent bond in which the bonding electrons are not shared equally between the two atoms. A bond between two atoms of different electronegativity is said to be polarized in the direction of the more electronegative atom, i.e. the electrons are drawn preferentially towards the atom. This leads to a small separation of charge and the development of a bond DIPOLE MOMENT as in, for example, hydrogen fluoride, represented as H→F or as $H^{\delta+}$–$F^{\delta-}$ (F is more electronegative).

The charge separation is much smaller than in ionic compounds; molecules in which bonds are strongly polar are said to display partial ionic character. The effect of the electronegative element can be transmitted beyond adjacent atoms, thus the C–C bonds in, for example, CCl_3CH_3 and CH_3CHO are slightly polar. *See also* intermolecular forces.

polarimeter /poh-lă-**rim**-ĕ-ter/ (**polariscope**) An instrument for measuring OPTICAL ACTIVITY.

polarizability /poh-lă-rȳ-ză-**bil**-ă-tee/ The ease with which an electron cloud is deformed (polarized). In ions, an increase in size or negative charge leads to an increase in polarizability. The concept is of particular use in the treatment of covalent contributions to predominantly ionic bonds embodied in Fajans' Rules. Thus ions, such as I^-, Se^{2-}, Te^{2-}, are especially prone to covalent character, particularly in combination with ions of small size and high positive charge (i.e. high ionic potential). Molecular polarizability is a measure of the ease with which electrons in the molecule are deformed, particularly by electromagnetic radiation.

polarization /poh-lă-ri-**zay**-shŏn/ **1.** The restriction of the vibrations in a transverse wave so that the vibration occurs in a single plane. Electromagnetic radiation, for instance, is a transverse wave motion. It can be thought of as an oscillating electric field and an oscillating magnetic field, both at right angles to the direction of propagation and at right angles to each other. Usually, the electric vector is considered since it is the electric field that interacts with charged particles of matter and causes the effects. In 'normal' unpolarized radiation, the electric field oscillates in all possible directions perpendicular to the wave direction. On reflection or on transmission through certain substances (e.g. Polaroid) the field is confined to a single plane. The radiation is then said to be *plane-polarized*. If the tip of the electric vector describes a circular helix as the wave propagates, the light is said to be *circularly polarized*.
2. *See* polarizability.
3. The reduction of current in a voltaic cell, caused by the build-up of products of the chemical reaction. Commonly, the cause is the build-up of a layer of bubbles (e.g. of hydrogen). This reduces the effective area of the electrode, causing an increase in the cell's internal resistance. It can also produce a back e.m.f. Often a substance such as manganese(IV) oxide (a *depolarizer*) is added to prevent hydrogen build-up.

polar molecule A molecule in which the individual polar bonds are not perfectly symmetrically arranged and are therefore not 'in balance'. Thus the charge separation in the bonds gives rise to an overall charge separation in the molecule as for example, in water. Such molecules possess a dipole moment.

polarography /poh-lă-**rog**-ră-fee/ An analytical method in which current is measured as a function of potential. A special type of cell is used in which there is a small easily polarizable cathode (the dropping mercury electrode) and a large non-polarizable anode (reference cell). The analytical reaction takes place at the cathode and is essentially a reduction of the cations, which are discharged according to the order of their electrode potential values. The data is expressed in the form of a *polarogram*, which is a plot of current against applied voltage. As the applied potential is increased a point is reached at which the ion is discharged. There is a step-wise increase in current, which levels off because of polarization effects. The potential at half the step height (called the *half-wave potential*) is used to identify the ion. Most elements can be determined by polarography. The optimum concentrations are in the range 10^{-2}–10^{-4}M; modified techniques allow determinations in the parts per million range.

polar solvent *See* solvent.

pollution Any damaging or unpleasant change in the environment that results from the physical, chemical, or biological side-effects of human industrial or social activities. Pollution can affect the atmosphere, rivers, seas, and the soil.

Air pollution is caused by the domestic and industrial burning of carbonaceous fuels, by industrial processes, and by car exhausts. Among recent problems are industrial emissions of sulfur(IV) oxide causing *acid rain*, and the release into the atmosphere of chlorofluorocarbons, used in refrigeration, aerosols, etc., has been linked to the depletion of ozone in the stratosphere. Carbon dioxide, produced by burning fuel and by car exhausts, is slowly building up in the atmosphere, which could result in an overall increase in the temperature of the atmosphere (greenhouse effect). Car exhausts also contain carbon monoxide and lead. The former has not yet reached dangerous levels, but vegetation near main roads contains a high proportion of lead and levels are sufficiently high in urban areas to cause concern about the effects on children. Lead-free gasoline is available. Photochemical smog, caused by the action of sunlight on hydrocarbons and nitrogen oxides from car exhausts, is a problem in several countries. Catalytic converters reduce harmful emissions from car exhausts.

Water pollutants include those that are *biodegradable*, such as sewage effluent, which cause no permanent harm if adequately treated and dispersed, as well as those which are nonbiodegradable, such as certain chlorinated hydrocarbon pesticides (e.g. DDT) and heavy metals, such as lead, copper, and zinc in some industrial effluents (causing *heavy-metal pollution*). When these accumulate in the environment they can become very concentrated in food chains. The pesticides DDT, aldrin, and dieldrin are now banned in many countries. Water supplies can become polluted by leaching of nitrates from agricultural land. The discharge of waste heat can cause thermal pollution of the environment, but this is reduced by the use of cooling towers. In the sea, oil spillage from tankers and the inadequate discharge of sewage effluent are the main problems.

Other forms of pollution are noise from aircraft, traffic, and industry and the disposal of radioactive waste.

polonium /pŏ-**loh**-nee-ŭm/ A radioactive metallic element belonging to group 16 of the periodic table. It occurs in very minute quantities in uranium ores. Over 30 radioisotopes are known, nearly all alpha-particle emitters. Polonium is a volatile metal and evaporates with time. It is also strongly radioactive; a quantity of polonium quickly reaches a temperature of a few hundred degrees C because of the alpha emission. For this reason it has been used as a lightweight heat supply in space satellites.

Symbol: Po; m.p. 254°C; b.p. 962°C; r.d. 9.32 (20°C); p.n. 84; stablest isotope ^{209}Po (half-life 102 years).

polyamide /pol-ee-**am**-ÿd, -id/ A synthetic polymer in which the monomers are linked by the group –NH–CO–. Nylon is an example of a polyamide.

polyatomic /pol-ee-ă-**tom**-ik/ Describing a molecule (or ion or radical) that consists of several atoms (three or more). Examples are benzene (C_6H_6) and methane (CH_4).

polybasic /pol-ee-**bay**-sik/ Describing an acid that has two or more replaceable hydrogen atoms. For example, phosphorus(V) acid, H_3PO_4, is tribasic.

polycarbonate /pol-ee-**kar**-bŏ-nayt/ A thermoplastic polymer consisting of polyesters of carbonic acid and dihydroxy compounds. They are tough and transparent, used for making soft-drink bottles and electrical connectors.

polychlorinated biphenyl /pol-ee-**klor**-ă-nay-tid, **-kloh**-ră- / (**PCB**) A type of compound based on biphenyl ($C_6H_5C_6H_5$), in which some of the hydrogen atoms have been replaced by chlorine atoms. They are used in certain polymers used for electrical insulators. PCBs are highly toxic and concern has been caused by the fact that they can accumulate in the food chain.

polychloroethene /pol-ee-klor-oh-**eth**-een, -kloh-roh-/ (**polyvinyl chloride**; PVC) A synthetic polymer made from chloroethene. It is a strong material with a wide variety of uses.

polycrystalline /pol-ee-**kris**-tă-lin, -lȳn/ Describing a substance composed of very many minute interlocking crystals that have solidified together.

polycyclic /pol-ee-**sȳ**-klik/ Describing a compound that has two or more rings in its molecules.

polydioxoboric(III) acid /pol-ee-dȳ-oks-oh-**bô**-rik/ *See* boric acid.

polyene /**pol**-ee-een/ An alkene with more than two double bonds in its molecules.

polyester /pol-ee-**es**-ter/ A synthetic polymer made by reacting alcohols with acids, so that the monomers are linked by the group –O–CO–. Synthetic fibers such as Dacron are polyesters.

polyethene /pol-ee-**eth**-een/ (**polyethylene**; **polythene**) A synthetic polymer made from ethene. It is produced in two forms – a soft material of low density and a harder, higher density form, which is more rigid. It can be made by the Ziegler process.

polyethylene /pol-ee-**eth**-ă-leen/ *See* polyethene.

polyhydric alcohol /pol-ee-**hȳ**-drik/ An alcohol that has several –OH groups in its molecules.

polymer /**pol**-i-mer/ A compound composed of very large molecules made up of repeating molecular units (*monomers*). A polymer has a repeated structural unit, known as a *mer*. Polymers do not usually have a definite relative molecular mass, as there are variations in the lengths of different chains. They may be natural substances (e.g. polysaccharides or proteins) or synthetic materials (e.g. nylon or polyethene). The two major classes of synthetic polymers are *thermosetting*, which harden irreversibly with increased temperature (e.g. Bakelite), and *thermoplastic*, which soften on heating (e.g. polyethene). *See also* polymerization.

polymerization /pol-i-mer-i-**zay**-shŏn/ The process in which one or more compounds react to form a POLYMER. *Homopolymers* are formed by polymerization of one monomer (e.g. the formation of polyethene from ethene). *Heteropolymers* or *copolymers* come from two or more monomers (e.g. the production of NYLON). Heteropolymers may be of different types depending on the arrangement of units. An *alternating copolymer* of two units A and B has an arrangement:

–A–B–A–B–A–B–

A *block copolymer* has an arrangement in which blocks of one monomer alternate with blocks of the other; for example:

–A–A–A–B–B–B–A–A–A–

In a *graft copolymer* there is a main choice of one monomer (–A–A–A–A–), with short side chains of the other monomer attached at regular intervals (–B–B–).

Stereospecific polymers have the subunit repeated along the chain in a regular way. These are *tactic polymers*. If one particular group is always on the same side of the chain, the polymer is said to be *isotactic*. If the group alternates in position along the chain the polymer is *syndiotactic*. If there is no regular pattern, the polymer is *atactic*.

Polymerization reactions are also classified according to the type of reaction. *Addition polymerization* occurs when the monomers undergo addition reactions, with no other substance formed. *Condensation polymerization* involves the elimination of small molecules in formation of the polymer. *See also* cross linkage.

polymethanal /pol-ee-**meth**-ă-nal/ *See* methanal.

polymethylmethacrylate /pol-ee-meth-ăl-meth-**ak**-ră-layt/ *See* Plexiglas.

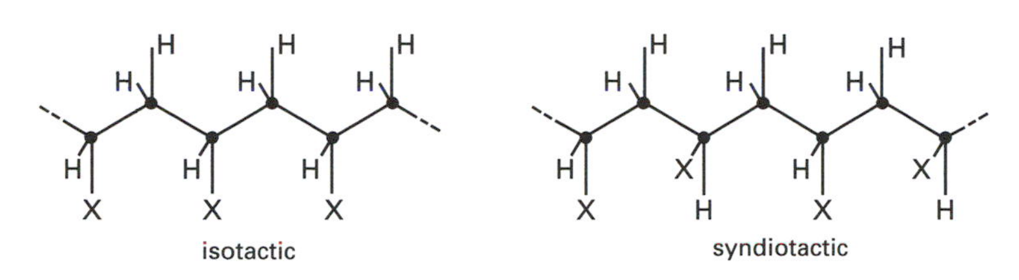

Addition polymers

polymorphism /pol-ee-**mor**-fiz-ăm/ The ability of certain chemical substances to exist in more than one physical form. For elements this is called allotropy. The existence of two forms is called *dimorphism*. Crystalline structures of compounds can vary with temperature as a result of different packing arrangements of the particles. There is a transition temperature between two forms and usually a marked change in density.

polynucleotide /pol-ee-**new**-klee-ŏ-tȳd/ *See* nucleotide.

polypeptide /pol-ee-**pep**-tȳd, -tid/ A PEPTIDE composed of a large number of amino-acid units. PROTEINS are polypeptides containing a few hundred amino-acid units.

polypropene /pol-ee-**proh**-peen/ (**polypropylene**) A synthetic polymer made from propene. It is similar in properties to polythene, but stronger and lighter. The propene is polymerized by the Ziegler process.

polypropylene /pol-ee-**proh**-pă-leen/ *See* polypropene.

polysaccharide /pol-ee-**sak**-ă-rȳd, -rid/ High-molecular-weight polymers of the monosaccharides or sugars. They contain many repeated units in their molecular structures. They can be broken down to smaller polysaccharides, disaccharides, and monosaccharides by hydrolysis or by the appropriate enzyme. Important polysaccharides are heparin, inulin, starch, glycogen (also known as animal starch), and cellulose. *See also* carbohydrates; sugar.

polystyrene /pol-ee-**stȳ**-reen/ A synthetic polymer made from styrene (phenylethene). Expanded polystyrene is a rigid foam used in packing and insulation.

polytetrafluoroethene /pol-ee-tet-ră-floo-ŏ-roh-**eth**-een/ (PTFE) A synthetic polymer made from tetrafluoroethene (i.e. CF_2:CF_2). It is able to withstand high temperatures without decomposing and also has a very low coefficient of friction, hence its use in non-stick pans, bearings, etc.

polyunsaturated /pol-ee-ŭn-**sach**-ŭ-ray-tid/ Describing a compound that has a number of C=C bonds in its compounds.

polyurethane /pol-ee-**yoor**-ă-thayn/ A synthetic polymer containing the group –NH–CO–O– linking the monomers. Polyurethanes are made by condensation of isocyanates (–NCO) with alcohols.

polyvalent /pol-ee-**vay**-lĕnt, pŏ-**liv**-ă-/ Describing an atom or group that has a valence of more than one.

polyvinyl acetate /pol-ee-**vȳ**-năl/ (**PVA**; **polyvinyl ethanoate**) A thermoplastic polymer made by the polymerization of vinyl ethanoate, CH_2:$CHOOCH_3$. It is used as a coating for paper and cloth and in adhesives.

polyvinyl chloride *See* polychloroethene.

polyvinyl ethanoate *See* polyvinyl acetate.

porcelain /**por**-sĕ-lin, **pohr**-/ A ceramic traditionally made from feldspar, kaolin (China clay), marble, and quartz.

porphyrins /**por**-fă-rinz/ Cyclic organic structures that have the important characteristic property of forming complexes with metal ions. Examples of such *metalloporphyrins* are the iron porphyrins (e.g. heme in hemoglobin) and the magnesium porphyrin, chlorophyll, the photosynthetic pigment in plants. In nature, the majority of metalloporphyrins are conjugated to proteins to form a number of very important molecules, e.g. hemoglobin, myoglobin, and the cytochromes.

potash *See* potassium carbonate; potassium hydroxide.

potash alum *See* alum; aluminum potassium sulfate.

potassamide /pŏ-**tass**-ă-mÿd/ *See* potassium monoxide.

potassium /pŏ-**tass**-ee-ŭm/ A soft reactive metal; the third member of the alkali metals (group 1 of the periodic table). The atom has the argon electronic configuration plus an outer $4s^1$ electron. The element gives a distinct lilac color to flames but, due to the intense yellow coloration from any trace impurity of sodium, the flame must be viewed through a cobalt glass. The ionization potential is low and the chemistry of potassium is largely the chemistry of the K^+ ion. Potassium accounts for 2.4% of the lithosphere and occurs in large salt deposits such as carnallite ($KCl.MgCl_2.6H_2O$) and arcanite (K_2SO_4). Large amounts also occur in mineral forms that are not of much use for recovery of potassium, e.g., orthoclase, $K_2Al_2Si_6O_{10}$. The commercial usage of potassium is much smaller than that of sodium; industrially the element is obtained by electrolysis of the fused hydroxide. Potassium hydroxide is obtained by electrolysis of carnallite solutions (the magnesium is initially precipitated as $Mg(OH)_2$), and both hydrogen and chlorine are recovered as by-products. The chemistry of potassium and its compounds is very similar to that of the other group 1 elements.

Particular points of distinction from sodium are:

1. Burning in air produces the superoxide KO_2.
2. Because of the larger size of the K^+ ion and hence lower lattice energies, potassium salts are often more soluble than corresponding sodium salts.
3. Potassium salts are usually less heavily hydrated than corresponding sodium salts.

The most abundant isotope of potassium is ^{39}K (93.1%), with ^{41}K also a stable isotope (6.8%), but there is a naturally occurring radioactive isotope ^{40}K (0.11%).

Like sodium, potassium forms a number of ionic alkyl- and aryl- derivatives of the type RK.

Symbol: K; m.p. 63.65°C; b.p. 774°C; r.d. 0.862 (20°C); p.n. 19; r.a.m. 39.0983.

potassium–argon dating A technique for dating rocks. It depends on the radioactive decay of the radioisotope ^{40}K to ^{40}Ar (half-life 1.27×10^{10} years) and is based on the assumption that argon has been trapped in the rock since the time that the rock cooled (i.e. since it formed). If the ratio $^{40}Ar/^{40}K$ is measured it is possible to estimate the age of the rock. *See also* radioactive dating.

potassium bicarbonate *See* potassium hydrogencarbonate.

potassium bromide (KBr) A white solid that is extremely soluble in water. It is formed by the action of bromine on hot potassium hydroxide solution or by the neutralization of the carbonate with hydrobromic acid. Potassium bromide forms colorless cubic crystals. It is used extensively in the manufacture of photographic plates, films, and papers and (formerly) as a sedative in medicine.

potassium carbonate (**pearl ash; potash;** K_2CO_3) A white deliquescent solid manufactured from potassium chloride by the Leblanc process or, more often, using the Precht process. It cannot be made by the

Solvay process. In the laboratory, it can be prepared by thermal decomposition of potassium hydrogencarbonate. Potassium carbonate is very soluble in water, its solutions being strongly alkaline due to salt hydrolysis. It is used in the laboratory as a drying agent and industrially in the manufacture of soft soap, hard glass, and in the dyeing industry. Potassium carbonate crystallizes out between 10 and 25°C as $K_2CO_3.3H_2O$; it dehydrates at 100°C to $K_2CO_3.H_2O$ and at 130°C to K_2CO_3.

potassium chlorate ($KClO_3$) A white soluble solid prepared by the electrolysis of a concentrated solution of potassium chloride. Industrially, it is prepared by the fractional crystallization of a solution containing sodium chlorate and potassium chloride. When heated, it decomposes to yield oxygen and potassium chloride. Potassium chlorate is a powerful oxidizing agent and is used in explosives, matches, weedkillers, and fireworks, and as a disinfectant. It oxidizes iodide ions to iodine when in an acidic medium. When heated just above its melting point, potassium perchlorate is formed.

potassium chloride (KCl) A white ionic solid prepared by neutralizing hydrochloric acid with potassium hydroxide solution. Potassium chloride occurs naturally as the minerals *sylvine* and *carnallite*. It is more soluble than sodium chloride in hot water but less soluble in cold water. On evaporation of an aqueous solution colorless cubic crystals similar to those of sodium chloride are produced. Potassium chloride is used as a fertilizer and in the manufacture of potassium hydroxide.

potassium chromate (K_2CrO_4) A bright yellow solid prepared by adding potassium hydroxide solution to a solution of potassium dichromate. It is extremely soluble in water. Addition of an acid to an aqueous solution of potassium chromate converts the chromate ions into dichromate ions. The salt is used as an indicator in silver nitrate titrations. The crystals are isomorphous with potassium sulfate.

potassium cyanide (KCN) A white ionic solid that is very soluble in water and extremely poisonous. It is made industrially by the Castner process and is used as a source of cyanide and hydrocyanic acid. Aqueous solutions of potassium cyanide are strongly hydrolyzed, the solutions being alkaline. On standing, these solutions slowly evolve hydrocyanic acid.

potassium dichromate ($K_2Cr_2O_7$) An orange-red solid prepared by adding potassium chloride solution to a concentrated solution of sodium dichromate and crystallizing out or by acidifying a solution of potassium chromate and evaporating. It is less soluble than sodium dichromate in cold water but more soluble in hot water. Potassium dichromate is used as an oxidizing agent both in volumetric analysis and in organic chemistry. The crystals are triclinic, anhydrous, and non-deliquescent. Addition of an alkali to a solution of the dichromate yields the chromate.

potassium hydride (KH) A white ionic solid prepared by passing hydrogen over heated potassium, the metal being suspended in an inert medium. It is an excellent reducing agent.

potassium hydrogencarbonate (**potassium bicarbonate**; $KHCO_3$) A white solid prepared by passing carbon dioxide through a saturated solution of potassium-carbonate. It occurs naturally in the mineral calcinite. Potassium hydrogen carbonate is more soluble in water than sodium hydrogencarbonate. When heated, it undergoes thermal decomposition to give the carbonate, water, and carbon dioxide. The salt forms monoclinic crystals. Its solutions are strongly alkaline as a result of salt hydrolysis.

potassium hydrogentartrate /hȳ-drŏ-jĕn-**tar**-trayt/ (**cream of tartar**; $HOOC(CHOH)_2COO^-K^+$) A white crystalline solid, an acid salt of tartaric acid that occurs in grape juice. It is used as the acid ingredient of BAKING POWDER.

potassium hydroxide (**caustic potash**; KOH) A white solid manufactured by the electrolysis of potassium chloride solution in a mercury-cathode cell. It can also be prepared by heating either potassium carbonate or potassium sulfate with slaked lime. The process closely resembles the Gossage process for making sodium hydroxide. In the laboratory, it can be made by reacting potassium, potassium monoxide, or potassium superoxide with water. Potassium hydroxide resembles sodium hydroxide but is more soluble in water and alcohol. It is preferred to sodium hydroxide for the absorption of carbon dioxide and sulfur(IV) oxide because of its greater solubility. Potassium hydroxide is used as an electrolyte in the Ni–Fe electric storage battery and in the production of soft soaps. It forms crystalline hydrates with 1, 1½, and 2 molecules of water.

potassium iodate /ÿ-ŏ-dayt/ (KIO_3) A white solid formed either by adding iodine to a hot concentrated solution of potassium hydroxide or by the electrolysis of potassium iodide solution. No hydrates are known. It is a source of iodide and iodic acid. When treated with a dilute acid and a reducing agent, the iodate ions are reduced to iodine.

potassium iodide (KI) A white ionic solid readily soluble in water. It is prepared by dissolving iodine in hot concentrated potassium hydroxide solution. Both the iodide and iodate are formed but the latter is removed by fractional crystallization. Potassium iodide has a sodium chloride cubic lattice. In solution with iodine it forms potassium tri-iodide. It is used in medicine, particularly in the treatment of goiter resulting from iodine deficiency. Dilute acidified solutions of potassium iodide can act as reducing agents, manganate(VII) ions being reduced to manganese(II) ions, copper(II) ions to copper(I) ions, and iodate ions to iodine.

potassium manganate(VII) (**potassium permanganate**; $KMnO_4$) A purple solid soluble in water. It is prepared by oxidizing potassium manganate(VI) with chlorine. Potassium permanganate is used in volumetric analysis as an oxidizing agent, as a bactericide, and as a disinfectant. In aqueous solution its behavior as an oxidizing agent depends on the pH of the solution.

potassium monoxide /mon-**oks**-ÿd, -id/ (K_2O) An ionic solid that is white when cold and yellow when hot. It is prepared by heating potassium with potassium nitrate. Potassium monoxide dissolves violently in water to form potassium hydroxide solution. The hydrate $K_2O.3H_2O$ is known. Potassium monoxide dissolves in liquid ammonia with the formation of potassium hydroxide and *potassamide* (KNH_2).

potassium nitrate (**saltpeter**; **niter**; KNO_3) A white solid, soluble in water, formed by fractional crystallization of sodium nitrate and potassium chloride solutions. It occurs naturally as niter (saltpeter) in rocks in India, South Africa, and Brazil. When heated it decomposes to give the nitrite and oxygen. Unlike sodium nitrate it is non-deliquescent. Potassium nitrate is used in gunpowder, fertilizers, and in the laboratory preparation of nitric acid.

potassium nitrite (KNO_2) A creamy deliquescent solid that is readily soluble in water. It reacts with cold dilute mineral acids to produce solutions of nitrous acid. Potassium nitrite is used in organic chemistry in the process of diazotization.

potassium permanganate *See* potassium manganate(VII).

potassium sulfate (K_2SO_4) A white solid prepared by the neutralization of either potassium hydroxide or potassium carbonate with dilute sulfuric acid. It occurs naturally in Stassfurt deposits as schönite. Potassium sulfate is soluble in water, forming a neutral solution. It is used as a fertilizer and in the chemical industry in the preparation of alums. The anhydrous salt crystallizes in the rhombic form.

potassium sulfide (K_2S) A yellowish-brown solid prepared by saturating an aqueous solution of potassium hydroxide with hydrogen sulfide and then adding an equal volume of potassium hydroxide. Industrially it is produced by heating potassium sulfate and carbon at high temperature. The sulfide crystallizes out from aqueous solution as $K_2S.5H_2O$. Its aqueous solutions undergo hydrolysis, the solution being strongly alkaline.

potassium superoxide (K_2O) A yellow paramagnetic solid prepared by burning potassium in excess oxygen. When treated with cold water or dilute mineral acids, hydrogen peroxide is produced. If heated strongly, it yields oxygen and potassium monoxide. Potassium superoxide is a powerful oxidizing agent.

potassium thiocyanate /th'ÿ-oh-**sÿ**-ă-nayt/ (KSCN) A colorless hygroscopic solid. Its solution is used in a test for iron(III) compounds, with which it turns a blood-red color.

potentiometric titration /pŏ-ten-shee-ŏ-**met**-rik/ A titration in which an electrode is used in the reaction mixture. The end point can be found by monitoring the electric potential of this during the titration.

praseodymium /pray-zee-oh-**dim**-ee-ŭm/ A soft ductile malleable silvery element of the lanthanoid series of metals. It occurs in association with other lanthanoids. Praseodymium is used in several alloys, as a catalyst, and in enamel and yellow glass for eye protection.

Symbol: Pr; m.p. 931°C; b.p. 3512°C; r.d. 6.773 (20°C); p.n. 59; r.a.m. 140.91.

precipitate /pri-**sip**-ă-tayt (*vb.*), pri-**sip**-ă-tayt, -tit (*n.*)/ A suspension of small particles of a solid in a liquid formed by a chemical reaction.

precursor A substance from which another substance is formed in a chemical reaction.

pressure Symbol: p The pressure on a surface due to forces from another surface or from a fluid is the force acting at 90° to unit area of the surface:

$$\text{pressure} = \text{force/area}$$

The unit is the pascal (Pa).

primary alcohol *See* alcohol.

primary amine *See* amine.

primary cell A voltaic cell in which the chemical reaction that produces the e.m.f. is not reversible. *Compare* accumulator.

primary standard A substance that can be used directly for the preparation of standard solutions without reference to some other concentration standard. Primary standards should be easy to purify, dry, capable of preservation in a pure state, unaffected by air or CO_2, of a high molecular weight (to reduce the significance of weighing errors), stoichiometric, and readily soluble. Any likely impurities should be easily identifiable.

prismane /**priz**-mayn/ *See* Ladenburg benzene.

producer gas (**air gas**) A mixture of carbon monoxide (25–30%), nitrogen (50–55%), and hydrogen (10–15%), prepared by passing air with a little steam through a thick layer of white-hot coke in a furnace or 'producer'. The air gas is used while still hot to prevent heat loss and finds uses in industrial heating, for example the firing of retorts and in glass furnaces. *Compare* water gas.

product *See* chemical reaction.

proline /**proh**-leen, -lin/ *See* amino acids.

promethium /prŏ-**mee**-th'ee-ŭm/ A radioactive element of the lanthanoid series of metals. It is not found naturally on Earth but can be produced artificially by the fission of uranium. It is used in some miniature batteries.

Symbol: Pm; m.p. 1168°C; b.p. 2730°C (approx.); r.d. 7.22 (20°C); p.n. 61; stablest isotope ^{145}Pm (half-life 18 years).

promoter (**activator**) A substance that improves the efficiency of a catalyst. It does not itself catalyze the reaction but assists the catalytic activity. For example, alumina or molybdenum promotes the catalytic activity of finely divided iron in the Haber process. The manner in which a promoter functions is not fully understood; no one theory covers all the examples. *See also* coenzymes.

proof A measure of the ethanol content of intoxicating drinks. In the United States, proof spirit contains 40% ethanol by volume. In the UK, it contains 49.28% of ethanol by weight (57.1% by volume at 16°C). The degree of proof gives the number of parts of proof spirit per 100 parts of the total. 100° of proof is 50% by volume, 80° of proof is 0.8 × 50%, etc.

propanal /**proh**-pă-nal/ (**propionaldehyde**; C_2H_5CHO) A colorless liquid aldehyde.

propane /**proh**-payn/ (C_3H_8) A gaseous alkane obtained either from the gaseous fraction of crude oil or by the cracking of heavier fractions. The principal use of propane is as a fuel for heating and cooking, since it can be liquefied under pressure, stored in cylinders, and transported easily. Propane is the third member of the homologous series of alkanes.

propanedioic acid /proh-payn-dÿ-**oh**-ik/ (**malonic acid**; $CH_2(COOH)_2$) A white crystalline dibasic carboxylic acid.

propane-1,2,3-triol (**glycerol**; **glycerine**; $CH_2(OH)CH(OH)CH_2(OH)$) A colorless viscous liquid obtained as a by-product from the manufacture of soap by the reaction of animal fats with sodium hydroxide. It is used as a solvent and plasticizer. *See also* glyceride.

propanoate /proh-pă-**noh**-ayt, -it/ A salt or ester of propanoic acid.

propanoic acid /proh-pă-**noh**-ik/ (**propionic acid**; C_2H_5COOH) A colorless liquid carboxylic acid.

propanol /**proh**-pă-nol, -nohl/ Either of two alcohols: propan-1-ol ($CH_3CH_2CH_2OH$) and propan-2-ol ($CH_3CH_2(OH)CH_3$). Both are colorless volatile flammable liquids.

propanone /**proh**-pă-nohn/ (**acetone**; $CH_3\text{-}COCH_3$) A colorless liquid ketone, used as a solvent and in the manufacture of methyl 2-methyl-propanoate (from which polymethylmethacrylate is produced). Propanone is manufactured from propene, either by the air-oxidation of propan-2-ol or, as a by-product from the cumene process.

propenal /**proh**-pĕ-năl/ (**acrolein**; $CH_2{:}CHCHO$) A colorless liquid unsaturated aldehyde with a pungent odor. It can be polymerized to make acrylate resins.

propene /**proh**-peen/ (**propylene**; C_3H_6) A gaseous alkene. Propene is not normally present in the gaseous crude-oil fraction but can be obtained from heavier fractions by catalytic cracking. This is the principal industrial source. Propene is the organic starting material for the production of propan-2-ol, required for the manufacture of propanone (acetone), and the starting material for the production of polypropene (polypropylene).

propenoate /**proh**-pĕ-nayt/ A salt or ester of propenoic acid.

propenoic acid /proh-pĕ-**noh**-ik/ (**acrylic acid**; $CH_2{:}CHCOOH$) An unsaturated liquid carboxylic acid with a pungent odor. The acid and its esters are used to make ACRYLIC RESINS.

propenonitrile /pro-pĕ-noh-**nÿ**-trăl, -tril, -trÿl/ (**acrylonitrile**; $H_2C{:}CH(CN)$) An organic compound from which acrylic-type polymers are produced.

propenyl group /**proh**-pĕ-năl/ (**allyl group**) The organic group, $CH_2{=}CH{-}CH_2{-}$, derived by removing one hydrogen atom from propene.

propionaldehyde /proh-pee-ŏ-**nal**-dĕ-hÿd/ *See* propanal.

propionic acid /proh-pee-**on**-ik/ *See* propanoic acid.

propylene /**proh**-pă-leen/ *See* propene.

propyl group /**proh**-păl/ The group $CH_3CH_2CH_2$–.

protactinium /proh-tak-**tin**-ee-ŭm/ A toxic radioactive element of the actinoid series of metals. It occurs in minute quantities in uranium ores as a radioactive decay product of actinium.

Symbol: Pa; m.p. 1840°C; b.p. 4000°C (approx.); r.d. 15.4 (calc.); p.n. 91; most stable isotope ^{231}Pa (half-life 32 500 years).

protein /**proh**-teen, -tee-in/ A naturally occurring compound found in all living matter, consisting of AMINO ACIDS joined into long chains by PEPTIDE links. The *primary structure* is the particular sequence of amino acids present. The protein also has a *secondary structure*, with the chains coiled in a helix or held together in pleated sheets. The *tertiary structure* is the way in which the chain or sheet is arranged in space. Secondary structures are held by hydrogen bonds between N–H and O=C groups. Tertiary structures are held by hydrogen bonds or cross linkages of the type –S–S– (cystine links).

proton An elementary particle with a positive charge ($+1.602\ 192 \times 10^{-19}$ C) and rest mass $1.672\ 614 \times 10^{-27}$ kg. Protons are nucleons, found in all nuclides.

proton number (**atomic number**) Symbol: Z The number of protons in the nucleus of an atom. The proton number determines the chemical properties of the element because the electron structure, which determines chemical bonding, depends on the electrostatic attraction to the positively charged nucleus.

Prussian blue *See* cyanoferrate.

prussic acid /**pruss**-ik/ See hydrocyanic acid.

pseudoaromatic /soo-doh-a-rŏ-**mat**-ik/ *See* aromatic compound.

pseudo-first order Describing a reaction that appears to exhibit first-order kinetics under special conditions, even though the 'true' order is greater than one. For example, in the hydrolysis of an ester in the presence of a *large* volume of water, the concentration of water remains approximately constant. The rate of reaction is thus found experimentally to be proportional to the concentration of the ester only (even though it also depends on the amount of water present). Such a reaction is described as 'bimolecular of the first order'.

pseudohalogens /soo-doh-**hal**-ŏ-jĕnz/ A small group of simple inorganic compounds with symmetrical molecules that resemble the halogens in some reactions and compounds. For example, $(CN)_2$ (cyanogen) has compounds analogous to halogen compounds (e.g. HCN, KCN, CH_3CN, etc.). Thiocyanogen, $(SCN)_2$, is another example.

PTFE *See* polytetrafluoroethene.

purine /**pyoor**-een, -in/ A simple nitrogenous organic molecule with a double ring structure. Members of the purine group include adenine and guanine, which are constituents of the nucleic acids, and certain plant alkaloids, such as caffeine and theobromine.

PVA *See* polyvinyl acetate.

PVC *See* polychloroethene.

pyranose /**pÿ**-ră-nohs, pi-**ran**-ohs/ A SUGAR that has a six-membered ring form (five carbon atoms and one oxygen atom).

pyrazine /**pÿ**-ră-zeen, -zin/ (**1,4-diazine**; $C_4H_4N_2$) A heterocyclic aromatic compound with a six-membered ring containing four carbon atoms and two nitrogen atoms.

pyrazole /**pȳ**-ră-zohl/ (**1,2-diazole**; $C_3H_4N_2$) A heterocyclic crystalline aromatic compound with a five-membered ring containing three carbon atoms and two nitrogen atoms.

pyrene /**pȳ**-reen/ ($C_{16}H_{10}$) A solid aromatic compound whose molecules consist of four benzene rings joined together. It is carcinogenic.

Pyrex /**pȳ**-reks/ (*Trademark*) A particularly strong, heat resistant, and chemically inert borosilicate glass commonly used for laboratory glassware.

pyridine /**pi**-ră-deen/ (C_5H_5N) An organic liquid of formula C_5H_5N. The molecules have a hexagonal planar ring and are isoelectronic with benzene. Pyridine is an example of an aromatic heterocyclic compound, with the electrons in the carbon–carbon pi bonds and the lone pair of the nitrogen delocalized over the ring of atoms. The compound is extracted from coal tar and used as a solvent and as a raw material for organic synthesis.

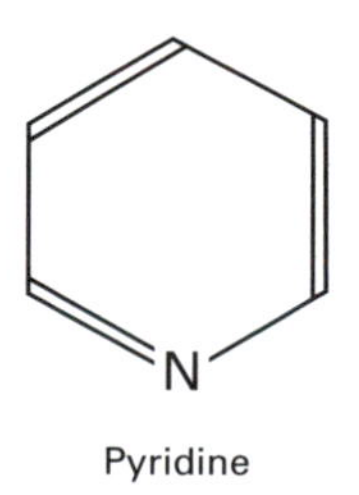

Pyridine

pyrimidine /pi-**rim**-ă-deen, **pi**-ră-mă-/ A simple nitrogenous organic molecule whose ring structure is contained in the pyrimidine bases cytosine, thymine, and uracil, which are constituents of the nucleic acids.

pyrites /pȳ-**rȳ**-teez, pi-, **pȳ**-rȳts/ A mineral sulfide of a metal; e.g. iron pyrites FeS_2.

pyrolusite /pȳ-rŏ-**loo**-sȳt/ See manganese(IV) oxide.

pyrolysis /pȳ-**rol**-ă-sis, pi-/ The decomposition of chemical compounds by subjecting them to very high temperature.

pyrometer /pȳ-**rom**-ĕ-ter, pi-/ An instrument used in the chemical industry to measure high temperature, e.g. in reactor vessels.

pyrone /**pȳ**-rohn, pi-**rohn**/ A compound having a six-membered ring of five carbon atoms and one oxygen, with one of the carbon atoms attached to a second oxygen in a carbonyl group. The pyrone ring system has two forms depending on the position of the carbonyl relative to the oxygen hetero atom. It occurs in many natural products.

pyrophoric /pȳ-rŏ-**fô**-rik, pi-/ **1.** Describing a compound that ignites spontaneously in air.
2. Describing a metal or alloy that gives sparks when struck. (Lighter flints are made of pyrophoric alloy.)

pyrophosphoric acid /pȳ-roh-fos-**fô**-rik/ *See* phosphoric(V) acid.

pyrosulfuric acid /pȳ-roh-sul-**fyoor**-ik/ *See* oleum.

pyrrhole /**pi**-rohl/ ($(CH)_4NH$) A heterocyclic liquid aromatic compound with a five-membered ring containing four carbon atoms and one nitrogen atom. It has important biochemical derivatives, including chlorophyll and heme.

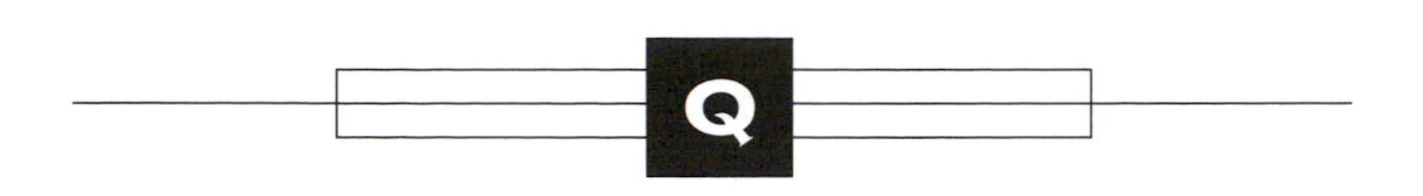

quadrivalent /kwod-ră-**vay**-lĕnt, kwod-**riv**-ă-/ (**tetravalent**) Having a valence of four.

qualitative analysis /**kwol**-ă-tay-tiv/ Analysis carried out with the purpose of identifying the components of a sample. Classical methods involved simple preliminary tests followed by a carefully devised scheme of systematic tests and procedures. Modern methods include the use of such techniques as infrared spectroscopy and emission spectrography. *Compare* quantitative analysis.

quanta /**kwon**-tă/ *See* quantum.

quantitative analysis /**kwon**-tă-tay-tiv/ Analysis carried out with the purpose of determining the concentration of one or more components of a sample. Classical wet methods include volumetric and gravimetric analysis. A wide range of more modern instrumental techniques are also used, including polarography and various types of chromatography and spectroscopy. *Compare* qualitative analysis.

quantized /**kwon**-tÿzd/ Describing a physical quantity that can take only certain discrete values, and not a continuous range of values. Thus, in an atom or molecule the electrons around the nucleus can have certain energies, E_1, E_2, etc., and cannot have intermediate values. Similarly, in atoms and molecules, the electrons have quantized values of spin angular momentum and orbital angular momentum.

quantum /**kwon**-tŭm/ (*pl.* **quanta**) A definite amount of energy released or absorbed in a process. Energy often behaves as if it were 'quantized' in this way. The quantum of electromagnetic radiation is the photon.

quantum electrodynamics The use of quantum mechanics to describe how particles and electromagnetic radiation interact.

quantum mechanics *See* quantum theory.

quantum number An integer or half integer that specifies the value of a quantized physical quantity (energy, angular momentum, etc.). *See* atom; Bohr theory; spin.

quantum states States of an atom, electron, particle, etc., specified by a unique set of quantum numbers. For example, the hydrogen atom in its ground state has an electron in the K shell specified by the four quantum numbers: $n = 1, l = 0, m = 0, m_s = \frac{1}{2}$. In the helium atom there are two electrons:

$$n = 1, l = 0, m = 0, m_s = \tfrac{1}{2}$$
$$n = 1, l = 0, m = 0, m_s = -\tfrac{1}{2}$$

quantum theory A mathematical theory originally introduced by Max Planck (1900) to explain the radiation emitted from hot bodies. Quantum theory is based on the idea that energy (or certain other physical quantities) can be changed only in certain discrete amounts for a given system. Other early applications were the explanations of the photoelectric effect and the Bohr theory of the atom.

Quantum mechanics is a system of mechanics that developed from quantum theory and is used to explain the behavior of atoms, molecules, etc. In one form it is based on de Broglie's idea that particles can have wavelike properties – this branch of quantum mechanics is called *wave mechanics*. *See* orbital.

quantum yield The average number of reactive events per absorbed photon in a photochemical reaction.

quartz A natural crystalline form of silica (SiO_2).

quasicrystal /kway-zÿ-**kris**-tăl,-sÿ-, **kwah**-zee-, -see-/ A type of crystal structure in which there is long-range order but in which the symmetry of the repeating unit is not one allowed in normal crystals. For example, in three dimensions, it would be possible to form a crystal structure made up of cubes, but not one made by joining icosahedra. However, there are examples of solids, such as the compound AlMn, that have icosahedral symmetry and a long-range repeating order in the stucture. These are known as quasicrystals.

quaternary ammonium compound A compound formed from an amine by addition of a proton to produce a positive ion. Quaternary compounds are salts, the simplest example being ammonium compounds formed from ammonia and an acid, for example:

$$NH_3 + HCl \rightarrow NH_4^+Cl^-$$

Other amines can also add protons to give analogous compounds. For instance, methylamine (CH_3NH_2) forms the compound

$$[CH_3NH_3]^+X^-$$

where X^- is an acid radical.

The formation of quarternary compounds occurs because the lone pair on the nitrogen atom can form a coordinate bond with a proton. This can also occur with heterogeneous nitrogen compounds, such as adenine, cytosine, thymine, and guanine. Such compounds are known as *nitrogenous bases*.

quenching A method used to alter the mechanical properties of metals. Hot metal is lowered quickly into a bath of oil, water, or brine and the rapid cooling results in a fine grain structure. This treatment increases the hardness of a metal but often makes it brittle.

quicklime *See* calcium oxide.

quinhydrone electrode /kwin-**hÿ**-drohn/ *See* quinone.

quinine /kwÿ-nÿn, kwi-**neen/** A poisonous ALKALOID found in the bark of the cinchona tree of South America. It is used in treating malaria.

quinol /kwin-ol. -ohl/ *See* benzene-1,4-diol.

quinoline /kwin-ŏ-leen, -lin/ (C_9H_7N) A colorless two-ring heterocyclic compound with an unpleasant odor, which acts as a base and forms salts with acids. First made from the alkaloid quinine, it is found in bone oil and coal tar and used for making drugs and dyestuffs.

quinone /kwă-**nohn**, **kwin**-ohn/ (**cyclohexadiene-1,4-dione; benzoquinone;** $C_6H_4O_2$) A yellow crystalline organic compound with a pungent odor. Its molecules contain a non-aromatic six-carbon ring and it behaves as an unsaturated diketone with conjugated double bonds. It is used in making dyestuffs. A platinum electrode in an equimolar solution of quinone and hydroquinone (benzene-1,4-diol, $C_6H_4(OH)_2$) is used as a standard electrode in electrochemistry. The reaction is:

$$C_6H_4(OH)_2 \rightleftharpoons C_6H_4O_2 + 2H^+ + 2e$$

This type of electrode is called a *quinhydrone electrode*.

quinquevalent /kwing-kwĕ-**vay**-lĕnt, kwing-**kwev**-ă-/ *See* pentavalent.

R

racemate /ray-**seem**-ayt, **-sem-**, ră-/ *See* optical activity.

racemic mixture /ray-**see**-mik, ră-/ *See* optical activity.

racemization /rass-ĕ-mi-**zay**-shŏn/ The conversion of an optical isomer into an equal mixture of isomers, which is not optically active.

rad A unit of absorbed dose of ionizing radiation, defined as being equivalent to an absorption of 10^{-2} joule of energy in one kilogram of material.

radian Symbol: rad The SI unit of plane angle; 2π radian is one complete revolution (360°).

radiation /ray-dee-**ay**-shŏn/ In general the emission of energy from a source, either as waves (light, sound, etc.) or as moving particles (beta rays or alpha rays).

radical /**rad**-ă-kăl/ A group of atoms in a molecule. *See also* free radical.

radioactive /ray-dee-oh-**ak**-tiv/ Describing an element or nuclide that exhibits natural radioactivity.

radioactive dating (**radiometric dating**) A technique for dating archaeological specimens, rocks, etc., by measuring the extent to which some radionuclide has decayed to give a product. *See* carbon dating; potassium–argon dating; rubidium–strontium dating; uranium–lead dating.

radioactive decay series The series of definite isotopes to which a given radioactive element is transformed as it decays.

radioactivity /ray-dee-oh-ak-**tiv**-ă-tee/ The disintegration or decay of certain unstable nuclides with emission of radiation. The emission of alpha particles, beta particles, and gamma waves are the three most important forms of radiation that occur during decay.

radiocarbon dating /ray-dee-oh-**kar**-bŏn/ *See* carbon dating.

radiochemistry /ray-dee-oh-**kem**-iss-tree/ The chemistry of radioactive isotopes of elements. Radiochemistry involves such topics as the preparation of radioactive compounds, the separation of isotopes by chemical reactions, the use of radioactive labels in studies of mechanisms, and experiments on the chemical reactions and compounds of transuranic elements.

radiogenic /ray-dee-oh-**jen**-ik/ Caused by radioactive decay.

radioisotope /ray-dee-oh-**ÿ**-sŏ-tohp/ A radioactive isotope of an element. Tritium, for instance, is a radioisotope of hydrogen. Radioisotopes are extensively used in research as souces of radiation and as tracers in studies of chemical reactions. Thus, if an atom in a compound is replaced by a radioactive nuclide of the element (a *label*) it is possible to follow the course of the chemical reaction. Radioisotopes are also used in medicine or diagnosis and treatment.

radiolysis /ray-dee-**ol**-ă-sis/ A chemical reaction produced by high-energy radiation (x-rays, gamma rays, or particles).

radiometric dating /ray-dee-oh-**met**-rik/ *See* radioactive dating.

radio waves A form of ELECTROMAGNETIC RADIATION with wavelengths greater than a few millimeters.

radium /**ray**-dee-ŭm/ A white radioactive luminescent metallic element of the alkaline-earth group. It has several short-lived radioisotopes and one long-lived isotope, radium-226 (half-life 1602 years). Radium is found in uranium ores, such as the oxides pitchblende and carnotite. It was formerly used in luminous paints and radiotherapy.

Symbol: Ra; m.p. 700°C; b.p. 1140°C; r.d. 5 (approx. 20°C); p.n. 88; r.a.m. 226.0254 (^{226}Ra).

radon /**ray**-don/ A colorless monatomic radioactive element of the rare-gas group, now known to form unstable compounds. It has 19 short-lived radioisotopes; the most stable, radon-222, is a decay product of radium-226 and itself disintegrates into an isotope of polonium with a half-life of 3.82 days. ^{222}Rn is sometimes used in radiotherapy. Radon occurs in uranium mines and is also detectable in houses built in certain areas of the country.

Symbol: Rn; m.p. –71°C; b.p. –61.8°C; d. 9.73 kg m^{-3} (0°C); p.n. 86.

raffinate /**raf**-ă-nayt/ The liquid remaining after the solvent extraction of a dissolved substance. *See* solvent extraction.

raffinose /**raf**-ă-nohs/ A SUGAR occurring in sugar beet. It is a trisaccharide consisting of fructose, galactose, and glucose units.

r.a.m. *See* relative atomic mass.

Raman effect /**rah**-măn/ A change in the frequency of electromagnetic radiation that occurs when a photon of radiation undergoes an inelastic collision with a molecule. The effect was first observed by the Indian physicists Sir Chandrasekhara Venkata Raman (1888–1970) and his colleague Sir Kariamanikkam Srinivasa Krishnan in 1928. It has been used extensively in *Raman spectroscopy* for the determination of molecular structure, particularly since the advent of the laser.

Raney nickel /**ray**-nee/ A catalytic form of nickel produced by treating a nickel–aluminum alloy with caustic soda. The aluminum dissolves (as aluminate) and Raney nickel is left as a spongy mass, which is pyrophoric when dry. It is used especially for catalyzing hydrogenation reactions.

Raoult's law /rah-**oolz**/ A relationship between the pressure exerted by the vapor of a solution and the presence of a solute. It states that the partial vapor pressure of a solvent above a solution (p) is proportional to the mole fraction of the solvent in the solution (X) and that the proportionality constant is the vapor pressure of pure solvent, (p_0), at the given temperature: i.e. $p = p_0X$. Solutions that obey Raoult's law are said to be *ideal.* There are some binary solutions for which Raoult's law holds over all values of X for either component. Such solutions are said to be *perfect* and this behavior occurs when the intermolecular attraction between molecules within one component is almost identical to the attraction of molecules of one component for molecules of the other (e.g. chlorobenzene and bromobenzene). Because of solvation forces this behavior is rare and in general Raoult's law holds only for dilute solutions.

For solutions that are ideal but not perfect the solute behavior is similar in that the partial pressure of the solute, p_s, is proportional to the mole fraction of the solute, X_s, but in this case the proportionality constant, p', is not the vapor pressure of the pure solute but must be determined experimentally for each system. This solute equivalent of Raoult's law has the form $p_s = p'X_s$, and is called HENRY'S LAW. Because of intermolecular attractions p' is usually less than p_0. The law is named for the French chemist Françoise-Marie Raoult (1830–1901). *See also* Babo's law.

rare earths *See* lanthanoids.

rare gases (**noble gases; inert gases; group 0 elements**) A group of of monatomic, low-boiling gases classified as belonging to group 0 of the periodic table and occupying a position between the highly electronegative group 17 elements and the highly electropositive group 1 elements. They are: helium (He), neon (Ne), argon (Ar), krypton (Kr), xenon (Xe), and radon (Rn). This complete filling of the s and p levels gives a closed-shell structure and is associated with a general lack of chemical reactivity and very high ionization energies. It is the acquisition of a stable rare-gas configuration that is associated with determining the valence of many covalent and ionic compounds.

Prior to 1962 the rare gases were frequently called *inert gases* as no chemical compounds were known (there were a few clathrates and 'hydrates'), but the realization that the ionization potential of xenon was sufficiently low to be accessible to chemical reaction led to the preparation of several fluorides, oxides, oxyfluorides, and a hexafluoroplatinate of xenon. Several unstable krypton and radon compounds have been synthesized.

Argon forms about 1% of the atmosphere but the other gases are present in only minute traces (excluding radon). They are obtained by fractionation of liquid air or (in the USA) of natural gas. Considerable amounts of argon are produced as a by-product from ammonia-plant tail gas. Their applications depend heavily on their inertness; for example, welding (Ar), filling light bulbs (Ar), gas-stirring in high temperature metallurgy (Ar), powder technology (Ar), discharge tubes (Ne), and chemical research as inert carriers (He, Ar). Neon has been proposed as an oxygen diluent for deep-sea diving because of its lower solubility in blood. Liquid helium is extensively used in low-temperature work and radon is used therapeutically as a source of α-particles.

The gases helium and argon are formed by radioactive decay in various minerals, for example argon by electron capture of ^{40}K. This mechanism may be applied to age-determination of minerals. Radon is generally obtained as ^{222}Rn from decay of radium. The half-life is only about 3.8 days.

Rashig process /**rash**-ig/ A method for the manufacture of chlorobenzene, and thence phenol, from benzene. Benzene vapor, hydrogen chloride, and air are passed over a copper(II) chloride catalyst (230°C):

$$2C_6H_6 + 2HCl + O_2 \rightarrow 2C_6H_5Cl + 2H_2O$$

The conversion to phenol is by a silicon catalyst (425°C):

$$C_6H_5Cl + H_2O \rightarrow HCl + C_6H_5OH$$

rate constant (**velocity constant; specific reaction rate**) Symbol: k The constant of proportionality in the rate expression for a chemical reaction. For example, in a reaction A + B → C, the rate may be proportional to the concentration of A multiplied by that of B; i.e.

$$\text{rate} = k[A][B]$$

where k is the rate constant for this particular reaction. The constant is independent of the concentrations of the reactants but depends on temperature; consequently the temperature at which k is recorded must be stated. The units of k vary depending on the number of terms in the rate expression, but are easily determined remembering that rate has the units s^{-1}.

rate-determining step (**limiting step**) The slowest step in a multistep reaction. Many chemical reactions are made up of a number of steps in which the one with the lowest rate is the one that determines the rate of the overall process.

The overall rate of a reaction cannot exceed the rate of the slowest step. For example, the first step in the reaction between acidified potassium iodide solution and hydrogen peroxide is the rate-determining step:

$$H_2O_2 + I^- \rightarrow H_2O + OI^- \text{ (slow)}$$
$$H^+ + OI^- \rightarrow HOI \text{ (fast)}$$
$$HOI + H^+ + I^- \rightarrow I_2 + H_2O \text{ (fast)}$$

rate of reaction A measure of the amount of reactant consumed in a chemical reaction in unit time. It is thus a measure of the number of effective collisions between reactant molecules. The rate at which a reaction proceeds can be measured by the rate the reactants disappear or by the rate at which the products are formed. The principal factors affecting the rate of reaction are temperature, pressure, concentration of reactants, light, and the action of a catalyst. The units usually used to measure the rate of a reaction are mol dm^{-3} s^{-1}. *See also* law of mass action.

rationalized units A system of units in which the equations have a logical form related to the shape of the system. SI units form a rationalized system of units. In it formulae concerned with circular symmetry contain a factor of 2π; those concerned with radial symmetry contain a factor of 4π.

raw material A substance from which other substances are made. In the chemical industry it may be simple (such as nitrogen from air used to make ammonia) or complex (such as coal and petroleum, used to make a wide range of products).

rayon /**ray**-on/ An artificial fiber formed from wood pulp (cellulose). There are two types. *Viscose rayon* is made by dissolving the cellulose in carbon disulfide and sodium hydroxide. The solution is forced through a fine nozzle into an acid bath, which regenerates the fibers. *Acetate rayon* is made by dissolving cellulose acetate in an organic solvent, and forcing the solution through a nozzle. The solvent is evaporated, and the cellulose acetate thus obtained as fibers.

reactant /ree-**ak**-tănt/ A compound taking part in a chemical reaction.

reaction *See* chemical reaction.

reactive dye A dye that 'sticks' to the fibers of a fabric by forming covalent chemical bonds with the substance of the fabric. The dyes used to color the cellulose fibers in rayon are examples of reactive dyes.

reagent /ree-**ay**-jĕnt/ A compound that reacts with another (the substrate). The term is usually used for common laboratory chemicals – sodium hydroxide, hydrochloric acid, etc. – used for experiment and analysis.

realgar /ree-**al**-ger/ The mineral form of arsenic(II) sulfide, As_4S_4. It is bright red in color and used as a pigment, in tanning, and in pyrotechnics.

real gas *See* gas laws.

rearrangement A reaction in which the groups of a compound rearrange themselves to form a different compound.

reciprocal proportions, law of *See* equivalent proportions, law of.

recrystallization /ree-kris-tă-li-**zay**-shŏn/ The repeated crystallization of a compound to ensure purity of the sample.

rectified spirit /**rek**-tă-fÿd/ A constant-boiling mixture of ethanol and water that contains about 6% water; no more water can be removed by further distillation. It is used as an industrial solvent.

red lead *See* dilead(II) lead(IV) oxide.

redox /**ree**-doks/ Relating to the process of oxidation and reduction, which are intimately connected in that during oxidation by chemical agents the oxidizing agent itself becomes reduced, and vice versa. Thus an oxidation process is always accompanied by a reduction process. In electrochemical processes this is equally true, oxidation taking place at the anode and reduction at the cathode. These systems are often called *redox systems*, particularly when the interest centers on both compounds.

Oxidizing and reducing power is indicated quantitatively by the *redox potential* or standard electrode potential, $E^{\ominus}$. Redox potentials are normally expressed as reduction potentials. They are obtained by electrochemical measurements and the values are referred to the H^+/H_2 couple for which $E^{\ominus}$ is set equal to zero. Thus increasingly negative potentials indicate increasing ease of oxidation or difficulty of reduction. Thus in a redox reaction the half reaction with the most positive value of $E^{\ominus}$ is the reduction half and the half reaction with the least value of $E^{\ominus}$ (or most highly negative) becomes the oxidation half. *See also* electrode potential.

red phosphorus *See* phosphorus.

reducing agent *See* reduction.

reduction /ri-**duk**-shŏn/ The gain of electrons by such species as atoms, molecules, or ions. It often involves the loss of oxygen from a compound, or addition of hydrogen. Reduction can be effected chemically, i.e. by the use of *reducing agents* (electron donors), or electrically, in which case the reduction process occurs at the cathode. For example,

$$2Fe^{3+} + Cu \rightarrow 2Fe^{2+} + Cu^{2+}$$

where Cu is the reducing agent and Fe^{3+} is reduced, and

$$2H_2O + SO_2 + 2Cu^{2+} \rightarrow 4H^+ + SO_4^{2-} + 2Cu^+$$

where SO_2 is the reducing agent and Cu^{2+} is reduced.

Tin(II) chloride, sulfur(IV) oxide, the hydrogensulfite ion, and hydroxylamine hydrochloride are common reducing agents. *See also* redox.

reduction potential *See* electrode potential.

refining The process of removing impurities from a substance or of extracting a substance from a mixture.

refluxing /ri-**fluks**-ing/ The process of boiling a liquid in a vessel connected to a condenser, so that the condensed liquid runs back into the vessel. By using a reflux condenser, the liquid can be maintained at its boiling point for long periods of time, without loss. The technique is a standard method of carrying out reactions in organic chemistry.

reforming The cyclization of straight-chain hydrocarbons from crude oil by heating under pressure with a catalyst, usually platinum on alumina. For example, the manufacture of methylbenzene from heptane:

$$C_7H_{16} \rightarrow C_6H_{11}CH_3$$

This first step is the production of methyl cyclohexane, which then loses six hydrogen atoms to give methylbenzene:

$$C_6H_{11}CH_3 \rightarrow C_6H_5CH_3 + 3H_2$$

See also steam reforming.

refractory /ri-**frak**-tŏ-ree/ Describing a compound (e.g. an inorganic oxide) that has a very high melting point.

Regnault's method A method used for the determination of the density of gases. A bulb of known volume is evacuated and weighed then the gas is admitted at a known pressure (from a vacuum line) and the bulb weighed again. The temperature is also noted and the data corrected to STP. The method is readily applicable to the determination of approximate relative molecular masses of gaseous samples. The method is named for the French physicist and chemist Henri Victor Regnault (1800–78).

relative atomic mass (r.a.m.) Symbol: A_r The ratio of the average mass per atom of the naturally occurring element to 1/12 of the mass of an atom of nuclide ^{12}C. It was formerly called *atomic weight*

relative density Symbol: d The ratio of the density of a given substance to the density of some reference substance. The relative densities of liquids are usually measured with reference to the density of water at 4°C. Relative densities are also specified for gases; usually with respect to air at STP. The temperature of the substance is stated or is understood to be 20°C. Relative density was formerly called *specific gravity*.

relative molecular mass Symbol: M_r The ratio of the average mass per molecule of the naturally occurring form of an element or compound to 1/12 of the mass of an atom of nuclide ^{12}C. This was formerly called *molecular weight*. It does not have to be used only for compounds that have discrete molecules; for ionic compounds (e.g.

NaCl) and giant-molecular structures (e.g. BN) the formula unit is used.

relaxation The process by which an excited species loses energy and falls to a lower energy level (such as the ground state).

rem /rem/ (*r*adiation *e*quivalent *m*an) A unit for measuring the effects of radiation dose on the human body. One rem is equivalent to an average adult male absorbing one rad of radiation. The biological effects depend on the type of radiation as well as the energy deposited per kilogram.

resin /**rez**-in/ A yellowish insoluble organic compound exuded by trees as a viscous liquid that gradually hardens on exposure to air to form a brittle amorphous solid. Synthetic resins are artificial polymers used in making adhesives, insulators and paints. *See also* rosin.

resolution (of racemates) The separation of a racemate into the two optical isomers. This cannot be done by normal methods, such as crystallization or distillation, because the isomers have identical physical properties. The main methods are:

1. Mechanical separation. Certain optically active compounds form crystals with distinct left- and right-handed shapes. The crystals can be sorted by hand.
2. Chemical separation. The mixture is reacted with an optical isomer. The products are then not optical isomers of each other, and can be separated by physical means.

For instance, a mixture of D- and L-forms of an acid, acting with a pure L-base, produces two salts that can be separated by fractional crystallization and then reconverted into the acids.

3. Biochemical separation. Certain organic compounds can be separated by using bacteria that feed on one form only, leaving the other.

resonance (**mesomerism**) The behavior of many compounds cannot be adequately explained by a single structure using simple single and double bonds. The bonding electrons of the compound have a different distribution in the molecules. The actual bonding in the molecule can be regarded as a hybrid of two or more conventional forms of the molecule, called *resonance forms* or *canonical forms*. The result is a resonance hybrid. For example, the carbonyl group in a ketone has negative charge on the oxygen atom. It can be described as a resonance hybrid, somewhere between =C=O, in which a pair of electrons is shared between the C and the O, and $=C^{+}–O^{-}$, in which the electrons are localized on the O atom. Note that the two canonical forms do not contribute equally in the hybrid. The bonding of benzene can be represented by a resonance hybrid of two Kekulé structures and, to a lesser extent, three Dewar structures.

resonance ionization spectroscopy (**RIS**) A type of spectroscopy that detects specific types of atoms using lasers. The frequency of the laser is chosen so that only the atoms of interest in a sample are excited by the laser.

resorcinol /rez-**or**-să-nol, -nohl/ *See* benzene-1,3-diol.

retort A piece of laboratory apparatus consisting of a glass bulb with a long narrow neck. In industrial chemistry, various metallic vessels in which distillations or reactions take place are called retorts.

retrosynthetic analysis /ret-roh-**sin**-th'ĕ-tik/ A technique for planning the synthesis of an organic molecule. The structure of the molecule is considered and it is divided into imaginary parts, which could combine by known reactions. These *disconnections* of the molecule suggest charged fragments, known as *synthons*, which give a guide to possible reagents for the synthesis.

reverberatory furnace /ri-**ver**-bĕ-ră-tor-ee, -toh-ree/ A furnace for smelting metals. It has a curved roof so that heat is reflected downwards, the fuel being in one part of the furnace and the ore in the other.

reverse osmosis A process by which water containing a salt is separated via a semipermeable membrane into pure water and salt water streams by subjecting the incoming water to pressures significantly greater than its OSMOTIC PRESSURE. This causes pure water to be forced through the membrane, leaving the salt behind. Reverse osmosis is used in the purification of sea water. *See* osmosis.

reversible change A change in the pressure, volume, or other properties of a system, in which the system remains at equilibrium throughout the change. Such processes could be reversed; i.e. returned to the original starting position through the same series of stages. They are never realized in practice. An isothermal reversible compression of a gas, for example, would have to be carried out infinitely slowly and involve no friction, etc. Ideal energy transfer would have to take place between the gas and the surroundings to maintain a constant temperature.

In practice, all real processes are *irreversible changes* in which there is not an equilibrium throughout the change. In an irreversible change, the system can still be returned to its original state, but not through the same series of stages. For a closed system, there is always an entropy increase involved in an irreversible change.

reversible reaction A chemical reaction that can proceed in either direction. For example, the reaction:

$$N_2 + 3H_2 \rightleftharpoons 2NH_3$$

is reversible. In general, there will be an equilibrium mixture of reactants and products.

***R*_F value** A measure of the relative distance traveled by a sample in a chromatography experiment. It is obtained by dividing the distance traveled by the solvent front into the distance traveled by the sample. Under standard conditions, the R_F value is characteristic of a particular substance. *See* paper chromatography; thin-layer chromatography.

rhenium /**ree**-nee-ŭm/ A rare silvery transition metal that usually occurs naturally with molybdenum (it is extracted from flue dust in molybdenum smelters). The metal is chemically similar to manganese and is used in alloys and catalysts.

Symbol: Re; m.p. 3180°C; b.p. 5630°C; r.d. 21.02 (20°C); p.n. 75; r.a.m. 186.207.

rheology /ree-**ol**-ŏ-jee/ The study of the ways in which matter can flow. This topic is of particular interest in the study of polymers.

rhodium /**roh**-dee-ŭm/ A rare silvery hard transition metal. It is difficult to work and highly resistant to corrosion. Rhodium occurs native but most is obtained from copper and nickel ores. It is used in protective finishes, alloys, and as a catalyst.

Symbol: Rh; m.p. 1966°C; b.p. 3730°C; r.d. 12.41 (20°C); p.n. 45; r.a.m. 102.90550.

rhombic crystal /**rom**-bik/ *See* crystal system.

ribonucleic acid /rȳ-boh-new-**klee**-ik/ *See* RNA.

ribose /**rȳ**-bohs/ ($C_5H_{10}O_5$) A monosaccharide; a component of RNA.

ring A closed loop of atoms in a molecule, as in benzene or cyclohexane. A *fused ring* is one joined to another ring in such a way that they share two atoms. Naphthalene is an example of a fused-ring compound.

ring closure A reaction in which one part of an open chain in a molecule reacts with another part, so that a ring of atoms is formed.

RIS *See* resonance ionization spectroscopy.

RNA (**ribonucleic acid**) A nucleic acid found mainly in the cytoplasm and involved in protein synthesis. It is a single polynucleotide chain similar in composition to a single strand of DNA except that the sugar ribose replaces deoxyribose and the pyrimidine base uracil replaces thymine. RNA is synthesized on DNA in the nucleus and exists in three forms. In certain viruses, RNA is the genetic material.

rock A definable part of the Earth's crust made up of a mixture of MINERALS. It may be hard (for example, granite), soft (chalk), consolidated (clay), or loose (sand).

rock salt (**halite**) A transparent naturally occurring mineral form of sodium chloride.

roentgen /**rent**-gĕn/ Symbol: R A unit of radiation, used for x-rays and gamma rays, defined in terms of the ionizing effect on air. One roentgen induces 2.58×10^{-4} coulomb of charge in one kilogram of dry air. The unit is named for the German physicist Wilhelm Konrad Roentgen (1845–1923).

roentgenium /**rent**-gĕn-ee-ŭm/ A radioactive element produced synthetically. It is named for Wilhelm Roentgen

Symbol: Rg; p.n. 111.

Rose's metal A fusible alloy containing 50% bismuth, 25–28% lead, and tin. Its low melting point (about 100°C) leads to its use in fire-protection devices.

rosin A brittle yellow or brown resin that remains after the distillation of turpentine. It is used as a flux in soldering and in making paints and varnishes. Powdered rosin gives a 'grip' to violin bows and boxers' shoes. *See also* resin.

rotary dryers Devices commonly used in the chemical industry for the drying, mixing, and sintering of solids. They consist essentially of a rotating inclined cylinder, which is longer in length than in diameter. Gases flow through the cylinder in either a countercurrent or cocurrent direction to regulate the flow of solids, which are fed into the end of the cylinder. Rotary dryers can be applied to both batch and continuous processes.

***R-S* convention** *See* optical activity.

rubber A natural or synthetic polymeric elastic material. Natural rubber is a polymer of methylbuta-1,3-diene (isoprene). Various synthetic rubbers are made by polymerization; for example chloroprene rubber (from 2-chlorobuta-1,3-diene) and silicone rubbers. *See also* vulcanization.

rubidium /roo-**bid**-ee-ŭm/ A soft silvery highly reactive element of the alkali-metal group. Naturally occurring rubidium comprises two isotopes, one of which, ^{87}Rb, is radioactive (half-life 5×10^{10} years). It is found in small amounts in several complex silicate minerals, including lepidolite. Rubidium is used in vacuum tubes, photocells, and in making special glass.

Symbol: Rb; m.p. 39.05°C; b.p. 688°C; r.d. 1.532 (20°C); p.n. 37; r.a.m. 85.4678.

rubidium–strontium dating A technique for dating rocks. It depends on the radioactive decay of the radioisotope ^{87}Rb to ^{87}Sr (half-life 4.7×10^{10} years). If the ratio $^{87}Sr/^{87}Rb$ is measured it is possible to estimate the age of the rock. *See also* radioactive dating.

ruby A gemstone variety of the mineral corundum (aluminum oxide, Al_2O_3) colored red by chromium impurities. Rubies can also be made synthetically. They are used in jewelry, as bearings in watches, and in some lasers.

rusting The corrosion of iron in air to form an oxide. Both oxygen and moisture must be present for rusting to occur. The process is electrolytic, involving electrochemical reactions on the wet iron:

$$Fe(s) \rightarrow Fe^{2+}(aq) + 2e^-$$

$$2H_2O + O_2(aq) + 4e^- \rightarrow 4OH^-(aq)$$

Iron(II) hydroxide precipitates and is oxidized to the red hydrated iron(III) oxide $Fe_2O_3.H_2O$.

ruthenium /roo-**th'ee**-nee-ŭm/ A transition metal that occurs naturally with platinum. It forms alloys with platinum that are used in electrical contacts. Ruthenium is also used in jewelry alloyed with palladium.

Symbol: Ru; m.p. 2310°C; b.p. 3900°C; r.d. 12.37 (20°C); p.n. 44; r.a.m. 101.07.

rutherfordium /ru*th*-er-**for**-dee-ŭm/ A radioactive metal not found naturally on earth. It is the first transactinide metal. Atoms of rutherfordium are produced by bombarding ^{249}Cf with ^{12}C or by bombarding ^{248}Cm with ^{18}O.

Symbol: Rf; m.p. 2100°C (est.); b.p. 5200°C (est.); r.d. 23 (est.); most stable isotope ^{261}Rf (half-life 65s).

rutile /**roo**-teel, -tÿl/ A naturally occurring reddish-brown form of titanium(IV) oxide, TiO_4. It is an ore of titanium and is also used in making ceramics. Some clear varieties of rutile are used as a semiprecious gemstone.

Rydberg constant /**rid**-berg/ A constant that occurs in formula for the frequencies of spectral lines in atomic spectra. For the hydrogen atom it has the value $1.0968 \times 10^7 m^{-1}$. The value of the Rydberg constant can be calculated from the BOHR THEORY of the hydrogen atom and from quantum mechanics. These calculations showed that:

$$R = M_0^2 m e^4 c^3 / 8h^3$$

where M_0 is the magnetic constant, m and e are the mass and charge respectively of an electron, h is the Planck constant and c is the speed of light in a vacuum. The constant is named for the Swedish physicist Johannes Robert Rydberg (1854–1919). *See also* hydrogen atom spectrum.

Sabatier–Senderens process A method of hydrogenating unsaturated vegetable oils to make margarine, using hydrogen and a nickel catalyst. It is named for the French chemists Paul Sabatier (1854–1941) and Jean-Baptiste Senderens (1856–1937). *See also* hardening.

saccharide /**sak**-ă-rÿd, -rid/ *See* sugar.

saccharin /**sak**-ă-rin/ ($C_7H_5NO_3S$) A white crystalline organic compound used as an artificial sweetener; it is about 550 times as sweet as sugar (sucrose). It is nearly insoluble in water and so generally used in the form of its sodium salt. Possible links with cancer in animals has restricted its use in some countries.

Sachse reaction /**sak**-sĕ/ A process for the manufacture of ethyne from natural gas (methane). Part of the methane is burned in two stages to raise the furnace temperature to about 1500°C. Under these conditions, the rest of the methane is converted to ethyne and hydrogen:

$$2CH_4 \rightarrow C_2H_2 + 3H_2$$

The process is important since it provides a source of ethyne from readily available natural gas, thus avoiding the expensive carbide process.

sacrificial protection A method of protection against electrolytic corrosion (especially rusting). In protecting steel pipelines, for instance, zinc or magnesium rods are buried in the ground at points along the line and connected to the pipeline. Rusting of iron is an electrochemical process in which Fe^{2+} ions dissolve in the water in contact with the surface. A more electropositive element, such as zinc, protects against this because Zn^{2+} ions dissolve in preference. In other words, the zinc rod sacrifices itself for the steel pipeline. The same effect occurs with galvanized iron. If the zinc coating is scratched, the iron exposed does not rust until all the zinc has dissolved away. (Note that tin, being less electropositive than iron, has the opposite effect.)

safety lamp (**Davy lamp**) A type of oil lamp for use in coal mines designed so that it will not ignite any methane (firedamp) present and cause an explosion. The flame is surrounded by a fine metal gauze which dissipates the heat from the flame but never reaches the ignition temperature of methane. If methane is present it burns inside the gauze; the lamp can be used as a test for pockets of methane. The lamp is also named for the British chemist Sir Humphry Davy (1778–1829).

sal ammoniac /sal ă-**moh**-nee-ak/ *See* ammonium chloride.

salicylate /**sal**-ă-sil-ayt, sal-ă-**sil**-ayt, să-**liss**-ă-layt/ (**hydroxybenzoate**) A salt or ester of salicylic acid.

salicylic acid /sal-ă-**sil**-ik/ (**2-hydroxybenzoic acid**; $C_6H_4(OH)COOH$) A crystalline aromatic carboxylic acid. It is used in medicines, as an antiseptic, and in the manufacture of azo dyes. Its ethanoyl (acetyl) ester is aspirin. *See* aspirin; methyl salicylate.

saline /**say**-lÿn/ Containing a salt, especially an alkali-metal halide such as sodium chloride.

salt A compound with an acidic and a basic radical, or a compound formed by total or partial replacement of the hydrogen in an acid by a metal. In general terms a salt is a material that has identifiable cationic and anionic components.

salt bridge An electrical contact between two half cells, used to prevent mixing. A glass U-tube filled with potassium chloride in agar is often used.

saltcake /**sawlt**-kayk/ An impure form of SODIUM SULFATE, formerly produced industrially as a by-product of the LEBLANC PROCESS.

salt hydrate A metallic salt in which the metal ions are surrounded by a fixed number of water molecules. The water molecules are bound to the metal ions by ion–dipole interactions. In the solid state the water molecules form part of the crystal structure. It is thus a clathrate.

Salt hydrates that contain several molecules of water for each metal ion can lose them progressively (effloresce) if the pressure of water vapor is kept below the dissociation pressure of the system; eventually an anhydrous salt is formed. Alternatively, if the vapor pressure is continuously raised the system will add molecules of water (deliquesce) until a saturated solution forms.

salting out The precipitation of a colloid (such as gelatin or soap) by adding an ionic salt to a solution. *See* gelatin; soap.

saltpeter /sawlt-**pee**-ter/ *See* potassium nitrate.

sal volatile /sal voh-**lat**-ă-lee/ *See* ammonium carbonate.

samarium /să-**mair**-ee-ŭm/ A silvery element of the lanthanoid series of metals. It occurs in association with other lanthanoids. Samarium is used in the metallurgical, glass, and nuclear industries.

Symbol: Sm; m.p. 1077°C; b.p. 1791°C; r.d. 7.52 (20°C); p.n. 62; r.a.m. 150.36.

sand A mineral form of silica (silicon(IV) oxide, SiO_2) consisting of separate grains of quartz. It results from erosion of quartz-containing rocks; the grains may be rounded by the tumbling action of water or wind. Sand is used in concrete and mortar and in making glass and other ceramics.

Sandmeyer reaction /**sand**-mȳ-er/ A method of producing chloro- and bromo-substituted derivatives of aromatic compounds by using the DIAZONIUM SALT with a copper halide. The reaction starts with an amine, which is diazotized at low temperature by using hydrochloric acid and sodium nitrite (to produce nitrous acid, HNO_2). For example:

$$C_6H_5NH_2 + NaNO_2 + HCl \rightarrow C_6H_5N_2^+ + Cl^- + OH^- + Na^+ + H_2O$$

The copper halide (e.g. CuCl) acts as a catalyst to give the substituted benzene derivative:

$$C_6H_5N_2^+ + Cl^- \rightarrow C_6H_5Cl + N_2$$

This reaction was discovered by the German chemist Traugott Sandmeyer (1854–1922) in 1884. A variation of the reaction, in which the catalyst is freshly precipitated copper powder, was reported in 1890 by the German chemist Ludwig Gatterman (1860–1920). This is known as the *Gatterman reaction* (or *Gatterman–Sandmeyer reaction*).

sandstone A sedimentary rock consisting of consolidated sand grains cemented together with calcium carbonate, clay, or oxides of iron. It is used mainly as a building stone.

sandwich compound A type of organometallic compound formed between transition metal ions and aromatic compounds, in which the metal ion is 'sandwiched' between two rings. Four, five, six, seven, and eight-membered rings are known to form complexes with a number of elements including V, Cr, Mn, Co, Ni, and FE. The bonding in such compounds is not between the metal ion and individual atoms on the ring; rather it involves bonding of *d* electrons of the ion with the pi-electron system of the ring as a whole. FERROCENE is the original example, having two parallel cyclopentadienyl rings sandwiching the iron ion. There are a number of related types of compound. Thus, *half-sandwich compounds* (sometimes known as *piano-stool compounds*) have only one ring bound to the central ion, with other ligands coordinating in the normal way. *Bent sandwich compounds* have two rings that are not parallel, and the ion is attached to other ligands. *Multidecker sandwich compounds* have more than two parallel rings (and more than two coordinating ions.

saponification /să-pon-ă-fă-**kay**-shŏn/ The process of hydrolyzing an ester with a hydroxide. The carboxylic acid forms a salt. For instance, with sodium hydroxide:

$$RCOOR' + NaOH \rightarrow NaOOCR + R'OH$$

The saponification of fats, which are esters of 1,2,3-trihydroxypropane with long-chain carboxylic acids, forms soaps.

sapphire A gemstone variety of the mineral corundum (aluminum oxide, Al_2O_3) colored blue (or other colors) by impurities. Sapphires can also be made synthetically.

saturated compound An organic compound that does not contain any double or triple bonds in its structure. A saturated compound will undergo substitution reactions but not addition reactions since each atom in the structure will already have formed its maximum possible number of single bonds. *Compare* unsaturated compound.

saturated solution A solution that contains the maximum equilibrium amount of solute at a given temperature. A solution is saturated if it is in equilibrium with its solute. If a saturated solution of a solid is cooled slowly, the solid may stay temporarily in solution; i.e. the solution may contain *more* than the equilibrium amount of solute. Such solutions are said to be *supersaturated.*

saturated vapor A vapor that is in equilibrium with the solid or liquid. A saturated vapor is at the maximum pressure (the saturated vapor pressure) at a given temperature. If the temperature of a saturated vapor is lowered, the vapor condenses. Under certain circumstances, the substance may stay temporarily in the vapor phase; i.e. the vapor contains *more* than the equilibrium concentration of the substance. The vapor is then said to be *supersaturated.*

s-block elements The elements of the first two groups of the periodic table; i.e. group 1 (H, Li, Na, K, Rb, Cs, Fr) and group 2 (Be, Mg, Ca, Sr, Ba, Ra). They are so called because their outer shells have the electronic configurations ns^1 or ns^2. The s-block excludes those with inner $(n - 1)$d levels occupied (i.e. it excludes transition elements, which also have s^2 and occasionally s^1 configurations).

scandium /**skan**-dee-ŭm/ A lightweight silvery element belonging to the first transition series. It is found in minute amounts in over 800 minerals, often associated with lanthanoids. Scandium is used in high-intensity lights and in electronic devices.

Symbol: Sc; m.p. 1541°C; b.p. 2831°C; r.d. 2.989 (0°C); p.n. 21; r.a.m. 44.955910.

scanning tunneling microscope (STM) An instrument that makes use of quantum-mechanical tunneling to investigate the surface structure of a sample of material. In this technique a sharp conducting tip is brought near a surface. This causes electrons to tunnel from the surface to the tip, with the probability of this occurring depending on both the distance between the surface and the tip and the electron density at the surface. The tunneling produces an electrical current which is kept constant by moving the tip up or down as it is moved over the surface. Single atoms can be observed using STM. *See also* atomic force microscope.

scavenger A compound or chemical species that removes a trace component from a system or that removes a reactive intermediate from a reaction.

Schiff's base /shiffs/ A type of compound formed by reacting an aldehyde or ketone (e.g. RCOR) with an aryl amine (e.g. $ArNH_2$). The product, an *N*-arylimide, which is usually crystalline, has the formula $R_2C{:}NAr$. The compound is named for the German chemist Hugo Schiff (1834–1915).

Schiff's reagent An aqueous solution of magenta dye decolorized by reduction with sulfur(IV) oxide. It is a test for aldehydes and ketones. Aliphatic aldehydes restore the color quickly; aliphatic ketones and aromatic aldehydes slowly; aromatic ketones give no reaction.

Schotten–Baumann reaction /**shot**-ĕn **bow**-mahn/ A method for the preparation of *N*-phenylbenzamide (benzanilide) in which a mixture of phenylamine (aniline) and sodium hydroxide is stirred while benzoyl chloride is added slowly. The *N*-phenylbenzamide precipitates and can be recrystallized from hot ethanol:

$$C_6H_5NH_2 + C_6H_5COCl \rightarrow C_6H_5NH.COC_6H_5 + HCl$$

Schottky defect /**shot**-kee/ *See* defect.

scrubber The part of a chemical plant that removes impurities from a gas by passing it through a liquid.

seaborgium /see-**bor**-gee-ŭm/ A radioactive metallic element not occurring naturally on Earth. It can be made by bombarding ^{249}Cf with ^{18}O nuclei using a cyclotron. Six isotopes are known.

Symbol: Sg; most stable isotope ^{266}Sg (half-life 27.3s).

second Symbol: s The SI base unit of time. It is defined as the duration of 9 192 631 770 cycles of a particular wavelength of radiation corresponding to a transition between two hyperfine levels in the ground state of the cesium-133 atom.

secondary alcohol *See* alcohol.

secondary amine *See* amine.

secondary cell *See* accumulator.

secondary structure *See* protein.

second-order reaction A reaction in which the rate of reaction is proportional to the product of the concentrations of two of the reactants or to the square of the concentration of one of the reactants; i.e.

$$\text{rate} = k[A][B]$$

or

$$\text{rate} = k[A]^2$$

For example, the hydrolysis by dilute alkali of an ester is a second-order reaction:

$$\text{rate} = k[\text{ester}][\text{alkali}]$$

The rate constant for a second-order reaction has the units $mol^{-1}\ dm^3\ s^{-1}$. Unlike a first-order reaction, the time for a definite fraction of the reactants to be consumed is dependent on the original concentrations.

sedimentation The settling of a suspension, either under gravity or in a centrifuge. The speed of sedimentation can be used to estimate the average size of the particles. This technique is used with an ultracentrifuge to find the relative molecular masses of macromolecules.

seed A small crystal added to a gas or liquid to assist solidification or precipitation from a solution. The seed, usually a crystal of the substance to be formed, enables particles to pack into predetermined positions so that a larger crystal can form.

selenium /si-**lee**-nee-ŭm/ A metalloid element existing in several allotropic forms and belonging to group 16 of the periodic table. It occurs in minute quantities in sulfide ores and industrial sludges. The common gray metallic allotrope is very light-sensitive and is used in photocells, solar cells, some glasses, and in xerography. The red allotrope is unstable and reverts to the gray form under normal conditions.

Symbol: Se; m.p. 217°C (gray); b.p. 684.9°C (gray); r.d. 4.79 (gray); p.n. 34; r.a.m. 78.96.

self-assembly *See* supramolecular chemistry.

Seliwanoff's test A test for ketose SUGARS in solution. The reagent used consists of resorcinol dissolved in hydrochloric acid. A few drops are added to the solution and a red precipitate indicates a ketonic sugar. The test is named for the Russian chemist F. F. Seliwanoff.

semicarbazide /sem-ee-**kar**-bă-zÿd/ *See* semicarbazone.

semicarbazone /sem-ee-**kar**-bă-zohn/ A type of organic compound containing the $C{:}N.NH.CO.NH_2$ grouping, formed by reaction of an aldehyde or ketone with semicarbazide ($H_2N.NH.CO.NH_2$). The compounds are crystalline solids with sharp melting points, which can be used to characterize the original aldehyde or ketone.

semiconductor /sem-ee-kŏn-**duk**-ter/ A crystalline material which conducts only under certain conditions. The conductivity, unlike in metals, increases with temperature because the highest occupied and lowest unoccupied energy levels are very close. A semiconductor can be formed by introducing small amounts of an impurity to an ultrapure material. If these impurities can withdraw electrons from the occupied energy level, leaving 'holes' which permit conduction, the resultant semiconduction is known as *p-type*. Alternatively, these impurities may donate electrons which enter the unoccupied energy level, thus forming an *n-type* semiconductor. Semiconductors are used extensively by the electronics industry in such devices as integrated circuits.

semipermeable membrane /sen-ee-**per**-mee-ă-băl/ A membrane that, when separating a solution from a pure solvent, permits the solvent molecules to pass through it but does not allow the transfer of solute molecules. Synthetic semipermeable membranes are generally supported on a porous material, such as unglazed porcelain or fine wire screens, and are commonly formed of cellulose or related materials. They are used in osmotic studies, gas separations, and medical applications.

Equilibrium is reached at a semipermeable membrane if the chemical potentials on both sides become identical; migration of solvent molecules towards the solution is an attempt by the system to reach equilibrium. The pressure required to halt this migration is the osmotic pressure.

semipolar bond /sem-ee-**poh**-ler/ A coordinate bond.

septivalent /sep-tă-**vay**-lĕnt, sep-**tiv**-ă-/ (**heptavalent**) Having a valence of seven.

sequence rule *See* CIP system.

sequestration /see-kwes-**tray**-shŏn/ The formation of a complex with an ion in solution, so that the ion does not have its normal activity. Sequestering agents are often chelating agents, such as edta.

serine /**se**-reen, -rin/ *See* amino acids.

sesqui- Prefix indicating a 2/3 ratio. A sesquioxide, for example, would have the formula M_2O_3. A sesquicarbonate is a mixed carbonate and hydrogencarbonate of the type $Na_2CO_3.NaHCO_3.2H_2O$, which contains 2 CO_3 units and 3 sodium ions.

sesquiterpene *See* terpene.

shell A group of electrons that share the same principal quantum number. Early work on x-ray emission studies used the terms K, L, M, and these are still sometimes used for the first three shells: $n = 1$ K-shell; $n = 2$ L-shell; $n = 3$ M-shell.

sherardizing /**sh'e**-răr-dÿ-zing/ A technique used to obtain corrosion-resistant articles of iron or steel by heating with zinc dust in a sealed rotating drum for several hours at about 375°C. At this temperature the two metals combine, forming internal layers of zinc–iron and zinc–steel alloys and an external layer of pure zinc. Sherardizing is principally used for small parts, such as springs, washers, nuts, and bolts.

short period *See* period.

SI *See* SI units.

side chain In an organic compound, an aliphatic group or radical attached to a longer straight CHAIN of atoms in an acyclic compound or to one of the atoms in the ring of a cyclic compound.

side reaction A chemical reaction that takes place to a limited extent at the same time as a main reaction. Thus the main product of a reaction may contain small amounts of other compounds.

siemens /**see**-mĕnz/ (**mho**) Symbol: S The SI unit of electrical conductance, equal to a conductance of one ohm^{-1}. The unit is named for the German electrical engineer Ernst Werner von Siemens (1816–92).

sievert /**see**-vert/ Symbol: Sv The SI unit of dose equivalent. It is the dose equivalent when the absorbed dose produced by ionizing radiation multiplied by certain dimensionless factors is 1 joule per kilogram ($1\ J\ kg^{-1}$). The dimensionless factors are used to modify the absorbed dose to take account of the fact that different types of radiation cause different biological effects. The unit is named for the Swedish physicist Rolf Sievert (1898–1966).

sigma orbital *See* orbital.

sigmatropic rearrangement /sig-mă-**trop**-ik/ *See* pericyclic reaction.

silane /**sil**-ayn/ Any of a small number of hydrides of silicon: SiH_4, Si_2H_6, Si_3H_8, etc. Silanes can be made by the action of acids on magnesium silicide (Mg_2Si). They are unstable compounds and ignite spontaneously in air. Only the first few members of the series are known and there is not the range of 'hydrosilicon' compounds to compare with the hydrocarbons.

silica /**sil**-ă-kă/ *See* silicon(IV) oxide.

silica gel A gel made by coagulating sodium silicate sol. The gel is dried by heating and used as a catalyst support and as a drying agent. The silica gel used in desiccators and in packaging to remove moisture is often colored with a cobalt salt to indicate whether it is still active (blue = dry; pink = moist).

silicates /**sil**-ă-kayts, -kits/ A large number of compounds containing metal ions and complex silicon–oxygen compounds. The negative ions in silicates are of the form SiO_4^{4-}, $Si_2O_7^{6-}$, etc., and many are polymeric, containing SiO_4 units linked in long chains, sheets, or three-dimensional arrays. *Aluminosilicates* and *borosilicates* are similar materials containing aluminum or boron atoms in the structure. Many silicates occur naturally in rocks and minerals.

silicic acid /să-**liss**-ik/ A colloidal or jelly-like hydrated form of silicon(IV) oxide (SiO_2), made by adding an acid (such as hydrochloric acid) to a soluble SILICATE.

silicide /**sil**-ă-sÿd/ A compound of silicon with a more electropositive element.

silicon /**sil**-ă-kŏn/ A hard brittle gray metalloid element; the second element in group 14 of the periodic table. It has the electronic configuration of neon with four additional outer electrons; i.e., $[Ne]3s^23p^2$.

Silicon accounts for 27.7% of the mass of the Earth's crust and occurs in a wide variety of silicates with other metals, clays, micas, and sand, which is largely SiO_2. The element is obtained on a small scale by the reduction of silicon(IV) oxide (SiO_2) by carbon or calcium carbide. For semiconductor applications very pure silicon is produced by direct reaction of silicon with an HCl/Cl_2 mixture to give silicon tetrachloride ($SiCl_4$), which can be purified by distillation. This is then decomposed on a hot wire in an atmosphere of hydrogen. For ultra-pure samples zone refining is used. Unlike carbon, silicon does not form allotropes but has only the diamond type of structure.

Silicon does not react with hydrogen except under extreme conditions in the presence of chlorine, in which case compounds such as trichlorosilane ($HSiCl_3$) are obtained. The hydride SiH_4 itself may be prepared in the laboratory by hydrolysis of magnesium silicide (a number of other hydrides up to Si_6H_{14} are obtained), or more conveniently by reduction of silicon tetrachloride with lithium tetrahydridoaluminate:

$$SiCl_4 + LiAlH_4 \rightarrow SiH_4 + LiCl + AlCl_3$$

The silanes are mildly explosive and the chloroalkyl silanes extremely so.

Silicon combines with oxygen when heated in air to form silicon(IV) oxide. The mineral silicates constitute a vast collection of materials with a wide range of structures including chains, rings, lattices, twisted chains, sheets, and varyingly cross-linked structures all based on different assemblies of $[SiO_4]^{2-}$ tetrahedra. Being a metalloid, silicon is somewhat amphoteric and consequently silica is weakly acidic, dissolving in fused alkalis to form the appropriate silicate.

Silicon also forms an immense number of organic derivatives. In many cases where single bonding only is involved these compounds are rather similar to organic analogs with the general distinction that nucleophilic attachment occurs at silicon (rather than at hydrogen) and bond angles at oxygen and nitrogen are often larger than the organic materials. Optical activity and inversion of activity is observed. The major differences are due to the type of 'double bond' available to carbon ($p\pi$–$p\pi$), and to silicon ($p \rightarrow d\pi$ or back bonding). Thus silicon does not form conventional double bonds and silicon bonds to oxygen are of the ether type (as in siloxanes) rather than the ketone type. The Si–OH bond is also very much less stable than the alcohol link, R–OH, ready condensation to the siloxane occurring in the silicon case:

$$2R_3SiOH \rightarrow R_3Si\text{–}O\text{–}SiR_3 + H_2O$$

Symbol: Si; m.p. 1410°C; b.p. 2355°C; r.d. 2.329 (20°C); p.n. 14; r.a.m. 28.0855

silicon carbide (**carborundum**; SiC) A black very hard crystalline solid made by heating silicon(IV) oxide (sand) with carbon (coke) in an electric furnace. It is used as an abrasive.

silicon(IV) chloride (**silicon tetrachloride**; $SiCl_4$) A colorless fuming liquid which rapidly hydrolyzes in moist air or water. It is used to produce pure silica for making special kinds of glass.

silicon dioxide /dÿ-**oks**-ÿd, -id/ *See* silicon(IV) oxide.

silicones /**sil**-ă-kohnz/ Polymeric synthetic silicon compounds containing chains of alternating silicon and oxygen atoms, with organic groups bound to the silicon atoms. Silicones are used as lubricants and water repellants and in waxes and varnishes. Silicone rubbers are superior to natural rubbers in their resistance to both high and low temperatures and chemicals. *See also* siloxanes.

silicon(IV) oxide (**silicon dioxide**; **silica**; SiO_2) A hard crystalline compound occurring naturally in three crystalline forms (quartz, tridymite, and crystobalite). Sand is mostly silicon(IV) oxide. Fused silica is a glassy substance used in laboratory apparatus. Silica is used in the manufacture of glass.

silicon tetrachloride /tet-ră-**klor**-ÿd, -id, -**kloh**-rÿd, -rid/ *See* silicon(IV) chloride.

siloxanes /să-**loks**-aynz/ Compounds containing Si–O–Si groups with organic groups bound to the silicon atoms. The silicones are polymers of siloxanes.

silver A transition metal that occurs native and as the sulfide (Ag_2S) and chloride (AgCl). It is extracted as a by-product in refining copper and lead ores. Silver darkens in air due to the formation of silver sulfide. It is used in coinage alloys, tableware, and jewelry. Silver compounds are used in photography.

Symbol: Ag; m.p. 961.93°C; b.p. 2212°C; r.d. 10.5 (20°C); p.n. 47; r.a.m. 107.8682.

silver(I) bromide (AgBr) A pale yellow precipitate obtained by adding a soluble bromide solution to a solution of silver(I) nitrate. The halide dissolves in concentrated ammonia solution but not in dilute ammonia solution. It is used extensively in photography for making the light-sensitive emulsions for plates and films. Silver(I) bromide crystallizes in a similar manner to sodium(I) chloride. Unlike silver(I) chloride and silver(I) iodide, silver(I) bromide does not absorb ammonia gas.

silver(I) chloride ($AgCl$) A compound prepared as a curdy white precipitate by the addition of a soluble chloride solution to a solution of silver(I) nitrate. It occurs in nature as the mineral 'horn silver'. Silver(I) chloride is insoluble in water but dissolves in concentrated hydrochloric acid and concentrated ammonia solution. It is rather unreactive but is affected by sunlight, being used extensively in photography. Silver(I) chloride crystallizes in the sodium chloride lattice pattern. It absorbs ammonia gas forming ammines, e.g. $AgCl.2NH_3$ and $AgCl.3NH_3$.

silver(I) iodide (AgI) A pale yellow precipitate prepared by the addition of a soluble iodide solution to a solution of silver(I) nitrate. Silver(I) iodide occurs in nature as iodoargyrite in the form of hexagonal crystals. It is virtually insoluble in ammonia solution and is used in the photographic industry. Silver(I) iodide will absorb ammonia gas. It is trimorphic and is found to contract when heated and expand on cooling.

silver-mirror test A test for the aldehyde group. A few drops of the sample are warmed with TOLLEN'S REAGENT. An aldehyde reduces Ag^+ to silver metal, causing a brilliant silver mirror to coat the inside wall of the tube.

silver(I) nitrate ($AgNO_3$) A white crystalline solid prepared by dissolving silver in dilute nitric acid and crystallizing the solution. Silver(I) nitrate is dimorphous, crystallizing in the orthorhombic and hexagonal systems. It undergoes thermal decomposition to yield silver, nitrogen(IV) oxide, and oxygen. Probably the most important silver salt, it is used in the medical industry for the treatment of warts and in the photographic industry. In the laboratory, it is used both as an analytical and volumetric reagent. Silver(I) nitrate in solution and in the solid state can oxidize organic compounds, e.g. aldehydes can be oxidized to carboxylic acids; the silver ions are reduced to silver.

silver(I) oxide (**argentous oxide**; Ag_2O) A brown amorphous solid formed by the addition of sodium or potassium hydroxide solution to a solution of silver nitrate. Silver(I) oxide turns moist red litmus blue and decomposes (at 160°C) to give silver and oxygen. Silver(I) oxide dissolves in concentrated ammonia solution, forming the complex ion $[Ag(NH_3)_2]^+$. In organic chemistry it is used when moist to convert alkyl halides to the corresponding alcohol and when dry to convert alkyl halides into ethers.

silver(II) oxide (**argentic oxide**; AgO) A black diamagnetic substance prepared by the oxidation of either silver or silver(I) oxide using ozone. Alternatively it can be prepared as a precipitate by the addition of a solution of potassium persulfate to one of silver(I) nitrate.

single bond A covalent bond between two elements that involves one pair of electrons only. It is represented by a single line, for example H–Br, and is usually a sigma bond, although it can be a pi bond. *Compare* multiple bond. *See also* orbital.

sintering /**sin**-ter-ing/ A process in which certain powdered substances (e.g. metals, ceramics) coagulate into a single mass when heated to a temperature below the substance's melting point. Sintered glass is a porous material used for laboratory filtration.

SI units (*S*ystème *I*nternational d'Unités) The internationally adopted system of units used for scientific purposes. It has seven base units and two dimensionless units, formerly called supplementary units. Derived units are formed by multiplication and/or division of base units. Standard prefixes are used for multiples and submulti-

BASE AND DIMENSIONLESS SI UNITS

Physical quantity	*Name of SI unit*	*Symbol for SI unit*
length	meter	m
mass	kilogram(me)	kg
time	second	s
electric current	ampere	A
thermodynamic temperature	kelvin	K
luminous intensity	candela	cd
amount of substance	mole	mol
*plane angle	radian	rad
*solid angle	steradian	sr

*supplementary units

DERIVED SI UNITS WITH SPECIAL NAMES

Physical quantity	*Name of SI unit*	*Symbol for SI unit*
frequency	hertz	Hz
energy	joule	J
force	newton	N
power	watt	W
pressure	pascal	Pa
electric charge	coulomb	C
electric potential difference	volt	V
electric resistance	ohm	Ω
electric conductance	siemens	S
electric capacitance	farad	F
magnetic flux	weber	Wb
inductance	henry	H
magnetic flux density	tesla	T
luminous flux	lumen	lm
illuminance (illumination)	lux	lx
absorbed dose	gray	Gy
activity	becquerel	Bq
dose equivalent	sievert	Sv

DECIMAL MULTIPLES AND SUBMULTIPLES USED WITH SI UNITS

Submultiple	*Prefix*	*Symbol*	*Multiple*	*Prefix*	*Symbol*
10^{-1}	deci-	d	10^{1}	deca-	da
10^{-2}	centi-	c	10^{2}	hecto-	h
10^{-3}	milli-	m	10^{3}	kilo-	k
10^{-6}	micro-	µ	10^{6}	mega-	M
10^{-9}	nano-	n	10^{9}	giga-	G
10^{-12}	pico-	p	10^{12}	tera-	T
10^{-15}	femto-	f	10^{15}	peta-	P
10^{-18}	atto-	a	10^{18}	exa-	E
10^{-21}	zepto-	z	10^{21}	zetta-	Z
10^{-24}	yocto-	y	10^{24}	yotta-	Y

ples of SI units. The SI system is a coherent rationalized system of units.

slag Glasslike compounds of comparatively low melting point formed during the extraction of metals when the impurities in an ore react with a flux. The fact that the slag is a liquid of comparatively low density means that it can be separated from the liquid metal on which it floats. In a blast furnace, the flux used is limestone ($CaCO_3$) and the main impurity is silica (SiO_2). The slag formed is mainly calcium silicate:

$$CaCO_3 + SiO_2 \rightarrow CaSiO_3$$

It is used as railroad ballast and in fertilizers and concrete.

slaked lime *See* calcium hydroxide.

slaking *See* calcium hydroxide.

slip planes The planes of weakness in a crystal, e.g. the boundaries between the octahedral layers in graphite.

slurry A thin paste of suspended solid particles in a liquid.

smectic crystal /**smek**-tik/ *See* liquid crystal.

smelting An industrial process for extracting metals from their ores at high temperatures. Generally the ore is reduced with carbon (for zinc and tin) or carbon monoxide (for iron). Copper and lead are obtained by reduction of the oxide with the sulfide; for example:

$$2Cu_2O + Cu_2S \rightarrow 6Cu + SO_2$$

A flux is also used to combine with impurities and form a slag on top of the molten metal (*see* slag).

smithsonite /**smith**-sŏ-nÿt/ *See* calamine.

SNG Substitute (or synthetic) natural gas. A mixture of hydrocarbons manufactured from coal or the naphtha fraction of petroleum for use as a fuel.

S_N1 reaction *See* nucleophilic substitution.

S_N2 reaction *See* nucleophilic substitution.

soap One of a number of sodium or potassium compounds of fatty acids that are commonly used to improve the cleansing properties of water. Soap was the earliest known DETERGENT. It is made by first reacting vegetable oils and animal fats with a strong solution of sodium or potassium hydroxide to produce organic acids, such as octadecanoic acid. The soap is then precipitated out as the salt, e.g. sodium octadecanoate, by adding excess sodium chloride. In water, soap molecules break up to produce ions, which are responsible for the cleansing properties.

soapstone (**steatite**) A soft type of talc which has a greasy feel and which is easy to carve to make ornaments. *See* talc.

soda *See* sodium carbonate; sodium hydroxide.

soda ash *See* sodium carbonate.

soda glass *See* glass.

soda lime A gray solid produced by adding sodium hydroxide solution to calcium oxide, to give a mixture of $Ca(OH)_2$ and NaOH. It is used in the laboratory as a drying agent and as an absorbent for dioxide.

sodamide /**soh**-dă-mÿd/ ($NaNH_2$) A white ionic solid formed by passing dry ammonia over sodium at 300–400°C. The compound reacts with water to give sodium hydroxide and ammonia. It reacts with red-hot carbon to form sodium cyanide and with nitrogen(I) oxide to form sodium azide. Sodamide is used in the Castner process and in the explosives industry.

soda water A solution of carbon dioxide dissolved in water under pressure, used in fizzy drinks. When the pressure is released, the liquid effervesces with bubbles of carbon dioxide gas. *See also* carbonic acid.

sodium /**soh**-dee-ŭm/ A soft reactive metal; the second member of the alkali metals (group 1 of the periodic table). It has the electronic configuration of a neon structure plus one additional outer 3s electron. Electronic excitation in flames or the familiar sodium lamps gives a distinctive yellow color arising from intense emission at the so called 'sodium-D' line pair. The ionization potential is rather low and the sodium atom readily loses its electron (i.e. the metal is strongly reducing). The chemistry of sodium is largely the chemistry of the monovalent Na^+ ion.

Sodium occurs widely as NaCl in seawater and as deposits of halite in dried up lakes etc. (2.6% of the lithosphere). The element is obtained commercially by electrolysis of NaCl melts in which the melting point is reduced by the addition of calcium chloride; sodium is produced at the iron cathode (the Downs cell). The metal is extremely reactive. It reacts vigorously with the halogens, and also with water to give hydrogen and sodium hydroxide. The chemistry of sodium is very similar to that of the other members of group 1.

Nearly all sodium compounds are soluble in water. Sodium hydroxide is produced commercially by the electrolysis of brine using diaphragm cells or mercury-cathode cells; chlorine is a coproduct.

Solutions of sodium metal in liquid ammonia are blue and have high electrical conductivities; the main current carrier of such solutions is the solvated electron. Such solutions are used in both organic and inorganic chemistry as efficient reducing agents. Sodium also forms a number of alkyl and aryl derivatives by reaction with the appropriate mercury compound, e.g.:

$$(CH_3)_2Hg + 2Na \rightarrow 2CH_3Na + Na/Hg$$

These materials are used as catalysts for polymerization reactions. The metal itself has a body-centered structure.

Symbol: Na; m.p. 97.81°C; b.p. 883°C; r.d. 0.971 (20°C); p.n. 11; r.a.m. 22.989768.

sodium acetate *See* sodium ethanoate.

sodium aluminate ($NaAlO_2$) A white solid produced by adding excess aluminum to a hot concentrated solution of sodium hydroxide. Once the reaction has been initiated, sufficient heat is liberated to keep the reaction going; hydrogen is also produced. In solution the aluminate ions have the structure $Al(OH)_4^-$ and $NaAl(OH)_4$ is known as sodium(I) tetrahydroxoaluminate(III). Addition of sodium hydroxide to an aluminum salt solution produces a white gelatinous precipitate of aluminum hydroxide, which dissolves in excess alkali to give a solution of sodium(I) tetrahydroxoaluminate(III).

sodium azide (NaN_3) A white solid prepared by passing nitrogen(I) oxide over heated sodamide. On heating, sodium azide decomposes to give sodium and nitrogen. It is used as a reagent in organic chemistry and in the preparation of lead azide (for use in detonators).

sodium benzenecarboxylate /ben-zeen-kar-**boks**-ă-layt/ (**sodium benzoate**; C_6H_5COONa) A white crystalline powder made by neutralizing benzoic acid with sodium hydroxide solution and then evaporating the solution. It is used as an urinary antiseptic, in the dyeing industry, and as a food preservative.

sodium benzoate /**ben**-zoh-ayt/ *See* sodium benzenecarboxylate.

sodium bicarbonate *See* sodium hydrogencarbonate.

sodium bisulfate *See* sodium hydrogensulfate.

sodium bisulfite *See* sodium hydrogensulfite.

sodium borate /**bor**-ayt, **boh**-rayt/ *See* disodium tetraborate decahydrate.

sodium bromide (NaBr) A white solid formed by the action of bromine on hot sodium hydroxide solution. Alternatively, it can be made by the action of dilute hydrobromic acid on sodium carbonate or hydroxide. Sodium bromide is soluble in water and forms crystals similar in shape to sodium chloride. It is used in medicine and photography. Sodium bromide forms a hydrate, $NaBr.2H_2O$.

sodium carbonate (soda ash; Na_2CO_3) A white amorphous powder, which aggregates on exposure to air owing to the formation of hydrates. Industrially it is prepared by the ammonia–soda (Solvay) process. On crystallizing from aqueous solution large translucent crystals of the decahydrate ($Na_2CO_3.10H_2O$) are formed (*washing soda*). These effloresce forming the monohydrate, $Na_2CO_3.H_2O$. Large quantities of sodium carbonate are used in the manufacture of sodium hydroxide. Washing soda is used as a domestic cleanser. The salt produces an alkaline solution in water by hydrolysis:

$$Na_2CO_3 + 2H_2O \rightarrow 2NaOH + H_2CO_3$$

It is used in volumetric analysis to standardize strong acids.

sodium chlorate(V) ($NaClO_3$) A white solid formed by the action of chlorine on hot concentrated sodium hydroxide solution, or by the electrolysis of a concentrated sodium chloride solution. It is soluble in water. Sodium chlorate is a powerful oxidizing agent. If heated it yields oxygen and sodium chloride. The chlorate is used in explosives, in matches, as a weed killer, and in the textile industry.

sodium chloride (common salt; salt; $NaCl$) A white solid prepared by the neutralization of hydrochloric acid with aqueous sodium hydroxide. It occurs in sea water and natural brines. Natural solid deposits of *rock salt* are also found in certain places. The compound dissolves in water with the absorption of heat, its solubility changing very little with temperature. It is used to season and preserve food. Industrially, it is used as a raw material in the manufacture of sodium carbonate (Solvay process), sodium hydroxide (electrolysis), and soap. Sodium chloride is almost insoluble in alcohol.

sodium-chloride structure A form of crystal structure that consists of a face-centered cubic arrangement of sodium ions

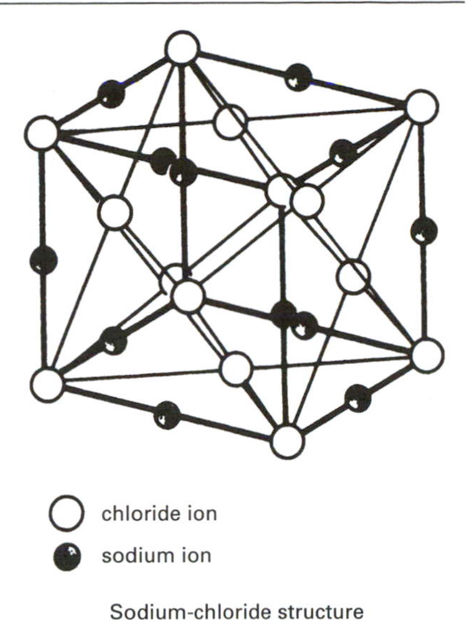

Sodium-chloride structure

with chloride ions situated at the middle of each edge and in the center of the cube. Electrostatic attraction holds the oppositely charged ions together.

The lattice can also be thought of as two interpenetrating face-centered cubes, one composed of sodium ions and the other of chloride ions. Other compounds (e.g. sodium bromide and potassium chloride) having their ions arranged in the same positions, are also described as having the sodium-chloride structure.

sodium cyanide ($NaCN$) A white solid formed either by the action of carbon on sodamide at high temperature or by passing ammonia over sodium at 300–400°C to form sodamide and reacting it with carbon. The industrial salt is prepared by the Castner process and purified by recrystallization from liquid ammonia. It is extremely poisonous and used as a source of cyanide and hydrocyanic acid. In the cyanide process it is used in the extraction of silver and gold. Its aqueous solutions are alkaline due to salt hydrolysis.

sodium dichromate(VI) ($Na_2Cr_2O_7$) An orange deliquescent solid that is very soluble in water. It is made from finely powdered chromite, which is heated with calcium oxide and sodium carbonate in a furnace. If an alkali is added to a solution of the dichromate, the chromate is produced. At high temperatures, sodium dichromate decomposes to give the chromate, chromic oxide, and oxygen. It is used as an oxidizing agent, particularly in organic chemistry and in volumetric analysis to estimate iron(II) ions and iodide ions.

sodium dihydrogen phosphate(V) /dȳ-**hȳ**-drŏ-jĕn/ (**sodium dihydrogen orthophosphate**; NaH_2PO_4) A white solid prepared by titrating phosphoric acid with sodium hydroxide solution using methyl orange as the indicator. On evaporation white crystals of the monohydrate are formed. It is used in some baking powders. The monohydrate crystallizes in the orthorhombic system.

sodium dioxide /dȳ-**oks**-ȳd, -id/ *See* sodium superoxide.

sodium ethanoate /eth-ă-**noh**-ayt/ (**sodium acetate**; CH_3COONa) A white solid prepared by the neutralization of ethanoic acid with either sodium carbonate or sodium hydroxide. Sodium ethanoate reacts with sulfuric acid to form sodium hydrogensulfate and ethanoic acid; with sodium hydroxide it gives rise to sodium carbonate and methane. Sodium ethanoate is used in the dyeing industry.

sodium fluoride (NaF) A white solid formed by the action of hydrofluoric acid on sodium carbonate or sodium hydroxide (the reaction between metallic sodium and fluorine is too violent). Sodium fluoride is soluble in water and reacts with concentrated sulfuric acid to yield hydrogen fluoride. It is used as a constituent of ceramic enamels, as an antiseptic, and as an agent to prevent fermentation.

sodium formate *See* sodium methanoate.

sodium hexafluoroaluminate /heks-ă-floo-ŏ-roh-ă-**loo**-mă-nayt/ (**cryolite**; Na_3AlF_6) A compound found in large quantities in South Greenland. It is white or colorless, but may be reddish or brown because of impurities. Sodium hexafluoroaluminate is used as a flux in the manufacture of aluminum. It crystallizes in the monoclinic system but in forms that closely resemble cubes and isometric octahedrals.

sodium hydride (NaH) A white crystalline solid prepared by passing a pure stream of dry hydrogen over sodium at 350°C; the sodium is usually suspended in an inert medium. Electrolysis of fused sodium hydride yields hydrogen at the anode, suggesting the presence of the H^- ion. Sodium hydride reacts with water to form sodium hydroxide solution and hydrogen. It is used as a powerful reducing agent to convert water to hydrogen, concentrated sulfuric acid to hydrogen sulfide, and iron(III) oxide to iron. Sodium hydride bursts into flames spontaneously on contact with the halogens at room temperature. It dissolves in liquid ammonia to give sodamide, $NaNH_2$.

sodium hydrogencarbonate (**sodium bicarbonate; baking soda**; $NaHCO_3$) A white solid formed either by passing an excess of carbon dioxide through sodium carbonate or hydroxide solution, or by precipitation when cold concentrated solutions of sodium chloride and ammonium hydrogencarbonate are mixed. Sodium hydrogencarbonate decomposes on heating to give sodium carbonate, carbon dioxide, and water. With dilute acids, it yields carbon dioxide. It is used as a constituent of baking powder, in effervescent beverages, and in fire extinguishers. Its aqueous solutions are alkaline as a result of salt hydrolysis. Sodium hydrogencarbonate forms monoclinic crystals.

sodium hydrogensulfate /hȳ-drŏ-jĕn-**sul**-fayt/ (**sodium bisulfate**; $NaHSO_4$) A white solid formed either by the partial neutralization of sodium hydroxide solution with sulfuric acid, by the action of concentrated sulfuric acid on sodium nitrate, or by heating equimolar quantities of sodium chloride and concentrated sulfuric acid. In aqueous solution sodium hydrogensulfate is strongly acidic. It crystallizes as the monohydrate ($NaHSO_4.H_2O$), which dehydrates on warming. If heated strongly the pyrosulfate ($Na_2S_2O_7$) is formed, which decomposes to give the sulfate and sulfur(VI) oxide. Sodium hydrogensulfate is used as a cheap source of sulfuric acid and in the dyeing industry.

sodium hydrogensulfite /hÿ-drŏ-jĕn-**sul**-fÿt/ (**sodium bisulfite**; $NaHSO_3$) A white powder prepared by saturating a solution of sodium carbonate with sulfur(IV) oxide. It is isolated from the aqueous solution by precipitation with alcohol. If heated it undergoes thermal decomposition to give sodium sulfate, sulfur(VI) oxide, and sulfur. Sodium hydrogensulfite is used to sterilize wine casks and in medicine as an antiseptic.

sodium hydroxide (**caustic soda**; NaOH) A white deliquescent slightly translucent solid. In solution it is a strong alkali and electrolyte. In the laboratory it can be prepared by reacting sodium, sodium monoxide, or sodium peroxide with water. Industrially it can be prepared by the electrolysis of sodium chloride using a mercury cathode (Castner–Kellner) or diaphragm cell. It can also be manufactured by the Gossage process. Sodium hydroxide dissolves readily in water with the evolution of heat. Its solutions have a soapy feel and are very corrosive. It is used to absorb acidic gases, such as carbon dioxide and sulfur(IV) oxide. Industrially it is used in soap and paper manufacture and in the purification of bauxite.

sodium iodide (NaI) A white solid formed by reacting sodium carbonate or sodium hydroxide with hydroiodic acid; the solution is then evaporated. Sodium iodide forms colorless crystals having a cubic shape. It is used as a source of iodine and in medicine and photography. Acidified solutions of sodium iodide exhibit reducing properties owing to the formation of hydroiodic acid.

sodium metasilicate /met-ă-**sil**-ă-kayt/ *See* sodium silicate.

sodium methanoate /meth-ă-**noh**-ayt/ (**sodium formate**; HCOONa) A white solid produced by reacting carbon monoxide with solid sodium hydroxide at 200°C and 10 atmospheres pressure. Alternatively it can be prepared by the neutralization of methanoic acid with sodium hydroxide solution.

sodium monoxide /mon-**oks**-ÿd, -id/ (Na_2O) A white solid formed by burning sodium in a deficiency of oxygen or alternatively by reducing sodium peroxide or sodium hydroxide with the requisite amount of sodium. It reacts violently with water to form sodium hydroxide and with acids to form solutions of their salts. Sodium monoxide forms cubic crystals. It dissolves in liquid ammonia to give a mixture of sodamide and sodium hydroxide.

sodium nitrate (**Chile saltpeter**; $NaNO_3$) A white solid formed by the neutralization of nitric acid with either sodium carbonate or sodium hydroxide. It occurs naturally in large quantities in South America. Impure industrial sodium nitrate is called *caliche*. Sodium nitrate is very soluble in water and on crystallization forms colorless deliquescent crystals. On heating it decomposes to give sodium nitrite and oxygen. When heated with concentrated sulfuric acid, nitric acid is produced. Sodium nitrate is used as a fertilizer and as a source of nitrates and nitric acid. The salt crystallizes in the rhombohedral system and is isomorphous with the iodate.

sodium nitrite ($NaNO_2$) A yellowish-white solid formed by the thermal decomposition of sodium nitrate. It is readily soluble in water. The anhydrous salt forms orthorhombic crystals. When treated with cold dilute hydrochloric acid, sodium nitrite forms nitrous acid. It is used in organic chemistry in the process of diazotization and industrially as a corrosion inhibitor.

sodium orthophosphate /or-thoh-**fos**-fayt/ *See* trisodium phosphate(V).

sodium peroxide (Na_2O_2) A yellowish-white ionic solid formed by the direct combination of burning sodium with excess oxygen. It reacts with water to form sodium hydroxide and hydrogen peroxide; the latter decomposes rapidly in the alkaline solution to give oxygen. Sodium peroxide is used as a bleaching agent and an oxidizing agent for such materials as wool and wood pulp. Sodium peroxide can convert nitrogen(II) oxide to sodium nitrate and iodine to sodium iodate.

sodium silicate (**sodium metasilicate;** $Na_2SiO_3.5H_2O$) A colorless crystalline solid used in detergents and other cleaning compounds, in refractories, and in cements for ceramics. A concentrated aqueous solution is known as *water glass* and is used as a sizing agent and for making precipated silica and silica gel.

sodium sulfate (Na_2SO_4) A white soluble solid formed by heating a mixture of sodium chloride and concentrated sulfuric acid. Industrially it is prepared by the first stage of the Leblanc process. Sodium sulfate forms two hydrates, the decahydrate ($Na_2SO_4.10H_2O$) and the heptahydrate ($Na_2SO_4.7H_2O$). Sodium decahydrate, known as *Glauber's salt*, is used in the manufacture of glass and as a purgative in medicine. It effloresces on exposure to air to form the anhydrous salt, which is used as a drying agent in organic chemistry. At room temperature sodium sulfate crystallizes in the orthorhombic system; around 250°C there is a transition to the hexagonal system. The decahydrate crystallizes in the monoclinic form. Glauber's salt is named for the German chemist Johann Rudolf Glauber (1604–68).

sodium sulfide (Na_2S) A yellow-red solid formed by the reduction of sodium sulfate using carbon (or carbon monoxide or hydrogen) at a high temperature. It is a corrosive material and deliquesces to release hydrogen sulfide. Sodium sulfide forms hydrates containing 4½, 5, and 9 H_2O. In aqueous solution it is readily hydrolyzed, the solution being strongly alkaline.

sodium sulfite (Na_2SO_3) A white solid formed by reacting the exact amount of sulfur(IV) oxide with either sodium carbonate or sodium hydroxide. It is readily soluble in water and crystallizes as colorless crystals of the heptahydrate ($Na_2SO_3.7H_2O$). When treated with dilute mineral acids, sulfur(IV) oxide is evolved. At high temperatures, sodium sulfite undergoes thermal decomposition to give sodium sulfate and sodium sulfide.

sodium superoxide (**sodium dioxide;** NaO_2) A pale yellow solid obtained by heating sodium peroxide in oxygen at 490°C. It reacts with water to give hydrogen peroxide, sodium hydroxide solution, and oxygen. Commercial sodium peroxide contains 10% sodium superoxide.

sodium thiosulfate(IV) ($Na_2S_2O_3$) A white solid prepared either by boiling sodium sulfite with flowers of sulfur or by passing sulfur(IV) oxide into a suspension of sulfur in boiling sodium hydroxide. Sodium thiosulfate is readily soluble in water and crystallizes as large colorless crystals of the pentahydrate ($Na_2S_2O_3.5H_2O$). It reacts with dilute acids to give sulfur and sulfur(IV) oxide. It is used in photography as 'hypo' and industrially as an antichlor. In volumetric analysis, solutions of sodium thiosulfate are usually prepared from the pentahydrate. On heating it disproportionates to give sodium sulfate and sodium sulfide.

soft iron Iron that has a low carbon content and is unable to form permanent magnets.

soft soap A liquid soap made by saponification with potassium hydroxide (rather than sodium hydroxide).

soft water *See* hardness (of water).

sol A COLLOID consisting of solid particles distributed in a liquid medium. A wide variety of sols are known; the colors often depend markedly on the particle size. The term *aerosol* is used for solid or liquid phases dispersed in a gaseous medium.

solder /**sod**-er/ An alloy used in joining metals. The molten solder wets the surfaces to be joined, without melting them, and solidifies on cooling to form a hard joint. The surfaces must be clean and free of oxide and a *flux* is often used to achieve this. *Soft solders* consist of lead with up to 60% tin and melt in the range 183–250°C. Soft-soldering is used, for example, in plumbing and making electrical contacts. *Brazing solders* are copper–zinc alloys that have higher melting points and produce stronger joints than soft solders; silver can be added to produce *silver solders*.

solid The state of matter in which the particles occupy fixed positions, giving the substance a definite shape. The particles are held in these positions by bonds. Three kinds of attraction fix the positions of the particles: ionic, covalent, and intermolecular. Since these bonds act over short distances the particles in solids are packed closely together. The strengths of these three types of bonds are different and so, therefore, are the mechanical properties of different solids.

solid solution A solid composed of two or more substances mixed together at the molecular level. Atoms, ions, or molecules of one component in the crystal are at lattice positions normally occupied by the other component. Certain alloys are solid solutions of one metal in another. Isomorphic salts can also sometimes form solid solutions, as in the case of crystalline alums.

solubility /sol-yŭ-**bil**-ă-tee/ The amount of one substance that could dissolve in another to form a saturated solution under specified conditions of temperature and pressure. Solubilities are stated as moles of solute per 100 grams of solvent, or as mass of solute per unit volume of solvent.

solubility product Symbol: K_s If an ionic solid is in contact with its saturated solution, there is a dynamic equilibrium between solid and solution:

$$AB(s) \rightleftharpoons A^+(aq) + B^-(aq)$$

The equilibrium constant for this is given by

$$[A^+][B^-]/[AB]$$

The concentration of undissolved solid [AB] is also constant, so

$$K_s = [A^+][B^-]$$

K_s is the solubility product of the salt (at a given temperature). For a salt A_2B_3, for instance:

$$K_s = [A^+]^2[B^-]^3, \text{etc.}$$

Solubility products are meaningful only for sparingly soluble salts. If the product of ions exceeds the solubility product, precipitation occurs.

solute /**sol**-yoot, **soh**-loot/ A material that is dissolved in a solvent to form a solution.

solution A liquid system of two or more species that are intimately dispersed within each other at a molecular level. The system is therefore totally homogeneous. The major component is called the solvent (generally liquid in the pure state) and the minor component is called the solute (gas, liquid, or solid).

The process occurs because of a direct intermolecular interaction of the solvent with the ions or molecules of the solute. This interaction is called solvation. Part of the energy of this interaction appears as a change in temperature on dissolution. *See also* solid solution; solubility.

solvation /sol-**vay**-shŏn/ The attraction of a solute species (e.g. an ion) for molecules of solvent. In water, for example, a positive ion will be surrounded by water molecules, which tend to associate around the ion because of attraction between the positive charge of the ion, and the negative part of the polar water molecule. The energy of this solvation (hydration in the case of water) is the 'force' needed to overcome the attraction between positive and negative ions when an ionic solid dissolves. The attraction of the dissolved ion for solvent molecules may extend for several layers. In the case of transition metal elements, ions may also form complexes by coordination to the nearest layer of molecules.

Solvay process /**sol**-vay/ (**ammonia–soda process**) An industrial process for making sodium carbonate. The raw materials are calcium carbonate and sodium chloride (with ammonia). The calcium carbonate is heated:

$$CaCO_3 \rightarrow CaO + CO_2$$

Carbon dioxide is bubbled into a solution of sodium chloride saturated with ammonia, precipitating sodium hydrogencarbonate and leaving ammonium chloride in solution. The sodium hydrogencarbonate is then heated:

$$2NaHCO_3 \rightarrow Na_2CO_3 + H_2O + CO_2$$

The ammonia is regenerated by heating the ammonium chloride with the calcium oxide:

$$2NH_4Cl + CaO \rightarrow CaCl_2 + 2NH_3 + H_2O$$

solvent /**sol**-věnt/ A liquid capable of dissolving other materials (solids, liquids, or gases) to form a solution. The solvent is generally the major component of the solution. Solvents can be divided into classes, the most important being:

Polar. A solvent in which the molecules possess a moderate to high dipole moment and in which polar and ionic compounds are easily soluble. Polar solvents are usually poor solvents for non-polar compounds. For example, water is a good solvent for many ionic species, such as sodium chloride or potassium nitrate, and polar molecules, such as the sugars, but does not dissolve paraffin wax.

Non-polar. A solvent in which the molecules do not possess a permanent dipole moment and consequently will solvate non-polar species in preference to polar species. For example, benzene and tetrachloromethane are good solvents for iodine and paraffin wax, but do not dissolve sodium chloride.

Amphiprotic. A solvent which undergoes self-ionization and can act both as a proton donator and as an acceptor. Water is a good example and ionizes according to:

$$2H_2O = H_3O^+ + OH^-$$

Aprotic. A solvent which can neither accept nor yield protons. An aprotic solvent is therefore the opposite to an amphiprotic solvent.

solvent extraction (**liquid–liquid extraction**) A method of removing a substance from solution by shaking it with and dissolving it in another (better) solvent that is imiscible with the original solvent.

solvent naphtha *See* naphtha.

solvolysis /sol-**vol**-ă-sis/ A reaction between a compound and the solvent in which it is dissolved. *See also* hydrolysis.

sorbitol /**sor**-bă-tol, -tohl/ ($HOCH_2(CHOH)_4CH_2OH$) A hexahydric alcohol that occurs in rose hips and rowan berries. It can be synthesized by the reduction of glucose. Sorbitol is used to make vitamin C (ascorbic acid) and surfactants. It is also used in medicines and as a sweetener (particularly in foods for diabetics). It is an isomer of mannitol.

sorption /**sorp**-shŏn/ Absorption of gases by solids.

specific /spě-**sif**-ik/ Denoting a physical quantity per unit mass. For example, volume (V) per unit mass (m) is called specific volume:

$$V = V_m$$

In certain physical quantities the term does not have this meaning: for example, *specific gravity* is more properly called relative density.

specific gravity *See* relative density.

specific heat capacity Symbol: c The amount of heat needed to raise a unit mass of a substance by one degree. It is measured in joules per kilogram per Kelvin ($J\ kg^{-1}K^{-1}$) in SI units.

specific rotatory power Symbol: α_m The rotation of plane-polarized light in degrees produced by a 10 cm length of solution containing 1 g of a given substance per milliliter of stated solvent. The specific rotatory power is a measure of the optical activity of substances in solution. It is measured at 20°C using the D-line of sodium.

spectra *See* spectrum.

spectral line A particular wavelength of electromagnetic radiation emitted or absorbed by an atom, ion, or molecule. *See* line spectrum.

spectral series A group of related lines in the absorption or emission spectrum of a substance. The lines in a spectral series occur when the transitions all occur between one particular energy level and a set of different levels. *See also* Bohr theory.

spectrograph /**spek**-trŏ-graf, -grahf/ An instrument for producing a photographic record of a spectrum.

spectrographic analysis /spek-trŏ-**graf**-ik/ A method of analysis in which the sample is excited electrically (by an arc or spark) and emits radiation characteristic of its component atoms. This radiation is passed through a slit, dispersed by a prism or a grating, and recorded as a spectrum, either photographically or photoelectrically. The photographic method was widely used for qualitative and semiquantitative work but photoelectric detection also allows wide quantitative application.

spectrometer /spek-**trom**-ĕ-ter/ **1.** An instrument for examining the different wavelengths present in electromagnetic radiation. Typically, spectrometers have a source of radiation, which is collimated by a system of lenses and/or slits. The radiation is dispersed by a prism or grating, and recorded photographically or by a photocell. There are many types for producing and investigating spectra over the whole range of the electromagnetic spectrum. Often spectrometers are called *spectroscopes*.
2. Any of various other instruments for analyzing the energies, masses, etc., of particles. *See* mass spectrometer.

spectrophotometer /spek-troh-foh-**tom**-ĕ-ter/ A form of spectrometer able to measure the intensity of radiation at different wavelengths in a spectrum, usually in the visible, infrared, or ultraviolet regions.

spectroscope /**spek**-trŏ-skohp/ An instrument for examining the different wavelengths present in electromagnetic radiation. *See also* spectrometer.

spectroscopy /spek-**tros**-kŏ-pee/ **1.** The production and analysis of spectra. There are many spectroscopic techniques designed for investigating the electromagnetic radiation emitted or absorbed by substances. Spectroscopy, in various forms, is used for analysis of mixtures, for identifying and determining the structures of chemical compounds, and for investigating energy levels in atoms, ions, and molecules. In the visible and longer wavelength ultraviolet, transitions correspond to electronic energy levels in atoms and molecules. The shorter wavelength ultraviolet corresponds to transitions in ions. In the X-ray region, transitions in the inner shells of atoms or ions are involved. The infrared region corresponds to vibrational changes in molecules, with rotational changes at longer wavelengths.
2. Any of various techniques for analysing the energy spectra of beams of particles or for determining mass spectra.

spectrum /**spek**-trŭm/ (*pl.* **spectra**) **1.** A range of electromagnetic radiation emitted or absorbed by a substance under particular circumstances. In an *emission spectrum*, light or other radiation emitted by the body is analyzed to determine the particular wavelengths produced. The emission of radiation may be induced by a variety of methods; for example, by high temperature, bombardment by electrons, absorption of higher-frequency radiation, etc. In an *absorption spectrum* a continuous flow of radiation is passed through the sample. The radiation is then analyzed to determine which wavelengths are absorbed. *See also* band spectrum; continuous spectrum; line spectrum.
2. In general, any distribution of a property. For instance, a beam of particles may have a spectrum of energies. A beam of ions may have a mass spectrum (the distribution of masses of ions). *See* mass spectrometer.

spelter Commercial zinc containing about 3% impurities, mainly lead.

spin A property of certain elementary particles whereby the particle acts as if it were spinning on an axis; i.e. it has an angular momentum. Such particles also have a magnetic moment. In a magnetic field the spins line up at an angle to the field direction and precess around this direction. Certain definite orientations to the field direction occur such that $m_s h/2\pi$ is the component of angular momentum along this direction. m_s is the spin quantum number, and for an electron it has values +1/2 and −1/2. *h* is the Planck constant.

spontaneous combustion The self-ignition of a substance that has a low ignition temperature. It occurs when slow oxidation of the substance (such as a heap of damp straw or oily rags) builds up sufficient heat for ignition to take place.

square-planar *See* complex.

stabilization energy The difference in energy between the delocalized structure and the conventional structure for a compound. For example, the stabilization energy of benzene is 150 kJ per mole, which represents the difference in energy between a Kekulé structure and the delocalized structure: the delocalized form being of lower energy is therefore more stable. The stabilization energy can be determined by comparing the experimental value for the heat of hydrogenation of benzene with that calculated for Kekulé benzene from bond-energy data.

stabilizer A substance added to prevent chemical change (i.e. a negative catalyst).

staggered conformation *See* conformation.

stainless steel *See* steel.

stalactites and stalagmites /**stal**-ăk-tÿts **stal**-ăg-mÿts, stă-**lak**-tÿts stă-**lag**-mÿys/ Pillars of calcium carbonate found hanging from the ceiling (stalactites) and standing on the floor (stalagmites) in limestone caverns. They form when water containing dissolved carbon dioxide forms weak carbonic acid, which is able to react with limestone rock (calcium carbonate) to give a solution of calcium hydrogencarbonate. The solution accumulates in drops on the roof of caves and as it evaporates, the reaction is reversed causing calcium carbonate to be deposited. Over a very long period, the deposits build up to give a stalactite hanging from the cavern roof. Water that drips onto the floor from the tip of the stalactite also evaporates to leave a deposit of calcium carbonate. This gradually grows to form a stalagmite. Stalagmites and stalactites can eventually meet to form a single rock pillar.

standard cell A voltaic cell whose e.m.f. is used as a standard. *See* Clark cell; Weston cadmium cell.

standard electrode A half cell used for measuring electrode potentials. The hydrogen electrode is the basic standard but, in practice, calomel electrodes are usually used.

standard pressure An internationally agreed value; a barometric height of 760 mmHg at 0°C; 101 325 Pa (approximately 100 kPa). This is sometimes called the *atmosphere* (used as a unit of pressure). The *bar*, used mainly when discussing the weather, is 100 kPa exactly.

standard solution A solution that contains a known weight of the reagent in a definite volume of solution. A standard flask or volumetric flask is used for this purpose. The solutions may be prepared by direct weighing for PRIMARY STANDARDS. If the reagent is not available in a pure form or is deliquescent the solution must be standardized by titration against another known standard solution.

standard state The standard conditions used as a reference system in thermodynamics: pressure is 101 325 Pa; temperature is 25°C (298.15 K); concentration is 1 mol. The substance under investigation must also be pure and in its usual state, given the above conditions.

standard temperature An internationally agreed value for which many measurements are quoted. It is the melting temperature of water, 0°C (273.15 K). *See also* STP.

stannane /**stan**-ayn/ (**tin(IV) hydride;** SnH_4) A colorless poisonous gas prepared by the action of lithium tetrahydroaluminate(III) on tin(II) chloride. It is unstable, decomposing immediately at 150°C. Stannane acts as a reducing agent.

stannate /**stan**-ayt/ *See* tin.

stannic chloride /**stan**-ik/ *See* tin(IV) chloride.

stannic compounds Compounds of tin(IV).

stannite /**stan**-ÿt/ *See* tin.

stannous compounds /**stan**-ŭs/ Compounds of tin(II).

starch A polysaccharide that occurs exclusively in plants. Starches are extracted commercially from maize, wheat, barley, rice, potatoes, and sorghum. The starches are storage reservoirs for plants; they can be broken down by enzymes to simple sugars and then metabolized to supply energy needs. Starch is a dietary component of animals.

Starch is not a single molecule but a mixture of amylose (water-soluble, blue color with iodine) and amylopectin (not water-soluble, violet color with iodine). The composition is amylose 10–20%, amylopectin 80–90%.

Stark effect The splitting of atomic spectral lines because of the presence of an external electric field. This phenomenon was discovered by the German physicist Johannes Stark (1874–1957) in 1913.

states of matter The three physical conditions in which substances occur: solid, liquid, and gas. The addition or removal of energy (usually in the form of heat) enables one state to be converted into another.

The major distinctions between the states of matter depend on the kinetic energies of their particles and the distances between them. In solids, the particles have low kinetic energy and are closely packed; in gases they have high kinetic energy and are very loosely packed; kinetic energy and separation of particles in liquids are intermediate.

Solids have fixed shapes and volumes, i.e. they do not flow, like liquids and gases, and they are difficult to compress. In solids the atoms or molecules occupy fixed positions in space. In most cases there is a regular pattern of atoms – the solid is crystalline.

Liquids have fixed volumes (i.e. low compressibility) but flow to take up the shape of the container. The atoms or molecules move about at random, but they are quite close to one another and the motion is hindered.

Gases have no fixed shape or volume. They expand spontaneously to fill the container and are easily compressed. The molecules have almost free random motion.

A plasma is sometimes considered to be a fourth state of matter.

stationary phase *See* chromatography.

statistical mechanics The branch of physics and chemistry associated with using statistical methods to relate the properties of a large number of microscopic particles, such as atoms and molecules, to the macroscopic properties of the system.

steam distillation A method of isolating or purifying substances by exploiting Dalton's law of partial pressures to lower the boiling point of the mixture. When two immiscible liquids are distilled, the boiling point will be lower than that of the more volatile component and consequently will be below 100°C if one component is water. The method is particularly useful for recovering materials from tarry mixtures.

steam reforming The conversion of a methane–steam mixture at 900°C with a nickel catalyst into a mixture of carbon monoxide and hydrogen. The mixture of

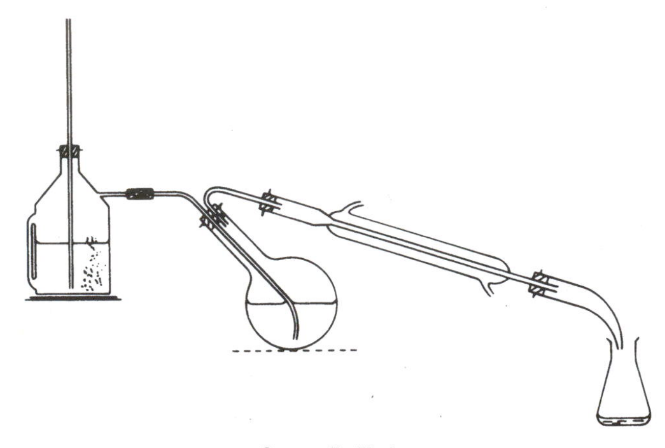

Steam distillation

gases (synthesis gas) provides a starting material in a number of processes, e.g. the manufacture of methanol.

stearate /**stee**-ă-rayt/ A salt or ester of stearic acid; an octadecanoate.

stearic acid /stee-**a**-rik/ *See* octadecanoic acid.

steatite /**stee**-ă-tÿt/ *See* soapstone.

steel An alloy of iron with small amounts of carbon and, often, other metals. *Carbon steels* contain 0.05–1.5% carbon – the more carbon, the harder the steel. *Alloy steels* contain carbon but also contain small amounts of certain other elements; for example chromium, manganese, and vanadium. Their properties depend on the composition. *Stainless steels*, for instance, are resistant to corrosion. The main nonferrous constituent is chromium (10–25%) with up to 0.7% carbon.

step An elementary stage in a chemical reaction, in which energy may be transferred from one molecule to another, bonds may be broken or formed, or electrons may be transferred. For example, in the reaction between hypochlorite ions and iodide ions in aqueous solution there are three distinct steps:

Step 1

$$OCl^-(aq) + H_2O(l) \rightarrow HOCl(aq) + OH^-(aq)$$

Step 2

$$I^-(aq) + HOCl(aq) \rightarrow HOI(aq) + Cl^-(aq)$$

Step 3

$$OH^-(aq) + HOI(aq) \rightarrow H_2O(l) + OI^-(aq)$$

steradian /sti-**ray**-dee-ăn/ Symbol: sr The SI unit of solid angle. The surface of a sphere, for example, subtends a solid angle of 4π at its center. The solid angle of a cone is the area intercepted by the cone on the surface of a sphere of unit radius.

stereochemistry /ste-ree-oh-**kem**-iss-tree, steer-ee-/ The branch of chemistry concerned with the shapes of molecules and the way these affect the chemical properties.

stereoisomerism /ste-ree-oh-ÿ-**som**-ĕ-riz-ăm, steer-ee-/ *See* isomerism.

stereospecific /ste-ree-oh-spĕ-**sif**-ik, steer-ee-/ Describing a chemical reaction involving an asymmetric atom (usually carbon) that results in the formation of only one geometric isomer. *See* isomerism; optical activity.

stereospecific polymer *See* polymerization.

steric effect /**ste**-rik, **steer**-ik/ An effect in which the shape of a molecule influences its reactions. A particular example occurs in molecules containing large groups, which hinder the approach of a reactant (*steric hindrance*).

steric hindrance *See* steric effect.

steroid /**steer**-oid/ Any member of a group of biochemical compounds having a complex basic ring structure. Examples are corticosteroid hormones (produced by the adrenal gland), sex hormones (progesterone, androgens, and estrogens), bile acids, and sterols (such as cholesterol).

sterol /**steer**-ol, -ohl/ A steroid with long aliphatic side chains (8–10 carbons) and at least one hydroxyl group. They are lipid-soluble and often occur in membranes. Examples are cholesterol and ergosterol.

still An apparatus for distillation.

stoichiometric coefficient /stoi-kee-ŏ-**met**-rik/ *See* chemical equation.

stoichiometry /stoi-kee-**om**-ĕ-tree/ The proportions in which elements form compounds. A *stoichiometric compound* is one in which the atoms have combined in small whole numbers.

storage battery *See* accumulator.

STP (NTP) Standard temperature and pressure. Conditions used internationally when measuring quantities that vary with both pressure and temperature (such as the density of a gas). The values are 101 325 Pa (approximately 100 kPa) and 0°C (273.15 K). *See also* standard pressure; standard temperature.

straight chain *See* chain.

Strecker synthesis /**strek**-er/ A method of synthesizing amino acids. Hydrogen cyanide forms an addition product (cyanohydrin) with aldehydes:

$$RCHO + HCN \rightarrow RCH(CN)OH$$

With ammonia further substitution occurs:

$$RCH(CN)OH + NH_3 \rightarrow RCH(CN)(NH_2) + H_2O$$

Acid hydrolysis of the cyanide group then produces the amino acid

$$RCH(CN)(NH_2) \rightarrow RCH(COOH)(NH_2)$$

The method is a general synthesis for α-amino acids (with the $-NH_2$ and $-COOH$ groups on the same carbon atom).

strong acid An acid that is almost completely dissociated into its component ions in solution. *Compare* weak acid.

strong base A base that is completely or almost completely dissociated into its component ions in solution.

strontia /**stron**-shee-ă/ *See* strontium oxide.

strontium /**stron**-shee-ŭm, -tee-/ A soft low-melting reactive metal; the fourth member of group 2 of the periodic table and a typical alkaline-earth element. The electronic configuration is that of krypton with two additional outer 5s electrons. Strontium and barium are both of low abundance in the Earth's crust, strontium occurring as strontianite ($SrCO_3$) and celestine ($SrSO_4$).

The element is produced industrially by roasting the carbonate to give the oxide (800°C) and then reducing with aluminum,

$$3SrO + 2Al \rightarrow Al_2O_3 + 2Sr$$

Strontium has a low ionization potential, is large, and is therefore very electropositive. The chemistry of strontium metal is therefore characterized by high reactivity of the metal. The properties of strontium fall into sequence with other alkaline earths. Thus it reacts directly with oxygen, nitrogen, sulfur, the halogens, and hydrogen to form respectively the oxide SrO, nitride Sr_3N_2, sulfide SrS, halides SrX_2, and hydride SrH_2, all of which are largely ionic in character. The oxide SrO and the metal react readily with water to form the hydroxide $Sr(OH)_2$, which is basic and mid-way between $Ca(OH)_2$ and $Ba(OH)_2$ in solubility. The carbonate and sulfate are both insoluble.

As the metal is very electropositive the salts are never much hydrolyzed in solution and the ions are largely solvated as $[Sr(H_2O)_6]^{2+}$.

Symbol: Sr; m.p. 769°C; b.p. 1384°C; r.d. 2.54 (20°C); p.n. 38; r.a.m. 87.62.

strontium bicarbonate *See* strontium hydrogencarbonate.

strontium carbonate ($SrCO_3$) A white insoluble solid that occurs naturally as the mineral strontianite. It can be prepared by passing carbon dioxide over strontium oxide or hydroxide or by passing the gas through a solution of a strontium salt. Strontium carbonate is then formed as a precipitate. It is used as a slagging agent in certain metal furnaces and to produce a red flame in fireworks.

strontium chloride ($SrCl_2$) A white solid obtained directly from strontium and chlorine or by passing chlorine over heated strontium oxide. It is used in fireworks to give a red flame. The hexahydrate ($SrCl_2.6H_2O$) is prepared by the neutralization of hydrochloric acid by strontium hydroxide or carbonate.

strontium hydrogencarbonate (**strontium bicarbonate**; $Sr(HCO_3)_2$) A compound present in solutions formed by the action of carbon dioxide on a suspension in cold water of strontium carbonate, to which it reverts on heating:

$$SrCO_3 + CO_2 + H_2O = Sr(HCO_3)_2$$

strontium hydroxide ($Sr(OH)_2$) A solid that normally occurs as the octahydrate ($Sr(OH)_2.8H_2O$), which is prepared by crystallizing an aqueous solution of strontium oxide. Strontium hydroxide is readily soluble in water and aqueous solutions are strongly basic. It is used in the purification of sugar.

strontium nitrate ($Sr(NO_3)_2$) A colorless crystalline compound, often used in pyrotechnics to produce a bright red flame.

strontium oxide (**strontia**; SrO) A grayish-white powder prepared by the decomposition of strontium carbonate, hydroxide, or nitrate. Strontium oxide is soluble in water, forming an alkaline solution because of the attraction between the oxide ions and protons from the water:

$$O^{2-} + H_2O \rightarrow 2OH^-$$

strontium sulfate ($SrSO_4$) A white sparingly soluble salt that occurs naturally as the mineral celestine. It can be prepared by dissolving strontium oxide, hydroxide, or carbonate in sulfuric acid. It is used as a substitute for barium sulfate in paints.

structural formula The formula of a compound showing the numbers and types of atoms present, together with the way in which these are arranged in the molecule. Often this can be done by grouping the atoms, as in the structural formula of ethanoic acid $CH_3.CO.OH$. *Compare* empirical formula; molecular formula.

structural isomerism *See* isomerism.

styrene /**stÿ**-reen/ *See* phenylethene.

sublimate /**sub**-lă-mayt/ A solid formed by sublimation.

sublimation /sub-lă-**may**-shŏn/ The conversion of a solid into a vapor without the solid first melting. For instance (at standard pressure) iodine, solid carbon dioxide, and ammonium chloride sublime. At certain conditions of external pressure and temperature an equilibrium can be established between the solid phase and vapor phase.

sub-shell A subdivision of an electron shell. It is a division of the orbitals that make up a shell into sets of orbitals, which are degenerate (i.e. have the same energy) in the free atom. For example, in the third shell or M-shell there are the 3s, 3p, and 3d sub-shells of which the latter two are 3- and 5-degenerate respectively.

substituent /sub-**stich**-û-ĕnt/ An atom or group substituted for another in a compound. Often the term is used for groups that have replaced hydrogen in organic compounds; for example, in benzene derivatives.

substitution reaction A reaction in which an atom or group of atoms in an organic molecule is replaced by another atom or group. The substitution of a hydrogen atom in an alkane by a chlorine atom is an example. Substitution reactions fall into three major classes depending upon the nature of the attacking substituent.

Nucleophilic substitution: the attacking substituent is a nucleophile. Such reactions are very common with alcohols and halogen compounds, in which the electron-deficient carbon atom attracts the nucleophile and the leaving group readily exists alone. Examples are the hydrolysis of a haloalkane and the chlorination of an alcohol:

$$C_2H_5Cl + OH^- \rightarrow C_2H_5OH + Cl^-$$
$$C_2H_5OH + HCl \rightarrow C_2H_5Cl + H_2O$$

Electrophilic substitution: the attacking substituent is an electrophile. Such reactions are common in aromatic compounds, in which the electron-rich ring attracts the electrophile. The nitration of benzene in which the electrophile is NO_2^+ is an example:

$$C_6H_6 + NO_2^+ \rightarrow C_6H_5NO_2 + H^+$$

Free-radical substitution: a free radical is the attacking substituent. Such reactions can be used with compounds that are inert to either nucleophiles or electrophiles, for instance the halogenation of an alkane:

$$CH_4 + Cl_2 \rightarrow CH_3Cl + HCl$$

The term 'substitution' is very general and several reactions that can be considered as substitutions are more normally given special names (e.g. esterification, hydrolysis, and nitration). *See also* electrophilic substitution; nucleophilic substitution.

substrate /**sub**-strayt/ A material that is acted on by a catalyst.

succinic acid /suk-**sin**-ik/ *See* butanedioic acid.

sucrose /**soo**-krohs/ (**cane sugar**; $C_{12}H_{22}O_{11}$) A sugar that occurs in many plants. It is extracted commercially from sugar cane and sugar beet. Sucrose is a disaccharide formed from a glucose unit and a fructose unit. It is hydrolyzed to a mixture of fructose and glucose by the enzyme invertase. Since this mixture has a different optical rotation (levorotatory) than the original sucrose, the mixture is called *invert sugar*.

sugar (**saccharide**) One of a class of sweet-tasting simple CARBOHYDRATES. Sugars have molecules consisting of a chain of carbon atoms with –OH groups attached, and either an aldehyde or ketone group. They can exist in a chain form or in a ring formed by reaction of the ketone or aldehyde group with an OH group. Monosaccharides are simple sugars that cannot be hydrolyzed to sugars with fewer carbon

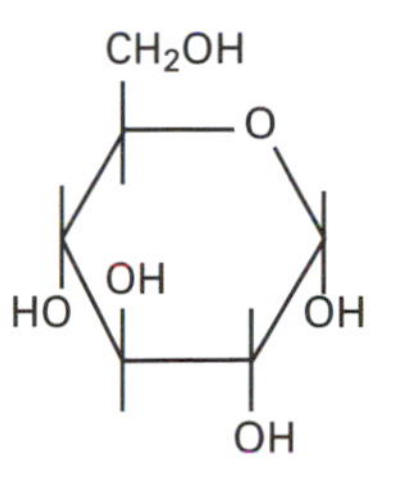

glucose – a pyranose ring

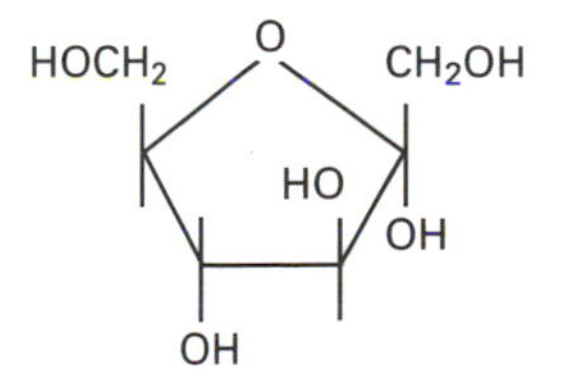

fructose – in a furanose ring form

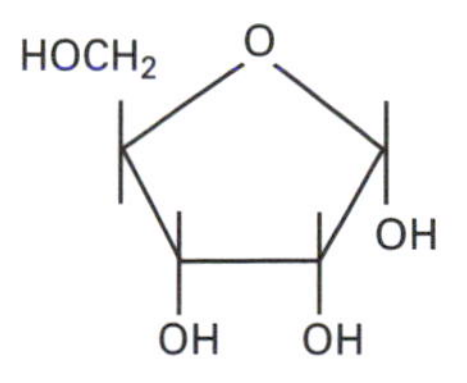

ribose – a furanose ring

Sugars

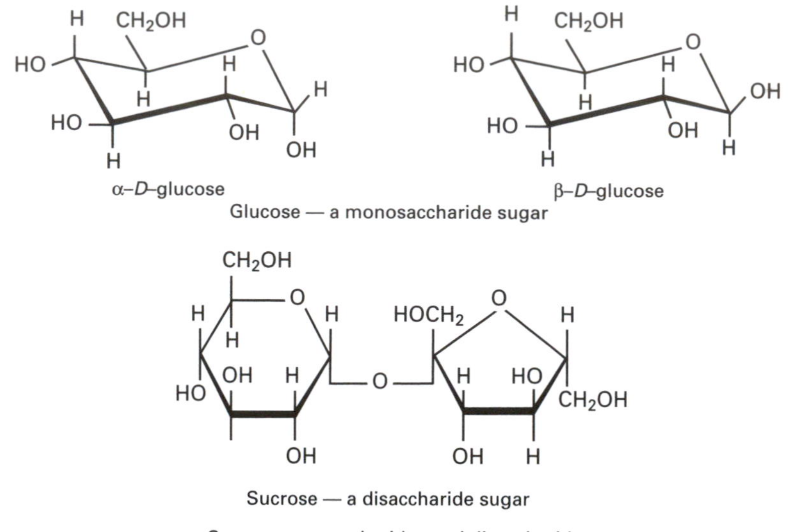

Glucose — a monosaccharide sugar

Sucrose — a disaccharide sugar

Sugar: monosaccharides and disaccharides

atoms. Two or more monosaccharide units can be linked in *disaccharides*, *trisaccharides*, etc.

Monosaccharides are also classified according to the number of carbon atoms: a *pentose* has five carbon atoms and a *hexose* six. Monosaccharides with aldehyde groups are *aldoses*; those with ketone groups are *ketoses*. Thus, an *aldohexose* is a hexose with an aldehyde group; a *ketopentose* is a pentose with a ketone group, etc.

The ring forms of monosaccharides are derived by reaction of the aldehyde or ketone group with one of the carbons at the other end of the chain. It is possible to have a six-membered (*pyranose*) ring or a five-membered (*furanose*) ring. *See also* polysaccharide.

sulfa drugs /**sul**-fă/ *See* sulfonamide.

sulfate /**sul**-fayt/ A salt or ester of sulfuric(VI) acid.

sulfide /**sul**-fÿd/ **1.** A compound of sulfur with a more electropositive element. Simple sulfides contain the ion S^{2-}. *Polysulfides* can also be formed containing ions with chains of sulfur atoms (S_x^{2-}). **2.** *See* thioether.

sulfinate /**sul**-fă-nayt, -nit/ *See* dithionite.

sulfinic acid /sul-**fin**-ik/ *See* dithionous acid.

sulfite /**sul**-fÿt/ A salt or ester of sulfurous acid.

sulfonamide /sul-**fon**-ă-mÿd/ A type of organic compound with the general formula $R.SO_2.NH_2$. Sulfonamides, which are amides of sulfonic acids, are active against bacteria, and some are used in pharmaceuticals ('sulfa drugs').

sulfonate /**sul**-fŏ-nayt/ A salt or ester of sulfonic acid. *See* sulfonic acid.

sulfonation /sul-fŏ-**nay**-shŏn/ A reaction introducing the $-SO_2OH$ (sulfonic acid) group into an organic compound. Sulfonation of aromatic compounds is usually accomplished by refluxing with concentrated sulfuric acid for several hours. The attacking species is SO_3 (sulfur(VI) oxide) and the reaction is an example of electrophilic substitution.

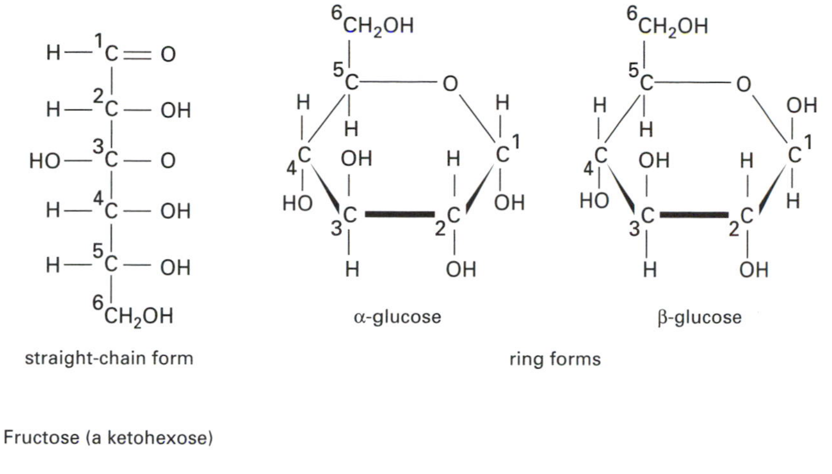

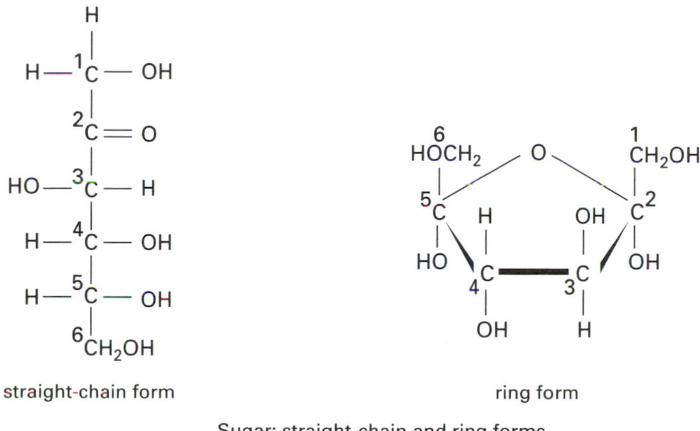

Sugar: straight-chain and ring forms

sulfonic acid /sul-**fon**-ik/ A type of organic compound containing the $-SO_2.OH$ group. The simplest example is benzenesulfonic acid ($C_6H_5SO_2OH$). Sulfonic acids are strong acids. Electrophilic substitution can introduce other groups onto the benzene ring; the $-SO_2.OH$ group directs substituents into the 3-position.

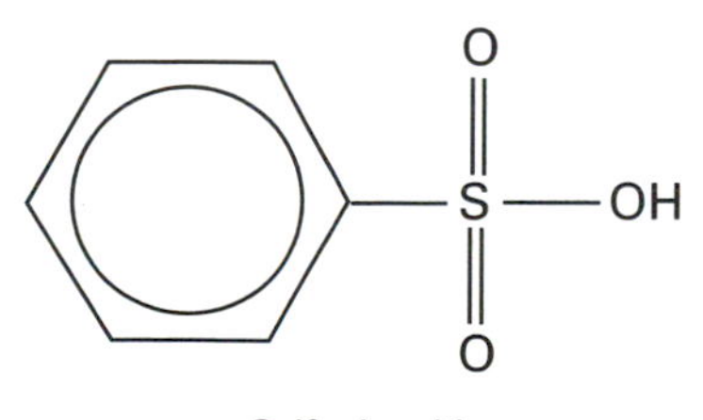

Sulfonic acid

sulfonium compound /sul-**foh**-nee-ŭm/ An organic compound of general formula R_3SX, where R is an organic radical and X is an electronegative radical or element; it contains the ion R_3S^+. An example is diethylmethylsulfonium chloride, $(C_2H_5)_2CH_3.S^+Cl^-$, made by reacting diethyl sulfide with chloromethane.

sulfoxide /sul-**foks**-ȳd, -id/ An organic compound of general formula RSOR′, where R and R′ are organic radicals. An example is dimethyl sulfoxide, $(CH_3)_2SO$, commonly used as a solvent.

sulfur /**sul**-fer/ A low melting non-metallic solid, yellow colored in its common forms; the second member of group 16 of the periodic table. It has the electronic configuration $[Ne]3s^2 3p^4$.

Sulfur occurs in the elemental form in Sicily and some southern states of the USA, and in large quantities in combined forms such as sulfide ores (FeS_2) and sulfate rocks ($CaSO_4$). It forms about 0.5% of the Earth's crust. Elemental sulfur is extracted commercially by the *Frasch process*, named for the German–American chemist Herman Frasch (1851–1914), in which superheated steam is forced down the outer of three concentric tubes leading into the deposit. This causes the sulfur to melt; then air is blown down the central tube causing molten sulfur to be forced out of the well to be collected and cooled at the well head. The large amount of sulfur obtained from smelting of sulfide ores or roasting of sulfates is not obtained as elemental sulfur but is used directly as SO_2/SO_3 for conversion to sulfuric acid (contact process).

Sulfur exhibits allotropy and its structure in all phases is quite complex. The common crystalline modification, *rhombic* sulfur, is in equilibrium with a *triclinic* modification above 96°C. Both have structures based on S_8-rings but the crystals are quite different. If molten sulfur is poured into water a dark red 'plastic' form is obtained in a semielastic form. The structure appears to be a helical chain of S atoms. Selenium and tellurium both have a gray 'metal-like' modification but sulfur does *not* have this form.

Sulfur reacts directly with hydrogen but the position of the equilibrium at normal temperatures precludes its use as a preparative method for hydrogen sulfide. The element burns readily in oxygen with a characteristic blue flame to form sulfur(IV) oxide (SO_2) and traces of sulfur(VI) oxide (SO_3). Sulfur also forms a large number of oxy-acid species, some containing peroxy-groups and some with two or more S-atoms. It forms four distinct fluorides, S_2F_2, SF_4, SF_6, S_2F_{10}; three chlorides, S_2Cl_2, SCl_2, SCl_4; a bromide S_2Br_2, but no iodide. Apart from SF_6, which is surprisingly stable and inert, the halides are generally susceptible to hydrolysis to give commonly SO_2, sometimes H_2S, and the hydrogen halide. The halides SF_4, S_2Cl_2, and SCl_2 are frequently used as selective fluorinating agents and chlorinating agents in organic chemistry. The chlorides are also industrially important for hardening rubbers. Sulfur hexafluoride may be prepared by direct reaction of the elements but SF_4 is prepared by fluorinating SCl_2 with NaF. In a limited supply of chlorine, sulfur combines to give sulfur monochloride, S_2Cl_2; in excess chlorine at higher temperatures the dichloride is formed,

$$2S + Cl_2 \rightarrow S_2Cl_2$$

$$S_2Cl_2 + Cl_2 \rightarrow 2SCl_2$$

In addition to these binary halides, sulfur forms two groups of oxyhalides; the sulfur dihalide oxides (thionyl halides), SOX_2 (F, Cl, and Br, but not I), and the sulfur dihalide dioxides (sulfuryl halides), SO_2X_2 (F and Cl, but not Br and I).

Most metals react with sulfur to form sulfides of which there are a large number of structural types. With electropositive elements of groups I and II the sulfides are largely ionic (S^{2-}) in the solid phase. Hydrolysis occurs in solution thus:

$$S^{2-} + H_2O \rightarrow SH^- + OH^-$$

Most transition metals form sulfides, which although formally treated as FeS, CoS, NiS, etc., are largely covalent and frequently non-stoichiometric.

In addition to the complexities of the binary compounds with metals, sulfur forms a range of binary compounds with elements such as Si, As, and P, which may be polymeric and generally of rather complex structure, e.g. $(SiS_4)_n$, $(SbS_3)_n$, N_4S_4, As_4S_4, P_4S_3.

Sulfur also forms a wide range of organic sulfur compounds, most of which have the typically revolting smell of H_2S.

Symbol: S; m.p. 112.8°C; b.p. 444.6°C; r.d. 2.07; p.n. 16; r.a.m. 32.066.

sulfur dichloride dioxide /dÿ-**oks**-ÿd, -id/ (**sulfuryl chloride**; SO_2Cl_2) A colorless fuming liquid formed by the reaction of chlorine with sulfur(IV) oxide in sunlight. It is used as a chlorinating agent.

sulfur dichloride oxide (**thionyl chloride**; $SOCl_2$) A colorless fuming liquid formed by passing sulfur(IV) oxide over phosphorus(V) chloride and distilling the mixture obtained. Sulfur dichloride oxide is used in organic chemistry to introduce chlorine atoms (for example, into ethanol to form monochloroethane), the organic product being easily isolated as the other products are gases.

sulfur dioxide /dÿ-**oks**-ÿd, -id/ *See* sulfur(IV) oxide.

sulfuretted hydrogen /sul-fyŭ-**ret**-id/ *See* hydrogen sulfide.

sulfuric(IV) acid /sul-**fyoor**-ik/ *See* sulfurous acid.

sulfuric(VI) acid (**oil of vitriol**; H_2SO_4) A colorless oily liquid manufactured by the CONTACT PROCESS. The concentrated acid is diluted by adding it slowly to water, with careful stirring. Concentrated sulfuric acid acts as an oxidizing agent, giving sulfur(IV) oxide as the main product, and also as a dehydrating agent. The diluted acid acts as a strong dibasic acid, neutralizing bases and reacting with active metals and carbonates to form sulfates.

Sulfuric acid is used in the laboratory to dry gases (except ammonia), to prepare nitric acid and ethene, and to absorb alkenes. In industry, it is used to manufacture fertilizers (e.g. ammonium sulfate), rayon, and detergents, to clean metals, and in vehicle batteries.

sulfur monochloride /mon-ŏ-**klor**-ÿd, -id, -**kloh**-rÿd, -rid/ *See* disulfur dichloride.

sulfurous acid /sul-**fyoor**-ŭs/ (**sulfuric(IV) acid**; H_2SO_3) A weak acid found only in solution, made by passing sulfur(IV) oxide into water. The solution is unstable and smells of sulfur(IV) oxide. It is a reducing agent, converting iron(III) ions to iron(II) ions, chlorine to chloride ions, and orange dichromate(VI) ions to green chromium(III) ions.

sulfur(IV) oxide (**sulfur dioxide**; SO_2) A colorless choking gas prepared by burning sulfur or heating metal sulfides in air, or by treating a sulfite with an acid. It is a powerful reducing agent, used as a bleach. It dissolves in water to form sulfurous acid (sulfuric(IV) acid) and combines with oxygen, in the presence of a catalyst, to form sulfur(VI) oxide. This latter reaction is important in the manufacture of sulfuric(VI) acid.

sulfur(VI) oxide (**sulfur trioxide**; SO_3) A fuming volatile white solid prepared by passing sulfur(IV) oxide and oxygen over hot vanadium(V) oxide (acting as a catalyst) and cooling the product in ice. Sulfur(VI) oxide reacts vigorously with water to form sulfuric acid. *See also* contact process.

sulfur trioxide /trÿ-**oks**-ÿd/ *See* sulfur(VI) oxide.

sulfuryl /**sul**-fŭ-răl, -yŭ-, -reel/ Describing a compound containing the group $=SO_2$.

sulfuryl chloride /**sul**-fŭ-răl, -yŭ-, -reel/ *See* sulfur dichloride dioxide.

superacid /soo-per-**ass**-id/ An acid with a high proton-donating ability. Superacids are substances such as $HF–SbF_3$ and $HSO_3–SbF_5$. Sometimes known as *magic acids*, they are able to produce carbenium ions from some saturated hydrocarbons.

superconduction /soo-per-kŏn-**duk**-shŏn/ The property of zero resistance exhibited by certain metals and metallic compounds at low temperatures. Metals conduct at temperatures close to absolute zero but a number of compounds are now known that conduct at higher (liquid-nitrogen) temperatures. Synthetic organic superconductors have also been produced.

superconductivity /soo-per-kon-duk-**tiv**-ă-tee/ The property of zero electrical resistance exhibited by certain metals and metallic compounds at low temperatures. Metals show superconductivity at temperatures close to absolute zero but a number of ceramic metal oxides are now known that are superconducting at higher (94K or better) temperatures. Synthetic organic superconductors have also been produced.

supercooling /soo-per-**koo**-ling/ The cooling of a liquid at a given pressure to a temperature below its melting temperature at that pressure without solidifying it. The liquid particles lose energy but do not spontaneously fall into the regular geometrical pattern of the solid. A supercooled liquid is in a metastable state and will usually solidify if a small crystal of the solid is introduced to act as a 'seed' for the formation of crystals. As soon as this happens, the temperature returns to the melting temperature until the substance has completely solidified.

superfluidity /soo-per-floo-**id**-ă-tee/ A property of liquid HELIUM at very low temperatures. At 2.186 K liquid helium makes a transition to a superfluid state, which has a high thermal conductivity and flows without friction.

superheating /soo-per-**hee**-ting/ The raising of a liquid's temperature above its boiling temperature, by increasing the pressure.

supernatant /soo-per-**nay**-tănt/ Denoting a clear liquid that lies above a sediment or a precipitate.

superoxide /soo-per-**oks**-ÿd/ An inorganic compound containing the O_2^- ion.

superphosphate /soo-per-**fos**-fayt/ A mixture – mainly calcium hydrogen phosphate and calcium sulfate – used as a fertilizer. It is made from calcium phosphate and sulfuric acid.

supersaturated solution /soo-per-**sach**-ŭ-ray-tit/ *See* saturated solution.

supersaturated vapor *See* saturated vapor.

supplementary units *See* dimensionless units.

supramolecular chemistry /soo-pră-mŏ-**lek**-yŭ-ler/ A branch of chemistry concerned with the synthesis and study of large structures consisting of molecules assembled together in a definite pattern. In a *supramolecule* the molecular units (which are sometimes known as *synthons*) are joined by intermolecular bonds – i.e. by hydrogen bonds or by ionic attractions. A particular interest in supramolecular research is the idea of *self-assembly* – i.e. that the molecules form well-defined structures spontaneously as a result of their geometry and chemical properties. In this way, supramolecular chemistry is 'chemistry beyond molecules'.

There are various different examples of supramolecular structures. For instance, single layers of carboxylic acid molecules held by hydrogen bonds may form two-dimensional crystalline structures with novel electrical properties. Large organic polymeric structures may be formed in which a number of individual chains radiate from a central point or region. These polymers are called *dendrimers* and they have a number of possible applications. Another type of supramolecule is a *helicate*, which has a double helix made of two chains of bipyridyl units held by copper ions along the axis. The structure is analogous to the double helix of DNA. *See also* host–guest chemistry.

supramolecule /soo-pră-**mol**-ĕ-kyool/ *See* supramolecular chemistry.

surfactant /ser-**fak**-tănt/ A substance that lowers surface tension and has properties of wetting, foaming, detergency, dispersion, and emulsification.

suspension A system in which small particles of a solid or liquid are dispersed in a liquid or gas.

symbol, chemical *See* chemical symbol.

syndiotactic polymer *See* polymerization.

syn-isomer /sin ÿ-sŏ-mer/ *See* isomerism.

synthesis /**sin**-th'ĕ-sis/ The preparation of chemical compounds from simpler compounds.

synthesis gas A mixture of carbon monoxide and hydrogen produced by steam reforming of natural gas.

$$CH_4 + H_2O \rightarrow CO + 3H_2$$

Synthesis gas is a useful starting material for the manufacture of a number of organic compounds.

synthon /**sin**-thon/ *See* retrosynthetic analysis.

Système International d'Unités /see-**stem** an-tair-nas-yo-**nal** doo-nee-**tay**/ *See* SI units.

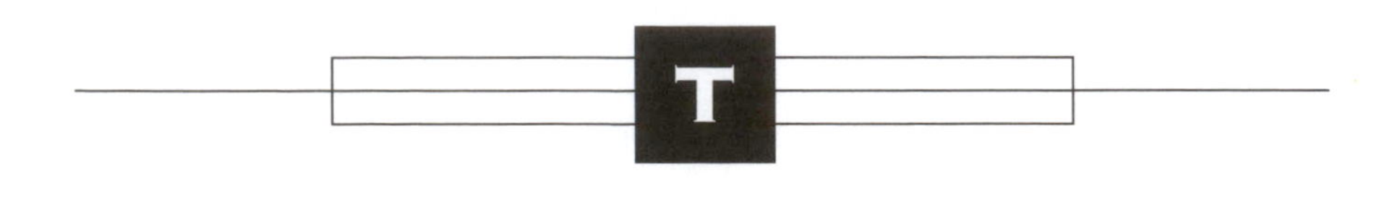

tactic polymer *See* polymerization.

talc /tal'k/ (**French chalk**) An extremely soft greenish or white mineral, a hydrous silicate of magnesium. Purified it is sold as talcum powder, and it is used as a lubricant, a filler (in paint and rubber) and an ingredient of some ceramics.

tannic acid /**tan**-ik/ *See* tannin.

tannin /**tan**-in/ Any of several yellow organic compounds found in vegetable sources such as bark of trees, oak galls, and tea. They are used in tanning animal skins to make leather and as mordants in dyeing. *Tannic acid* (a type of tannin) is a white solid heterocyclic organic acid extracted from oak galls and used for making dyes and inks.

tantalum /**tan**-tă-lŭm/ A silvery transition element. It is strong, highly resistant to corrosion, and is easily worked. Tantalum is used in turbine blades and cutting tools and in surgical and dental work.

Symbol: Ta; m.p. 2996°C; b.p. 5425 ± 100°C; r.d. 16.654 (20°C); p.n. 73; r.a.m. 180.9479.

tar (**bitumen**) A dark oily viscous liquid obtained by the destructive distillation of coal or the fractionation of PETROLEUM. Tars are mixtures consisting mainly of high-molecular weight hydrocarbons and phenols.

tartar emetic /**tar**-ter i-**met**-ik/ (**potassium antimonyl tartrate**; $KSbO(C_4H_4O_6).\frac{1}{4}H_2O$) A poisonous compound of antimony, which is used as an emetic in medicine and as a mordant in dyeing and as an insecticide.

tartaric acid /tar-**ta**-rik/ A crystalline hydroxy carboxylic acid with the formula:

$$HOOC(CHOH)_2COOH$$

Its systematic name is 2,3-dihydroxybutanedioic acid. *See also* optical activity.

tartrate /**tar**-trayt/ A salt or ester of tartaric acid.

tautomerism /taw-**tom**-ĕ-riz-ăm/ Isomerism in which each isomer can convert into the other, so that the two isomers are in equilibrium. The isomers are called *tautomers*. Tautomerism often results from the migration of a hydrogen atom. *See* keto–enol tautomerism.

TCCD *See* dioxins.

technetium /tek-**nee**-shee-ŭm/ A transition metal that does not occur naturally on Earth. It is produced artificially by bombarding molybdenum with neutrons and also during the fission of uranium. It is radioactive.

Symbol: Tc; m.p. 2172°C; b.p. 4877°C; r.d. 11.5 (est.); p.n. 43; r.a.m. 98.9063 (^{99}Tc); most stable isotope ^{98}Tc (half-life 4.2×10^6 years).

Teflon (*Trademark*) The synthetic polymer polytetrafluoroethane.

telluride /**tel**-yŭ-rȳd, -rid/ A compound of TELLURIUM and another element.

tellurium /te-**loor**-ee-ŭm/ A brittle silvery metalloid element belonging to group 16 of the periodic table. It is found native and in combination with metals. Tellurium is used mainly as an additive to improve the qualities of stainless steel and various metals.

Symbol: Te; m.p. 449.5°C; b.p. 989.8°C; r.d. 6.24 (20°C); p.n. 52; r.a.m. 127.6.

temperature scale A practical scale for measuring temperature. A temperature scale is determined by fixed temperatures (*fixed points*), which are reproducible systems assigned an agreed temperature. On the Celsius scale the two fixed points are the temperature of pure melting ice (the *ice temperature*) and the temperature of pure boiling water (the *steam temperature*). The difference between the fixed points is the *fundamental interval* of the scale, which is subdivided into temperature units. The International Temperature Scale has 11 fixed points that cover the range 13.81 kelvin to 1337.58 kelvin.

temporary hardness A type of water hardness caused by the presence of dissolved calcium, iron, and magnesium hydrogencarbonates. This form of hardness can be removed by boiling the water. Temporary hardness arises because rainwater combines with carbon dioxide from the atmosphere to form weak carbonic acid. This reacts with carbonate rocks, such as limestone, to produce calcium hydrogencarbonate, which then goes into solution. When this solution is heated, the complete reaction sequence is reversed so that calcium carbonate is precipitated. If this is not removed, it will accumulate in boilers and hot-water pipes and reduce their efficiency. *Compare* permanent hardness. *See also* water softening.

tera- Symbol: T A prefix denoting 10^{12}. For example, 1 terawatt (TW) = 10^{12} watts (W).

terbium /**ter**-bee-ŭm/ A soft ductile malleable silvery rare element of the lanthanoid series of metals. It occurs in association with other lanthanoids. One of its few uses is as a dopant in solid-state devices.

Symbol: Tb; m.p. 1356°C; b.p. 3123°C; r.d. 8.229 (20°C); p.n. 65; r.a.m. 158.92534.

terephthalic acid /te-ref-**thal**-ik/ *See* benzene-1,4-dicarboxylic acid.

ternary compound /**ter**-nă-ree/ A chemical compound formed from three elements; e.g. Na_2SO_4 or $LiAlH_4$.

terpene /**ter**-peen/ Any of a class of natural unsaturated hydrocarbons with formulae $(C_5H_8)_n$, found in plants. Terpenes consist of isoprene units,

$$CH_2{=}C(CH_3)CH{=}CH_2.$$

Monoterpenes have two units ($C_{10}H_{16}$), *diterpenes* four units ($C_{20}H_{32}$), *triterpenes* three units ($C_{30}H_{48}$), etc. *Sesquiterpenes* have three isoprene units ($C_{15}H_{24}$).

tertiary alcohol *See* alcohol.

tertiary amine *See* amine.

tertiary structure *See* protein.

tervalent /ter-**vay**-lĕnt/ (**trivalent**) Having a valence of three.

Terylene /**te**-ră-leen/ (*Trademark*) A polymer made by condensing benzene-1,4-dicarboxylic acid (terephthalic acid) and ethane-1,2-diol (ethylene glycol), used for making fibers for textiles.

tesla /**tess**-lă/ Symbol: T The SI unit of magnetic flux density, equal to a flux density of one weber of magnetic flux per square meter. 1 T = 1 Wb m^{-2}. The unit is named for the Croatian–American physicist Nikola Tesla (1856–1943).

tetracarbonyl nickel(0) /tet-ră-**kar**-bŏ-năl/ *See* nickel carbonyl.

tetrachloroethene /tet-ră-klor-oh-**eth**-een, -kloh-roh-/ (**ethylene tetrachloride; tetrachloroethylene**; $CCl_2{:}CCl_2$) A colorless poisonous liquid organic compound (a haloalkene) used as a solvent in dry cleaning and as a de-greasing agent.

tetrachloroethylene /tet-ră-klor-oh-**eth**-ă-leen, -kloh-roh-/ *See* tetrachloroethene.

tetrachloromethane /tet-ră-klor-oh-

meth-ayn/ (**carbon tetrachloride**; CCl_4) A colorless nonflammable liquid made by the chlorination of methane. Its main use is as a solvent.

tetraethyl lead /tet-ră-**eth**-ăl/ *See* lead tetraethyl.

tetrafluoroethene /tet-ră-floo-ŏ-roh-**eth**-een/ (CF_2:CF_2) A gaseous organic compound (a fluorocarbon and a haloalkene) used to make the plastic polytetrafluoroethene (PTFE). *See* polytetrafluoroethene.

tetragonal crystal /te-**trag**-ŏ-năl/ *See* crystal system.

tetrahedral compound /tet-ră-**hee**-drăl/ A compound such as methane, in which an atom has four bonds directed towards the corners of a regular tetrahedron. The angles between the bonds in such a compound are about 109°.

tetrahydrate /tet-ră-**hÿ**-drayt/ A crystalline hydrated compound containing four molecules of water of crystallization per molecule of compound. *See also* complex.

tetrahydrofuran /tet-ră-hÿ-droh-**fyoo**-ran, -fyoo-**ran**/ (**THF**, C_4H_8O) A colorless liquid widely used as a solvent and for making polymers.

tetravalent /tet-ră-**vay**-lĕnt, te-**trav**-ă-/ (**quadrivalent**) Having a valence of four.

thallium /**thal**-ee-ŭm/ A soft malleable grayish metallic element belonging to group 13 of the periodic table. It is found in lead and cadmium ores, and in pyrites (FeS_2). Thallium is highly toxic and was used previously as a rodent and insect poison. Various compounds are now used in photocells, infrared detectors, and low-melting glasses.

Symbol: Tl; m.p. 303.5°C; b.p. 1457°C; r.d. 11.85 (20°C); p.n. 81; r.a.m. 204.3833.

thermal dissociation The decomposition of a chemical compound into component atoms or molecules by the action of heat. Often it is temporary and reversible.

thermite /**ther**-mÿt/ (**thermit**) A mixture of aluminum powder and (usually) iron(III) oxide. When ignited the oxide is reduced; the reaction is strongly exothermic and molten iron is formed:

$$2Al + Fe_2O_3 \rightarrow Al_2O_3 + 2Fe$$

Thermite is used in incendiary bombs and for welding steel. *See also* Goldschmidt process.

thermochemistry /ther-moh-**kem**-iss-tree/ The branch of chemistry concerned with heats of reaction, solvation, etc.

thermodynamics /ther-moh-dÿ-**nam**-iks/ The study of heat and other forms of energy and the various related changes in physical quantities such as temperature, pressure, density, etc.

The *first law of thermodynamics* states that the total energy in a closed system is conserved (constant). In all processes energy is simply converted from one form to another, or transferred from one system to another.

A mathematical statement of the first law is:

$$\delta Q = \delta U + \delta W$$

Here, δQ is the heat transferred to the system, δU the change in internal energy (resulting in a rise or fall of temperature), and δW is the external work done *by* the system.

The *second law of thermodynamics* can be stated in a number of ways, all of which are equivalent. One is that heat cannot pass from a cooler to a hotter body without some other process occurring. Another is the statement that heat cannot be totally converted into mechanical work, i.e. a heat engine cannot be 100% efficient.

The *third law of thermodynamics* states that the entropy of a substance tends to zero as its thermodynamic temperature approaches zero.

Often a zeroth law of thermodynamics is given: that if two bodies are each in thermal equilibrium with a third body, then they are in thermal equilibrium with each other. This is considered to be more fundamental than the other laws because they assume it. *See also* Carnot cycle; entropy.

thermodynamic temperature Symbol: T A temperature measured in kelvins. *See also* absolute temperature.

thermoplastic polymer /ther-moh-**plas**-tik/ *See* polymer.

thermosetting polymer /ther-moh-**set**-ing/ *See* polymer.

THF *See* tetrahydrofuran.

thiazine /**th'ÿ**-ă-zeen/ (C_4H_4NS) Any of a group of heterocyclic organic compounds that have a six-membered ring containing four carbon atoms, one nitrogen atom, and one sulfur atom.

thiazole /**th'ÿ**-ă-zohl/ (C_3H_3NS) A colorless volatile liquid, a beterocyclic compound with a five-membered ring containing three carbon atoms, one nitrogen atom, and one sulfur atom. It resembles PYRIDINE in its reactions.

thin-layer chromatography A technique widely used for the analysis of mixtures. Thin-layer chromatography employs a solid stationary phase, such as alumina or silica gel, spread evenly as a thin layer on a glass plate. A base line is carefully scratched near the bottom of the plate using a needle, and a small sample of the mixture is spotted onto the base line using a capillary tube. The plate is then stood upright in solvent, which rises up to the base line and beyond by capillary action. The components of the spot of the sample will dissolve in the solvent and tend to be carried up the plate. However, some of the components will cling more readily to the solid phase than others and will not move up the plate so rapidly. In this way, different fractions of the mixture eventually become separated. When the solvent has almost reached the top, the plate is removed and quickly dried. The plate is developed to locate the positions of colorless fractions by spraying with a suitable chemical or by exposure to ultraviolet radiation. The components are identified by comparing the distance they have traveled up the plate with standard solutions that have been run simultaneously, or by computing an R_F value.

thio alcohol /**th'ÿ**-oh/ *See* thiol.

thiocarbamide /th'ÿ-oh-**kar**-bă-mÿd, -mid/ (**thiourea**; NH_2CSNH_2) A colorless crystalline organic compound (the sulfur analog of urea). It is converted to the inorganic compound ammonium thiocyanate on heating. It is used as a sensitizer in photography and in medicine.

thioether /th'ÿ-oh-**ee**-ther/ An organic sulfide; a compound of the type RSR′.

thiol /**th'ÿ**-ol, -ohl/ (**thio alcohol; mercaptan**) A compound of the type RSH, similar to an alcohol with the oxygen atom replaced by sulfur.

thionyl /**th'ÿ**-ŏ-năl/ Describing a compound containing the group =SO.

thionyl chloride /**th'ÿ**-ŏ-năl/ *See* sulfur dichloride oxide.

thiophene /**th'ÿ**-ŏ-feen/ (C_4H_4S) A colorless liquid that smells like benzene, a heterocyclic compound with a five-membered ring containing four carbon atoms and one sulfur atom. It occurs as an impurity in commercial benzene and is used as a solvent and in organic syntheses.

thiosulfate /th'ÿ-oh-**sul**-fayt/ A salt containing the ion $S_2O_3^{2-}$. *See* sodium thiosulfate(IV).

thiourea /th'ÿ-oh-yû-**ree**-ă/ *See* thiocarbamide.

thixotropy /thiks-**ot**-rŏ-pee/ The change of viscosity with movement. Fluids that undergo a decrease in viscosity with increasing velocity are said to be thixotropic. Non-drip paint is an example of a thixotropic fluid.

thoria /**thor**-ee-ă, **thoh**-ree-/ *See* thorium dioxide.

thorium /**thor**-ee-ŭm, **thoh**-ree-/ A toxic radioactive element of the actinoid series that is a soft ductile silvery metal. It has several long-lived radioisotopes found in a variety of minerals including monazite. Thorium is used in magnesium alloys, incandescent gas mantles, and nuclear fuel elements.

Symbol: Th; m.p. 1780°C; b.p. 4790°C (approx.); r.d. 11.72 (20°C); p.n. 90; r.a.m. 232.0381.

thorium dioxide /dў-**oks**-ўd, -id/ (**thoria**; ThO_2) A white insoluble compound, used as a refractory, in gas mantles, and as a replacement for silica in some types of optical glass.

threonine /**three**-ŏ-neen, -nin/ *See* amino acids.

thulium /**thoo**-lee-ŭm/ A soft malleable ductile silvery element of the lanthanoid series of metals. It occurs in association with other lanthanoids.

Symbol: Tm; m.p. 1545°C; b.p. 1947°C; r.d. 9.321 (20°C); p.n. 69; r.a.m. 168.93421.

thymidine /**th'ў**-mă-din, -deen/ The NUCLEOSIDE formed when thymine is linked to D-ribose by a β-glycosidic bond.

thymine /**th'ў**-min, -meen/ A nitrogenous base found in DNA. It has a PYRIMIDINE ring structure.

tin A white lustrous metal of low melting point; the fourth member of group 14 of the periodic table. Tin itself is the first distinctly metallic element of the group even though it retains some amphoteric properties. Its electronic structure has outer s^2p^2 electrons ($[Kr]4d^{10}5s^25p^2$). The element is of low abundance in the Earth's crust (0.004%) but is widely distributed, largely as cassiterite (SnO_2). The metal has been known since early bronze age civilizations when the ores used were relatively rich but currently worked ores are as low as 1–2% and considerable concentration must be carried out before roasting. The metal itself is obtained by reduction using carbon,

$$SnO_2 + C \rightarrow Sn + CO_2$$

Tin is an expensive metal and several processes are used for recovering tin from scrap tin-plate. These may involve chlorination (dry) to the volatile $SnCl_4$, or electrolytic methods using an alkaline electrolyte:

$$Sn + 4OH^- \rightarrow Sn(OH)_4^{2-} + 2e^- \text{ (anode)}$$
$$Sn(OH)_4^{2-} \rightarrow Sn^{2+} + 4OH^- \text{ (cathode)}$$
$$Sn^{2+} + 2e \rightarrow Sn \text{ (cathode)}$$

Tin does not react directly with hydrogen but an unstable hydride, SnH_4, can be prepared by reduction of $SnCl_4$. The low stability is due to the rather poor overlap of the diffuse orbitals of the tin atom with the small H-orbitals. Tin forms both tin(II) oxide and tin(IV) oxide. Both are amphoteric, dissolving in acids to give tin(II) and tin(IV) salts, and in bases to form *stannites* and *stannates*,

$$SnO + 4OH^- \rightarrow [SnO_3]^{4-} + 2H_2O$$
stannite (relatively unstable)
$$SnO_2 + 4OH^- \rightarrow [SnO_4]^{4-} + 2H_2O$$
stannate

The halides, SnX_2, may be prepared by dissolving tin metal in the hydrogen halide or by the action of heat on SnO plus the hydrogen halide. Tin(IV) halides may be prepared by direct reaction of halogen with the metal. Although tin(II) halides are ionized in solution their melting points are all low suggesting considerable covalence in all but the fluoride. The tin(IV) halides are volatile and essentially covalent with slight polarization of the bonds. Tin(II) compounds are readily oxidized to tin(IV) compounds and are therefore good reducing agents for general laboratory use.

Tin has three crystalline modifications or allotropes, α-tin or 'gray tin' (diamond structure), β-tin or 'white tin', and γ-tin; the latter two are metallic with close packed structures. Tin also has several isotopes. It is used in a large number of alloys including Babbit metal, bell metal, Britannia metal, bronze, gun metal, and pewter as well as several special solders.

Symbol: Sn; m.p. 232°C; b.p. 2270°C; r.d. 7.31 (20°C); p.n. 50; r.a.m. 118.710.

tincal /tink-ăl/ A naturally occurring form of borax. *See* disodium tetraborate decahydrate.

tin(II) chloride (**stannous chloride**; $SnCl_2$) A transparent solid made by dissolving tin in hydrochloric acid. Tin(II) chloride is a reducing agent, it combines with ammonia, and forms hydrates with water. It is used as a mordant.

tin(IV) chloride (**stannic chloride**; $SnCl_4$) A colorless fuming liquid. Tin(IV) chloride is soluble in organic solvents but is hydrolyzed by water. It dissolves sulfur, phosphorus, bromine, and iodine, and it dissolves in concentrated hydrochloric acid to give the anion $SnCl_6^{2-}$.

tincture A solution in which alcohol (ethanol) is the solvent (e.g., tincture of iodine).

tin(IV) hydride *See* stannane.

tin(II) oxide (**stannous oxide**; SnO) A dark green or black solid. It can also be obtained in an unstable red form, which turns black on exposure to air. Tin(II) oxide can be made by precipitating the hydrated oxide from a solution containing tin(II) ions and dehydrating the product at 100°C.

tin(IV) oxide (**stannic oxide**; **tin dioxide**; SnO_2) A colorless crystalline solid, which is usually discolored owing to the presence of impurities. It exists as hexagonal or rhombic crystals or in an amorphous form. Tin(IV) oxide is insoluble in water. It is used for polishing glass and metal.

tin(II) sulfide (**stannous sulfide**; SnS) A gray solid that can be prepared from tin and sulfur. Above 265°C tin(II) sulfide slowly turns into tin(IV) sulfide and tin:

$$2SnS(s) \rightarrow SnS_2(s) + Sn(s)$$

tin(IV) sulfide (**stannic sulfide**; SnS_2) A yellowish solid that can be precipitated by reacting hydrogen sulfide with a solution of a soluble tin(IV) salt. A crystalline form sometimes known as *mosaic gold* is obtained by heating a mixture of tin filings, sulfur, and ammonium chloride. Tin(IV) sulfide is used as a pigment.

titania /tÿ-**tay**-nee-ă, ti-/ *See* titanium(IV) oxide.

titanic chloride /tÿ-**tan**-ik/ *See* titanium(IV) chloride.

titanic oxide *See* titanium(IV) oxide.

titanium /tÿ-**tay**-nee-ŭm, ti-/ A silvery transition metal that occurs in various ores as titanium(IV) oxide and also in combination with iron and oxygen. It is extracted by conversion of titanium(IV) oxide to the chloride, which is reduced to the metal by heating with sodium. Titanium is reactive at high temperatures. It is used in the aerospace industry as it is strong, resistant to corrosion, and has a low density. It forms compounds with oxidation states +4, +3, and +2, the +4 state being the most stable.

Symbol: Ti; m.p. 1660°C; b.p. 3287°C; r.d. 4.54 (20°C); p.n. 22; r.a.m. 47.867.

titanium(IV) chloride (**titanic chloride**; **titanium tetrachloride**; $TiCl_4$) A volatile colorless liquid formed by heating titanium(IV) oxide with carbon in a stream of dry chlorine at 700°C. The product is purified by fractional distillation. Titanium(IV) chloride fumes in moist air forming the oxychlorides of titanium. It undergoes hydrolysis in water but this can be prevented by the presence of excess hydrochloric acid. Crystallization of such a solution gives crystals of the di- and pentahydrates. Titanium(IV) chloride is used in the preparation of pure titanium and as an intermediate in the production of titanium compounds from minerals.

titanium dioxide /dÿ-**oks**-ÿd/ *See* titanium(IV) oxide.

titanium(IV) oxide (**titanic oxide**; **titanium dioxide**; TiO_2) A white ionic solid that occurs naturally in three crystalline forms: rutile (tetragonal), brookite (orthorhombic), and anatase (tetragonal). It reacts slowly with acids and is attacked by the halogens at high temperatures. On heating it decomposes to give titanium(III) oxide and oxygen. Titanium(IV) oxide is amphoteric; it is used as a white pigment. With hot concentrated sulfuric acid, it forms titanyl sulfate, $TiOSO_4$. When fused with alkalis, it forms titanates.

titanium tetrachloride /tet-ră-**klor**-ÿd, -id, -**kloh**-rÿd, -rid/ *See* titanium(IV) chloride.

titrant /**tÿ**-trănt/ *See* titration.

titration /tÿ-**tray**-shŏn/ A procedure in VOLUMETRIC ANALYSIS in which a solution of known concentration (called the *titrant*) is added to a solution of unknown concentration from a buret until the equivalence point or end point of the titration is reached.

TNT *See* trinitrotoluene.

Tollen's reagent /**tol**-ĕn/ A solution of the complex ion $Ag(NH_3)_2^+$ produced by precipitation of silver oxide from silver nitrate with a few drops of sodium hydroxide solution, and subsequent dissolution of the silver oxide in aqueous ammonia. Tollen's reagent is used in the SILVER-MIRROR TEST for aldehydes, where the Ag^+ ion is reduced to silver metal. It is also a test for alkynes with a triple bond in the 1-position. A yellow precipitate of silver carbide is formed in this case.

$$RCCH + Ag^+ \rightarrow RCC^-Ag^+ + H^+$$

The reagent is named for Bernhard Tollens (1841–1918).

toluene /**tol**-yoo-een/ *See* methylbenzene.

toluidine /tŏ-**loo**-ă-deen, -din/ (**methylaniline; aminotoluene**; $CH_3C_6H_4NH_2$) An aromatic amine used in making dyestuffs and drugs. There are three isomers; the 1,2- (ortho-aminotoluene) and 1,3- (meta-) forms are liquids, the 1,4- (para-) isomer is a solid.

tonne /tun/ Symbol: t A unit of mass equal to 10^3 kilograms (i.e. one megagram).

topaz A hydrous aluminosilicate mineral containing some fluoride ions, used for making refractories, glasses, and glazes. A clear pale yellow or brown form is used as a gemstone.

torr A unit of pressure equal to a pressure of 101 325/760 pascals (133.322 Pa). One torr is equal to one mmHg. The unit is named for the Italian physicist Evangelista Toricelli (1608–47).

tracer An ISOTOPE of an element used to investigate chemical reactions or physical processes (e.g. diffusion).

trans- Designating an isomer with groups that are on opposite sides of a bond or structure. *See* isomerism.

transactinide elements /tran-**sak**-tă-nÿd, -**zak**-, -nid/ Those elements following ^{103}Lw (i.e. following the actinide series). To date, element 104 (rutherfordium) through to element 112 have been synthesized; so far, elements 110, 111, and 112 are unnamed. All are unstable and have very short half-lives. There is speculation that islands of stable nuclei with very high mass numbers exist since theoretical calculations predict unusual stability for elements 114 and 184.

trans-fat *See* hardening.

transient species A short-lived intermediate in a chemical reaction.

transition elements /tran-**zish**-ŏn/ A class of elements occurring in the periodic table in four series: from scandium to zinc; from yttrium to cadmium; from lanthanum to mercury; and from actinium to element 112. The transition elements are all metals. They owe their properties to the presence of d electrons in the atoms. In the first transition series, calcium has the configuration $3s^23p^64s^2$. The next element scandium has $3s^23p^63d^14s^2$, and the series is formed by filling the d levels up to zinc at $3d^{10}4s^2$. The other transition series occur by filling of the 4d, 5d, and 6d levels. Often the elements scandium, yttrium, and lanthanum are considered with the lanthanoids rather than the transition elements, and zinc, cadmium, and mercury are also treated as a separate group. Transition elements have certain characteristic properties resulting from the existence of d levels:

1. They have variable valences – i.e. they can form compounds with the metal in different oxidation states (Fe^{2+} and Fe^{3+}, etc.).
2. They form a vast number of inorganic complexes, in which coordinate bonds are formed to the metal atom or ion.
3. Compounds of transition elements are often colored.

See also d-block elements; lanthanoids; metals; transactinide elements; zinc group.

transition state (**activated complex**) Symbol: ‡ A short-lived high-energy molecule, radical, or ion formed during a reaction between molecules possessing the necessary activation energy. The transition state decomposes at a definite rate to yield either the reactants again or the final products. The transition state can be considered to be at the top of the energy profile.

For the reaction,

$$X + YZ = X \ldots Y \ldots Z^{\ddagger} \rightarrow XY + Z$$

the sequence of events is as follows. X approaches YZ and when it is close enough the electrons are rearranged producing a weakening of the bond between Y and Z. A partial bond is now formed between X and Y producing the transition state. Depending on the experimental conditions, the transition state either breaks down to form the products or reverts back to the reactants.

transition temperature A temperature at which some definite physical change occurs in a substance. Examples of such transitions are change of state, change of crystal structure, and change of magnetic behavior.

transmutation /trans-myoo-**tay**-shŏn, tranz-/ A change of one element into another by radioactive decay or by bombardment of the nuclei with particles.

transport number Symbol: *t* In an electrolyte, the transport number of an ion is the fraction of the total charge carried by that type of ion in conduction.

transuranic elements /trans-yû-**ran**-ik, tranz-/ Those elements of the actinoid series that have higher atomic numbers than uranium. The transuranic elements are formed by adding neutrons to the lower actinoids by high-energy bombardment and they generally have such short half-lives that macroscopic isolation is impossible. Neptunium and plutonium are formed in trace quantities by natural neutron bombardment from spontaneous fission of ^{235}U.

triaminotriazine /trÿ-am-ă-noh-**trÿ**-ă-zeen/ *See* melamine.

triammonium phosphate /trÿ-ă-**moh**-nee-ŭm/ *See* ammonium phosphate.

triatomic /trÿ-ă-**tom**-ik/ Denoting a molecule, radical, or ion consisting of three atoms. For example, O_3 and H_2O are triatomic molecules.

triazine /**trÿ**-ă-zeen, -zin/ ($C_3H_3N_3$) A heterocyclic organic compound with a six-membered ring containing three carbon atoms and three nitogen atoms. There are three isomers, used as dyestuffs and herbicides. *See also* melamine.

triazole /**trÿ**-ă-zohl/ ($C_2H_3N_3$) A heterocyclic organic compound with a five-membered ring containing two carbon atoms and three nitrogen atoms. There are two isomers.

tribasic acid /trÿ-**bay**-sik/ An ACID with three replaceable hydrogen atoms (such as phosphoric(V) acid, H_3PO_4).

tribromomethane /trÿ-broh-moh-**meth**-ayn/ (**bromoform**; $CHBr_3$) A colorless liquid compound. *See* haloform.

trichloroacetic acid /trÿ-klor-oh-ă-**see**-tik, -**set**-ik, -kloh-roh-/ *See* chloroethanoic acid.

trichloroethanal /trȳ-klor-oh-**eth**-ă-nal, -kloh-roh-/ (**chloral**; CCl_3CHO) A colorless liquid aldehyde made by chlorinating ethanal. It was used to make the insecticide DDT. It can be hydrolyzed to give *2,2,2-trichloroethanediol* (chloral hydrate, $CCl_3CH(OH)_2$). Most compounds with two –OH groups on the same carbon atom are unstable. However, in this case the effect of the three chlorine atoms stabilizes the compound. It is used as a sedative.

2,2,2-trichloroethanediol /trȳ-klor-oh-eth-ayn-**dȳ**-ol, -ohk, -kloh-roh-/ *See* trichloroethanal.

trichloroethanoic acid /trȳ-klor-oh-eth-ă-**noh**-ik, -kloh-roh-/ *See* chloroethanoic acid.

trichloromethane /trȳ-klor-oh-**meth**-ayn, -kloh-roh-/ (**chloroform**; $CHCl_3$) A colorless volatile liquid formerly used as an anesthetic. Now its main use is as a solvent and raw material for making other chlorinated compounds. Trichloromethane is made by reacting ethanal, ethanol, or propanone with chlorinated lime.

triclinic crystal /trȳ-**klin**-ik/ *See* crystal system.

triglyceride /trȳ-**gliss**-ĕ-rȳd/ A GLYCERIDE in which esters are formed with all three –OH groups of glycerol.

trigonal bipyramid /**trig**-ŏ-năl bȳ-**pi**-ră-mid/ *See* complex.

trigonal crystal *See* crystal system.

trihydrate /trȳ-**hȳ**-drayt/ A crystalline hydrated compound that contains three molecules of water of crystallization per molecule of compound.

trihydric alcohol /trȳ-**hȳ**-drik/ *See* triol.

triiodomethane /trȳ-ȳ-ŏ-doh-**meth**-ayn/ (**iodoform**; CHI_3) A yellow crystalline compound made by warming ethanal with an alkaline solution of an iodide:

$$CH_3CHO + 3I^- + 4OH^- \rightarrow CHI_3 + HCOO^- + 3H_2O$$

The reaction also occurs with all ketones of general formula CH_3COR (R is an alkyl group) and with secondary alcohols $CH_3CH(OH)R$. Iodoform is used as a test for such reactions (the *iodoform reaction*).

triiron tetroxide /trȳ-**ȳ**-ern tet-**roks**-ȳd/ (**ferrosoferric oxide**; **magnetic iron oxide**; Fe_3O_4) A black solid prepared by passing either steam or carbon dioxide over red-hot iron. It may also be prepared by passing steam over heated iron(II) sulfide. Triiron tetroxide occurs in nature as the mineral magnetite. It is insoluble in water but will dissolve in acids to give a mixture of iron(II) and iron(III) salts in the ratio 1:2. Generally it is chemically unreactive; it is, however, a fairly good conductor of electricity.

trimer /**trȳ**-mer/ A molecule (or compound) formed by addition of three identical molecules. *See* ethanal; methanal.

trimethylaluminum /trȳ-meth-ăl-ă-**loo**-mă-nŭm/ (**aluminum trimethyl**; $(CH_3)_3Al$) A colorless liquid produced by the sodium reduction of dimethyl aluminum chloride. It ignites spontaneously on contact with air and reacts violently with water, acids, halogens, alcohols, and amines. Aluminum alkyls are used in the Ziegler process for the manufacture of high-density polyethene.

trimolecular /trȳ-mŏ-**lek**-yŭ-ler/ Describing a reaction or step that involves three molecules interacting simultaneously with the formation of a product. For example, the final step in reaction between hydrogen peroxide and acidified potassium iodide is trimolecular:

$$HOI + H^+ + I^- \rightarrow I_2 + H_2O$$

It is uncommon for reactions to take place involving trimolecular steps. The oxidation of nitrogen(II) oxide to nitrogen(IV) oxide,

$$2NO + O_2 \rightarrow 2NO_2$$

is often classified as a trimolecular reaction but many believe it to involve two bimolecular reactions.

trinitroglycerine /trÿ-nÿ-troh-**gliss**-ĕ-rin, -reen/ *See* nitroglycerine.

trinitrophenol /trÿ-nÿ-troh-**fee**-nol, -nohl/ *See* picric acid.

trinitrotoluene /trÿ-nÿ-troh-**tol**-yoo-een/ (**TNT**; $CH_3C_6H_2(NO_2)_3$) A yellow crystalline solid. It is a highly unstable substance, used as an explosive. The compound is made by nitrating methylbenzene and the nitro groups are in the 2, 4, and 6 positions.

triol /**trÿ**-ol, -ohl/ (**trihydric alcohol**) An alcohol that has three hydroxyl groups (-OH) per molecule of compound.

triose /**trÿ**-ohs/ A SUGAR that contains three carbon atoms.

trioxoboric(III) acid /trÿ-oks-ŏ-**bô**-rik/ *See* boric acid.

trioxosulfuric(IV) acid / trÿ-oks-oh- sul-**fyoor**-ik / *See* sulfurous acid.

trioxygen /trÿ-**oks**-ă-jĕn/ *See* ozone.

triple bond A covalent bond formed between two atoms in which three pairs of electrons contribute to the bond. One pair forms a sigma bond (equivalent to a single bond) and two pairs give rise to two pi bonds. It is conventionally represented as three lines, thus H–C≡C–H. *See* multiple bond.

triple point The only point at which the gas, solid, and liquid phases of a substance can coexist in equilibrium. The triple point of water (273.16 K at 101 325 Pa) is used to define the kelvin.

trisodium phosphate(V) /trÿ-**soh**-dee-ŭm/ (**sodium orthophosphate**; Na_3PO_4) A white solid prepared by adding sodium hydroxide to disodium hydrogenphosphate. On evaporation white hexagonal crystals of the dodecahydrate ($Na_3PO_4.12H_2O$) may be obtained. These crystals do not effloresce or deliquesce. They dissolve readily in water to produce alkaline solutions owing to salt hydrolysis. Trisodium phosphate is used as a water softener.

triterpene *See* terpene.

tritiated compound A compound in which one or more ^1H atoms have been replaced by tritium (^3H) atoms.

tritium /**trit**-ee-ŭm, **trish**-/ Symbol: T, ^3H A radioactive isotope of hydrogen of mass number 3. The nucleus contains 1 proton and 2 neutrons. Tritium decays with emission of low-energy beta radiation to give ^3He. The half-life is 12.3 years. It is useful as a tracer in studies of chemical reactions. Compounds in which ^3H atoms replace the usual ^1H atoms are said to be *tritiated*. A positive tritium ion, T^+, is a *triton*.

triton /**trÿ**-tŏn/ *See* tritium.

trivalent /trÿ-**vay**-lĕnt, **triv**-ă-/ (**tervalent**) Having a valence of three.

tropylium ion /trŏ-**pil**-ee-ŭm/ The positive ion $C_7H_7^+$, having a symmetrical seven-membered ring of carbon atoms. The tropylium ion ring shows non-benzenoid aromatic properties.

tryptophan /**trip**-tŏ-fan/ *See* amino acids.

tungsten /**tung**-stĕn/ A transition metal occurring naturally in wolframite ($(Fe,Mn)WO_4$) and scheelite ($CaWO_4$). It was formerly called *wolfram*. It is used as the filaments in electric lamps and in various alloys.

Symbol: W; m.p. 3410 ± 20°C; b.p. 5650°C; r.d. 19.3 (20°C); p.n. 74; r.a.m. 183.84.

tungsten carbide Either of two carbides (W_2C and WC) produced by heating powdered tungsten with carbon. The carbides are extremely hard and are used in industry to make cutting tools or as an abrasive. WC has a very high melting point (2770°C) and will conduct electricity. W_2C also has a very high melting point (2780°C) but is a less efficient conductor of electricity. It is very resistant to chemical attack and behaves in a manner very similar to that of tungsten. It is strongly attacked by chlorine to give tungsten hexachloride, WCl_6.

turpentine /**ter**-pĕn-tÿn/ (**pine-cone oil**) A yellow viscous resin obtained from coniferous trees. It can be distilled to produce turpentine oil (also known simply as turpentine), used in medicine and as a solvent in paints, polishes, and varnishes. *See* resin.

turquoise /**ter**-koiz, -kwoiz/ A naturally occurring hydrated copper aluminum phosphate, used as a green-blue gemstone.

tyrosine /**tÿ**-rŏ-seen, -sin, **ti**-/ *See* amino acids.

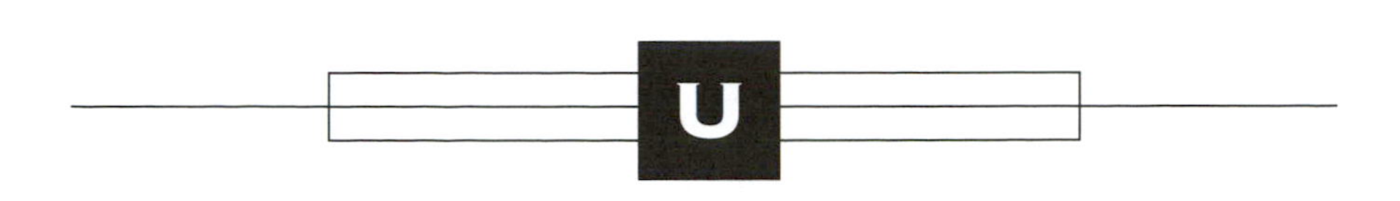

ultracentrifuge /ul-tră-**sen**-tră-fyooj/ A high-speed centrifuge used for separating out very small particles. The sedimentation rate depends on the particle size, and the ultracentrifuge can be used to measure the mass of colloidal particles and large molecules (e.g. proteins).

ultrahigh vacuum *See* vacuum.

ultramarine /ul-tră-mă-**reen**/ *See* lazurite.

ultraviolet /ul-tră-**vȳ**-ŏ-lĕt/ (UV) A form of ELECTROMAGNETIC RADIATION, shorter in wavelength than visible light. Ultraviolet wavelengths range between about 1 nm and 400 nm. Ordinary glasses are not transparent to these waves; quartz is a much more effective material for making lenses and prisms for use with ultraviolet. Like light, ultraviolet radiation is produced by electronic transitions between the outer energy levels of atoms. However, having a higher frequency, ultraviolet photons carry more energy than those of light and can induce photolysis of compounds.

unimolecular /yoo-nă-mŏ-**lek**-yŭ-ler/ Describing a reaction (or step) in which only one molecule is involved. For example, radioactive decay is a unimolecular reaction:

$$Ra \rightarrow Rn + \alpha$$

Only one atom is involved in each disintegration.

In a unimolecular chemical reaction, the molecule acquires the necessary energy to become activated and then decomposes. The majority of reactions involve only uni- or bimolecular steps. The following reactions are all unimolecular:

$$N_2O_4 \rightarrow 2NO_2$$
$$PCl_5 \rightarrow PCl_3 + Cl_2$$
$$CH_3CH_2Cl \rightarrow C_2H_4 + HCl$$

unit A reference value of a quantity used to express other values of the same quantity. *See also* SI units.

unit cell The smallest group of atoms, ions, or molecules that, when repeated at regular intervals in three dimensions, will produce the lattice of a crystal system. There are seven basic types of unit cells, which result in the seven crystal systems.

unit processes (**chemical conversions**) The recognized steps used in chemical processes, e.g. alkylation, distillation, hydrogenation, pyrolysis, nitration, etc. Industrial processing and the economics, design, and use of the equipment are based on these unit processes rather than consideration of each reaction separately.

univalent /yoo-nă-**vay**-lĕnt, yoo-**niv**-ă-/ (**monovalent**) Having a valence of one.

universal indicator (**multiple-range indicator**) A mixture of indicator dyestuffs that shows a gradual change in color over a wide pH range. A typical formulation contains methyl orange, methyl red, bromothymol blue, and phenolphthalein and changes through a red, orange, yellow, green, blue, and violet sequence between pH 3 and pH 10. Several commercial preparations are available as both solutions and test papers.

unnil- A prefix used in the names of new chemical elements. Synthetic elements with high proton numbers were given temporary names based on the proton number, pending official agreement on the actual name to be used. These names were formed from the word elements in the next table.

Element	*Number*	*Symbol*
nil	0	n
un	1	u
bi	2	b
tri	3	t
quad	4	q
pent	5	p
hex	6	h
sept	7	s
oct	8	0

The suffix -ium was also used (because the elements were metals). So element 104, for example, had the temporary systematic name unnilquadium (un + nil + quad + ium). The names for elements using this system are shown in the table below.

unsaturated compound An organic compound that contains at least one double or triple bond between two of its carbon atoms. The multiple bond is relatively weak; consequently unsaturated compounds readily undergo addition reactions to form single bonds. *Compare* saturated compound.

unsaturated solution *See* saturated solution.

unsaturated vapor *See* saturated vapor.

UPVC Unplasticized PVC; a hard-wearing form of PVC used in building work (e.g. for window frames).

uracil /**yoor**-ă-săl/ A nitrogenous base that is found in RNA, replacing the thymine of DNA. It has a pyrimidine ring structure.

uraninite /**yoor**-ă-nÿt/ *See* pitchblende.

uranium /yû-**ray**-nee-ŭm/ A toxic radioactive silvery element of the actinoid series of metals. Its three naturally occurring radioisotopes, ^{238}U (99.283% in abundance), ^{235}U (0.711%), and ^{234}U (0.005%), are found in numerous minerals including the uranium oxides pitchblende, uraninite, and carnotite. The readily fissionable ^{235}U is a major nuclear fuel and nuclear explosive, while ^{238}U is a source of fissionable ^{239}Pu.

Symbol: U; m.p. 1132.5°C; b.p. 3745°C; r.d. 18.95 (20°C); p.n. 92; r.a.m. 238.0289.

uranium hexafluoride /heks-ă-**floo**-ō-

P.N.	*Temporary name*	*Symbol*	*Name*	*Symbol*
101	unnilunium	Unu	mendelevium	Md
102	unnilbium	Unb	nobelium	No
103	unniltrium	Unt	lawrencium	Lr
104	unnilquadium	Unq	rutherfordium	Rf
105	unnilpentium	Unq	dubnium	Db
106	unnilhexium	Unh	seaborgium	Sg
107	unnilseptium	Uns	bohrium	Bh
108	unniloctium	Uno	hassium	Hs
109	unnilennium	Une	meitnerium	Mt
110	ununnilium	Uun	darmstadtium	Ds
111	unununium	Uuu	roentgenium	Rg
112	ununbium	Uub	unnamed	
113	ununtrium	Uut	unnamed	
114	ununquadium	Uuq	unnamed	
115	ununpentium	Uup	unnamed	
116	ununhexium	Uuh	unnamed	

rÿd/ (UF_6) A crystalline volatile compound, used in separating uranium isotopes by differences in the rates of gas diffusion.

uranium–lead dating A technique for dating certain rocks depending on the decay of the radioisotope ^{238}U to ^{206}Pb (half-life 4.5×10^9 years) or of ^{235}U to ^{207}Pb (half-life 7.1×10^8 years). The decay of ^{238}U releases alpha particles and an estimate of the age of the rock can be made by measuring the amount of helium present. Another method is to measure the amount of radioactive lead present to the amount of nonradioactive lead. *See also* radioactive dating.

uranium(IV) oxide (**uranium dioxide; urania**; UO_2) An extremely poisonous radioactive black crystalline solid. It occurs in pitchblende (uraninite) and is used in ceramics, photographic chemicals, pigments, and as a nuclear fuel.

uranium(VI) oxide (**uranium trioxide**; UO_3) An extremely poisonous radioactive orange solid, used as a pigment in ceramics and in uranium refining.

uranyl /yoor-ă-năl/ The inorganic radical UO_2^{2+}, as in uranyl nitrate $UO_2(NO_3)_2$.

urea /yû-**ree**-ă/ (**carbamide**; $CO(NH_2)_2$) A white crystalline compound made from ammonia and carbon dioxide. It is used in the manufacture of urea–formaldehyde (methanal) resins. Urea is the end product of metabolism in many animals and is present in urine.

urethane /yoor-ĕ-thayn/ (**ethyl carbamate**; $CO(NH_2)OC_2H_5$) A poisonous flammable organic compound, used in medicine, as a solvent, and as an intermediate in the manufacture of polyurethane resins.

uric acid /yoor-ik/ A nitrogen compound produced from purines. In certain animals (e.g. birds and reptiles), it is the main excretory product resulting from breakdown of amino acids. In humans, uric acid crystals in the joints are the cause of gout.

uridine /yoor-ă-din, -deen/ The nucleoside formed when uracil is linked to D-ribose by a β-glycosidic bond.

UV *See* ultraviolet.

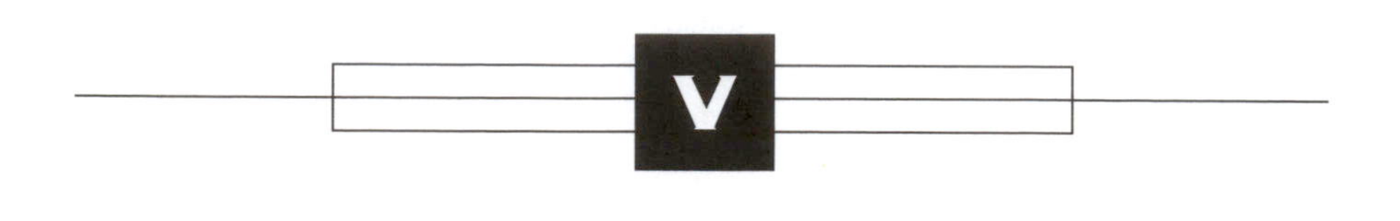

vacancy *See* defect.

vacuum (*pl.* **vacuums** or **vacua**) A space containing gas below atmospheric pressure. A perfect vacuum contains no matter at all, but for practical purposes *soft* (*low*) *vacuum* is usually defined as down to about 10^{-2} pascal, and *hard* (*high*) *vacuum* as below this. *Ultrahigh vacuum* is lower than 10^{-7} pascal.

vacuum distillation The distillation of liquids under a reduced pressure, so that the boiling point is lowered. Vacuum distillation is a common laboratory technique for purifying or separating compounds that would decompose at their 'normal' boiling point.

vacuum flask *See* Dewar flask.

valence /**vay**-lĕns/ (**valency**) The combining power of an element or radical, equal to the number of hydrogen atoms that will combine with or displace one atom of the element. For simple covalent molecules the valence is obtained directly, for example C in CH_4 is tetravalent; N in NH_3 is trivalent. For ions the valence is regarded as equivalent to the magnitude of the charge; for example Ca^{2+} is divalent, CO_3^{2-} is a divalent radical. The rare gases are zero-valent because they do not form compounds under normal conditions. As the valence for many elements is constant, the valence of some elements can be deduced without reference to compounds formed with hydrogen. Thus, as the valence of chlorine in HCl is 1, the valence of aluminum in $AlCl_3$ is 3; as oxygen is divalent (H_2O) silicon in SiO_2 is tetravalent. The product of the valence and the number of atoms of each element in a compound must be equal. For example, in Al_2O_3 for the two aluminum atoms (valence 3) the product is 6 and for the three oxygen atoms (valence 2) the product is also 6.

The valence of an element is generally equal to either the number of valence electrons or eight minus the number of valence electrons. Transition metal ions display variable valence.

valence band The highest ENERGY LEVEL in a SEMICONDUCTOR or nonconductor that can be occupied by electrons.

valence-bond theory A technique for calculating the electronic structure of molecules. In valence-bond theory the starting point is to consider the atoms in a molecule and the ways in which valence electrons in the atoms can pair up in chemical bonds. Each way of pairing up the electrons is called a *valence bond structure*. The overall structure for the molecule is determined by a linear combination of the wavefunctions for the valence-bond structures. *See* resonance.

valence electron An outer electron in an atom that can participate in forming chemical bonds.

valency /**vay**-lĕns/ *See* valence.

valeric acid /vă-**le**-rik, -**leer**-ik/ *See* pentanoic acid.

valine /**val**-een, -in/ *See* amino acids.

vanadium /vă-**nay**-dee-ŭm/ A silvery transition element occurring in complex ores in small quantities. It is used in alloy steels. Vanadium forms compounds with oxidation states +5, +4, +3, and +2.

It forms colored ions.

Symbol: V; m.p. 1890°C; b.p. 3380°C; r.d. 6.1 (20°C); p.n. 23; r.a.m. 50.94.

vanadium(V) oxide (**vanadium pentoxide**; V_2O_5) An oxide of vanadium extensively used as a catalyst in oxidation processes, as in the contact process.

van der Waals equation /**van**-der-wawlz/ An equation of state for real gases. For *n* moles of gas the equation is

$$(p + n^2a/V^2)(V - nb) = nRT$$

where p is the pressure, V the volume, and T the thermodynamic temperature. a and b are constants for a given substance and R is the gas constant. The equation gives a better description of the behavior of real gases than the perfect gas equation ($pV = nRT$).

The equation contains two corrections: b is a correction for the non-negligible size of the molecules; a/V^2 corrects for the fact that there are attractive forces between the molecules, thus slightly reducing the pressure from that of an ideal gas. The equation is named for the Dutch physicist Johannes Diderik van der Waals (1837–1923). *See also* gas laws; kinetic theory.

van der Waals force An intermolecular force of attraction, considerably weaker than chemical bonds and arising from weak electrostatic interactions between molecules (the energies are often less than 1 J mol^{-1}).

The van der Waals interaction contains contributions from three effects; permanent dipole–dipole interactions found for any polar molecule; dipole–induced dipole interactions, where one dipole causes a slight charge separation in bonds that have a high polarizability; and dispersion forces, which result from temporary polarity arising from an asymmetrical distribution of electrons around the nucleus. Even atoms of the rare gases exhibit dispersion forces.

van't Hoff factor /vant-**hoff**/ Symbol: i The ratio of the number of particles present in a solution to the number of undissociated molecules added. It is used in studies of colligative properties, which depend on the number of entities present. For example, if n moles of a compound are dissolved and dissociation into ions occurs, then the number of particles present will be in. Osmotic pressure (π), for instance, will be given by the equation

$$\pi V = inRT$$

The factor is named for the Dutch theoretical chemist Jacobus Henricus Van't Hoff (1852–1911).

van't Hoff isochore The equation:

$$d(\log_e K)/dT = \Delta H/RT^2$$

showing how the equilibrium constant, K, of a reaction varies with thermodynamic temperature, T. ΔH is the enthalpy of reaction and R is the gas constant.

vapor A gas formed by the vaporization of a solid or liquid. Some particles near the surface of a liquid acquire sufficient energy in collisions with other particles to escape from the liquid and enter the vapor; some particles in the vapor lose energy in collisions and re-enter the liquid. At a given temperature an equilibrium is established, which determines the vapor pressure of the liquid at that temperature.

vapor density The ratio of the mass of a certain volume of a vapor to the mass of an equal volume of hydrogen (measured at the same temperature and pressure). Determination of vapor densities is one method of finding the relative molecular mass of a compound (equal to twice the vapor density). Victor Meyer's method, Dumas' method, or Hofmann's method can be used.

vaporization The process by which a liquid or solid is converted into a gas or vapor by heat. Unlike boiling, which occurs at a fixed temperature, vaporization can occur at any temperature. Its rate increases as the temperature rises.

vapor pressure The pressure exerted by a vapor. The *saturated vapor pressure* is the pressure of a vapor in equilibrium with its liquid or solid. It depends on the nature of the liquid or solid and the temperature.

vat dyes A class of insoluble dyes applied by first reducing them to derivatives that are soluble in dilute alkali. In this condition they have a great attraction for certain fibers, such as cotton. The solution is applied to the material and the insoluble dye is regenerated in the fibers by atmospheric oxidation. Indigo and indanthrene are examples of vat dyes.

verdigris /**ver**-di-grees, -gris, -gree/ Any of various greenish basic salts of copper. True verdigris is basic copper acetate, a green or blue solid of variable composition used as a paint pigment. The term is often applied to green patinas formed on metallic copper. Copper cooking vessels can form a coating of basic copper carbonate, $CuCO_3.Cu(OH)_2$. The verdigris formed on copper roofs and domes is usually basic copper sulfate, $CuSO_4.3Cu(OH)_2.H_2O$, while basic copper chloride, $CuCl_2.Cu(OH)_2$, may form in regions near the sea.

vermilion *See* mercury(II) sulfide.

vesicant /**ves**-ă-kănt/ A substance that causes blistering of the skin. Mustard gas is an example.

vicinal positions /**vis**-ă-năl/ Positions in a molecule at adjacent atoms. For example, in 1,2-dichloroethane the chlorine atoms are in vicinal positions, and this compound can thus be named *vic*-dichloroethane.

Victor Meyer's method /**vik**-ter **mÿ**-erz/ A method for determining vapor densities in which a given weight of sample is vaporized and the volume of air displaced by it is measured. In practice, a bulb in a heating bath is connected via a fairly long tube to a water-bath gas-collection arrangement. The system is brought to equilibrium and the sample is then added (without opening the apparatus to the atmosphere). As the air displaced by gas is collected over water a correction for the vapor pressure of water is necessary and the method may fail if the vapor is soluble in water. The method is named for the German chemist Victor Meyer (1848–97). *See also* Dumas' method; Hofmann's method.

vinegar A dilute solution (about 4% by volume) of ETHANOIC ACID (acetic acid), often with added coloring and flavoring such as caramel. Natural vinegar is produced by the bacterial fermentation of cider or wine; it can also be made synthetically.

vinylation /vÿ-nă-**lay**-shŏn/ A catalytic reaction in which a compound adds across the triple bond of ethyne (acetylene) to form an ethenyl (vinyl) compound. For example, an alcohol can add as follows:

$$ROH + HC{\equiv}CH \rightarrow RHC{=}CH(OH)$$

vinyl benzene /**vÿ**-năl/ *See* phenylethene.

vinyl chloride *See* chloroethene.

vinyl group The group $CH_2{:}CH–$.

viscose rayon /**viss**-kohs/ *See* rayon.

viscosity /vis-**koss**-ă-tee/ Symbol: η The resistance to flow of a fluid.

visible radiation *See* light.

vitreous /**vit**-ree-ŭs/ Resembling glass, or having the structure of a glass.

volatile Easily converted into a vapor.

volt Symbol: V The SI unit of electrical potential, potential difference, and e.m.f., defined as the potential difference between two points in a circuit between which a constant current of one ampere flows when the power dissipated is one watt. 1 V = 1 J C^{-1}. The unit is named for the Italian physicist Count Alessandro Volta (1745–1827).

voltaic cell /vol-**tay**-ik/ *See* cell.

volume strength *See* hydrogen peroxide.

volumetric analysis /vol-yŭ-**met**-rik/ One of the classical wet methods of quantitative analysis. It involves measuring the volume of a solution of accurately known concentration that is required to react with a solution of the substance being determined. The solution of known concentration (the standard solution) is added in small portions from a buret. The process is called a titration and the equivalence point is called the *end point*. End points are observed with the aid of indicators or by instrumental methods, such as conduction or light absorption. Volumetric analysis can also be applied to gases. The gas is typically held over mercury in a graduated tube, and volume changes are measured on reaction or after absorption of components of a mixture.

VSEPR (valence-shell electron-pair repulsion) A method of predicting the shape of molecules. In this theory one atom is taken to be the central atom and pairs of valence electrons are drawn round the central atom. The shape of the molecule is determined by minimizing the Coulomb repulsion between pairs and the repulsion between pairs due to the Pauli exclusion principle. Thus, for three pairs of electrons an equilateral triangle is favored and for four pairs of electrons a tetrahedron is favored. The VSEPR theory has had considerable success in predicting molecular shapes. *See* lone pair.

vulcanite /**vul**-kă-nÿt/ (**ebonite**) A hard black insulator made by vulcanizing rubber with a large amount of sulfur.

vulcanization /vul-kă-ni-**zay**-shŏn/ A process of improving the quality of rubber (hardness and resistance to temperature changes) by heating it with sulfur (about 150°C). Accelerators are used to speed up the reaction. Certain sulfur compounds can also be used for vulcanization.

Wacker process /**wak**-er/ An industrial process for making ethanal (and other carbonyl compounds). To produce ethanal, ethene and air are bubbled through an acid solution of palladium(II) chloride and copper(II) chloride (20–60°C and moderate pressure):

$$C_2H_4 + Pd^{2+} + O_2 \rightarrow CH_3CHO + Pd + 2H^+$$

The reaction involves an intermediate complex between palladium(II) ions and ethene. The purpose of the copper(II) chloride is to oxidize the palladium back to Pd^{2+} ions:

$$Pd + 2Cu^{2+} \rightarrow Pd^{2+} + 2Cu^+$$

The copper(I) ions spontaneously oxidize to copper(II) ions in air. The process provides a cheap source of ethanal (and, by oxidation, ethanoic acid) from the readily available ethene.

Walden inversion /**wawl**-dĕn/ A reaction in which an optically active compound reacts to give an optically active product in which the configuration has been inverted. This happens in the S_N2 mechanism. The inversion is named for the Russian–German chemist Paul Walden (1863–1957). *See* nucleophilic substitution.

washing soda *See* sodium carbonate.

water (H_2O) A colorless liquid that freezes at 0°C and, at atmospheric pressure, boils at 100°C. In the gaseous state water consists of single H_2O molecules. Due to the presence of two lone pairs the atoms do not lie in a straight line, the angle between the central oxygen atom and the two hydrogen atoms being 105°; the distance between each hydrogen atom and the oxygen atom is 0.099 nm. When ice forms, hydrogen bonds some 0.177 nm long develop between the hydrogen atom and oxygen atoms in adjacent molecules, giving ice its tetrahedral crystalline structure with a density of 916.8 kg m^{-3} at STP. Different ice structures develop under higher pressures. When ice melts to form liquid water, the tetrahedral structure breaks down, but some hydrogen bonds continue to exist; liquid water consists of groups of associated water molecules, $(H_2O)_n$, mixed with some monomers and some dimers. This mixture of molecular species has a higher density than the open-structured crystals. The maximum density of water, 999.97 kg m^{-3}, occurs at 3.98°C. This accounts for the ability of ice to float on water and for the fact that water pipes burst on freezing.

Although water is predominantly a covalent compound, a very small amount of ionic dissociation occurs ($H_2O \rightleftharpoons H^+ + OH^-$). In every liter of water at STP there is approximately 10^{-7} mole of each ionic species. It is for this reason that, on the pH scale, a neutral solution has a value of 7.

As a polar liquid, water is the most powerful solvent known. This is partly a result of its high dielectric constant and partly its ability to hydrate ions. This latter property also accounts for the incorporation of water molecules into some ionic crystals as water of crystallization.

Water is decomposed by reactive metals (e.g. sodium) when cold and by less active metals (e.g. iron) when steam is passed over the hot metal. It is also decomposed by electrolysis.

water gas A mixture of carbon monoxide and hydrogen produced when steam is passed over red-hot coke or made to combine with hydrocarbons, e.g.

$$C(s) + H_2O(g) \rightarrow CO(g) + H_2(g)$$

$$CH_4(g) + H_2O(g) \rightarrow CO(g) + 3H_2(g)$$

The production of water gas using methane is an important step in the preparation of hydrogen for ammonia synthesis. *Compare* producer gas.

water glass *See* sodium silicate.

water of crystallization Water present in definite proportions in crystalline compounds. Compounds containing water of crystallization are called *hydrates*. Examples are copper(II) sulfate pentahydrate ($CuSO_4.5H_2O$) and sodium carbonate decahydrate ($Na_2CO_3.10H_2O$). The water can be removed by heating.

When hydrated crystals are heated the water molecules may be lost in stages. For example, copper(II) sulfate pentahydrate changes to the monohydrate ($CuSO_4.H_2O$) at 100°C, and to the anhydrous salt ($CuSO_4$) at 250°C. The water molecules in crystalline hydrates may be held by hydrogen bonds (as in $CuSO_4.H_2O$) or, alternatively, may be coordinated to the metal ion as a complex aquo ion.

water softening The removal from water of dissolved calcium, magnesium, and iron compounds, thus reducing the hardness of the water. The compounds are potentially damaging because they can accumulate in pipes and boilers and they react with, and therefore waste, soap. Temporary hardness can be removed by boiling the water. Permanent hardness can be removed in a number of ways: by distillation; by the addition of sodium carbonate, which causes dissolved calcium, for example, to precipitate out as calcium carbonate; and by the use of ion-exchange products such as Permutit and Calgon. *See also* hardness (of water).

watt /wot/ Symbol: W The SI unit of power, defined as a power of one joule per second. 1 W = 1 J s^{-1}. The unit is named for the British instrument maker and inventor James Watt (1736–1819).

wavefunction /**wayv**-fung-shŏn/ *See* quantum theory; orbital.

waveguides /**wayv**-gÿdz/ *See* microwaves.

wavelength Symbol: λ The distance between the ends of one complete cycle of a wave. Wavelength is related to the speed (c) and frequency (v) thus:

$$c = v\lambda$$

wave mechanics *See* quantum theory.

wave number Symbol: σ The reciprocal of the wavelength of a wave. It is the number of wave cycles in unit distance, and is often used in spectroscopy. The unit is the $meter^{-1}$ (m^{-1}). The circular wave number (symbol: k) is given by:

$$k = 2\pi\sigma$$

wax A soft amorphous water-repellant organic material. The term is particularly used for fatty-acid esters of monohydric alcohols.

weak acid (or base) An acid or base that is not fully dissociated in solution.

weak base A base that is only partly or incompletely dissociated into its component ions in solution.

weber /**vay**-ber/ Symbol: Wb The SI unit of magnetic flux, equal to the magnetic flux that, linking a circuit of one turn, produces an e.m.f. of one volt when reduced to zero at uniform rate in one second. 1 Wb = 1 V s. The unit is named for the German physicist Wilhelm Eduard Weber (1804–91).

Weston cadmium cell A standard cell that produces a constant e.m.f. of 1.0186 volts at 20°C. It consists of an H-shaped glass vessel containing a negative cadmium-mercury amalgam electrode in one leg and a positive mercury electrode in the other. The electrolyte – saturated cadmium sulfate solution – fills the horizontal bar of the vessel to connect the two electrodes. The e.m.f. of the cell varies very little with temperature, being given by the equation $E = 1.0186 - 0.000\,037\,(T - 293)$, where T is the thermodynamic temperature.

wet cell A type of cell, such as a car battery, in which the electrolyte is a liquid solution. *See also* Leclanché cell.

white arsenic *See* arsenic(III) oxide.

white lead *See* lead(II) carbonate hydroxide.

white phosphorus *See* phosphorus.

white spirit A liquid hydrocarbon resembling kerosene obtained from petroleum, used as a solvent and in the manufacture of paints and varnishes.

Williamson's continuous process A method of preparing simple ethers by dehydration of alcohols with concentrated sulfuric acid. The reaction is carried out at 140°C under reflux with an excess of the alcohol:

$$2ROH \rightarrow ROR + H_2O$$

The concentrated sulfuric acid both catalyzes the reaction and displaces the equilibrium to the right. The product, ether, is termed 'simple', because the R groups are identical.

Williamson's synthesis A method for the preparation of mixed ethers by nucleophilic substitution. A haloalkane is refluxed with an alcoholic solution of sodium alkoxide (from sodium dissolved in alcohol):

$$R'Cl + RO^-Na^+ \rightarrow R'OR + NaCl$$

The product ether is termed 'mixed' if the alkyl groups R and R′ are different. This synthesis can produce both simple and mixed ethers. The synthesis is named for the British chemist Alexander William Williamson (1824–1904).

will-o'-the-wisp *See* ignis fatuus.

witherite /**wi***th*-ĕ-rÿt/ A mineral form of barium carbonate, $BaCO_3$.

Wöhler's synthesis /**voh**-lerz/ A synthesis of urea from inorganic compounds (Friedrich Wöhler, 1828). The urea was produced by evaporating NH_4NCO (ammonium cyanate). The experiment was important as it demonstrated that 'organic' chemicals (i.e. ones produced in living organisms) could be made from inorganic starting materials. The synthesis is named for the German chemist Friedrich Wöhler (1800–82).

wolfram /**wûlf**-răm/ A former alternative name for tungsten.

wood alcohol *See* methanol.

Wood's metal A fusible alloy containing 50% bismuth, 25% lead, and 12.5% tin and cadmium. Its low melting point (70°C) leads to its use in fire-protection devices. The alloy is named for the English scientist William Wood (1671–1730).

work function *See* photoelectric effect.

wrought iron A low-carbon steel obtained by refining the iron produced in a blast furnace. Wrought iron is processed by heating and hammering to reduce the slag content and to ensure its even distribution.

Wurtz reaction /voorts/ A reaction for preparing alkanes by refluxing a haloalkane (RX) with sodium metal in dry ether:

$$2RX + 2Na \rightarrow RR + 2NaX$$

The reaction involves the coupling of two alkyl radicals. The *Fittig reaction* is a similar process for preparing alkyl-benzene hydrocarbons by using a mixture of halogen compounds. For example, to obtain methylbenzene:

$$C_6H_5Cl + CH_3Cl + 2Na \rightarrow C_6H_5CH_3 + 2NaCl$$

In this mixed reaction phenylbenzene ($C_6H_5C_6H_5$) and ethane (CH_3CH_3) are also produced by side reactions. The Wurtz reaction is named for the French chemist Charles Adolphe Wurtz (1817–84) and the Fittig reaction is named for the German organic chemist Rudolph Fittig (1835–1910).

xanthate /**zan**-thayt/ (-SCS(OR)) A salt or ester of XANTHIC ACID (where R is an organic group). Cellulose xanthate is used to make RAYON.

xanthene /**zan**-theen/ ($CH_2(C_6H_4)_2O$) A yellow crystalline organic compound, used in making dyestuffs and fungicides.

xanthic acid /**zan**-thik/ (HSCS(OR)) Any of several unstable organic acids (where R is an organic group), whose esters and salts have various industrial applications. *See* xanthate.

xanthine /**zan**-theen, -th'ÿn/ (**2,6-dioxypurine;** $C_5H_4N_2O_2$) A poisonous colorless crystalline organic compound that occurs in blood, coffee beans, potatoes, and urine. It is used as a chemical intermediate.

xanthone /**zan**-thohn/ (**dibenzo-4-pyrone;** $CO(C_6H_4)_2O$) A colorless crystalline organic compound found as a pigment in gentians and other flowers. It is used as an insecticide and in making dyestuffs.

xenon /**zen**-on/ A colorless odorless monatomic element of the rare-gas group. It occurs in trace amounts in air. Xenon is used in electron tubes and strobe lighting.

Symbol: Xe; m.p. –111.9°C; b.p. –107.1°C; d. 5.8971 (0°C) kg m^{-3}; p.n. 54; r.a.m. 131.29.

x-radiation An energetic form of electromagnetic radiation. The wavelength range is 10^{-11} m to 10^{-8} m. X-rays are normally produced by absorbing high-energy electrons in matter. The radiation can pass through matter to some extent (hence its use in medicine and industry for investigating internal structures). It can be detected with photographic emulsions and devices like the Geiger–Müller tube.

X-ray photons result from electronic transitions between the inner energy levels of atoms. When high-energy electrons are absorbed by matter, an x-ray line spectrum results. The structure depends on the substance and is thus used in x-ray spectroscopy. The line spectrum is always formed in conjunction with a continuous background spectrum. The minimum (cutoff) wavelength λ_0 corresponds to the maximum x-ray energy, W_{max}. This equals the maximum energy of electrons in the beam producing the x-rays. Wavelengths in the continuous spectrum above λ_0 are caused by more gradual energy loss by the electrons, in the process called *bremsstrahlung* (braking radiation).

x-ray crystallography The study of the internal structure of crystals using the technique of x-ray diffraction.

x-ray diffraction A technique used to determine crystal structure by directing x-rays at the crystals and examining the diffraction patterns produced. At certain angles of incidence a series of spots are produced on a photographic plate; these spots are caused by interaction between the x-rays and the planes of the atoms, ions, or molecules in the crystal lattice. The positions of the spots are consistent with the Bragg equation $n\lambda = 2d\sin\theta$.

x-rays *See* x-radiation.

xylene /**zÿ**-leen/ *See* dimethylbenzene.

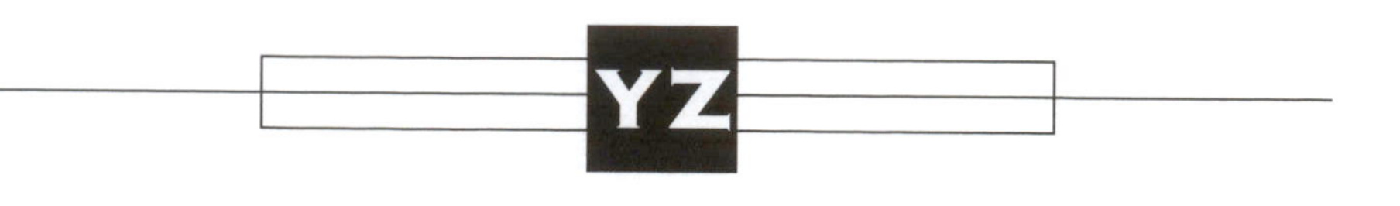

yellow phosphorus *See* phosphorus.

yocto- Symbol: y A prefix denoting 10^{-24}. For example, 1 yoctometer (ym) = 10^{-24} meter (m).

yotta- Symbol: Y A prefix denoting 10^{24}. For example, 1 yottameter (Ym) = 10^{24} meter (m).

ytterbium /i-**ter**-bee-ŭm/ A soft malleable silvery element having two allotropes and belonging to the lanthanoid series of metals. It occurs in association with other lanthanoids. Ytterbium has been used to improve the mechanical properties of steel.

Symbol: Yb; m.p. 824°C; b.p. 1193°C; r.d. 6.965 (20°C); p.n. 70; r.a.m. 173.04.

yttrium /**it**-ree-ŭm/ A silvery metallic element belonging to the second transition series. It is found in almost every lanthanoid mineral, particularly monazite. Yttrium is used in various alloys, in yttrium–aluminum garnets used in the electronics industry and as gemstones, as a catalyst, and in superconductors. A mixture of yttrium and europium oxides is widely used as the red phosphor on television screens.

Symbol: Y; m.p. 1522°C; b.p. 3338°C; r.d. 4.469 (20°C); p.n. 39; r.a.m. 88.90585.

Zeeman effect /**zay**-mahn/ The splitting of atomic spectral lines by an external applied magnetic field. It was first reported by the Dutch physicist Pieter Zeeman (1865–1943) in 1896.

Zeisel reaction /**zȳ**-sĕl/ The reaction of an ether with excess concentrated hydroiodic acid. On refluxing, a mixture of iodoalkanes is formed:

$$ROR' + 2HI \rightarrow H_2O + RI + R'I$$

Analysis to identify the iodoalkanes gives information about the composition of the original ether.

Zeise's salt /**zȳ**-sĕz, **tsȳ**-/ A complex of platinum and ethane (ethylene) first synthesized by W. C. Zeise in 1827. It has the formula $PtCl_3(CH_2CH_2)$ and was the first complex in which the metal ion coordinated to a pi-electron system rather than to individual atoms.

zeolite /**zee**-ŏ-lȳt/ A member of a group of hydrated aluminosilicate minerals, which occur in nature and are also manufactured for their ion-exchange and selective-absorption properties. They are used for water softening and for sugar refining. The zeolites have an open crystal structure and can be used as MOLECULAR SIEVES. *See also* ion exchange.

zepto- Symbol: z A prefix denoting 10^{-21}. For example, 1 zeptometer (zm) = 10^{-21} meter (m).

zero order Describing a chemical reaction in which the rate of reaction is independent of the concentration of a reactant; i.e.

$$\text{rate} = k[X]^0$$

The concentration of the reactant remains constant for a period of time although other reactants are being consumed. The hydrolysis of 2-bromo-2-methylpropane using aqueous alkali has a rate expression,

$$\text{rate} = k[\text{2-bromo-2-methylpropane}]$$

i.e. the reaction is zero order with respect to the concentration of the alkali (it does not depend on the alkali concentration). The rate constant for a zero reaction has the units mol dm^{-3} s^{-1}.

zero point energy The energy possessed by the atoms and molecules of a substance at absolute zero (0 K).

zetta- Symbol: Z A prefix denoting 10^{21}. For example, 1 zettameter (Zm) = 10^{21} meter (m).

Ziegler process /**zee**-gler/ A method for the manufacture of high-density polyethene using a catalyst of titanium(IV)] chloride and triethyl aluminum ($Al(C_2H_5)_3$) under slight pressure. The mechanism involves formation of titanium alkyls

$$TiCl_3(C_2H_5)$$

which coordinate the σ-orbitals of titanium with the π bond of ethene. The chain-length, and hence the density, of the polymer can be controlled. The process is named for the German chemist Karl Ziegler (1898–1973).

zinc A bluish-white transition metal occurring naturally as the sulfide (zinc blende) and carbonate (smithsonite). It is extracted by roasting the ore in air and then reducing the oxide to the metal using carbon. Zinc is used to galvanize iron, in alloys (e.g. brass), and in dry batteries. It reacts with acids and alkalis but only corrodes on the surface in air. Zinc ions are identified in solution by forming white precipitates with sodium hydroxide or ammonia solution, both precipitates being soluble in excess reagent.

Symbol: Zn; m.p. 419.58°C; b.p. 907°C; r.d. 7.133 (20°C); p.n. 30; r.a.m. 65.39.

zincate /**zink**-ayt/ A salt containing the ion ZnO_2^{2-}.

zinc-blende (sphalerite) structure A form of crystal structure that consists of a zinc atom surrounded by four sulfur atoms arranged tetrahedrally; each sulfur atom is similarly surrounded by four atoms of zinc. Zinc sulfide crystallizes in the cubic system. Covalent bonds of equal strength and length result in the formation of a giant molecular structure. If the zinc and sulfur atoms are replaced by carbon atoms the diamond structure is produced. Wurtzite (another form of zinc sulfide) is similar but belongs to the hexagonal crystal system.

zinc-blende structure A form of crystal structure that consists of a zinc atom surrounded by four sulfur atoms arranged tetrahedrally; each sulfur atom is similarly surrounded by four atoms of zinc. Zinc sulfide crystallizes in the cubic system. Covalent bonds of equal strength and length result in the formation of a giant molecular structure. If the zinc and sulfur atoms are replaced by carbon atoms the diamond structure is produced. Wurtzite (another form of zinc sulfide) is similar but belongs to the hexagonal crystal system.

zinc carbonate *See* calamine.

zinc chloride ($ZnCl_2$) A white crystalline solid prepared by the action of dry hydrogen chloride or chlorine on heated zinc. The anhydrous product undergoes sublimation. The hydrated salt may be prepared by the addition of excess zinc to dilute hydrochloric acid and crystallization of the solution. Hydrates with 4, 3, 5/2, 3/2, and 1 molecule of water exist. Zinc chloride is extremely deliquescent and dissolves easily in water. It is used in dentistry, Leclanché cells, as a flux, and as a dehydrating agent in organic reactions.

zinc group A group of metallic elements in the periodic table, consisting of zinc (Zn), cadmium (Cd), and mercury (Hg). The elements all occur at the ends of the three transition series and all have outer $d^{10}s^2$ configurations; Zn $[Ar]3d^{10}4s^2$, Cd $[Kr]4d^{10}5s^2$, $Hg[Xn]5d^{10}6s^2$. The elements all use only outer s-electrons in reaction and in combination with other elements unlike the coinage metals, which immediately precede them. Thus all form dipositive ions, and oxidation states higher than +2 are not known. In addition mercury forms the mercurous ion, sometimes written as Hg(I]], but actually the ion $^+Hg\text{-}Hg^+$. Although the elements fall naturally at the end of the transition series their properties are more like those of main-group elements than of transition metals:

1. The melting points of the transition elements are generally high whereas those of the zinc group are low (Zn 419.5°C, Cd 329.9°C, Hg –38.87°C).
2. The zinc-group ions are diamagnetic and colorless whereas most transition elements have colored ions and many paramagnetic species.
3. The zinc group does not display the variable valence associated with transition metal ions. However the group does have the transition-like property of forming many complexes or coordination compounds, such as $[Zn(NH_3)_4]^{2+}$ and $[Hg(CN]]_4]^{2-}$.

Within the group, mercury is anomalous because of the Hg_2^{2+} ion mentioned above and its general resistance to reaction. Zinc and cadmium are electropositive and will react with dilute acids to release hydrogen whereas mercury will not. When zinc and cadmium are heated in air they burn to give the oxides, MO, which are stable to further strong heating. In contrast mercury does not react readily with oxygen (slow reaction at the boiling point), and mercuric oxide, HgO, decomposes to the metal and oxygen on further heating. All members of the group form dialkyls and diaryls, R_2M.

zincite /**zink**-ÿt/ (**spartalite**) A red-orange mineral form of ZINC OXIDE, ZnO, often containing also some manganese. It is an important ore of zinc.

zinc oxide (ZnO) A compound prepared by the thermal decomposition of zinc nitrate or carbonate; it is a white powder when cold, yellow when hot. Zinc oxide is an amphoteric oxide, almost insoluble in water, but dissolves readily in both acids and alkalis. If mixed with powdered carbon and heated to red heat, it is reduced to the metal. Zinc oxide is used in the paint and ceramic industries. Medically it is used in zinc ointment.

zinc sulfate ($ZnSO_4$) A white crystalline solid prepared by the action of dilute sulfuric acid on either zinc oxide or zinc carbonate. On crystallization the hydrated salt is formed. The heptahydrate ($ZnSO_4.7H_2O$) is formed below 30°C, the hexahydrate ($ZnSO_4.6H_2O$) above 30°C, and the monohydrate ($ZnSO_4.H_2O$) at 100°C; at 450°C the salt is anhydrous. Zinc sulfate is extremely soluble in water. It is used in the textile industry.

zinc sulfide (ZnS) A compound that can be prepared by direct combination of zinc and sulfur. Alternatively it can be prepared as a white amorphous precipitate by bubbling hydrogen sulfide through an alkaline solution of a zinc salt or by the addition of ammonium sulfide to a soluble zinc salt solution. Zinc sulfide occurs naturally as the mineral zinc blende. It dissolves in dilute acids to yield hydrogen sulfide. Impure zinc sulfide is phosphorescent. It is used as a pigment and in the coatings on luminescent screens.

zircon /**zer**-kon/ *See* zirconium.

zirconium /zer-**koh**-nee-ŭm/ A hard lustrous silvery transition element that occurs in a gemstone, zircon ($ZrSiO_4$). It is used in some strong alloy steels.

Symbol: Zr; m.p. 1850°C; b.p. 4380°C; r.d. 6.506 (20°C); p.n. 40; r.a.m. 91.224.

zwitterion /**tsvit**-er-ÿ-on, -ŏn, **zwit**-/ (**ampholyte ion**) An ion that has both a positive and negative charge on the same species. Zwitterions occur when a molecule contains both a basic group and an acidic group; formation of the ion can be regarded as an internal acid-base reaction. For example, amino-ethanoic acid (glycine) has the formula $H_{2+}N.CH_2.COOH$. Under neutral conditions it exists as the zwitterion $H_3N.CH_2.COO^-$, which can be formed by transfer of a proton from the carboxyl group to the amine group. At low pH (acidic conditions) the ion $^+H_3N.CH_2.COOH$ is formed; at high pH (basic conditions) $H_2N.CH_2.COO^-$ is formed. *See also* amino acid.

APPENDIXES

Appendix I

Carboxylic Acids

In the examples below, the systematic name is given first, followed by the trivial (common) name.

Simple saturated monocarboxylic acids:

methanoic	formic	$HCOOH$
ethanoic	acetic	CH_3COOH
propanoic	proprionic	C_2H_5COOH
butanoic	butyric	C_3H_7COOH
pentanoic	valeric	C_4H_9COOH
hexanoic	caproic	$C_5H_{11}COOH$
heptanoic	enathic	$C_6H_{13}COOH$
octanoic	caprylic	$C_7H_{15}COOH$
nonanoic	pelargonic	$C_8H_{17}COOH$
decanoic	capric	$C_9H_{19}COOH$

Other simple saturated acids are found in naturally occurring glycerides. They all contain even numbers of carbon atoms:

dodecanoic	lauric	$C_{11}H_{23}COOH$
tetradecanoic	myristic	$C_{13}H_{27}COOH$
hexadecanoic	palmitic	$C_{15}H_{31}COOH$
octadecanoic	stearic	$C_{17}H_{35}COOH$
eicosanoic	arachidic	$C_{19}H_{39}COOH$
docosanoic	behenic	$C_{21}H_{43}COOH$
tetracosanoic	lignoceric	$C_{23}H_{47}COOH$
hexacosanoic	cerotic	$C_{25}H_{51}COOH$

Certain important unsaturated monocarboxylic acids occur naturally:

octadec-9-enoic	oleic	$C_8H_{17}CH{=}CHC_7H_{14}COOH$
octadeca-9,12- dienoic	linoleic	$C_5H_{11}CH{=}CHCH_2CH{=}CHC_7H_{14}COOH$
octadeca-9,12,15- trienoic	linolenic	$C_2H_5CH{=}CHCH_2CH{=}CHCH_2CH{=}CH\text{-}C_7H_{14}COOH$

There are a number of common saturated dicarboxylic acids:

ethanedioic	oxalic	$HOOCCOOH$
propanedioic	malonic	$HOOCCH_2COOH$
butanedioic	succinic	$HOOCC_2H_4COOH$
pentanedioic	glutaric	$HOOCC_3H_6COOH$
hexanedioic	adipic	$HOOCC_4H_8COOH$

Examples of unsaturated dicarboxylic acids are:

cis-butenedioic	maleic	HOOCCH=CHCOOH
trans-butenedioic	fumaric	HOOCCH=CHCOOH

Certain hydroxy acids occur naturally:

hydroxyethanoic	glycolic	$CH_2(OH)COOH$
2-hydroxypropanoic	lactic	$CH_3CH(OH)COOH$
hydroxybutanedioic	malic	$CH(OH)CH_2(COOH)_2$
2-hydroxy-propane-1,2,3-tricarboxylic	citric	$(CH_2)_2C(OH)(COOH)_3$

Naturally occurring amino carboxylic acids are shown in Appendix II.

Appendix II

Amino Acids

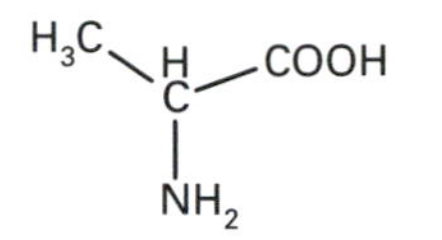

alanine

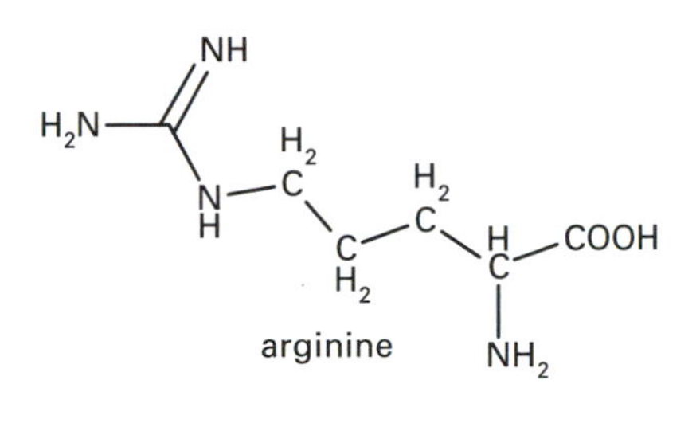

arginine

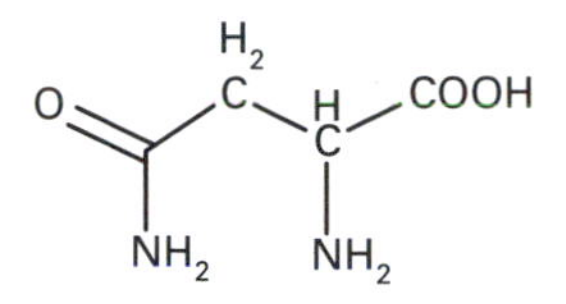

asparagine

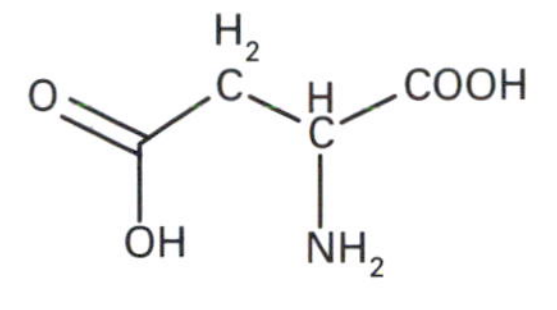

aspartic acid

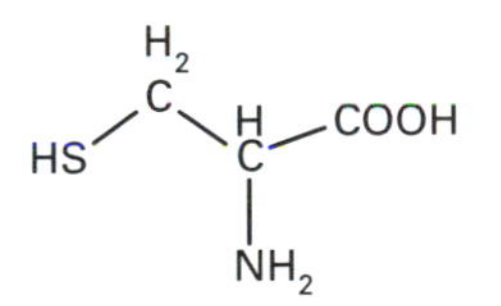

cysteine

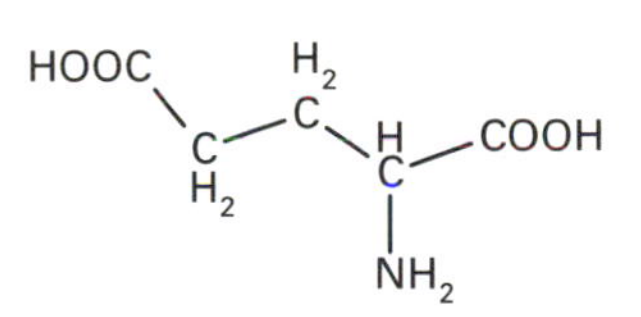

glutamic acid

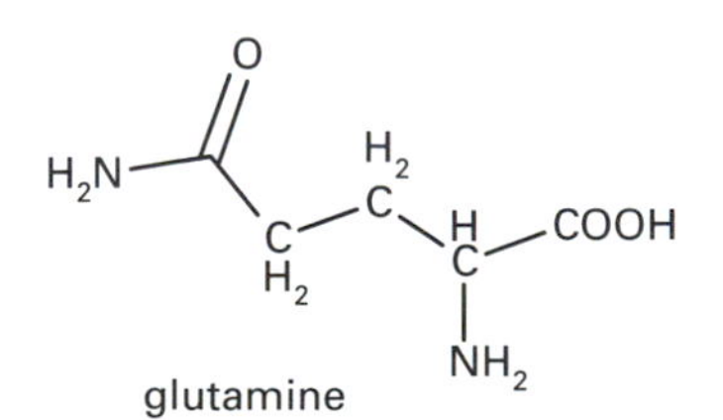

glutamine

COOH
H_2C
NH_2

glycine

Amino Acids

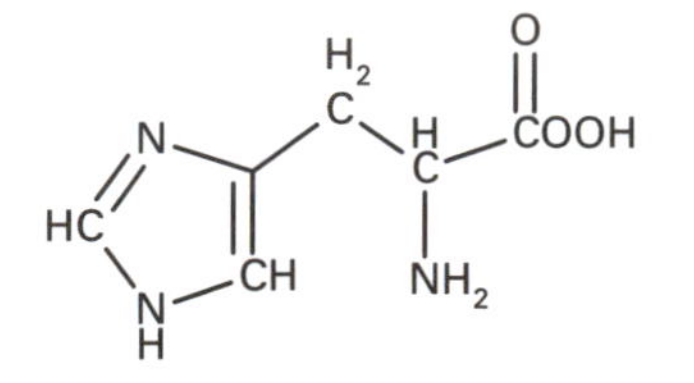

histidine

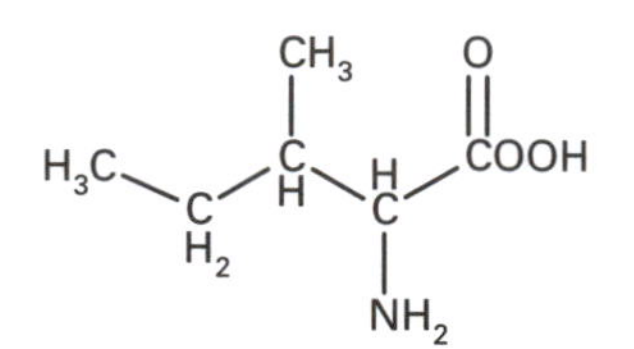

isoleucine

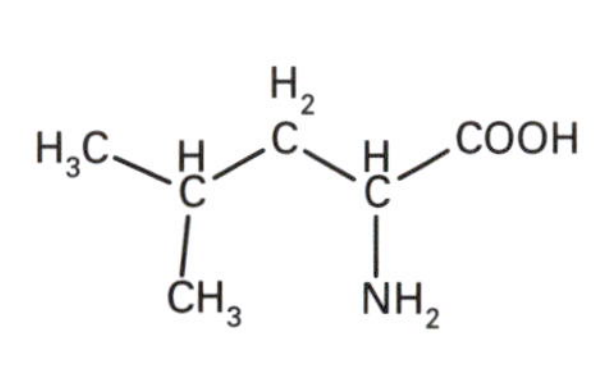

leucine

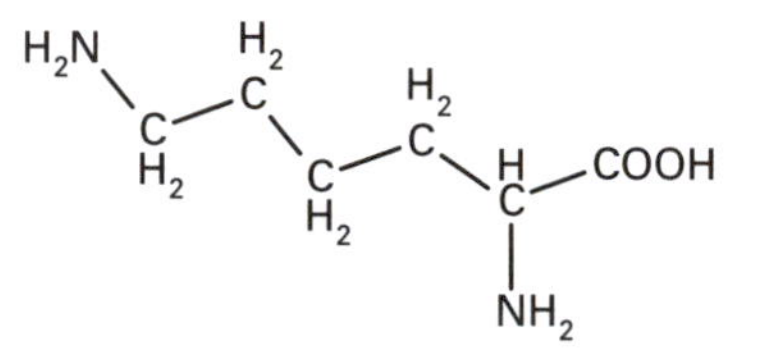

lysine

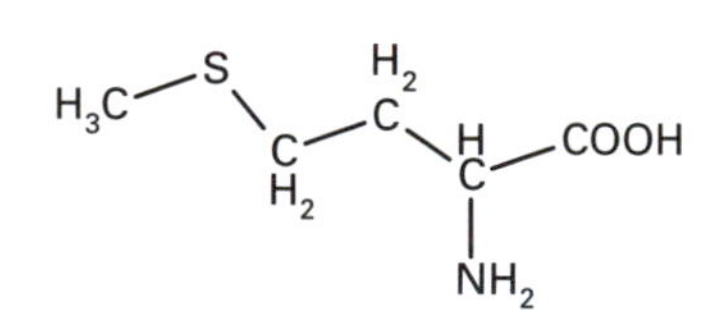

methionine

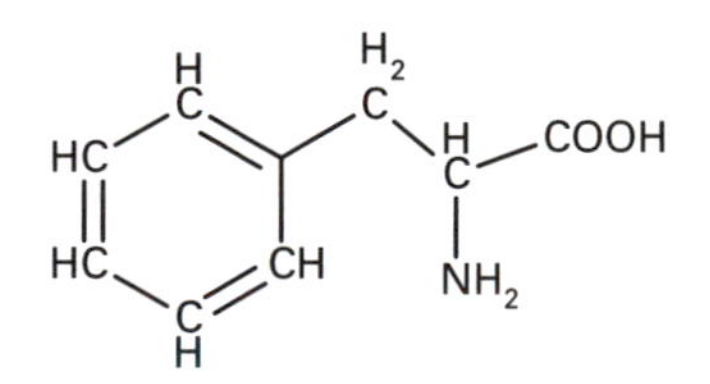

phenylalanine

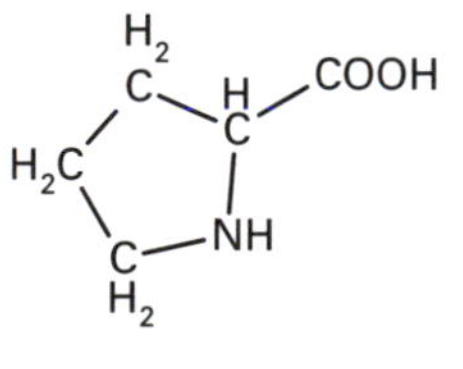

proline

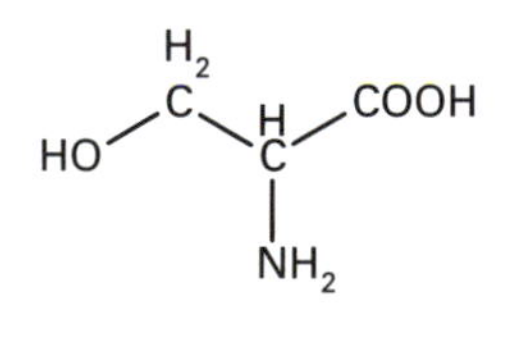

serine

Amino Acids

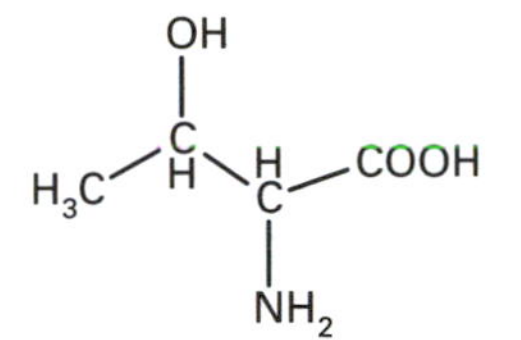

threonine

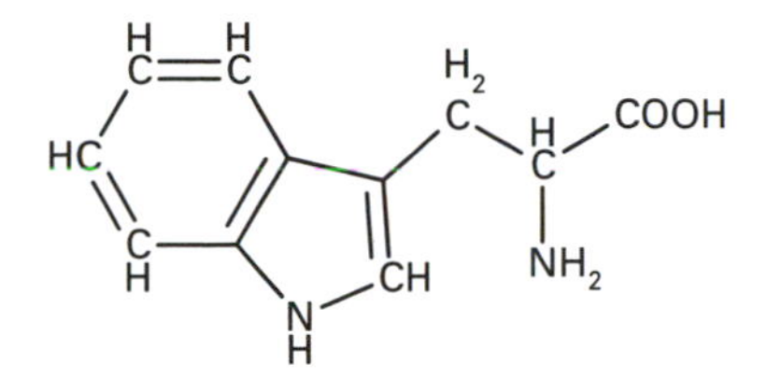

tryptophan

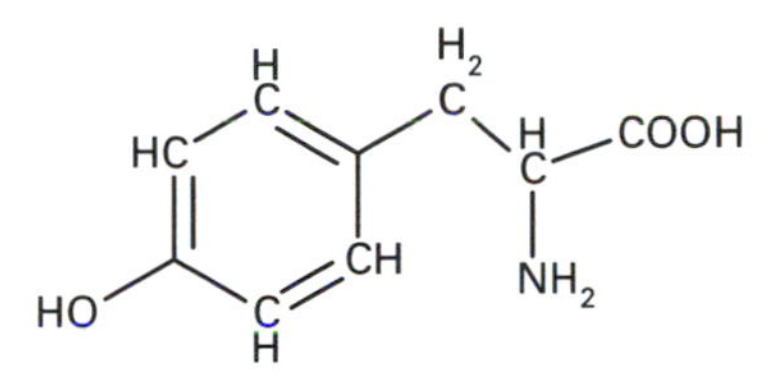

tyrosine

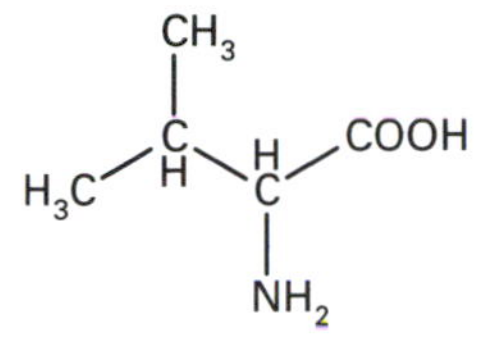

valine

Appendix III

Sugars

Some simple monosaccharides. The β-D-form is shown in each case

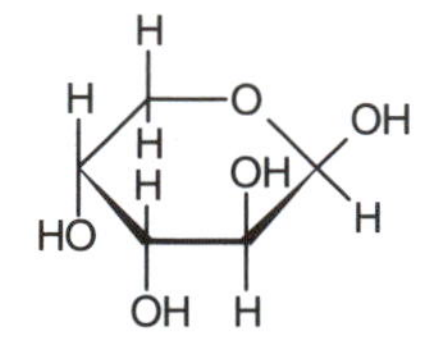

arabinose

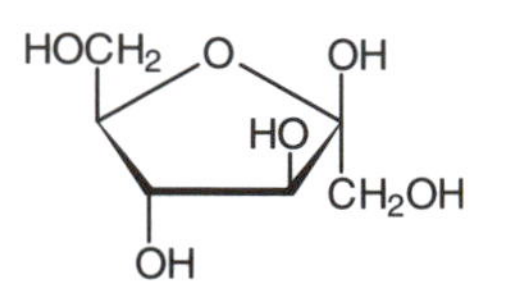

fructose

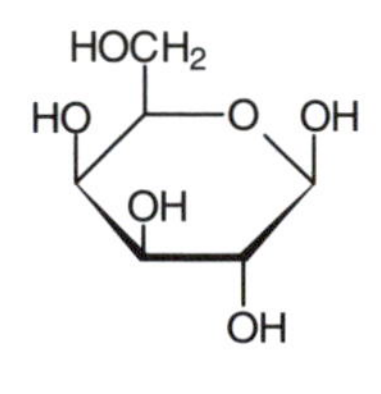

galactose

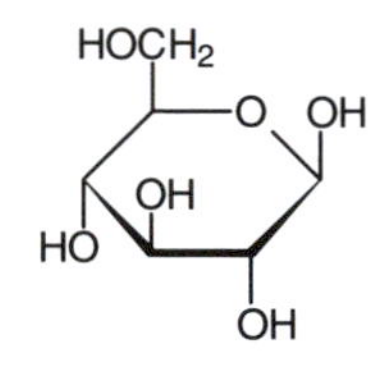

glucose

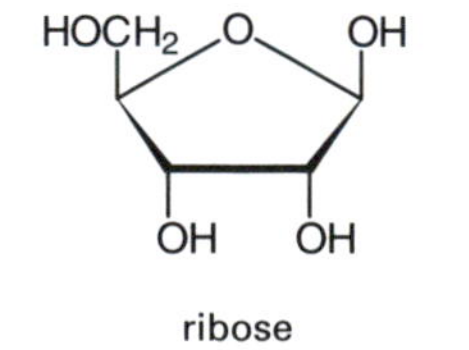

ribose

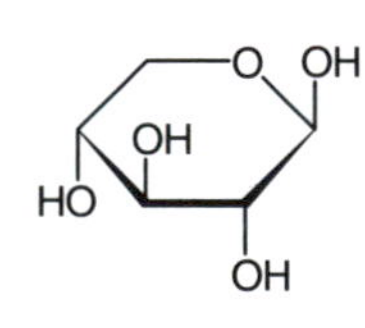

xylose

Appendix IV

Nitrogenous Bases and Nucleosides

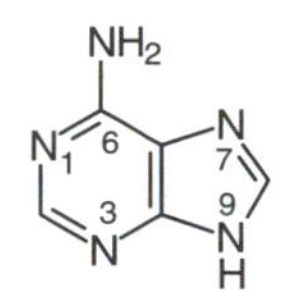

adenine

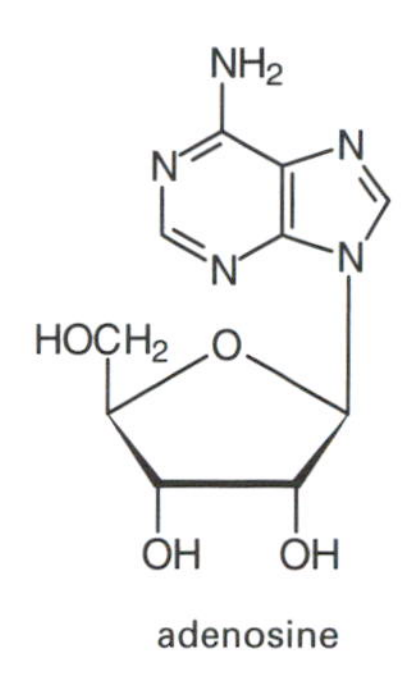

adenosine

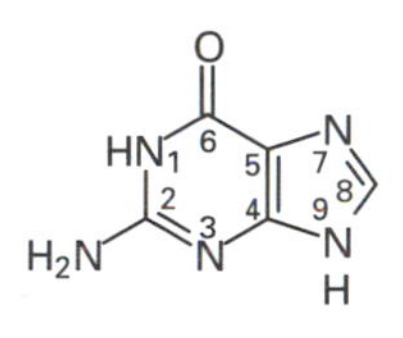

guanine

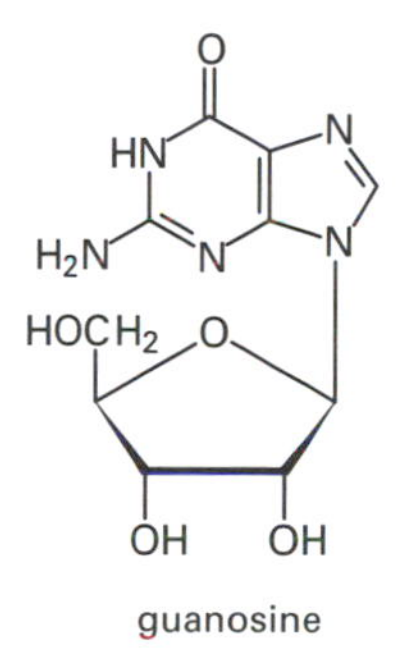

guanosine

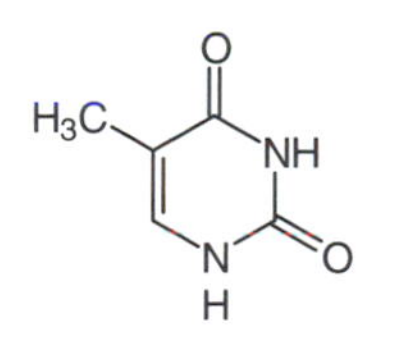

thymine

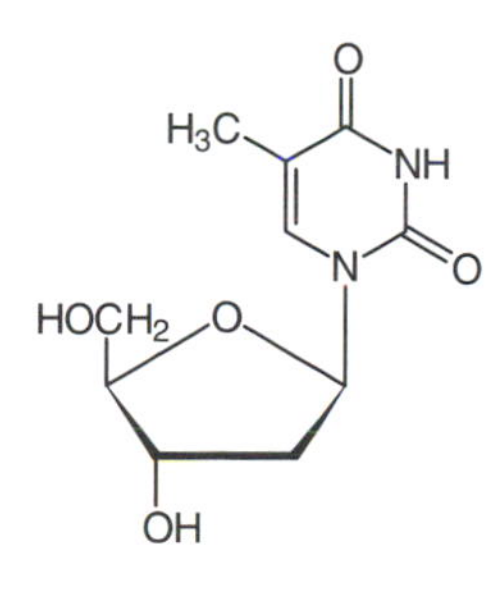

thymidine

Nitrogenous Bases and Nucleosides

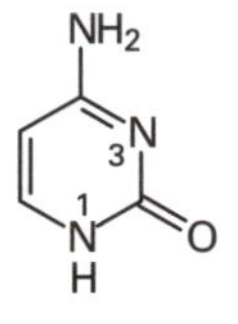

cytosine

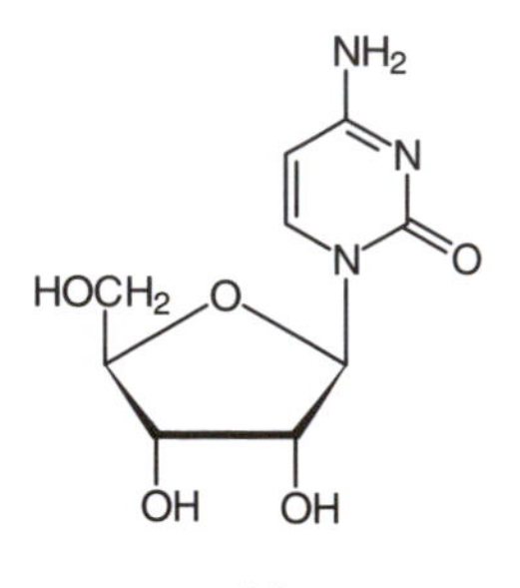

cytidine

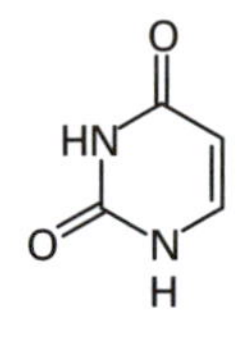

uracil

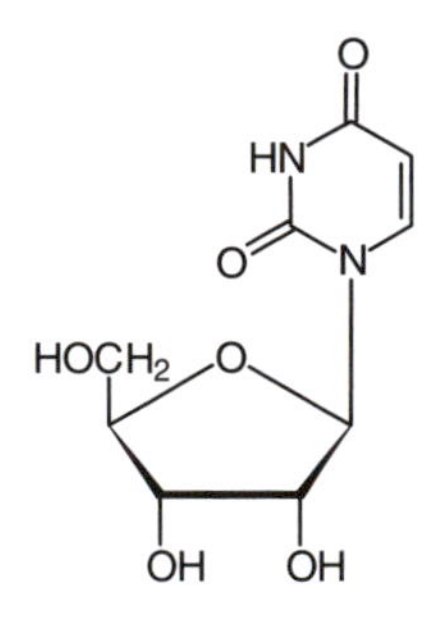

uridine

Appendix V

The Chemical Elements

(* *indicates the nucleon number of the most stable isotope*)

Element	*Symbol*	*p.n.*	*r.a.m*	*Element*	*Symbol*	*p.n.*	*r.a.m*
actinium	Ac	89	227*	europium	Eu	63	151.965
aluminum	Al	13	26.982	fermium	Fm	100	257*
americium	Am	95	243*	fluorine	F	9	18.9984
antimony	Sb	51	112.76	francium	Fr	87	223*
argon	Ar	18	39.948	gadolinium	Gd	64	157.25
arsenic	As	33	74.92	gallium	Ga	31	69.723
astatine	At	85	210	germanium	Ge	32	72.61
barium	Ba	56	137.327	gold	Au	79	196.967
berkelium	Bk	97	247*	hafnium	Hf	72	178.49
beryllium	Be	4	9.012	hassium	Hs	108	265*
bismuth	Bi	83	208.98	helium	He	2	4.0026
bohrium	Bh	107	262*	holmium	Ho	67	164.93
boron	B	5	10.811	hydrogen	H	1	1.008
bromine	Br	35	79.904	indium	In	49	114.82
cadmium	Cd	48	112.411	iodine	I	53	126.904
calcium	Ca	20	40.078	iridium	Ir	77	192.217
californium	Cf	98	251*	iron	Fe	26	55.845
carbon	C	6	12.011	krypton	Kr	36	83.80
cerium	Ce	58	140.115	lanthanum	La	57	138.91
cesium	Cs	55	132.905	lawrencium	Lr	103	262*
chlorine	Cl	17	35.453	lead	Pb	82	207.19
chromium	Cr	24	51.996	lithium	Li	3	6.941
cobalt	Co	27	58.933	lutetium	Lu	71	174.967
copper	Cu	29	63.546	magnesium	Mg	12	24.305
curium	Cm	96	247*	manganese	Mn	25	54.938
darmstadtium	Ds	110	269*	meitnerium	Mt	109	266*
dubnium	Db	105	262*	mendelevium	Md	101	258*
dysprosium	Dy	66	162.50	mercury	Hg	80	200.59
einsteinium	Es	99	252*	molybdenum	Mo	42	95.94
erbium	Er	68	167.26	neodymium	Nd	60	144.24

The Chemical Elements

Element	*Symbol*	*p.n.*	*r.a.m*	*Element*	*Symbol*	*p.n.*	*r.a.m*
neon	Ne	10	20.179	selenium	Se	34	78.96
neptunium	Np	93	237.048	silicon	Si	14	28.086
nickel	Ni	28	58.69	silver	Ag	47	107.868
niobium	Nb	41	92.91	sodium	Na	11	22.9898
nitrogen	N	7	14.0067	strontium	Sr	38	87.62
nobelium	No	102	259*	sulfur	S	16	32.066
osmium	Os	76	190.23	tantalum	Ta	73	180.948
oxygen	O	8	15.9994	technetium	Tc	43	99*
palladium	Pd	46	106.42	tellurium	Te	52	127.60
phosphorus	P	15	30.9738	terbium	Tb	65	158.925
platinum	Pt	78	195.08	thallium	Tl	81	204.38
plutonium	Pu	94	244*	thorium	Th	90	232.038
polonium	Po	84	209*	thulium	Tm	69	168.934
potassium	K	19	39.098	tin	Sn	50	118.71
praseodymium	Pr	59	140.91	titanium	Ti	22	47.867
promethium	Pm	61	145*	tungsten	W	74	183.84
protactinium	Pa	91	231.036	ununbium	Uub	112	285*
radium	Ra	88	226.025	ununtrium	Uut	113	284*
radon	Rn	86	222*	ununquadium	Uuq	114	289*
rhenium	Re	75	186.21	ununpentium	Uup	115	288*
rhodium	Rh	45	102.91	ununhexium	Uuh	116	292*
roentgenium	Rg	111	272*	uranium	U	92	238.03
rubidium	Rb	37	85.47	vanadium	V	23	50.94
ruthenium	Ru	44	101.07	xenon	Xe	54	131.29
rutherfordium	Rf	104	261*	ytterbium	Yb	70	173.04
samarium	Sm	62	150.36	yttrium	Y	39	88.906
scandium	Sc	21	44.956	zinc	Zn	30	65.39
seaborgium	Sg	106	263*	zirconium	Zr	40	91.22

Appendix VI

Periodic Table of the Elements - giving group, atomic number, and chemical symbol

Period	1	2	3	4	5	6	7	8	9	10	11	12	13	14	15	16	17	18
1	1 H																	2 He
2	3 Li	4 Be											5 B	6 C	7 N	8 O	9 F	10 Ne
3	11 Na	12 Mg											13 Al	14 Si	15 P	16 S	17 Cl	18 Ar
4	19 K	20 Ca	21 Sc	22 Ti	23 V	24 Cr	25 Mn	26 Fe	27 Co	28 Ni	29 Cu	30 Zn	31 Ga	32 Ge	33 As	34 Se	35 Br	36 Kr
5	37 Rb	38 Sr	39 Y	40 Zr	41 Nb	42 Mo	43 Tc	44 Ru	45 Rh	46 Pd	47 Ag	48 Cd	49 In	50 Sn	51 Sb	52 Te	53 I	54 Xe
6	55 Cs	56 Ba	57-71 La-Lu	72 Hf	73 Ta	74 W	75 Re	76 Os	77 Ir	78 Pt	79 Au	80 Hg	81 Tl	82 Pb	83 Bi	84 Po	85 At	86 Rn
7	87 Fr	88 Ra	89-103 Ac-Lr	104 Rf	105 Db	106 Sg	107 Bh	108 Hs	109 Mt	110 Ds	111 Rg	112 Uub	113 Uut	114 Uuq	115 Uup	116 Uuh		

Period																
6	Lanthanides	57 La	58 Ce	59 Pr	60 Nd	61 Pm	62 Sm	63 Eu	64 Gd	65 Tb	66 Dy	67 Ho	68 Er	69 Tm	70 Yb	71 Lu
7	Actinides	89 Ac	90 Th	91 Pa	92 U	93 Np	94 Pu	95 Am	96 Cm	97 Bk	98 Cf	99 Es	100 Fm	101 Md	102 No	103 Lr

The above is the modern recommended form of the table using 18 groups. Older group designations are shown below.

Modern form	1	2	3	4	5	6	7	8	9	10	11	12	13	14	15	16	17	18
European convention	IA	IIA	IIIA	IVA	VA	VIA	VIIA	VIII (or VIIIA)			IB	IIB	IIIB	IVB	VB	VIB	VIIB	0 (or VIIIB)
N. American convention	IA	IIA	IIIB	IVB	VB	VIB	VIIB	VIII (or VIIIB)			IB	IIB	IIIA	IVA	VA	VIA	VIIA	VIIIA (or 0)

Appendix VII

The Greek Alphabet

Α	α	alpha	Ν	ν	nu
Β	β	beta	Ξ	ξ	xi
Γ	γ	gamma	Ο	ο	omikron
Δ	δ	delta	Π	π	pi
Ε	ε	epsilon	Ρ	ρ	rho
Ζ	ζ	zeta	Σ	σ	sigma
Η	η	eta	Τ	τ	tau
Θ	θ	theta	Υ	υ	upsilon
Ι	ι	iota	Φ	φ	phi
Κ	κ	kappa	Χ	χ	chi
Λ	λ	lambda	Ψ	ψ	psi
Μ	μ	mu	Ω	ω	omega

Appendix VIII

Web Sites

Chemical society web sites:

American Chemical Society	www.chemistry.org
Royal Society of Chemistry	www.rsc.org
The International Union of Pure and Applied Chemistry	www.iupac.org

Information on nomenclature:

Queen Mary College, London	www.chem.qmul.ac.uk/iupac
Advanced Chemistry Development, Inc	www.acdlabs.com/iupac/nomenclature

An extensive set of organic chemistry links:

WWW Virtual Library, Chemistry Section	www.liv.ac.uk/Chemistry/Links

The definitive site for the chemical elements:

WebElements Periodic Table	www.webelements.com

Other resources:

Chemical Education Resource Shelf, Journal of Chemical Education, University of Missouri-St. Louis	www.umsl.edu/~chemist/books
Molecule of the Month	www.chm.bris.ac.uk/motm/motm.htm
The Chemistry Hypermedia Project at Virginia Tech	www.chem.vt.edu/chem-ed/vt-chem-ed.html

Biographies:

Nobel Prizes	http://nobelprize.org
WWW Virtual Library	www.liv.ac.uk/Chemistry/Links/refbiog.html

Bibliography

Comprehensive texts covering organic chemistry:

Carey, Francis and Richard J. Sundberg. *Advanced Organic Chemistry: Structure and Mechanism*. 4th ed. New York: Plenum, 2000.

Carey, Francis and Richard J. Sundberg. *Advanced Organic Chemistry: Reactions*. 4th ed. New York: Plenum, 2000.

Clayden, Jonathon; Nick Greeves; Stuart Warren; and Peter Wothers. *Organic Chemistry*. Oxford, U.K.: Oxford University Press, 2000.

McMurray, John. *Organic Chemistry*. 6th ed. Pacific Grove, Calif.: Brooks/Cole, 2004.

March, Jerry. *Advanced Organic Chemistry: Reactions, Mechanisms, and Structure*. 4th ed. New York: Wiley, 1992.

Volhardt, Peter K. and Neil E. Schore. *Organic Chemistry: Structure and Function*. 4th ed. New York: W. H. Freeman, 2003.

Comprehensive texts covering inorganic chemistry:

Cotton, F. A.; C. Murillo; G. Wilkinson; M. Bochmann; and R. Grimes. *Advanced Inorganic Chemistry*. 6th ed. New York: Wiley, 1999.

Greenwood, N. N. and A. Earnshaw. *Chemistry of the Elements*. Oxford, U.K.: Butterworth-Heinemann, 1997.

Shriver, D. F. and P. W. Atkins. *Inorganic Chemistry*. 3rd ed. Oxford, U.K.: Oxford University Press, 1999.

Additional useful sources:

Emsley, J. *Nature's Building Blocks – An A-Z Guide to the Elements*. Oxford, U.K.: Oxford University Press, 2001.

King, R.B. (ed.) *Encyclopedia of Inorganic Chemistry*. New York: Wiley, 1994.

An advanced text on biochemistry:

Nelson, David L. and Michael M. Cox. *Lehninger Principles of Biochemistry*. 3rd ed. New York: Worth Publishers, 2000.

The definitive book for physical chemistry:

Atkins, Peter and Julio de Paula. *Atkins' Physical Chemistry*, 7th ed. Oxford, UK: Oxford University Press, 2004.